RECENT PROGRESS *in* MANY-BODY THEORIES

Series on Advances in Quantum Many-Body Theory

Edited by R. F. Bishop, C. E. Campell, J. W. Clark and S. Fantoni
(International Advisory Committee for the Series of International Conferences on Recent Progress in Many-Body Theories)

Published

Vol. 1: Proceedings of the Ninth International Conference on Recent Progress in Many-Body Theories
Edited by D. Neilson and R. F. Bishop

Vol. 3: Proceedings of the Tenth International Conference on Recent Progress in Many-Body Theories
Edited by R. F. Bishop, K. A. Gernoth, N. R. Walet and Y. Xian

Vol. 4: Microscopic Approaches to Quantum Liquids in Confined Geometries
E. Krotscheck and J. Navarro

Vol. 5: 150 Years of Quantum Many-Body Theory
A Festschrift in Honour of the 65th Birthdays of John W Clark, Alpo J Kallio, Manfred L Ristig and Sergio Rosati
Raymond F. Bishop, Klaus A. Gernoth and Niels R. Walet

Vol. 6: Proceedings of the Eleventh International Conference on Recent Progress in Many-Body Theories
Edited by Raymond F. Bishop, Tobias Brandes, Klaus A. Gernoth, Niels R. Walet and Yang Xian

Vol. 7: Introduction to Modern Methods of Quantum Many-Body Theory and Their Applications
Adelchi Fabrocini, Stefano Fantoni and Eckhard Krotscheck

Vol. 9: Proceedings of the Twelfth International Conference on Recent Progress in Many-Body Theories
Edited by J. A. Carlson and G. Ortiz

Vol. 10: Proceedings of the Thirteenth International Conference on Recent Progress in Many-Body Theories
Edited by S. Hernández and H. Cataldo

Forthcoming

Vol. 2: Microscopic Approaches to the Structure of Light Nuclei
Edited by R. F. Bishop and N. R. Walet

Vol. 8: Pairing in Fermionic Systems: Basic Concepts and Modern Applications
Edited by S. Armen, M. Alford and J. W. Clark

Series on Advances in Quantum Many-Body Theory – Vol. 10

Proceedings of the 13th International Conference

RECENT PROGRESS *in* MANY-BODY THEORIES

Buenos Aires, Argentina 5 – 9 December 2005

Editors

Susana Hernández

Horacio Cataldo

University of Buenos Aires, Argentina

World Scientific

NEW JERSEY • LONDON • SINGAPORE • BEIJING • SHANGHAI • HONG KONG • TAIPEI • CHENNAI

Published by

World Scientific Publishing Co. Pte. Ltd.
5 Toh Tuck Link, Singapore 596224
USA office: 27 Warren Street, Suite 401-402, Hackensack, NJ 07601
UK office: 57 Shelton Street, Covent Garden, London WC2H 9HE

British Library Cataloguing-in-Publication Data
A catalogue record for this book is available from the British Library.

RECENT PROGRESS IN MANY-BODY THEORIES
Proceedings of the 13th International Conference

ISBN 981-270-035-8

Printed in Singapore by World Scientific Printers (S) Pte Ltd

CONTENTS

PREFACE

The Thirteenth International Conference on *Recent Progress in Many-Body Theories* (RPMBT-13) was held at the University of Buenos Aires, Buenos Aires, Argentina, 5-9 December 2005. The present volume contains most of the invited talks plus a selection of excellent poster presentations.

This conference series is now firmly established as one of the premier series of international meetings in the field of Many-Body physics. The first official RPMBT meeting was held in Trieste in 1978, in response to several precursor meetings that accentuated the need for a continuing series. The most important of these, which can be regarded as RPMBT-0, is the 1972 conference on *The Nuclear Many-Body Problem* organized by F. Calogero and C. Cioffi degli Atti in Rome. Additionally, there were two very significant workshops held in 1975 and 1977 at the University of Illinois, Urbana, with Vijay Pandharipande as the chief organizer. The quantum many–body community suffered a profound loss with Pandharipande's untimely death in early 2006. Later conferences in the series have been the 1981 RPMBT-2 meeting in Oaxtepec, Mexico; the 1983 RMPBT-3 meeting in Altenberg, Germany; the 1985 RPMBT-4 meeting in San Francisco, USA; the 1987 RPMBT-5 meeting in Oulu, Finland; the 1989 RPMBT-6 meeting in Arad, Israel; the 1991 RPMBT-7 meeting in Minneapolis, USA; the 1994 RPMBT-8 meeting in Schloss Seggau, Styria, Austria; the 1997 RPMBT-9 meeting in Sydney, Australia; the 1999 RPMBT-10 meeting in Seattle, USA; the 2001 RPMBT-11 meeting in Manchester, UK; the 2004 RPMBT-12 meeting in Santa Fe, USA, and the present 2005 meeting in Buenos Aires. Highlights of past meetings can be found in earlier volumes of this series.

This conference series is also responsible for awarding the prestigious Eugene Feenberg Memorial Medal in Many-Body Physics. This medal, first presented in 1985, is designated for work that is firmly established and that can be demonstrated to have significantly advanced the field of many-body physics. Past recipients have included David Pines (1985), John W. Clark (1987), Malvin H. Kalos (1989), Walter Kohn (1991), David M. Ceperley (1994), Lev P. Pitaevskii (1997), Anthony J. Leggett (1999), Philippe Noziéres (2001), and Spartak T. Belyaev and Lev P. Gor'kov (2004). Professors Kohn and Leggett received the Nobel Prize in 1998 and 2003 respectively. We are pleased that the Tenth Feenberg Medal was awarded at this conference to Professors Hermann Kümmel and Raymond Bishop for their development and application of the Coupled-Cluster Method to diverse physical problems. During the past half century, the application of this quantum many-body physics approach has contributed substantially to our understanding of challenging problems in condensed matter physics (including electrons in solids, nuclear matter, quantum liquids and gases, and quantum magnetism), atomic and molecular physics, nuclear physics, and subnuclear physics/quantum field theory (see R.F. Bishop and H. G. Kümmel, "The Coupled-Cluster Method," *Physics Today*, March 1987). The Coupled-Cluster Method is one of very few *ab initio* methods that has an enduring impact in quantum chemistry. The presentation was made by Charles

E. Campbell, Chair of the Feenberg Medal Selection Committee. The text of the tribute as well as the responses of the Medal recipients are included in this volume.

The conference also hosted a special session to celebrate the 70th birthdays of John Clark and Manfred Ristig, whose very substantial contributions to many-body physics were warmly recalled by Charles Campbell and Leszek Szybisz. These reminiscences, as well as Clark's and Ristig's lectures, are collected in this volume.

The current conference maintains the tradition of covering the entire spectrum of theoretical tools developed to tackle important and current quantum many-body problems with the aim of fostering the exchange of ideas and techniques among physicists working in diverse subfields of physics such as nuclear and sub-nuclear physics, astrophysics, atomic and molecular physics, quantum chemistry, complex systems, quantum field theory, strongly correlated electronic systems, magnetism, quantum fluids and condensed matter physics. The highlights of the conference included state-of-the-art contributions to the understanding of supersolid Helium, BEC-BCS crossover, fermionic BEC, quantum phase-transition, computing, simulations, as well as the latest results on more traditional topics of liquid helium, droplets, nuclear and electronic systems. The conference continues to demonstrate the vitality and the fundamental importance of many-body theories, techniques, and applications in understanding diverse and novel phenonomena at the cutting-edge of physics.

The Local Organizing and Program Committees deserve great thanks in creating a well-run and productive meeting, with an exciting program of talks and poster presentations. We also thank the speakers and participants for making the meeting a lively mutual learning experience. It is a special pleasure to thank the local organizers, Susana Hernández, Leszek Szybisz, Guillermo Dussel, Ana María Llois and Horacio Cataldo, for their tireless efforts in making the conference not only a success, but a memorable occasion. We are all indebted to Susana's resourcefulness, the hospitality and charm of Buenos Aires, and the resilience of the Argentinian economy.

Siu A. Chin
Chair, International Advisory Committee
for the Series of International Conferences on
Recent Progress in Many-Body Theories
College Station, USA

SERIES EDITORIAL BOARD AND OTHER COMMITTEES

K. Hallberg (Argentina)
E. Krotscheck (Austria)
D. Neilson (Italy)
J. P. Paz (Argentina)
A. Polls (Spain)
M. Ueda (Japan)
M. Weissmann (Argentina)

LOCAL ORGANIZING COMMITTEE FOR THE THIRTEENTH CONFERENCE

Susana Hernández (Chair)
Leszek Szybisz
Guillermo Dussel
Ana María Llois
Horacio Cataldo

all at University of Buenos Aires, Argentina

FOREWORD BY THE EDITORS

The *Thirteenth International Conference on Recent Progress in Many–Body theories* took place at the University of Buenos Aires, Buenos Aires, Argentina, from December 5 to 9, 2005. The Conference Programme consisted of 43 oral presentations plus near 30 posters. This book contains a majority of those invited talks and a selection of short write-ups corresponding to posters. We warmly thank Ray Bishop and Hermann Kümmel for their valuable advice in order to collect short contributed papers for publication in this volume. Notwithstanding, the final decision regarding the contents of this book corresponds to the Editors. We regret not being able to include material from several outstanding talks, and warmly thank all authors appearing in this volume for joining us in the effort of offering these proceedings to the Quantum Many-Body international community, for the extreme care in preparing their contributions and for the high quality of the final product. A special word of gratitude is due to Siu Chin, Chair of the International Advisory Committee of the Conference Series, for enriching this volume with an enlightening preface, to Charles Campbell for his vivid remembrance of John Clark's achievements and scientific life in his 70th birthday, and to Leszek Szybisz for his warm recollection of Manfred Ristig's life and career at the turn of his seventh decade.

As in previous volumes of this series, the articles are collected into groups that intend to reflect the spirit of the meeting, on the understanding that both the grouping and the headings are somehow arbitrary and that many papers could appear in more than one group, or under a different classification. The present scheme aims at orienting the reader to locate the material delivered at the conference, in a broad sense.

As members of the Local Organizing Committee and Editors of this volume, we feel largely gratified for the outcome, that to our belief meets the highest standard of the series.

Susana Hernández
Horacio Cataldo
Buenos Aires, Argentina
April 30, 2006

FEENBERG MEMORIAL MEDAL PRESENTATION

RAYMOND BISHOP AND HERMANN KÜMMEL: FEENBERG MEDALISTS 2005 THE COUPLED CLUSTER METHOD

C. E. CAMPBELL

School of Physics and Astronomy, University Minnesota
Minneapolis, Mn 55455, U.S.A.
campbell@umn.edu

S. A. CHIN

Department of Physics, Texas A&M University
College Station, TX 77843, USA
chin@physics.tamu.edu

J. W. CLARK

Department of Physics, Washington University
St. Louis, MO 63130, USA
jwc@wuphys.wustl.edu

1. The Eugene Feenberg Memorial Medal

Raymond Bishop and Hermann Kümmel have been named as the recipients of the The Tenth Eugene Feenberg Award in Many-Body Physics, in recognition of their key roles in the development and application of the powerful coupled cluster method of quantum many-body theory. Feenberg Medals were presented to the awardees on December 5, 2005 at the Thirteenth International Conference on Recent Progress in Many-Body Theories, held in Buenos Aires, Argentina.

The Eugene Feenberg Memorial Medal was established in 1983 by the quantum many-body physics community to honor Eugene Feenberg, a pioneer in theoretical nuclear physics, for his unique and enduring contributions to the development of the microscopic quantum many-body theory of strongly interacting systems. The award commemorates and celebrates Feenberg's wise stewardship of a field that permeates all branches of physics, his deep physical insights and great formal achievements, his dedicated service as teacher and mentor, and his exemplary integrity in his professional and personal life.

The selection of Feenberg Medalists is conducted under the auspices of the International Advisory Committee for the Series of International Conferences on Recent Progress in Many-Body Theories. The previous recipients of the award are David Pines (1985), John W. Clark (1987), Malvin H. Kalos (1989), Walter Kohn (1991),

Hermann Kümmel and Raymond Bishop

David M. Ceperley (1994), Lev P. Pitaevskii (1997), Anthony J. Leggett (1999), Philippe Nozières (2001); and Spartak T. Belyaev and Lev P. Gor'kov (2004).

2. Eugene Feenberg

As we near the centenary of his birth, it is especially appropriate to include memories of Eugene Feenberg on this occasion. Born on October 19, 1906 in Fort Smith, Arkansas, Feenberg received a B.A. in physics and M.A. in mathematics in 1929 from the University of Texas, Austin after three years of study. While a Harvard doctoral student he spent a year and a half in Europe as a Parker Traveling Fellow, visiting the institutes of Sommerfeld, Pauli, and Fermi. His time in Europe was cut short by the turmoil accompanying the Nazi ascent to power in Germany. Returning to Harvard, he completed his Ph.D. in 1933 under the direction of E. C. Kemble, founder of one of the earliest significant schools of theoretical physics in the United States. As a second mentor, John van Vleck also had a strong influence on Feenberg's early professional development. His Ph.D. thesis, on quantum scattering theory, contained the first statement and proof of the quantum optical theorem.

In the ensuing years, Feenberg held appointments as Instructor or Fellow at Harvard, Wisconsin, and the Institute for Advanced Studies, collaborating with Eugene Wigner, Gregory Breit, John Bardeen, and Melba Phillips, among others. In 1936, back-to-back papers in *The Physical Review* by Cassen and Condon and Breit and Feenberg presented evidence and arguments for the charge independence of nuclear forces and formally identified the first internal symmetry (isospin) of particle physics.[1,2] After eight years on the faculty of New York University, intermixed with four years in defense work at Sperry Gyroscopic during World War II, Feenberg joined the faculty of Washington University in St. Louis as an associate professor. In 1964 he succeeded Edward Condon as the Wayman Crow Professor of Physics, a chair previously held by Arthur H. Compton, and subsequently by Edwin T. Jaynes and currently by John Clark, continuing the very high standards and traditions of this important chair.

Much of Feenberg's research during the first phase of his career was concerned with nuclear theory, in work that culminated in publication of *Shell Theory of the Nucleus* by Princeton University Press 50 years ago.[3] In the mid-1950's, a career-long interest in perturbation theory stimulated his early entry into the new field of quantum many-body theory, coincident with pioneering developments in Russia and the West driven by advances in field theory. This engagement brought a shift of Feenberg's primary focus from nuclear problems to the theory of quantum fluids, most notably the helium liquids, a subject to which he would contribute very importantly for the next two decades until his death in 1977. Along with his students, he developed the method of correlated basis functions (CBF) in order to cope with the strong, short-range repulsion between helium atoms – a feature that defies mean-field approaches and makes the theory of the helium fluids virtually intractable using ordinary perturbation theory. The early part of this research is the subject of his now-classic monograph *Theory of Quantum Fluids.*[4]

Eugene Feenberg's courage in confronting the challenge of strong, short-range correlations by conceiving and developing a comprehensive *ab initio* framework to deal with them quantitatively from first principles, was characteristic of his research style. This body of work is the foundation of much of modern theoretical research on strongly correlated quantum fluids. His personal integrity and his high standards of professional achievement and behavior continue to inspire and guide his former students and colleagues.

The esteem in which Eugene Feenberg is held by those who knew and learned from him is eloquently captured in an excerpt from the commencement address "Giants Must Stand on Solid Ground" delivered in 1991 at the Massachusetts Institute of Technology by Walter Massey, President of Morehouse College and former Director of the U.S. National Science Foundation:[5]

> So, where do young researchers first learn the rules of the game that protect the integrity of their endeavors? For generations, the community has relied on the unique mentor-apprentice relationship that develops during the process of doctoral and postdoctoral research to teach these important lessons. When all goes well, a bond of trust develops between the professor and his or her students which is grounded in intellectual curiosity, a desire to discover new knowledge, and a common commitment to truth. What is conveyed and what is learned is more than simply a body of facts; it is an approach to understanding, an appreciation of standards, and a set of values – in effect, an ethic. I was very fortunate in my own graduate training to have had a research director from whom I learned a great deal of physics and much more besides. Professor Eugene Feenberg (now deceased) was a prominent physicist, well-respected among his peers, a man of integrity and unyielding honesty. All of us who were his students saw these traits in him and profited from working with him. To come to class ill-prepared or not take time to listen to his students and their problems; to publish a result prematurely simply to gain priority; or to put his name on a paper by one of his students when he had not shared in the work himself – all these things were anathema to Gene Feenberg. Those of us who were his students realize how lucky we are to have had an almost ideal mentor-apprentice relationship.

Those who would like to know more about Eugene Feenberg and his work are invited to read George Pake's biographical memoir for the National Academy of Sciences.[6]

Presentation of Feenberg Medals at successive conferences in the series on Recent Progress in Many-Body Theories serves both to remind us of and to perpetuate this man's extraordinary influence on physics and physicists. Our two laureates, Hermann Kümmel and Raymond Bishop, exemplify the steadfast integrity and uncommon commitment to the highest of standards that were so visible in the professional career and personal life of Eugene Feenberg.

3. Hermann Kümmel

The Tenth Feenberg Memorial award cites Hermann Kümmel for his role in the creation and early development of the coupled cluster method, and pioneering high-accuracy applications of it to problems in nuclear and subnuclear physics.

The volume of the World Scientific Series on Advances in Quantum Many-Body Theory devoted to the proceedings of MB11 contains an excellent biography of Hermann Kümmel, in a tribute written by Raymond Bishop in celebration of Kümmel's 80th birthday.[7] From that source we learn that Kümmel was born in 1922 in Berlin, and that all of his formal education took place in that city. He received the Diplom in 1950 from Humboldt University in the East Zone, and his Ph.D. in theoretical physics in 1952 from the Free University in West Berlin, where he continued with his research until moving to the U.S. to work at the University of Iowa. It was during his two years as a research associate in Iowa City that, in consultation with Fritz Coester, the foundations were laid for the coupled cluster method (CCM) (more commonly known in the early days as the Exponential-S Theory). Kümmel and Coester presented the seminal ideas of CCM in a 1960 paper published in *Nuclear Physics*.[8]

After a brief period at the University of Tübingen, Kümmel spent several years in Mainz with joint appointments at the Max Planck Institute for Nuclear Chemistry and the University of Mainz. Then came two years back in the U.S. as Professor of Physics at Oklahoma State University. Returning to Mainz as a senior scientist at the Max Planck Institute and adjunct professor at the University, he built a strong research group in nuclear theory.

In 1969 Kümmel moved with his entire research group to take a Chair in Physics at the newly founded Ruhr University in Bochum (RUB), Germany. His efforts soon established Bochum as one of the world's leading centers in quantum many-body theory. Significantly, it was at this time that he turned his attention in earnest to the theoretical and computational development of CCM. Although nuclear systems remained the primary focus of applications, important results were also obtained for the electron gas, a basic problem in solid-state physics and a testbed for many-body theories. The level of sophistication achieved in Bochum under his leadership, in both formal and computational strength, has rarely been matched. He retired in 1988 at the mandatory age, but has remained active in research at RUB as Professor Emeritus.

An authoritative and insightful account of Hermann Kümmel's pioneering contributions to the formal development of CCM and its subsequent application in a host of problem settings may be found in his paper entitled "A Biography of the Coupled Cluster Method".[9] Kümmel was both a co-founder of the CCM and the first to appreciate and exploit its power in the context of nuclear physics. Arguably, the earliest applications of coupled cluster theory by his group to medium-mass atomic nuclei stand to this day as never having been surpassed. Considering the computing power that was available at the time, the results obtained for both closed-shell nuclei and open-shell nuclei comprising one or two valence particles or holes outside closed shells, are breathtaking in their sophistication and accuracy. They were undoubtedly the first fully microscopic calculations for nuclei to be demonstrably converged for a selection of the best available, realistic, two- and three-nucleon interaction potentials, for both ground- and excited-state energies and properties.

Based on his successes in applying coupled cluster theory to finite nuclei, Kümmel next turned his attention to various strongly correlated systems in subnuclear physics and quantum field theory. In particular, his treatment of the deuteron from the field-theoretic standpoint of pions and nucleons interacting via the standard pseudoscalar, isovector coupling, may be recognized as a landmark calculation. It provides a rare example of a fully converged calculation for a non-trivial, strongly interacting quantum field theory. Throughout his long and fruitful career, Hermann Kümmel has been a pioneer, a leader, an inspiration, and a mentor in the international community of many-body theorists.

4. Raymond Bishop

The Tenth Feenberg Memorial Medal award cites Raymond Bishop for his development of the coupled cluster method toward a comprehensive ab initio approach, and innovative applications across the full spectrum of subfields of quantum many-body physics.

Raymond Bishop was born in London in 1945. After earning his B.A. degree at the Queen's College, Oxford in 1966 and a Ph.D. in theoretical physics at Stanford University in 1971, he returned to England as Science Research Council Research Fellow. Subsequently, he was Senior Research Associate and Lecturer in the Department of Physics, Manchester University, simultaneously being associated with the Theory Group at the Science Research Council, Daresbury Laboratory.

Bishop returned to the U.S. and spent several years at the University of California, Berkeley as Staff Scientist at the Lawrence Berkeley Laboratory and Lecturer in the Department of Physics. In 1979 he joined the faculty of the Department of Mathematics at the University of Manchester Institute of Science and Technology (UMIST). He was appointed Professor of Theoretical Physics at UMIST in 1988 and Head of the Department of Mathematics in 1991. Theoretical Physics moved to the Department of Physics in 1995, where Bishop became Head in 1996. He is currently Professor of Theoretical Physics in the School of Physics and Astronomy of the combined University of Manchester formed in the merger of UMIST with Manchester University in 2004.

Bishop initiated his research program in the coupled cluster method during lengthy visits to Bochum in the mid-to-late-1970s. While at Bochum, he began a fruitful collaboration with Karl-Heinz Lührmann, in which they applied CCM theory exhaustively, and with great success, to that touchstone problem of many-body theory, the electron gas. From that point on, Bishop was responsible for a series of pivotal formal developments in coupled cluster theory, often with the close participation of the late Jouko Arponen. He has systematically and energetically developed the CCM into the elegant, consistent, flexible, and accurate tool for microscopic applications that we know today. Indeed, he has been the driving force behind the 25-year evolution of coupled cluster theory from a promising framework for *ab initio* treatment of quantum many-particle systems into its mature form as a powerful and universal method for quantitative investigation of ground states, excitations, and phase transitions. Applications abound in nuclei and nuclear matter, atoms and molecules, quantum fluids and electronic systems in condensed matter, lattice-spin and lattice-gauge models, quantum optics, and quantum field theory.

In more detail, his numerous significant achievements include: (a) incisive analysis of correlations in the electron gas at high, intermediate (metallic), and low densities; (b) extensive development of the extended coupled cluster formalism, its uses to map arbitrary quantum field or quantum many-body theories exactly onto classical Hamiltonian mechanics, and for its applications to the treatment of excited states and the zero-temperature hydrodynamics of strongly-interacting condensed Bose fluids; (c) formulation and implementation of a translationally-invariant coupled cluster theory for light nuclei; (d) pathbreaking studies of spin-lattice models yielding new insights into quantum antiferromagnets; (e) exploration of Hamiltonian lattice-gauge theories in the coupled cluster framework; and (f) publication of comprehensive and authoritative reviews on the important ramifications of coupled cluster theory within both chemistry and physics.

At Manchester, Bishop has established one of the world's foremost groups in quantum many-body physics and has directed an intensive computational effort exploiting automated symbolic manipulation. At the same time he has developed and leads what amounts to an informal international school of many-body physics research through his extensive collaborations throughout the world and his development and application of CCM to an enormously broad range of theoretical physics subdisciplines.

5. Selected significant publications

Hermann Kümmel

F. Coester and H. Kümmel, Short-range correlations in nuclear wave functions, *Nucl. Phys.* **17**, 477-485 (1960).

H. Kümmel, Theory of many-body wave functions with correlations, *Nucl. Phys.* **A176**, 205-218 (1971).

H. Kümmel and K. H. Lührmann, Equations for linked clusters and the energy variational principle, *Nucl. Phys.* **A191**, 525-534 (1972).

R. Offermann, W. Ey and H. Kümmel, Degenerate many-fermion theory in exp(S)–form, *Nucl. Phys.* **A273**, 349 (1976)

H. Kümmel, K. H. Lürmann and J. G. Zabolitzky, Many fermion theory in the exp S–(or coupled cluster) form, *Phys. Rep.* **36C**, 1-63 (1978)

Raymond Bishop

R. F. Bishop and K. H. Lührmann, Electron correlations. II. Ground- state results at low and metallic densities, *Phys. Rev.* **B26**, 5523-5557 (1982).

J. S. Arponen, R. F. Bishop, and E. Pajanne, Extended coupled cluster method. I. Generalized coherent bosonization as a mapping of quantum theory into classical Hamiltonian mechanics, *Phys. Rev.* **A36**, 2519-2538 (1987).

R. F. Bishop, M. F. Flynn, M. C. Boscá, E. Buendia, and R. Guardiola, Translationally invariant coupled-cluster theory for simple finite systems, *Phys. Rev.* **C42**, 1341-1360 (1990).

R. F. Bishop, J. B. Parkinson, and Y. Xian, Coupled-cluster treatments of correlations in quantum antiferromagnets, *Phys. Rev.* **B44**, 9425-9443 (1991).

R. F. Bishop, An overview of coupled cluster theory and its applications in physics, *Theor. Chim. Acta* **80**, 95-148 (1991)

References

1. B. Cassen and E. U. Condon, *Phys. Rev.* **50**, 846 (1936)
2. G. Breit and E. Feenberg, *Phys. Rev.* **50**, 850 (1936)
3. E. Feenberg, *Shell Theory of the Nucleus* (Princeton University Press, Princeton, NJ 1955)
4. E. Feenberg, *Theory of Quantum Fluids* (Academic Press, New York, 1969)
5. W. Massey, in *The Tech* **111**, Number 28, p. 2 (June 26, 1991); available from: http://www-tech.mit.edu/V111/N28/massey.28n.html. Accessed November 15, 2005.
6. G. Pake, in *Biographical Memoirs,* (The National Academies Press); available from: http://www.nap.edu/readingroom/books/biomems/efeenberg.html. Accessed November 12, 2005; see also K. A. Brueckner, C. E. Campbell, J. W. Clark, and H. Primakoff, *Nuclear Physics* **A317**, pp. i-vii (1979).
7. R. F. Bishop, in *Proceedings of the 11th International Conference on Recent Progress in Many-Body Theories,* ed R.F. Bishop, T. Brandes, K. A. Gernoth, N. R. Walet, and

Y. Xian (World Scientific, Singapore, 2002) pp. 319-333; R. F. Bishop, *International Journal of Modern Physics* **B17**, 5295(2003)

8. F. Coester and H. Kümmel, *Nucl. Phys.* **17**, 477-485 (1960).
9. H. Kümmel, in *Proceedings of the 11th International Conference on Recent Progress in Many-Body Theories,* ed R.F. Bishop, T. Brandes, K. A. Gernoth, N. R. Walet, and Y. Xian (World Scientific, Singapore, 2002) pp. 334-348; , H. Kümmel, *International Journal of Modern Physics* **B17**, 5311 (2003)

QUANTUM THEORY "WITHOUT MEASUREMENT"

HERMANN G. KÜMMEL
Institut für Theoretische Physik II, Ruhr-Universität
44780 Bochum, Germany
hermann.kuemmel.tp2@ruhr-uni-bochum.de

Attempting to describe the "collapse of the wave function" of the orthodox quantum theory as a physical process, a natural assumption is that the measurement apparatus is subject to the same physical laws as the measured object, leading to a theory without rules for measurement. This - together with the loss of classical-quantum correspondence for chaotic systems - enforces the inclusion of the environment. I very shortly describe the present state of the art in this field - a challenge for the quantum many body community.

Keywords: decoherence; environment; measurement

1. Prologue: Why is the Orthodox Quantum Theory Incomplete?

This meeting is about applied quantum theory. We all are convinced that there are good reasons to work in this field. Rarely in science exist results as exact as those obtained in quantum theory. Also, we know that there are many beautiful experiments exploring and testing the most sophisticated aspects of quantum physics - and that their results are always in agreement with the predictions from this theory. Quantum computing, cryptography, teleportation and so on are the keywords which matter here. Just because we don't have the least doubts that quantum theory in the "orthodox" (i.e. the Copenhagen or Bohr-Born) form correctly describes nature, we rarely worry about the fundamental problems involved. We more or less consider them as a nuisance, something esoteric, even irrelevant, and of interest more for philosophers or for retired and quite old people like myself rather than for active physicists. Yet quite a few of our famous colleagues like Einstein, Schrödinger, Wigner, Feynman, Bell and others never believed that quantum physics is fully understood, or complete, or correct. I quote Feynman as a representative of the younger (post-Einstein) generation: *"I think I can safely say that nobody today understands quantum physics"*. Bell even believed that *"Quantum mechanics carries the seeds of its own destruction"*, he said *"It is rotten"*. Nevertheless all these people fully recognized the utilitarian success of the theory - they had to. I believe, however, that now the situation has changed: a consistent and complete theory is in sight. Thus is may be a good idea to present the new developments to the quantum mechanics practitioner.

Let me first write down the well known rules of the orthodox interpretation about which almost everybody agrees that they are correct for applying quantum theory. I leave out some aspects which are irrelevant for the points I wish to make, even though some of them have been hotly debated in the past. From now on "object" means the physical system we want to investigate, and "apparatus" is the measurement apparatus we use to do this. For lack of space I don't go into the rather delicate (almost philosophical) questions of what constitutes these two entities, or what are the boundaries between them, or what an "observable" really means.

The rules are:

1. Attach to any observable O a Hermitian operator $\mathbf{O} = \sum \mathbf{o_i}|\varphi_i\rangle\langle\varphi_i|$ (for instance pairs of generalized momenta and coordinates $(\mathbf{P}_j, \mathbf{Q}_j)$ with the usual commutation relations). This implies the existence of a Hilbert space for the object.

2. Invent a suitable Hermitian Hamilton operator $\mathbf{H}$ characteristic for the object at hand.

3. The measurement of an observable O puts an observed object *"suddenly"* into an eigenstate $|\varphi_i\rangle$ of the Hermitian operator $\mathbf{O}$ related to this observable ("collapse or reduction of the wave function"). The value of this observable attached to the object is the eigenvalue o_i. The apparatus is macroscopic and the measured value of the observable is imprinted on it as a macroscopic observable.

4. The probability for the occurrence of o_i is given by $p_i = |\langle\varphi_i|\psi\rangle|^2$ with ψ as the state of the object before the measurement ("Borns rule").

5. Before and after measurements the state of the object moves according to the Schrödinger equation driven by the unitary operator $exp(-i/(\mathbf{H}t)$.

To rules 1 and 2: Here one is free to define physical objects (Hamiltonians) and observables and their hermitian operators. Discussing the fundamentals of quantum physics we have to assume that all relevant operators are known some way or another.

Rule 4 should be interpreted as follows: p_i is approximately the fraction of objects found in state $|\varphi_i\rangle$ of a large finite number of objects investigated. It is standard to introduce a density matrix ϱ representing the ensemble of the quantum objects. For such an ensemble one obtains $p_i = Tr(\varrho|\varphi_i\rangle\langle\varphi_i|)$. Then rules 3 and 4 applied to an ensemble of equally prepared objects all in the state ψ transforms the original density matrix $|\psi\rangle\langle\psi|$ suddenly into one consisting of the sum of projectors:

$$\varrho_{before} \equiv |\psi\rangle\langle\psi| \Rightarrow \varrho_{after} \equiv \sum p_i|\varphi_i\rangle\langle\varphi_i|, \tag{1}$$

where the arrow does *not* represent a unitary transformation. ϱ_{after} is interpreted as describing an ensemble with the fraction p_i of objects in the state $|\varphi_i\rangle$. According to rule (3) what is called a "measurement" in quantum theory actually is "putting the system into a state". Moreover, after the measurement on a single object we have learned almost nothing about the status of the object as it was before the measurement. Even if we repeat the measurement as often as possible on as many as possible objects - all of them prepared as well as possible in the same way - we merely know (approximately) the set $|\langle\varphi_i|\psi\rangle|^2$ and not the wave function before the measurement. Our daily experience and the belief that we can learn something about the history of the universe from its present state is not compatible with this rigor inherent in the orthodox quantum theory. I shall come back to this problem later. On the other hand, after a (perfect) measurement we know everything about the state of the object immediately after this measurement - and forever if we know the Hamiltonian and can use it (rule (5)).

After these comments on the established fundamental laws of quantum physics let me come to the central problem: the theory in this form is incomplete since for the sudden change of the wave function there is no equation of motion. In this interpretation the mechanism of the measurement remains unexplained. As we shall see, the search for something very general describing this process, i.e. for some universal features for all kinds of measurement apparatuses, leads to a series of more problems.

In any case, as soon as the apparatus is considered to be more than a "black box" one arrives at the following four alternatives for a complete theory:

I. *Both object and apparatus move according to the Schrödinger equation.*

II. *Object and apparatus are quantum systems not treated on the same footing* ("Consistent histories", "many worlds" and other interpretations).

III. *Everything is classical - although described by waves instead of particles.* Quantum phenomena appear only if one does not look - or is not able to look - carefully enough. The "hidden parameter" theories are of this kind. Due to Bells famous papers it is known that they necessarily are non-local and as a consequence violate special relativity.

IV. *A "super"-theory, encompassing everything including general relativity.* String theories as theories of everything come into our mind. However, the question how the measurement apparatus is embedded in it to my knowledge never has been considered. Such ideas at present are very far from concrete.

I follow a minimalist approach preserving as much of the known quantum theory as possible, especially the quantum equation of motion. This is alternative I of above. In other words: both microphysics and macrophysics including the measurement apparatus will be described as ordinary (if appropriate many body) quantum objects subject to the Schrödinger equation. In the end such a quantum theory would have no rules for the measurement - like in classical physics. Using this approach recently much progress has been made and conclusions have been reached on which very likely a final theory can be built.

2. Object and Apparatus Coupled together

Once one has decided to treat also the measurement apparatus as a quantum object one is bound to construct a wave function containing the degrees of freedom of object and apparatus. Thus one necessarily has to deal with product as well as with entangled states: before the measurement the state of object plus apparatus is the product of the wave functions $|\psi\rangle = \sum c_i|\varphi_i\rangle$ (state of the object) and $|A_0\rangle$ (initial state of the apparatus). The interpretation of this measurement as a unitary

(Hamiltonian driven) time development then leads to an entangled state consisting of the wave functions $|\varphi_i\rangle$ and the apparatus in the pointer states A_i. Explicitly:

$$|\psi\rangle|A_0\rangle = \sum c_i|\varphi_i\rangle|A_0\rangle \to \sum c_i|\varphi_i\rangle|A_i\rangle, \tag{2}$$

where ***the arrow now is a shortcut for this unitary transformation.*** Hamiltonians containing an object-apparatus interaction $\mathbf{H}_{OA}$ which perform (2) can easily be constructed, see below. This approach - due to John v. Neumann[1] - seems to be the only one possible as long as one sticks to the idea to have a common entity "object+apparatus" described by the time development of quantum dynamics and separated from the rest of the world, i.e. as a closed system. And one is tempted to be happy by observing that if the apparatus is seen in the state $|A_i\rangle$ the object certainly is in the state $|\varphi_i\rangle$. However, this cannot be the final solution. First of all: what does "seen in the state $|A_i\rangle$" mean? Indeed, one has shifted the measurement problem of the object to the measurement problem of the apparatus. This is no wonder since the apparatus now is a quantum system as well as the object we are investigating and this implies a certain symmetry between both. And the system object+apparatus is not necessarily in one of the states $|\varphi_i\rangle|A_i\rangle$ after the measurement as required by the laws of quantum physics. In the final density matrix

$$\varrho_{final} = \sum_{ij} c_i c_j^* |\varphi_i\rangle\langle\varphi_j||A_i\rangle\langle A_j| \tag{3}$$

there are off-diagonal elements and not just projectors $|\varphi_i\rangle\langle\varphi_i||A_i\rangle\langle A_i|$ onto the individual states of the object+apparatus with probabilities $p_i = |c_i|^2$ like in equation Eq. (1). In other words: the transition into one of the states $|\varphi_i\rangle|A_i\rangle$ has not been achieved. There are some more problems with this approach: one easily shows that one can change the basis such that the measured observables and even the roles of object and apparatus can be exchanged. This means that we neither know what we measure nor what is the object and what is the apparatus. This is referred to as the ***"basis ambiguity"***. Both effects are due to the above mentioned symmetry between object and apparatus in the interaction $\mathbf{H}_{OA}$. This can be made explicit by considering a simple model, a two state object and a two state apparatus. Call the states of system and apparatus $|0\rangle_O, |1\rangle_O$ and $|0\rangle_A, |1\rangle_A$. Define a complementary basis for both by the general form $|\pm\rangle = 1/\sqrt{2}(|0\rangle \pm |1\rangle)$. Then the unitary transformation $exp(-it/\hbar\mathbf{H}_{OA})$ with

$$\mathbf{H}_{OA} = g[|1\rangle\langle 1|]_O[|-\rangle\langle -|]_A \tag{4}$$

- corresponding to the arrow in (2) if g is chosen as $g = \pi\hbar/t$ - can be written in terms of the two bases either as

$$(a) \quad \mathbf{H}_{OA} = g/2[|1\rangle\langle 1|]_O\Big[\mathbf{1}_A - [|0\rangle\langle 1|]_A - [|1\rangle\langle 0|]_A\Big],$$

or as

$$(b) \qquad \mathbf{H}_{OA} = g/2[|-\rangle\langle -|]_A\Big[\mathbf{1}_O - [|-\rangle\langle +|]_O - [|+\rangle\langle -|]_O\Big].$$

Clearly both the role of apparatus and object as well as the observables corresponding to the two basis sets $|+\rangle, |-\rangle$ and $|1\rangle, |0\rangle$ have been exchanged in (a) and (b). Note that $\mathbf{H}_{OA}$ commutes with all projectors of the object states. This guarantees that after the measurement the state does not change due to this interaction. ($\mathbf{H}_{OA}$ - also known as "controlled shift" or "cnot" - is the prototype for the operators used to control the wave functions in quantum computers etc. and has been realized in various ways by the experimentalists working in this field.) An additional unitary time change due to the internal Hamiltonians of object and apparatus poses no problems and thus has been ignored here. The Hamiltonian (4) can be generalized to infinite dimensional systems and the same kind of ambiguity always appears if some of the $|c_i|$ in the wave function ψ are equal.

Finally, the existence of classical chaotic systems presents another problem. Especially if one is aware of the fact that chaotic motion is more the rule than the exception. One knows since a long time that closed macroscopic quantum objects driven by a Hamiltonian allowing classical chaos loose "quantum-classical correspondence". This refers to a breakdown of the well known Bohr correspondence principle describing the similarities between classical and quantum dynamics.

The effect of chaos on quantum systems has been investigated since about 1978 and it is a research field on its own. I cannot go into details - I am not enough of an expert for this. Let me nevertheless try to make plausible how this loss of correspondence comes about. Assume we have a chaos prone Hamiltonian. Patches of chaotic classical phase space distributions stretch exponentially - with $exp(+\Lambda t)$ - in instable directions and shrink exponentially - with $exp(-\Lambda t)$ - in stable directions, where $\Lambda > 0$ is the Lyapunov exponent. But Heisenberg uncertainty requires for instance that shrinking in the momentum (p) direction generates immediate stretching in the coordinate (x) direction. Thus quantum mechanics rapidly spoils the classical distribution. It delocalizes the wave function and finally extends the distribution over the whole available space. Thus it looses all traces of classicality. This happens after a surprisingly short time, as can be seen as follows: The compression of the original momentum distribution Δp_0 is

$$\Delta p = \Delta p_0 exp(-\Lambda t).$$

Via the Heisenberg uncertainty relation the spatial region is stretched as follows:

$$\Delta x = (\hbar/\Delta p_0) exp(+\Lambda t).$$

The available space with the size of the order L is then filled with the wave function after the time

$$t_\hbar = \Lambda^{-1} ln(\Delta p_0 L/\hbar).$$

Since Δp_0 is smaller or equal to the total momentum range P it follows

$$t_\hbar \leq \Lambda^{-1} ln(PL/\hbar),$$

and we have an upper limit for the time it takes to fill the whole available space. (PL is of the order of magnitude of the classical action I). After this time the wave function will fold in itself. The normal spreading time due to the Schödinger equation is of the order of $I/\hbar$, i.e. typically much larger than $t_\hbar$. W.H. Zurek[2] cites as an example the Saturn moon Hyperion. It has a prolate shape like a potato. Its gravitational quadrupole interaction with the tidal field of Saturn leads to chaotic tumbling. He then estimates - using all known data about this object - that after about 20 years Hyperion would be delocalized, i.e. be in a coherent superposition of different orientations - a very macroscopic quantum effect (something like a "Schrödinger cat state" on an astronomical scale!). This of course never has been observed. Even if one doubts the validity of the estimates about Hyperion the problem remains that "almost all" closed macroscopic systems (all those with chaos prone Hamiltonians) *necessarily* must show macroscopic quantum effects some way or another - contrary to experience. Then we are in trouble if we insist on describing everything as closed quantum system.

Finally we know that we can get complete information about the macroscopic past of a macroscopic object by just solving the equations of motion backwards in time. As mentioned before, orthodox quantum theory does not give us any tool for dealing with this situation. Clearly this problem is very much interwoven with the description of the measurement process. But the inclusion of the apparatus alone does not help in any way.

Attempting to describe the measurement process with both object and macroscopic apparatus as a quantum system we thus arrive at the following **preliminary list of problems**:

1. Remove the *basis ambiguity*.

2. *Describe macroscopic objects* - including the measurement apparatus - *via quantum theory* and remove the problems with *chaotic motion*.

3. Select the basis of states for the object related to the desired observable and the stable pointer states, i.e. justify the structure of $\mathbf{H}_{OA}$ (*"preferred basis problem"*).

4. Describe - via the Schrödinger equation - the *measurement as fast transition* of object and apparatus into a new state by destroying the coherent superposition $|\psi\rangle \equiv \sum c_i|\varphi_i\rangle$ (*"problem of definite outcome"*). This must happen with a probability according to *Born's rule*.

Fortunately the fact that the problems with chaos occur only for closed systems forces a way out: treat object and apparatus as open system - i.e include the environment. This idea has been promoted for some time, but only recently has been made more concrete and acceptable.

3. Object, Apparatus and Environment Coupled together

One "opens" the system object+apparatus by generalizing (2) to the system object+apparatus+environment in a rather straight forward way. The object is assumed to have contact with the apparatus first, and then the apparatus to interact with the environment.[2] With $|\epsilon_i\rangle$ as the states of the environment we thus have

$$\sum c_i|\varphi_i\rangle|A_0\rangle|\epsilon_0\rangle \to \sum c_i|\varphi_i\rangle|A_i\rangle|\epsilon_0\rangle \to \sum c_i|\varphi_i\rangle|A_i\rangle|\epsilon_i\rangle \equiv |\Phi\rangle. \tag{5}$$

Here both arrows describe unitary transformations via suitably constructed Hamiltonians. The states $|\epsilon_i\rangle$ represent a (certainly incomplete) set of (not necessarily orthogonal) states of the environment. In addition to the interaction between object+apparatus $\mathbf{H}_{OA}$ the Hamiltonian now contains an interaction $\mathbf{H}_{AE}$ between apparatus and environment. Again it is possible to formally construct such Hamiltonians. The density matrix is

$$\varrho_{OAE} = |\Phi\rangle\langle\Phi| = \sum_{ij} c_i c_j^* |\varphi_i\rangle\langle\varphi_j||A_i\rangle\langle A_j||\epsilon_i\rangle\langle\epsilon_j|. \tag{6}$$

Summing (tracing) over all states $|\epsilon_i\rangle$ of the environment we obtain *if and only if*

$$\langle\epsilon_i|\epsilon_j\rangle = \delta_{ij} \tag{7}$$

the density matrix

$$\varrho_{OA} = \sum_i |c_i|^2 |\varphi_i\rangle\langle\varphi_i||A_i\rangle\langle A_i|. \tag{8}$$

This is the desired reduced (or "marginal") density matrix consisting of the projections on the states of the system object+apparatus with the correct probabilities $p_i = |c_i|^2$. We justify the trace by saying that we don't observe (measure) the environment. Of course it may be sufficient "for all practical purposes" (Bell) if these relations are only approximately true. We thus have made a big step in the right direction contributing to solving problem 4.

There is one remarkable feature in this approach: The measurement is monitored by the environment, the results are inscribed in the latter via the states $|\epsilon_i\rangle$. This says something about the objective existence of states. Remember that in the Copenhagen (orthodox) interpretation a single measurement on an object tells us practically nothing about the state before this measurement. But a careful analysis shows that in the present theory the information about the former state is inscribed in the environment and remains so. Inspection of the environment in principle may yield information about it, i.e. about the past. As mentioned before, this is a desirable side result.

We still cannot claim that problem 4 of above has been solved completely: although the density matrix has the correct form, the question how the object and apparatus manage to end up in one of the states (given by one of the elements of the now diagonal density matrix (8)) is still open. After all the density matrix is only a convenient computational tool: the final diagonal form tells in which state the object *may* end up, not that it really is in this state. Note that the diagonal form as such does not necessarily correspond to an ensemble of objects in the states occurring in the matrix elements: after all one may diagonalise any density matrix.

Also, we neither have a justification of the Hamiltonians nor of the environment states $|\epsilon_i\rangle$, especially their orthogonality. We merely know that $\mathbf{H}_{OA}$ must be of the general form

$$\mathbf{H}_{AE} = \sum_{klm} g_{klm} |A_k\rangle\langle A_k||\epsilon_l\rangle\langle\epsilon_m\rangle, \tag{9}$$

to guarantee that the states $|\varphi_i\rangle|A_i\rangle$ remain intact. This is a necessary condition since the apparatus must stay in the state belonging to a given measured result. The pointer, counter etc. "show" the results forever if we let them alone.

Problem 1 is solved: the basis ambiguity has been put away with. The proof is rather formal, but the result is quite plausible because now the interaction chain $\mathbf{H}_{OA}$ - $\mathbf{H}_{AE}$ treats all three constituents in different ways.

Introducing the apparatus-environment interaction (9) one thus has selected pointer states and a "preferred basis" of object states is fixed (in contrast to the ambiguity we had without the environment). This is some kind of superselection called environment induced selection (einselection).

Problem 2 has been partially solved: Due to the openness of the system object+apparatus+environment the strange behaviour of chaotic systems now does not *necessarily* occur. One still has to make sure that "in most cases" this behaviour indeed does not show up. (Note that there are exceptional well understood objects - for instance buckyballs[3] or objects in the realm of superconductivity - where some macroscopic quantum effects are well known.)

The **list of problems** of above now looks like this:

1. Basis ambiguity - solved

2. *Describe macroscopic objects* - including the measurement apparatus - *via quantum theory* and remove the problems with *chaotic motion*. Although we know that the "Schrödinger cat states" due to the environment do not *necessarily* occur, one still has to prove that in most cases they indeed do not show up.

3. Select the basis of states for the object related to the desired observable, the stable pointer states and states of the environment, i.e. justify the structure of $\mathbf{H}_{OA}$ and $\mathbf{H}_{AE}$(*"preferred basis problem"*).

4. Describe - via the Schrödinger equation - the *measurement as fast transition* of object and apparatus into a new state destroying the coherent superposition $\sum c_i|\varphi_i\rangle$(*"problem of definite outcome"*). This must happen with a probability according to *Born's rule*.

Thus problems 3 and 4 of the former list survive in essentially the same form, whereas problem 2 has been partially solved due to the "openess" of the system. Realistic environments and environment-apparatus interactions have to be introduced in order to attack the remaining problems. There is a large amount of papers[2,4] on these subjects, mostly using models and plausibility arguments. Not surprisingly, since extensive use has to be made of the complicated environment, at present there are no final results. The environment has to produce the change from the coherent quantum states to the non-coherent classical ones. In other words it has to produce "decoherence". All published attempts to describe this replace the complexity of the environment by some rather simple models for the interaction between apparatus and environment $\mathbf{H}_{AE}$ and for the environment states $|\epsilon_i\rangle$. In this context often the density matrix in the form of the "Wigner distribution"

$$W(x,p) = 1/(2\pi\hbar) \int dyexp(ipy/\hbar)\langle x - y/2|\psi\rangle\langle\psi|x + y/2\rangle \tag{10}$$

has been used. It yields momentum and coordinate distributions via $\int dxW(x,p) = |\psi(p)|^2$ and $\int dpW(x,p) = |\psi(x)|^2$. But, since $W(x,p)$ is not positive definite it cannot be interpreted as density distribution in phase space. However, introducing into the (Schrödinger) equation for $W(x,p)$ a coupling to various -sometimes semi-realistic - model environments of finite temperature it develops *almost instantaneously* into an approximately positive definite quantity subject to the Liouville equations, i.e. to a classical density in phase space. In a similar way one may show that no unrealistic macroscopic quantum phenomena occur because the fast decoherence initiated by the environment overcomes the effect of the Heisenberg uncertainty relation. I don't have enough space to be more explicit. Therefore let me describe only a simple example which demonstrates the feasibility of the ideas behind what has been named the "decoherence program".[4]

4. A Model

This simplest imaginable model (due to Zurek[2]) illustrates quite nicely the strategies one has to apply. It consists of (1) a two state object with states $|+\rangle$ and $|-\rangle$, (2) a two state apparatus with states $|\Uparrow\rangle$ and $|\Downarrow\rangle$ and (3) an environment state of N two state systems with states $|\uparrow\rangle_i$ and $|\downarrow\rangle_i$ $(i = 1..N)$ originally in the state

$$|\epsilon_0\rangle = \prod^{N} (\alpha_i|\uparrow\rangle_i + \beta_i|\downarrow\rangle_i). \tag{11}$$

The Hamiltonian is

$$\mathbf{H}_{AE} = \big(|\Uparrow\rangle\langle\Uparrow| - |\Downarrow\rangle\langle\Downarrow|\big) \sum_k g_g \big(|\uparrow\rangle\langle\uparrow| - |\downarrow\rangle\langle\downarrow|\big)_k. \tag{12}$$

It is easily seen that this Hamiltonian does the job of the second arrow in Eq. (5) - it has been constructed for this purpose. If the system object+apparatus at the time $t = 0$ is in the state $a|+\rangle|\Uparrow\rangle + b|-\rangle|\Downarrow\rangle$ (obtained via the first arrow in Eq. (5)) the wave function of object, apparatus and environment is

$$\Phi(0) = \big(a|+\rangle|\Uparrow\rangle + b|-\rangle|\Downarrow\rangle\big)|\epsilon_0\rangle. \tag{13}$$

For $t > 0$ one obtains

$$\Phi(t) = a|+\rangle|\Uparrow\rangle|\epsilon_\uparrow\rangle + b|-\rangle|\Downarrow\rangle|\epsilon_\downarrow\rangle \tag{14}$$

with

$$|\epsilon_\uparrow(t)\rangle = \prod_i \big(\alpha_i exp(ig_k t)|\uparrow\rangle_i + \beta_i exp(-ig_k t)|\downarrow\rangle_i\big) = |\epsilon_\downarrow(-t)\rangle. \tag{15}$$

The exact reduced density matrix is obtained by summing over all states of the environment:

$$\varrho_{OAE} = |a|^2|+\rangle\langle+||\Uparrow\rangle\langle\Uparrow| + |b|^2|-\rangle\langle-||\Downarrow\rangle\langle\Downarrow| + \left(ab^*r^*(t)|+\rangle\langle-||\Uparrow\rangle\langle\Downarrow| + c.c.\right), \quad (16)$$

with

$$r(t) = \langle\epsilon_\uparrow(t)|\epsilon_\downarrow(t)\rangle. \quad (17)$$

However, for $tg_k \to \infty$ $|r(t)| \to 0$ as 2^{-n} (excluding some exceptional values of α_i and β_i). For reasonable g_k this happens in an extremely short ("decoherence") time. Thus for large N the desired reduced density matrix

$$\varrho_{OAE} = |a|^2|+\rangle\langle+||\Uparrow\rangle\langle\Uparrow| + |b|^2|-\rangle\langle-||\Downarrow\rangle\langle\Downarrow| \quad (18)$$

is obtained, although as an approximation only. It consists only of the two projectors on the object+apparatus states as they may come out of measurements of the object. Thus this model reproduces (approximately) all steps from Eqs. (5) to (8). The environment has selected the pointer states $|+\rangle|\Uparrow\rangle$ and $|-\rangle|\Downarrow\rangle$ out of the general set of states $a|+\rangle|\Uparrow\rangle + b|-\rangle|\Downarrow\rangle$ i.e. it has produced decoherence. The Hamiltonian $\mathbf{H}_{AE}$ - designed for einselection and decoherence - may look a bit strange. No relation to any realistic interaction is visible - a rather typical situation in this field. For other models producing decoherence see Refs. 2 and 4.

5. Summary

After a long struggle now one seems to be on the way to a final understanding of the foundations of quantum theory. To get away from relying on models and qualitative arguments the following tasks are waiting to be taken care of:

i) justifying the interactions between object, apparatus and environment in terms of the standard many body Hamiltonians (kinetic energy and two body interaction and possibly external fields)

ii) understanding (deriving) exactly (" for all practical purposes") those *universal* features of the environment which are essential for decoherence.

These are problems to be solved with the methods of quantum many body theory. This is why I present this overview to the quantum many body community, hoping that more people with the necessary expertise get involved in this field.

References

1. J.v.Neumann, *Mathematische Grundlagen der Quantenmechanik* (Springer, Berlin, 1932); reprinted 1981; English translation, 1955: Mathematical Foundations of Quantum Mechanics (Princeton University, Princeton, NJ).
2. W.H.Zurek, *Rev.Mod.Phys.* **75**, 715 (2003). In this paper there is a rather complete list of further references.
3. L.Hackermüller, K. Hornberger, B.Brezger, A.Zeilinger and M. Arndt, *Nature* **427**, 711 (2004).
4. M.Schlosshauer, *Rev.Mod.Phys.* **76**, 1267 (2004). This paper contains another exhaustive list of references.

TOWARDS A COUPLED-CLUSTER TREATMENT OF SU(N) LATTICE GAUGE FIELD THEORY

RAYMOND F. BISHOP, NORBERT LIGTERINK,* AND NIELS R. WALET

School of Physics and Astronomy, The University of Manchester, Sackville Street Building, P.O. Box 88, Manchester M60 1QD, UK
[email: raymond.bishop@manchester.ac.uk]

A consistent approach to Hamiltonian SU(N) lattice gauge field theory is developed using the maximal-tree gauge and an appropriately chosen set of angular variables. The various constraints are carefully discussed, as is a practical means for their implementation. A complete set of variables for the colourless sector is thereby determined. We show that the one-plaquette problem in SU(N) gauge theory can be mapped onto a problem of N fermions on a torus, which is solved numerically for the low-lying energy spectra for $N \leq 5$. We end with a brief discussion of how to extend the approach to include the spatial (inter-plaquette) correlations of the full theory, by using a coupled-cluster method parametrisation of the full wave functional.

Keywords: gauge fixing; maximal-tree gauge; Hamiltonian approach

1. Introduction

Largely due to its conceptual elegance and simplicity, most calculations in lattice gauge-field theory (LGFT) have been performed within the Lagrangian formulation, which is based on a path-integral approach to the imaginary-time propagator. A great deal of work has been done to improve the accuracy of the Monte Carlo methods used in the numerical implementation of the Lagrangian approach, further increasing its viability.

Nevertheless, the much less studied Hamiltonian approaches have at least four important advantages. Firstly, since it is based on an imaginary-time evolution, the Lagrangian approach does not allow easy access to the vacuum wave functional. By contrast, such a wave functional is at the core of any Hamiltonian approach. Once the wave functional is known, most important properties of theories such as QCD, including, for example, confinement and chiral symmetry, should follow automatically. Secondly, time-dependent phenomena can only be discussed in a real-time (Hamiltonian) setting. Thirdly, the physical interpretation of the variables is much more transparent in the Hamiltonian framework, where, for example, electric and magnetic operators have their classical meanings. Lastly, the Hamiltonian approach

*Present address: Universiteit Twente, Faculteit EWI/Elektrotechniek, Postbus 217, 7500 AE Enschede, The Netherlands.

may be studied both with analytical techniques and numerically. Thus, for example, we can study low-lying excitations with a harmonic approximation and we can disentangle the dependence of observables on the parameters in the approximation.

In view of these advantages, we develop here a consistent approach to Hamiltonian LGFT using the maximal-tree gauge, which is further formulated in terms of a set of angular variables. The various constraints in the theory are discussed and implemented, and an independent and complete set of variables is determined for the colourless sector. We describe a general scheme to construct the eigenstates of the electric energy operator using an efficient symbolic method. It is shown how the one-plaquette problem for SU(N) LGFT can be mapped onto an N-fermion problem for arbitrary values of N. The low-lying energy spectra are investigated numerically, and explicit solutions are shown for SU(2), SU(3), SU(4), and SU(5). Previous attempts to use the coupled cluster method to include multi–plaquette correlations have largely been confined to the ground–state energy and low–lying excitation gaps (glueball masses) in the pure gauge (gluon) sector (i.e., without fermions) of the U(1) and SU(2) cases in one and two spatial dimensions.[1,2] We conclude with a brief discussion on how the new maximal-tree formulation might profitably be used to make further progress for the non-Abelian SU(N) gauge theories with $N \geq 2$ in three spatial dimensions where, as we show, significant new complications arise.

2. Hamiltonian Lattice Gauge Theory

The original Abelian gauge theory of electromagnetism was extended by Yang and Mills[3] using gauge fields of more complicated structure, which included internal degrees of freedom. This generates self-interactions since the gauge fields do not commute, but are chosen to obey the commutation relations of a specific Lie algebra.

2.1. *Lagrangian*

We shall concentrate on a gauge-field Lagrangian where the field $\boldsymbol{A}_\mu$ is an element of the Lie algebra su(N),

$$\mathcal{L} = \frac{1}{2g^2}\mathrm{Tr}[\boldsymbol{F}_{\mu\nu}\boldsymbol{F}^{\mu\nu}] \quad . \tag{1}$$

Here the field tensor $\boldsymbol{F}_{\mu\nu}$ is defined as

$$\boldsymbol{F}_{\mu\nu} \equiv \partial_\mu \boldsymbol{A}_\nu - \partial_\nu \boldsymbol{A}_\mu - i[\boldsymbol{A}_\mu, \boldsymbol{A}_\nu] \quad , \tag{2}$$

and the field variable $\boldsymbol{A}_\mu$ is an element of the algebra, conveniently parametrised as

$$\boldsymbol{A}_\mu \equiv g\frac{1}{2}\lambda^a A^a_\mu \quad . \tag{3}$$

The λ^a are the $N^2 - 1$ generators of the Lie algebra, satisfying the commutation relations

$$[\lambda^a, \lambda^b] = 2if^{abc}\lambda^c \quad . \tag{4}$$

The index a thus runs from 1 to N^2-1. The λ^a can be represented by traceless $N \times N$ matrices, normalised such that their squares have trace 2, as can be seen from the anticommutation relations

$$\{\lambda^a, \lambda^b\} = 2d^{abc}\lambda^c + \frac{4}{N}\delta_{ab}I \quad . \tag{5}$$

In Eq. (3) we have absorbed the coupling constant g into the field $\boldsymbol{A}_\mu$, so that we can interpret the fields geometrically, since the field tensor is now the curvature that follows from the covariant derivative

$$\boldsymbol{d}_\mu \equiv \partial_\mu - i[\boldsymbol{A}_\mu, \cdot] \quad , \tag{6}$$

from which follows the relation

$$[\boldsymbol{d}_\mu, \boldsymbol{d}_\nu] = -i[\boldsymbol{F}_{\mu\nu}, \cdot] \quad . \tag{7}$$

As we are interested in the Hamiltonian, we perform the standard equal-time quantisation and reformulate the Lagrangian in terms of generalised electric and magnetic fields. This is strictly speaking a 3-dimensional result, since this interpretation requires the use of three-dimensional algebra. We shall nonetheless use the result below for other numbers of spatial dimensions as well. We find

$$\mathcal{L} = \frac{1}{g^2}\mathrm{Tr}\left[\sum_{k=1}^{D}(\boldsymbol{E}_k^2 - \boldsymbol{B}_k^2)\right] \quad , \tag{8}$$

where $\boldsymbol{B}_i \equiv -\frac{1}{2}\epsilon_{ijk}\boldsymbol{F}_{jk}$ and $\boldsymbol{E}_i \equiv \boldsymbol{F}_{i0}$. Since we wish to impose the temporal gauge $\boldsymbol{A}_0 = 0$, we separate the Lagrangian into two parts, thereby isolating the $\boldsymbol{A}_0$ dependent part,[4]

$$\mathcal{L} = \frac{1}{g^2}\mathrm{Tr}\left[\sum_{k=1}^{D}(\boldsymbol{E}_k^2 - \boldsymbol{B}_k^2)\right]_{\boldsymbol{A}_0=0} + \frac{1}{g^2}\mathrm{Tr}\left[\boldsymbol{A}_0\boldsymbol{\mathcal{G}} + \boldsymbol{A}_0\boldsymbol{\mathcal{X}}(\boldsymbol{A}_0)\right] \quad , \tag{9}$$

where we have added a total divergence. The function $\boldsymbol{\mathcal{X}}$ is second order in $\boldsymbol{A}_0$ and does not contribute to the equations of motion or to the constraint equations in the temporal gauge ($\boldsymbol{A}_0 = 0$) discussed below.

Since the time-derivative of $\boldsymbol{A}_0$ does not occur in the Lagrangian, equation of motion for $\boldsymbol{A}_0$ is a time-independent algebraic equation, and thus A_0 is constant. This set of equations (one for each colour index) is the non-Abelian analogue of the Gauss' law constraint, and in the absence of colour charges they take the simple form

$$\mathcal{G}^a(\mathbf{x}) = 0 \quad , \tag{10}$$

where

$$\mathcal{G}^a(\mathbf{x}) = \sum_{i=1}^{D}[\partial_i E_i^a(\mathbf{x}) + g f^{abc} A_i^b(\mathbf{x}) E_i^c(\mathbf{x})] = \sum_{i=1}^{D}\boldsymbol{d}_i E_i^a \quad . \tag{11}$$

The components of the fields can be obtained via the relation

$$A^a_\mu = \frac{1}{g}\mathrm{Tr}[\boldsymbol{A}_\mu \lambda^a] \quad , \qquad E^a_\mu = \frac{1}{g}\mathrm{Tr}[\boldsymbol{E}_\mu \lambda^a] \quad . \tag{12}$$

The constraints obey the same commutation relations as the generators of the gauge group. Thus, Gauss' law cannot be implemented as a strict operator condition as it leads to contradictions, since the non-commuting constraints cannot all be diagonalised simultaneously. However, within the physical (in this case colourless) subspace defined by

$$\mathcal{G}^a(\mathbf{x})|\mathrm{Phys}\rangle = 0 \quad , \tag{13}$$

no such problem arises, since the eigenvalue of the commutators is also 0. The space of states consists of wave functionals, taking values on the SU(N) group manifold. From Eq. (13) we find $N^2 - 1$ functional conditions on each wave functional, consisting of functions on the group manifold at each space point.

As is well known, quantisation of problems involving redundant degrees of freedom (i.e., where some of the equations of motion are constraints) is quite involved. The two main techniques used are Dirac and BRS quantisation, and they require a large amount of additional analysis. For more details one can consult the seminal work by Dirac,[5,6] as well as Refs. [4,7,8,9,10]. If we are able to work within the physical subspace only, we can ignore these formal problems and define the quantisation of the canonical momenta, $\Pi^a_i \equiv \partial_0 A^a_i$, by

$$\Pi^a_i(\mathbf{x}) = \hat{E}^a_i = -i\frac{\delta}{\delta A^a_i(\mathbf{x})} \quad , \tag{14}$$

which involves a functional derivative[11,12] with respect to the field variables.

Since A^a_0 is not dynamical, we cannot associate a canonical momentum with it. We therefore use the temporal gauge, $A^a_0 = 0$, which leaves us with a residual gauge freedom $\boldsymbol{\phi}(\mathbf{x})$ independent of the time coordinate, such that under the transformation $\boldsymbol{A}_\mu(\mathbf{x}) \to {}^{\boldsymbol{\phi}}\boldsymbol{A}_\mu(\mathbf{x})$, where

$${}^{\boldsymbol{\phi}}\boldsymbol{A}_\mu(\mathbf{x}) = \boldsymbol{\phi}(\mathbf{x})\boldsymbol{A}_\mu(\mathbf{x})\boldsymbol{\phi}^{-1}(\mathbf{x}) + i[\partial_\mu\boldsymbol{\phi}(\mathbf{x})]\boldsymbol{\phi}^{-1}(\mathbf{x}) \quad , \tag{15}$$

$${}^{\boldsymbol{\phi}}\boldsymbol{F}_{\mu\nu}(\mathbf{x}) = \boldsymbol{\phi}(\mathbf{x})\boldsymbol{F}_{\mu\nu}(\mathbf{x})\boldsymbol{\phi}^{-1}(\mathbf{x}) \quad , \tag{16}$$

with $\boldsymbol{\phi} \in$ SU(N), the Lagrangian is invariant.

2.2. *Discretisation and the Hamiltonian*

Many quantum field theories suffer from singularities, both in the infrared and ultraviolet limits. In many interesting cases, such as QCD,[13] these are renormalisable. Rather than dealing directly with the continuum, we shall regularise the problem by introducing a simple hypercubic lattice in the D-dimensional space, with lattice spacing a. Since we are pursuing a Hamiltonian approach, time will remain continuous. We shall concentrate here only on the pure gauge theory.

As is by now well known (and see our full paper[14] for further details), the system is described by a set of gauge fields (or chromo-vector-potentials)

$$A_l = \frac{1}{2} g \sum_{a=1}^{N^2-1} \lambda^a A_l^a \quad , \tag{17}$$

that are now defined on the links l of the lattice. They are Hermitian, since λ^a is Hermitian. The chromo-electric fields, E_l^a, are the corresponding canonical momenta obeying the commutation relation,

$$[E_l^a, A_{l'}^{a'}] = -i\delta_{ll'}\delta_{aa'} \quad . \tag{18}$$

The group elements are the link variables, $U_l \equiv \exp\{iaA_l\}$, which are thus $N \times N$ matrices. They correspond to parallel transporters on the lattice. Since A_l is an element of the su(N) Lie algebra, U_l is an SU(N) matrix. The product of four such group elements around a primitive square on the lattice (usually called a plaquette or Wilson plaquette) defines the corresponding plaquette operator, $U_p \equiv U_1 U_2 U_3^{-1} U_4^{-1}$. Their traces form the simplest gauge-invariant quantities on the lattice.

Ignoring temporarily problems with overcompleteness of the variables, one can derive the Kogut-Susskind Hamiltonian,[9]

$$\begin{aligned} H_{\mathrm{KS}} &= \frac{g^2}{2a^{D-2}} \left(\sum_l \sum_{a=1}^{N^2-1} E_l^a E_l^a + \lambda \sum_p \mathrm{Tr}[2 - U_p - U_p^{-1}] \right) \\ &\equiv \frac{g^2}{2a^{D-2}} (H_E + \lambda H_M) \quad , \end{aligned} \tag{19}$$

where D is the number of spatial dimensions, and $\lambda \equiv a^{2D-6}/g^4$. The sum on l in the electric term H_E runs over all links, while the sum on p in the magnetic term H_M runs over all plaquettes on the lattice.

Our physical states are now gauge-invariant in the vacuum sector. So far, the price we have paid for adopting the Hamiltonian approach is twofold. Firstly, explicit Lorentz invariance has been broken and, secondly, the problem remains of determining the physical subspace.

2.3. *A first look at handling the constraints*

As we have seen above, the gauge freedom leads to constrained dynamics. The residual gauge symmetry involves all time-independent local gauge transformations. These gauge freedoms thus generally prove an obstacle to establishing a proper set of variables in which the wave functional may be expressed. Before proceeding let us do some simple counting of the number of *independent* (i.e., unconstrained) degrees of freedom in our formulation. The primitive variables are the set E_l^a defined on the set of N_l links. Hence, the total number of degrees of freedom is $(N^2 - 1)N_l$, where the first factor is simply the number of SU(N) group generators. However,

Table 1. The number of degrees of freedom on a hypercubic lattice of size $n \times n \times \cdots n$, for different numbers of space dimensions, D. [The $D =$ "1" case comprises n plaquettes on a line.]

Dimensionality, D	"1"	2	3	$D \geq 2$
number of sites, N_s	$2(n+1)$	$(n+1)^2$	$(n+1)^3$	$(n+1)^D$
number of links, N_l	$3n+1$	$2n(n+1)$	$3n(n+1)^2$	$Dn(n+1)^{D-1}$
number of plaquettes, N_p	n	n^2	$3n^2(n+1)$	$\frac{1}{2}D(D-1)n^2(n+1)^{D-2}$

there are constraints between them due to Gauss' law. Thus, we can readily derive the lattice versions $\mathcal{G}_i^a$ of the continuum generators $\mathcal{G}^a(\mathbf{x})$, for each of the N_s lattice sites i. Hence, the number of (unconstrained) independent degrees of freedom is

$$N_u = (N^2 - 1)(N_l - N_s + 1) \quad , \tag{20}$$

where the additional unity term in the second factor in this expression arises due to the overall global gauge degree of freedom that would finally still remain.

As shown in Table 1, we see that the total number of plaquettes, N_p on the lattice is given by $N_p = N_l - N_s + 1$ for the cases $D = 1$ and $D = 2$, but this relation is not true for $D > 2$. The construction of Gauss' law makes it clear why it is so attractive to work with plaquette variables or, more generally, with traces of products of group operators U_l around closed loops (i.e., Wilson loops), since these variables are automatically gauge-invariant. Hence, for $D = 1, 2$ the plaquette variables form a complete (i.e., neither over- nor under-complete) set of unconstrained variables. We turn below to the vexed question of what variables to choose for $D > 2$.

3. Gauge Fixing and Maximal Trees

As we have seen above, it is the gauge freedom that leads to the constrained dynamics and the fact that the set of link variables must, in general, be over-complete. However, since the link variables are still one of the most attractive sets to use, we are motivated to fix the gauge as much as possible. We actually choose to fix the gauge fully (apart from an overall global gauge transformation that we cannot fix) by separating all of the links on the spatial lattice into two sets. One set is chosen to be just sufficient to connect any two lattice sites in a unique way. Any such set of links is called a *maximal tree*.[15,16] Its choice for a given lattice is clearly not unique, but one such choice is shown in Fig. 1 for $D = 3$. The case shown corresponds to the union of all links on the x-axis for $y = z = 0$, and all links in the y-direction for $z = 0$, and all links in the z-direction, with an obvious labelling for the axes.

Thus, all SU(N) link variables U_l are thereby divided into two non-overlapping sets, $\{U_l\} \to \{V_l \,|\, l \notin \text{tree}; W_l \,|\, l \in \text{tree}\}$. The variables $\{W_l\}$ are now associated with the irrelevant gauge degrees of freedom; they are essentially the irrelevant *longitudinal electric fields*. Conversely, the variables $\{V_l\}$ form our basis for the relevant degrees of freedom, the magnetic variables. However, these latter link variables are not by themselves invariant under local gauge transformations. Hence the wave

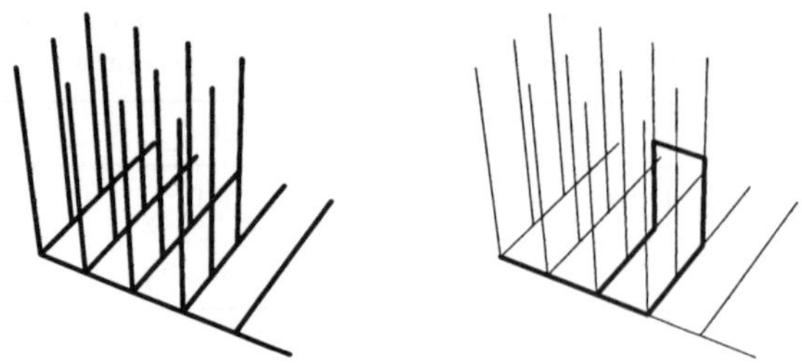

Fig. 1. A particular choice of maximal tree for $D = 3$ is shown on the left; and the path associated with a typical variable X_l for a particular link not on this tree is shown on the right.

function cannot depend directly on them since, gauge invariance would then be violated. Instead, we transform the links not on the tree into closed loops by combining them with (unique) paths on the maximal tree joining each end of the link from and to the origin, as is illustrated in Fig. 1. In this way we make a unique mapping, $V_l \leftrightarrow X_l; \forall l \notin$ tree, where X_l simply comprises a product of the corresponding U_l or U_l^{-1} operators on the links comprising the closed loop, where the inverse operators appear on links pointing back to the origin, in a convention that orients the links on the maximal tree along the direction away from the origin. Thus, all the variables X_l transform in the same way under local gauge transformations with the gauge transformation at the origin, and are invariant under all other local gauge changes. We know that when we fix the gauge we cannot fix a global gauge transformation, and we are thus led to identify this with the one at the origin.

We have shown[14] that the electric field operator E_l^a associated with the link l on the maximal tree can be transformed by a body-fixed frame rotation, such that when it acts on one of the X_m-variables one of the following relations holds, depending on the position of the link,

$$E_l^a X_m = -\frac{1}{2}\lambda^a X_m\,, \quad E_l^a X_m = \frac{1}{2} X_m \lambda^a\,, \quad E_l^a X_m = -\frac{1}{2}\lambda^a X_m + \frac{1}{2} X_m \lambda^a\,. \tag{21}$$

In this last equation the first result holds if link l is part of the path leading up to link m from the origin, the second result holds if it is part of the path leading back back from link m to the origin, and the third result holds if it is part of both paths. In this way the electric field operator generates long-range interactions between two link variables X_l and $X_{l'}$.

The links W_l on the maximal tree change under local gauge transformations, and any function in the physical subspace must hence be a function only of the gauge-invariant X_l variables. The gauge is effectively fixed by setting the SU(N) matrix operators W_l to be the unit operator, so that we have $\{V_l; W_l\} \to \{X_l; 1\}$. It is now a matter of simple counting to show that the X_l variables are precisely the (unconstrained) canonical variables of our Hamiltonian theory in the colourless sector, as summarised in Table 2. On a finite lattice of length n in D dimensions there are $N_s = (n+1)^D$ lattice points, and $N_l = Dn(n+1)^{D-1}$ links. The maximal tree contains $N_t = n+n(n+1)+n(n+1)^2+\cdots+n(n+1)^{D-1} = (n+1)^D-1 = N_s-1$

Table 2. The number of degrees of freedom in the maximal tree approach on a hypercubic lattice of size $n \times n \times \cdots n$, for different number of space dimensions, D.

Dimensionality, D	2	3	$D \geq 2$
number of sites, N_s	$(n+1)^2$	$(n+1)^3$	$(n+1)^D$
number of links, N_l	$2n(n+1)$	$3n(n+1)^2$	$Dn(n+1)^{D-1}$
number of links on tree, N_t	$n+n(n+1)$	$(n+1)^3-1$	$(n+1)^D-1$
number of independent X_l variables, $N_l - N_t$	n^2	$2n^3+3n^2$	$Dn(n+1)^{D-1} -(n+1)^D+1$

links, as can easily be seen from (the D-dimensional generalisation of) our explicit choice of maximal tree. Therefore there are $N_l - N_t = Dn(n+1)^{D-1} - (n+1)^D + 1$ remaining links, and the same number of variables X_l. Each of these variables has $N^2 - 1$ degrees of freedom. Hence, the number of degrees of freedom of the X_l variables is precisely equal to N_u, the number of unconstrained canonical degrees of freedom in the theory, as given by Eq. (20).

3.1. *The Hamiltonian in terms of the* X_l *variables*

We now start with the Kogut-Susskind Hamiltonian of Eq. (19) and assume that it will act on a function of the variables X_l. For the electric piece of the Hamiltonian, H_E, we make use of the result in Eq. (21). In the magnetic part, H_M, we can set all of the links on the maximal tree to unity. After some algebra we arrive at a final expression for our Hamiltonian given entirely in terms of our complete set of (unconstrained) canonical variables, X_l. We do not quote the somewhat unedifying expression here. The interested reader is referred to Ref. [14] for further details and the full expression.

4. Colour Neutrality and Further Constraints

Although we have solved the major problem above of the constrained dynamics due to the gauge freedom, we still have to face a further problem of overcompleteness related to the number of degrees of freedom in the *traces* of SU(N) matrices. This leads to the existence of dependences among the traces of the Wilson loop variables, usually known as Mandelstam constraints.[17]

Our wave function in the pure gauge sector studied here should be a colour singlet, and the above complications arise when we impose colour neutrality on the wave function. Thus, the natural choice for us is now to work with traces of products of the variables X_l, as discussed above. A suitable approach would be to construct a basis of eigenstates of the electric part of the Hamiltonian, H_E, and calculate matrix elements of the magnetic energy between these states. Such an approach is a quite natural calculational scheme for the Hamiltonian approach. One can also use the method inherent in the Lagrangian calculations, which is based on invariant integration over the full group.[18] However, for a proper Hamiltonian approach this discards many of the advantages of the method.

To find eigenstates of the electric operator, one can resort to three general approaches. Firstly, group theory gives us, in principle, a way to construct general eigenstates, the group characters. However, for a large basis, and $N > 2$, this is extremely involved,[19] unless it can be automated, and we see no easy way to do this.

A second approach is based on integrating configurations, and constructing orthogonal combinations from them. In this case one must start off with much larger overcomplete sets of configurations, and at increasing orders the integration, based on Creutz's integration method,[20,21] tends to become more and more involved.[22]

The third approach is based on the action of the electric operator itself, which leads to a block-diagonal matrix which has to be diagonalised to recover the eigenstates. In combination with a symbolic method explained more fully elsewhere,[14] this seems to be the most powerful approach, which allows one to tackle any arbitrary SU(N) group. This is the method that we have developed and that we prefer to use with the maximal tree approach, although we note that it also has wider applicability. for further details the reader is referred to Ref. [14]

Our whole approach so far has been designed for applications in mind using a universal method of microscopic quantum many-body theory in the Hamiltonian formulation, such as the coupled cluster method (CCM).[23] We do not have the scope in the present paper to discuss in any detail such many-body applications. Instead, we consider below the simpler limit where all plaquettes appear independently, and in so doing we introduce an extremely useful set of angular variables that considerably aid and simplify the analysis. We show that they serve to map an SU(N) one-loop problem onto an N-fermion problem on a torus.

5. The One-Plaquette Problem

If all of the loop variables are combined in such a way that they form plaquettes,[21] we may consider our wave functional as depending only on the plaquette variables. Since the trace of a matrix is invariant under similarity transformations, we may write

$$\mathrm{Tr}[U^m] = \mathrm{Tr}\left[\left(VUV^{-1}\right)^m\right] = \sum_{j=1}^{N} \mathrm{e}^{im\phi_j}\ ; \quad m \in [1, 2, \cdots, N] \quad , \tag{22}$$

where V diagonalises the SU(N) unitary matrix U of the one-plaquette problem. Hence, out of the N^2-1 degrees of freedom for SU(N) the trace depends only on its N eigenvalues $\exp(i\phi_j)$, all of which lie on the unit circle in the complex plane. The angular variables ϕ_j satisfy the constraint $\sum_{j=1}^{N}\phi_j = 0$ coming from the SU(N) condition, $\det U = 1$.

The wave functions of the one-plaquette problem can be expressed in a basis of group characters[24] labelled by the standard SU(N) partitions $\lambda \equiv [\lambda_1, \lambda_2, \cdots, \lambda_N]$. These are simply proportional to the eigenstates of the electric operator (which is just the quadratic Casimir operator) for the single plaquette. We denote the group characters of SU(N) as $\chi_\lambda(\phi)$, where we use the shorthand notation $\phi \equiv [\phi_1, \phi_2, \cdots, \phi_N]$. They may be represented as

$$\chi_\lambda(\phi) = \frac{\epsilon(\lambda_1, \lambda_2, \cdots, \lambda_N)}{\epsilon((N-1),(N-2),\cdots,2,1,0)} \; ; \quad \lambda_1 > \lambda_2 > \cdots > \lambda_N = 0 \, , \tag{23}$$

where $\epsilon(\lambda)$ is the determinant of the $N \times N$ matrix M with elements given by $M_{kl} = \exp(i\lambda_k\phi_l)$. The determinant in the denominator is a Vandermonde determinant of the eigenvalues of U, namely $\Delta \equiv \epsilon((N-1),(N-2),\cdots,2,1,0) = \prod_{k<l=1}^{N}[\exp(i\phi_k) - \exp(i\phi_l)]$. Its presence and its antisymmetry properties in the angular-variable eigenvalues allow us to map the original one-plaquette problem, which is symmetric under the interchange of the eigenvalues (i.e., an effective bosonic problem) but which has a complicated integration measure, into a much simpler fermionic problem by mapping $\Psi_\lambda \to \Phi_\lambda \equiv \Delta\Psi_\lambda$. In the electric part of the Hamiltonian we write correspondingly,

$$\sum_{a=1}^{N^2-1} E^a E^a \to \Delta \left(\sum_{a=1}^{N^2-1} E^a E^a \right) \frac{1}{\Delta} = D_N - \frac{N(N^2-1)}{24} \quad , \tag{24}$$

where the differential operator D_N, given by

$$D_N = -\frac{1}{2}\sum_{i=1}^{N} \frac{\partial^2}{\partial\phi_i^2} + \frac{1}{2}\left(\frac{1}{\sqrt{N}} \sum_{i=1}^{N} \frac{\partial}{\partial\phi_i} \right)^2 \quad , \tag{25}$$

acts on the antisymmetric wave function Φ.

When the the term Tr[U]+Tr[U^{-1}] in the magnetic part, H_M, of the Hamiltonian in Eq. (19), acts on a wave function, it now reduces to a multiplication[24] of group characters,

$$[\chi_1(\phi) + \chi_1(-\phi)]\epsilon_\lambda = \sum_{\lambda'=\lambda+1} \epsilon_{\lambda'} + \sum_{\lambda'=\lambda-1} \epsilon_{\lambda'} \quad , \tag{26}$$

where the symbols $\lambda \pm 1$ in the sums denote the inclusion of all possible partitions in which one and only one of the $\lambda_i \to \lambda_i \pm 1$ (with $1 \leq i \leq N$), while still fulfilling the requirements in Eq. (23). The operators Tr[U] and Tr[U^{-1}] thus act as raising and lowering operators, which act in a simple way on the Young tableaux corresponding to the group characters or wave functions with specific symmetry properties.[14]

We see from the above results that for a single SU(N) loop comprising L links (where $L = 4$ for a plaquette), the electric operator, H_E, in the Hamiltonian essentially yields L times the difference between the total kinetic energy of the N particles (now fermions) on a torus (represented by the angular variables ϕ_i; $i = 1, 2, \cdots, N$) and their centre-of-mass energy. It is very naturally expressed in terms of the momentum operators $p_j \equiv -i\frac{\partial}{\partial\phi_j}$. The magnetic term, H_M, for the single plaquette is also easily seen to be given as $2\lambda\sum_{j=1}^{N}(1-\cos\phi_j)$. The constraint det $U = 1$ now reduces to a constraint on the centre-of-mass motion, $\Phi \equiv N^{-1}\sum_{j=1}^{N}\phi_j = 0$. We need

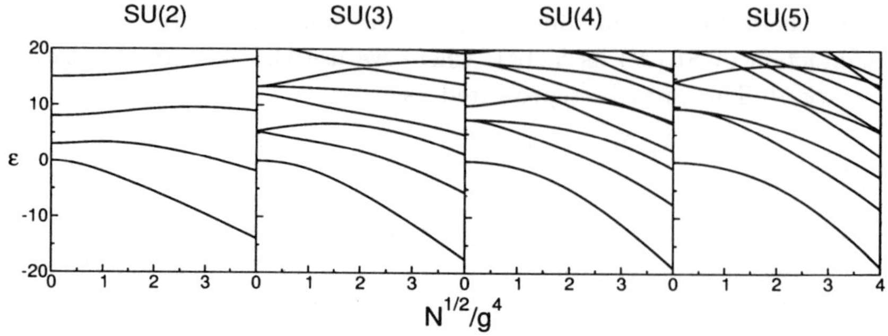

Fig. 2. The lowest levels of the one-plaquette energy spectra for SU(2) to SU(5).

also to impose the constraint on the centre-of-mass momentum, $P \equiv \sum_{j=1}^{N} p_j = 0$, since Φ is an unphysical variable for the SU(N) problem. A straightforward application of the Dirac quantisation procedure to handle the constraints then easily leads to the following realisation of the one-plaquette Hamiltonian,

$$H = \frac{g^2}{2a}\left(2\sum_{j=1}^{N}[p_j - P/N]^2 + 2\lambda\sum_{j=1}^{N}[1-\cos(\phi_j - \Phi)]\right) \quad , \tag{27}$$

which we have written for the physical case $D = 3$ for which $\lambda \equiv g^{-4}$. In this form the Hamiltonian is now manifestly translationally invariant. In the weak-coupling limit ($g \to 0$) (or, equivalently, $\lambda \to \infty$) we may readily use the harmonic approximation for the magnetic ("potential energy") term of the N-fermion problem. In this limit the entire centre-of-mass energy factorises. This great simplification allows for the easy implementation of the constraint $\det U = 1$, and the degeneracies ν_n of the nth level of the equidistant spectrum are now given by $\nu_n = P_N(n) - P_N(n-1)$, where $P_m(n)$ is the number of different ways the positive integer n can be partitioned as the sum of m positive integers.

5.1. *Numerical results*

We now investigate the one-plaquette spectra of a few of the relevant SU(N) gauge theories, namely for $2 \leq N \leq 5$. The method we use to solve the problem is first to work in a basis of eigenstates of the electric Hamiltonian, and then to evaluate the action of $\mathrm{Tr}[U]$ and $\mathrm{Tr}[U^{-1}]$ on these states. For the low-lying spectra that we calculate we solve the linear eigenvalue equations numerically. From the form of the Hamiltonian in Eq. (19) or Eq. (27), we see that it is convenient to define a scaled energy ε as follows,

$$E = \frac{g^2}{2a}\varepsilon + \frac{N}{g^2 a} \quad , \tag{28}$$

which we use for representing the numerical results.

For SU(2), the one-plaquette Schrödinger equation reduces to the Mathieu equation and the spectrum is given by its odd characteristic values, as shown in the left panel of Fig. 2. No such closed-form analytical solutions seem to exist for SU(N) with $N > 2$, and we resort to the numerical procedure described above to obtain the remaining results shown in Fig. 2. We note that the spectra for $N > 2$ are much richer than for $N = 2$. They include not only many avoided crossings but also, very interestingly, what appear to be several real crossings in the cases $N = 4$ and $N = 5$. We have verified that the distances between the respective two levels in these latter cases are equal to zero within our numerical accuracy. Such real crossings, of course, are a reflection of some (otherwise hidden) symmetry in these theories. They clearly deserve further investigation.

Finally, we note that the region of the coupling constant shown in Fig. 2 is insufficient to observe the asymptotic convergence to the harmonic approximation discussed above. However, we have checked numerically, by going to values $g^{-4} \geq 50$, that our stated results for the degeneracy factors for the equidistant harmonic oscillator spacings in this limit are correct.

5.2. *Independent-plaquette wave functionals*

The results for the one-plaquette problem have more consequences for more general (e.g., variational or CCM) wave functionals than one might at first suspect. Thus, if the trace variables, $\xi_m \equiv m^{-1}\mathrm{Tr}(U^m)$, of the one-plaquette matrix are used, the wave functional is a function of the group characters only. The specific wave functional that is the sum of one-plaquette functions,

$$\langle\{\phi_j^\alpha\}|\Sigma\rangle = \sum_{\text{plaquettes }\alpha} F\left(\{\phi_j^\alpha\}_{j=1}^N\right) \quad , \tag{29}$$

naturally leads to the sum of one-plaquette problems, leading to total energies which are the sum of one-plaquette energies. However, the corresponding product wave functional

$$\langle\{\phi_j^\alpha\}|\Pi\rangle = \prod_{\text{plaquettes }\alpha} F\left(\{\phi_j^\alpha\}_{j=1}^N\right) \quad , \tag{30}$$

also leads to the same result as we now show. This absence of correlations between nearest-neighbour plaquettes, follows in our case from the symmetry of (the original bosonic) wave functional, Ψ_λ, in the angular variables,

$$\Psi_\lambda(\cdots\phi_i\cdots\phi_j\cdots) = \Psi_\lambda(\cdots\phi_j\cdots\phi_i\cdots) \quad , \tag{31}$$

where we decompose

$$F(\phi_1\cdots\phi_N) = \sum_\lambda c_\lambda\Psi_\lambda(\phi_1\cdots\phi_N) \quad . \tag{32}$$

Therefore the cross-product term from the electric operator vanishes,

$$\begin{aligned} &D_N\Psi'_\lambda(\phi_1^\alpha\cdots\phi_N^\alpha)\Psi'_{\lambda'}(\phi_1^\beta\cdots\phi_N^\beta) \\ &= \Psi'_{\lambda'}(\phi_1^\beta\cdots\phi_N^\beta)D_N\Psi'_\lambda(\phi_1^\alpha\cdots\phi_N^\alpha) + \Psi'_\lambda(\phi_1^\alpha\cdots\phi_N^\alpha)D_N\Psi'_{\lambda'}(\phi_1^\beta\cdots\phi_N^\beta) \quad , \end{aligned} \tag{33}$$

where α and β are plaquettes containing the link l, and $\Psi'_\lambda = \det(J)\Psi_\lambda$, where $\det(J) = |\Delta|^2$ is the Jacobian of the transformation from trace variables to angular variables, $J_{mj} \equiv \partial\xi_m/\partial\phi_j = i\exp(im\phi_j)$. The differential operator D_N contains both sets of angular operators, $\partial_{\phi_i} = \partial_{\phi_i^\alpha} + \partial_{\phi_i^\beta}$. Therefore the Hamiltonian, when acting on the product wave functional $|\Pi\rangle$, also reduces to the sum of one-plaquette Hamiltonians.

6. Summary and Whither Next?

In summary, we have shown how to fix the gauge by using the maximal-tree gauge, which is specific to the lattice versions of gauge field theories. It does not suffer from the typical problems of gauge fixing in the continuum. We then introduced a very natural set of angular variables , simply related to the trace (or Wilson loop) variables for a fundamental plaquette, in terms of which the SU(N) one-loop problem can be mapped onto an N-fermion problem on a torus. Exact solutions were obtained for a single plaquette in both the weak-coupling ($g \to 0$) and strong-coupling ($g \to \infty$) limits. The eigenstates in the strong-coupling limit are the group characters of the corresponding group, and in the weak-coupling limit the harmonic approximation to the fermion problem leads to explicit formulas for the degeneracies of the equidistant energy levels. In terms of the group characters (i.e., the eigenstates of the electric piece of the Hamiltonian), the ground and low-lying energy states of the single plaquette can be determined numerically for arbitrary values of the coupling constant, since the Hamiltonian reduces to a simple linear equation in terms of these group characters. The terms in the magnetic piece of the Hamiltonian are simple raising and lowering operators in this basis, and hence the energy spectra are found by a simple configuration-interaction method technique of diagonalising the resultant block-diagonal Hamiltonian matrix in a restricted basis. The procedure can be automated using the theory of multiplication of group characters, in terms of the Young tableaux of the corresponding group.

Another, somewhat unexpected, advantage of our approach is the fact that correlations between spatially distinct trace variables cancel. This leads to the intriguing possibility that spatial correlations in the full lattice problem might be weak in our approach, which would in turn tend to suggest that successive approximations in a fully consistent microscopic many-body approach based on our approach might be expected to converge rapidly. Clearly, the simplicity of the angular variables formulation is promising for more elaborate wave functionals. An obvious extension to include correlations is to employ the coupled cluster method (CCM),[23] which has been extensively applied with great success to a wide variety of quantum many-body systems and quantum field theories. Typical applications of the CCM to strongly interacting *continuum* quantum field theories include ϕ^4 field theory[25,26] and a model field theory of pions and nucleons.[27] There have been many applications of the CCM to various spin-lattice models in quantum magnetism.[28] Other applications to *lattice* field theories include the O(4) nonlinear sigma model as a model of meson field theory with a phase transition due to chiral symmetry breaking.[29]

Previous attempts to use the CCM in lattice gauge theory have largely been confined to the ground and low-lying excited states in the pure gauge (gluon) sectors of the U(1) and SU(2) cases in one and two spatial dimensions.[1,2] As we have seen, for $D < 3$ the plaquette variables U_p form a natural complete set. However, they are overcomplete for $D \geq 3$. One of our key aims here has been to find a natural complete set of variables for this case. We have shown that the variables X_l fulfil this role.

At the heart of the CCM is the parametrisation of the ket-state many-body wave function as an exponential of a correlation operator formed from a linear combination of mutually commuting creation (or excitation) operators with respect to a model or reference state $|\Phi\rangle$,

$$|\Psi\rangle = \mathrm{e}^{S}|\Phi\rangle \quad ; \quad S = \sum_{I} c_I |I\rangle\langle\Phi| \quad . \tag{34}$$

The creation operators here excite from the model state to an arbitrary excited state $|I\rangle$, and do not act between different excited states on overlapping lattice-site configurations.

From this vantage point the key question is then the choice of the form of the model state $|\Phi\rangle$ and of the states in which to expand the many-body Hilbert space. Clearly, the simplest choice for the model state is just the chromo-electric vacuum. As for the remainder, this is precisely the question that we have addressed here, where we have established the viability for such purposes of the set of independent maximal-tree variables, X_l, and their conjugate variables, E_l^a. Thus, for the colourless (pure gauge) sector we have to use closed contours, which are traces over products of X_l variables, since only these variables are invariant under gauge transformations generated by Gauss' law.

We conclude with some final comments. Firstly, we note that the form of our SU(N) Hamiltonian in the X_l variables is very different from the naive Kogut-Susskind form of Eq. (19). This is likely to have an important effect on the role of correlations in the ensuing analysis. Secondly, a price that we pay for using the maximal-tree gauge is that the tree, and hence our Hamiltonian, has a preferred direction and we thereby lose explicit translational invariance. Although this is presumably restored in a full CCM calculation carried out to all orders, in practice we need to make approximations via a systematic hierarchy of choices for which excited states $|I\rangle$ to include in the CCM ket-state correlation operator S in Eq. (34), and it is difficult to know in advance how serious the loss of translational invariance might be in practice at attainable levels of implementation. An obvious next step in this regard would be to investigate the possibility of designing different, more symmetric, choices for the maximal tree in order to circumvent this problem. On the other hand, although the current choice might turn out to be problematic for a study of the vacuum sector, it would seem to be a natural choice for the study of the interaction between fixed sources on the lattice, which explicitly breaks translational invariance. Such studies are themselves an obvious next step in the extension of the current approach away from the pure gauge sector to include fermions. We note that some very preliminary steps in this direction have already been taken in the much simpler case of the Schwinger model of quantum electrodynamics in one spatial dimension, whose Hamiltonian form on the lattice has been investigated, using CCM techniques, with the inclusion of massive staggered fermions.[30]

References

1. R. F. Bishop and Y. Xian, *Acta Phys. Pol.* **B24**, 541 (1993); R. F. Bishop, A. S. Kendall, L. Y. Wong, and Y. Xian, *Phys. Rev. D* **48**, 887 (1993); R. F. Bishop and Y. Xian, in *Condensed Matter Theories*, Vol. **9**, eds. J. W. Clark, K. A. Shoaib, and A. Sadiq, (Nova Science Publ., Commack, N.Y., 1994), p. 433; S. J. Baker, R. F. Bishop, and N. J. Davidson, *Phys. Rev. D* **53**, 2610 (1996).
2. C. H. Llewellyn Smith and N. J. Watson, *Phys. Lett.* **B302**, 463 (1993).
3. C. N. Yang and R. L. Mills, *Phys. Rev.* **96**, 191 (1954).
4. L. D. Faddeev and A. A. Slavnov, *Gauge Fields, Introduction to Quantum Theory* (Benjamin Cummings, Reading, Mass., 1980).
5. P. A. M. Dirac, *Principles of Quantum Mechanics*, 4th ed., (Oxford University Press, Oxford, 1967).
6. P. A. M. Dirac, *Lectures on Quantum Mechanics*, (Belfer Graduate School of Science, Yeshiva University, New York, 1964); *Lectures on Quantum Field Theory*, (Belfer Graduate School of Science, Yeshiva University, New York, 1966).
7. T. Muta, *Foundations of Quantum Chromodynamics*, (World Scientific, Singapore, 1987).
8. M. Henneaux and C. Teitelboim, *Quantization of Gauge Systems*, (Princeton University Press, Princeton, 1992).
9. J. Kogut and L. Susskind, *Phys. Rev. D* **11**, 395 (1975).
10. J. B. Kogut, *Rev. Mod. Phys.* **51**, 659 (1979).
11. F. A. Berezin, *The Method of Second Quantization* (Academic Press, New York, 1966).
12. B. Felsager, *Geometry, Particles and Fields* (Odense University Press, Odense, 1981).
13. G. 't Hooft and M. Veltman, *Nucl. Phys.* **B44**, 189 (1972).
14. N. E. Ligterink, N. R. Walet, and R. F. Bishop, *Ann. Phys. (N.Y.)* **284**, 215 (2000).
15. V. F. Müller and W. Rühl, *Nucl. Phys.* **B230**, 49 (1984).
16. J. B. Bronzan, *Phys. Rev. D* **31**, 2020 (1985); *ibid.* **37**, 1621 (1988); *ibid.* **38**, 1994 (1988); J. B. Bronzan and T. E. Vaughan, *ibid.* **43**, 3499 (1991); *ibid.* **47**, 3543 (1993).
17. S. Mandelstam, *Phys. Rev. D* **19**, 2391 (1979).
18. D. Horn and M. Karliner, *Nucl. Phys.* **B235**,135 (1984).
19. M. A. B. Bég and H. Ruegg, *J. Math. Phys.* **6**, 677 (1965).
20. M. Creutz, *J. Math. Phys.* **19**, 2043 (1974).
21. M. Creutz, *Quarks, Gluons, and Lattices*, (Cambridge Univ. Press, Cambridge, 1983).
22. C. R. Leonard, Ph.D. thesis, Melbourne University, 1998.
23. R. F. Bishop, *Theor. Chim. Acta* **80**, 95, (1991); in *Microscopic Quantum Many-Body Theories and Their Applications*, eds. J. Navarro and A. Polls, Lecture Notes in Physics, Vol. **510**, (Springer-Verlag, Berlin, 1998), p. 1.
24. H. Weyl, *The Classical Groups*, 2nd ed., (Princeton Univ. Press, Princeton, 1946); J. E. Hetrick, *Int. J. Mod. Phys. A* **9**, 3153 (1994); J. Hallin, *Class. Quant. Grav.* **11**, 1615 (1994).
25. U. Kaulfuss, *Phys. Rev. D* **32**, 1421 (1985); C. S. Hsue, H. Kümmel, and P. Ueber-

holz, *ibid.* **32**, 1435 (1985); M. Altenbokum and H. Kümmel, *ibid.* **32**, 2014 (1985); M. Funke, U. Kaulfuss, and H. Kümmel, *ibid.* **35**, 621 (1987).
26. A. H. Rezaeian and N. R. Walet, *J. High Energy Phys.* **12**, 040 (2003); *Phys. Lett.* **B570**, 129 (2003).
27. H. Kümmel, *Phys. Rev. C* **27**, 765 (1983); G. Hasberg and H. Kümmel, *ibid.* **33**, 1367 (1986).
28. R. F. Bishop, J. B. Parkinson, and Y. Xian, *Phys. Rev. B* **44**, 9425 (1991); C. Zheng, D. J. J. Farnell, and R. F. Bishop, *J. Stat. Phys.* **60**, 327 (1998); J. Rosenfeld, N. E. Ligterink, and R. F. Bishop, *Phys. Rev. B* **60**, 4030 (1999); S. E. Krüger, J. Richter, J. Schulenburg, D. J. J. Farnell, and R. F. Bishop, *ibid.* **61**, 14607 (2000); D. J. J. Farnell, R. F. Bishop, and K. A. Gernoth, *J. Stat. Phys.* **108**, 401 (2002); D. J. J. Farnell and R. F. Bishop, in *Quantum Magnetism*, eds. U. Schollwöck, J. Richter, D. J. J. Farnell and R. F. Bishop, Lecture Notes in Physics, Vol. **645**, (Springer-Verlag, Berlin, 2004), p. 307.
29. N. E. Ligterink, N. R. Walet, and R. F. Bishop, *Ann. Phys. (N.Y.)* **267**, 97 (1998); *Phys. Rev. E* **63**, 037103 (2001).
30. R. McDonald and N. R. Walet, *Int. J. Mod. Phys. B* **17**, 5393 (2003).

A TRIBUTE TO SEVEN DECADES: PHYSICS AND MUCH MORE

A TRIBUTE TO JOHN WALTER CLARK ON HIS 70th BIRTHDAY

CHARLES E. CAMPBELL

School of Physics and Astronomy, University of Minnesota, Minneapolis, MN 55455, USA
campbell@umn.edu

The 13th International Conference on Recent Progress in Many-Body Theories afforded participants the opportunity to celebrate the 70th birthdays of two esteemed colleagues who have contributed enormously to many-body theoretical physics during five decades: John Clark and Manfred Ristig.

It is my honor and privilege to offer a few words on this occasion about my dear friend, former teacher, academic brother, and collaborator, John Walter Clark. For over forty years John has been an international leader, pioneer, and outstanding contributor to many-body theory and to an impressively broad set of physical problems falling under that umbrella. He has deepened our understanding of the physics of systems with many degrees of freedom while addressing a wide variety of physical systems, from nuclear physics, astrophysics, and the physics of condensed matter quantum fluids to neural networks, computational neuroscience, and quantum control theory.

John was born on April 7, 1935, in Lockhart, Texas. He received his B.S. (1955) and M.A. (1957) in Physics from the University of Texas, Austin. He travelled north for his Ph.D. in Physics, receiving it in 1959 at Washington University in St. Louis, where his advisor was Eugene Feenberg. This was the period when the foundations were being laid for quantum many-body physics. John was the first of Feenberg's students to do research in many-body theory. Together they were the pioneers of a way of approaching strongly interacting many-body systems non-perturbatively from a microscopic, first principles level. John's Ph.D. dissertation was one of the first in microscopic many-body physics.

John returned to Washington University in 1963 as Assistant Professor of Physics, where he was a close friend and colleague of Feenberg until his death in 1977. John was promoted to associate professor in 1966 and full professor in 1972. He served as interim chair of the Department of Physics during 1996-1997, and has been chair since 2002. In 1999 John was appointed Wayman Crow Professor of Physics, a storied chair whose previous occupants included Arthur Holly Compton, Edward Uhler Condon, Eugene Feenberg, and Edwin T. Jaynes.

In the years between his Ph.D. and his return to St. Louis, John was successively an NSF Postdoctoral Fellow with Eugene Wigner at Princeton University, an asso-

John W. Clark

ciate research scientist at the Martin Company, Denver, and a NATO Postdoctoral Fellow both at the University of Birmingham, England, with Sir Rudolph Peierls, and the French nuclear research establishment in Saclay, with Claude Bloch. In 1965 he was awarded an Alfred P. Sloan Foundation Fellowship, and was elected Fellow of the American Physical Society in 1972.

John was the recipient of the second Eugene Feenberg Medal for Many Body Physics presented at the Fifth International Conference on Recent Progress in Many Body Theory, in Oulu, Finland, in 1987, for his fundamental developments and high-precision applications of the method of correlated basis functions toward a universal microscopic approach for the study of strongly interacting quantum many-body systems.[1]

John has long been an international leader of physics. He has chaired, been on program and advisory committees, and/or co-edited the proceedings of dozens of international conferences and workshops. In particular, for more than thirty years he has been a key leader of the series of International Workshops on Condensed Matter Theories, perhaps the oldest international, annual workshop in the field.[2] For a similar length of time he has provided leadership, and was chair for six years of the International Advisory Committee, for the International Conferences on Recent Progress in Many-body Theories, of which the present conference is the thirteenth edition. He has contributed in important ways to the International Forum of the American Physical Society, including chairing the John Wheatley Award Committee.

John has been a mentor for many young physicists throughout the world. An examination of his published papers reveals more than 100 co-authors, with a strong international flavor. He has supervised the Ph.D. dissertation research of more than 25 students and has hosted a similarly large number of postdocs and visitors for extended visits to his research group.

John Clark's contributions to the development of the method of correlated basis functions (CBF) have provided much of the foundation of our theoretical understanding of strongly interacting many-body systems such as nuclear matter and neutron matter, for both finite nuclei and astrophysical objects, and terrestrial quantum fluids such as the helium liquids and spin polarized deuterium fluids. The strong interactions between the constituent particles precluded the direct application of ordinary perturbation theory to these systems. Clark and Feenberg recognized that the use of Jastrow variational wavefunctions to tame these strong, short-range interactions was just the first step in a complete, *ab initio* theory wherein the complete wavefunction basis was not the non-interacting particle basis, but was built instead out of that basis by appending a correlation factor–a Jastrow function or, in principle, the exact boson ground state wavefunction–to create the correlated basis, with which multi-configurational methods could be applied to improve upon the ground state and study low excited states of the systems of interest. The seminal work in this area was Clark's Ph.D.thesis, his 1959 paper with Eugene Feenberg, and his collaborations with his student Paul Westhaus. These early papers are listed in the section of selected publications below.

Subsequent applications of this theory included insightful work on superfluidity of nucleonic matter, with important astrophysical applications. We will return to that subject below.

In Clark's long and productive collaboration with Manfred Ristig, they developed–with their students–the first systematic inclusion of state-dependent correlations in the theory of quantum fluids, which was a particularly important advance in the theory of nuclear and neutron matter where state dependent interactions cannot be ignored in a quantitatively accurate theory. Clark and Ristig collaborated on a number of other problems of importance to quantum fluids and nucleon systems, co-authoring more than 50 papers. Also particularly noteworthy was their research on the one-body density matrix and momentum distribution of both boson and fermion quantum fluids.

While Clark's early CBF work focused on the systematic formulation of cluster expansion methods to evaluate matrix elements of the Hamiltonian and identity operator within the correlated basis, he and Eckhard Krotscheck and collaborators extended the FHNC resummation method–shortly after it was developed by Fantoni and Rosati and independently by Krotscheck and Ristig–to a much more complete evaluation of these matrix elements, including off-diagonal elements, necessary for dealing with the more strongly correlated helium three liquid. Subsequently, they and their collaborators extended the CBF program to correlated versions of RPA, with one consequence being the recognition of the systematics of the more widely studied weakly interacting systems.

In such a brief reflection it is impossible to do justice to the importance, breadth, and depth of John Clark's research. Ray Bishop's history and tribute to John when he was presented the second Feenberg Medal provides a much more thorough and authoritative review of Clark's research that preceded the award.[1]

There have been a number of other significant thrusts of John Clark's research program during the last twenty-five years. He and his students, together with several distinguished collaborators, have developed an enormously promising program of neural-network models and advanced methods of machine learning, which they have applied, e.g., to existing data alone to predict the properties of nuclides such as the mass formula and one- and two-nucleon separation energies. A number of Clark's reviews and monographs listed below are devoted to this area of research.

The 1983 paper by Huang, Tarn, and Clark, listed below, is a foundational work in quantum control theory in which they extend classical control theory to the quantum domain. During the past 20 years, John and his collaborators have continued to make important contributions to this exciting area. A recent paper by Clark, Lucarelli, and Tarn provides a brief primer on the subject and notes its implications for quantum information theory and quantum computation.[3]

Early on, John and his collaborators contributed significantly to the theoretical study of nucleonic superfluidity in neutron stars (e.g., the paper with Chao and Yang listed in the selected papers list below). As with much of his other work, one of the major challenges that was overcome was the strong, short-range interaction, in this earliest case the S-wave channel, between neutrons. More recently, Clark has collaborated with Khodel, Khodel, and Zverev to investigate triplet pairing in the neutronic component of a neutron star, where the strong short-range correlations suppress the singlet S-wave pairing. By using the recently developed "separation" method for solving the gap equation, they avoid the traditional–and in this case entirely inappropriate–assumption of a constant gap. Rich and surely very important results emerge from these remarkable calculations.

It is wonderful to observe that, at the age of 70, John Clark's research program is richer than ever in its breadth and depth, despite his heavy and selfless commitment to administration and service to the international physics community. His current research programs–on microscopic quantum many-body theory; on computational neuroscience; on global statistical models for predicting nuclear properties far from stability; and on quantum control, quantum entanglement, and quantum phase transitions– promise many new and exciting research results in the future.

Selected publications of John W. Clark (chosen from over 250)

- J. W. Clark and E. Feenberg, Simplified treatment for strong-short range repulsions in N particle systems, *Phys. Rev.* **113**, 388-399 (1959).
- J. W. Clark and P. Westhaus, Method of correlated basis functions, *Phys. Rev.* **141**, 833-857 (1966); **149**, 990 (1966).
- J. W. Clark and P. Westhaus, Cluster expansions in many-fermion theory, I. Factor-cluster formalisms, *J. Math. Phys.* **9**, 131-148(1968); P. Westhaus and J. W. Clark, Cluster expansions in many-fermion theory; II. Rearrangements of primitive decomposition equations, *J. Math. Phys.* **9**, 149-154 (1968).
- N.-C. Chao, J. W. Clark, and C.-H. Yang, Proton superfluidity in neutron-star matter, *Nucl. Phys.* **A179**, 320-332(1972).
- J. W. Clark, C.-G. Källman, C.-H. Yang, and D. A. Chakkalakal, Effect of polarization on superfluidity in low-density neutron matter, *Phys. Lett.* **61B**, 331-334(1976).
- M. L. Ristig and J. W. Clark Density matrix of quantum fluids, *Phys. Rev.* **B 14**, 2875-2887(1976).
- Garng M. Huang, T. J. Tarn, and J. W. Clark, On the controllability of quantum-mechanical systems, *J. Math. Phys.* **24**, 2608-2618(1983).
- J. M. C. Chen, J. W. Clark, R. D. Dave, and V. V. Khodel, Pairing gaps in nucleonic superfluids, *Nucl. Phys.* **A555**, 59-89(1993).
- V. A. Khodel, V. V. Khodel, and J. W. Clark, Solution of the gap equation in neutron matter, *Nucl. Phys.* **A598**, 390-417(1996); M. V. Zverev, J. W. Clark, and V. A. Khodel, (2003) $^3P_2 - ^3F_2$ pairing in dense neutron matter. The spectrum of solutions, *Nucl. Phys.* **720**, 20-42.
- J. W. Clark, V. A. Khodel, and M. V. Zverev (2005) Anomalous Low-

Temperature Behavior of strongly correlated Fermi systems, *Phys. Rev.* **B 71**, 012401: 1-4.

Monographs and reviews by John W. Clark

- J. W. Clark, Variational Theory of Nuclear Matter, in *Progress in Particle and Nuclear Physics,* ed. D. H. Wilkinson (Pergamon, Oxford, 1979), Vol. 2, pp. 89-199.
- J. W. Clark, J. Rafelski, and J. V. Winston, Brain Without Mind: Computer Simulation of Neural Networks with Modifiable Neuronal Interactions, *Physics Reports* **123**(4), pp. 215-273 (1985)
- J. W. Clark, Statistical Mechanics of Neural Networks, *Physics Reports* **158**, pp. 91-158 (1988)
- J. W. Clark, Introduction to Neural Networks, in *Nonlinear Phenomena in Complex Systems,* ed. A. N. Proto (Elsevier, Amsterdam, 1989).
- J. W. Clark, Neural network modelling, *Physics in Medicine and Biology* **36**, pp. 1259-1317 (1991)
- J. W. Clark, Neural networks: New tools for modeling and data analysis in science, in *Scientific Applications of Neural Nets, Springer Lecture Notes in Physics*, Vol. 522, ed. J. W. Clark, T. Lindenau, and M. L. Ristig (Springer-Verlag, Berlin, 1999), pp. 1-96.

References

1. R. F. Bishop, in *Recent Progress in Many-Body Theories, Volume 1* ed. A. J. Kallio, E. Pajanne, and R.F. Bishop,(Plenum Press, New York, 1987), pp. 385-393
2. J.W. Clark in *Condensed Matter Theories, Volume 9* ed. J. W. Clark, K. A. Shoaib, and A. Sadiq (Nova Science Publishers, Commack, NY, 1994) pp. xiii–xv.
3. J. W. Clark, D. G. Lucarelli, and T. J. Tarn, in *Recent Progress in Many-Body Theories, Volume 6* ed. R.F. Bishop, T. Brandes, K. A. Gerrnoth, N. R. Walet, and Y. Xian (World Scientific, Singapore, 2002), pp. 411-424.

APPLICATION OF SUPPORT VECTOR MACHINES TO GLOBAL PREDICTION OF NUCLEAR PROPERTIES

JOHN W. CLARK and HAOCHEN LI

Physics Department, Washington University, St. Louis, Missouri 63130, USA

Advances in statistical learning theory present the opportunity to develop statistical models of quantum many-body systems exhibiting remarkable predictive power. The potential of such "theory-thin" approaches is illustrated with the application of Support Vector Machines (SVMs) to global prediction of nuclear properties as functions of proton and neutron numbers Z and N across the nuclidic chart. Based on the principle of structural-risk minimization, SVMs learn from examples in the existing database of a given property Y, automatically and optimally identify a set of "support vectors" corresponding to representative nuclei in the training set, and approximate the mapping $(Z, N) \to Y$ in terms of these nuclei. Results are reported for nuclear masses, beta-decay lifetimes, and spins/parities of nuclear ground states. These results indicate that SVM models can match or even surpass the predictive performance of the best conventional "theory-thick" global models based on nuclear phenomenology.

Keywords: global nuclear modeling; machine learning; database mining.

1. Introduction

Consider a class of quantum many-body problems such that each member of the class is characterized by X_1 particles of type 1 and X_2 particles of a different type 2, and by the interactions, both pairwise and multiparticle, that operate among any subset of particles. With the interactions in place, all of the states and all of the observables of any exemplar of the class are determined by the integers X_1 and X_2. Specifically, for any observable Y defined for the class, there exists a physical mapping from (X_1, X_2) to Y. This notion is trivially extended to more than two type of constituents.

Quite obviously, the nuclear many-body problem defines a class of this kind, with X_1 and X_2 taken as the numbers Z and N of protons and neutrons in a nuclide. Other examples coming easily to mind: ^{3}He-^{4}He clusters, binary and ternary alloys, etc.

Approaches to calculation – or prediction – of the properties of individual systems belonging to such a class span a broad spectrum from pure *ab initio* microscopic treatments to phenomenological models having few or many adjustable parameters, with hybrid macroscopic/microscopic and density-functional methods in between. These approaches are "theory-thick" to varying degrees, with the *ab initio* ones based in principle on exact theory and the phenomenological ones invoking physical intuition, semi-classical pictures, and free parameters.

Thinking in the spirit of Edwin Jaynes, inventor of the MaxEnt method and charismatic proponent of Bayesian probability,[1] it becomes of special interest to go all the way in the "theory-thin" direction and ask the question:

> *To what extent does the existing data on property Y across the members of a system class, and only the data, determine the mapping $(X_1, X_2) \to Y$?*

In general, this mapping takes one of two forms, depending on whether Y is a continuous variable (e.g., the nuclear mass excess or quadrupole moment) or a discrete variable (e.g., the nuclear spin and parity). The former case defines a problem of function approximation, while the latter defines a classification problem. During the past three decades, powerful new methods have been developed for attacking such problems. Chief among these are advanced techniques of statistical learning theory, or "machine learning" with artificial neural networks as a subclass. Considering the concrete example of the mapping $(Z, N) \to M$ that determines the nuclear (i.e., atomic) mass, a learning machine consists of (i) an input interface where Z and N are fed to the device in coded form, (ii) a system of intermediate processing elements, and (iii) an output interface where an estimate of the mass appears for decoding. Given a body of training data to be used as exemplars of the desired mapping (consisting of input "patterns," also called vectors, and their associated outputs), a suitable learning algorithm is used to adjust the parameters of the machine, e.g., the weights of the connections between the processing elements in the case of a neural network. These parameters are adjusted in such a way that the learning machine (a) generates responses at the output interface that reproduce, or closely fit, the masses of the training examples, and (b) serves as a reliable predictor of the masses of test nuclei absent from the training set. This second requirement is a strong one – the system should not merely serve as a lookup table for masses of known nuclei; it should also perform well in the much more difficult task of prediction or *generalization*.

The most widely applied learning machine is the Multilayer Perceptron (MLP), consisting of a feedforward neural network with at least one layer of "hidden neurons" between input and output interfaces.[2] MLPs are usually trained by the back-propagation algorithm,[2–4] essentially a gradient-descent procedure for adjusting weight parameters incrementally to improve performance on a set of training examples. A significant measure of success has been achieved in constructing global models of nuclear properties based on such networks, with applications to atomic masses, neutron and proton separation energies, spins and parities of nuclear ground states, stability vs. instability, branching ratios for different decay modes, and beta-decay lifetimes. (Reviews and original references may be found in Ref. 5.)

The Support Vector Machine (SVM),[2,6,7] a versatile and powerful approach to problems in classification and nonlinear regression, entered the picture in the 1990s. Rooted in the strategy of structural-risk minimization,[2] it has become a standard tool in statistical modeling. Although Multilayer Perceptrons as well as Support Vector Machines are in principle universal approximators, SVMs eliminate much of the guesswork of MLPs, since they incorporate an automatic process for determining the architecture of the learning machine. Not surprisingly, they have become the method of choice for a wide variety of problems.

Our selection of global nuclear systematics as a concrete example for the application of advanced machine-learning algorithms is neither accidental nor academic. There exists a large and growing body of excellent data on nuclear properties for thousands of nuclides, providing the raw material for the construction of robust and accurate statistical models. Moreover, interest in this classic problem in nuclear physics has never been greater.

The advent of radioactive ion-beam facilities, and the promise of the coming generation epitomized by the Rare Isotope Accelerator (RIA), have given new impetus to the quest for a unified, global understanding of the structure and behavior of nuclei across a greatly expanded nuclear chart. The creation of hundreds of new nuclei far from stability opens exciting prospects for discovery of exotic phenomena, while presenting difficult challenges for nuclear theory. Following the pattern indicated above, traditional methods for theoretical postdiction or prediction of the properties of known or unknown nuclides include *ab initio* many-body calculations employing the most realistic nuclear Hamiltonians[8] and, more commonly, density functional approaches and semi-phenomenological models. Since computational barriers limit *ab initio* treatment to light nuclei, viable global models inevitably contain parameters that are adjusted to fit experimental data on certain reference nuclei. Global models currently representing the state of the art are hybrids of microscopic theory and phenomenology; most notably, they include the macroscopic/microscopic droplet models of Möller et al.[9,10] and the density functional theories employing Skyrme, Gogny, or relativistic mean-field Lagrangian parametrizations of self-consistent mean-field theory.[11–13] From the standpoint of data analysis, these approaches are inherently theory-thick, since their formulation rests on a deep knowledge of the problem domain.

It is evident that data-driven, "theory-thin", statistical models built with machine-learning algorithms can never compete with traditional global models in providing new physical insights. Nevertheless, in several respects they can be of considerable value in complementing the traditional methods, especially in the present climate of accelerated experimental and theoretical exploration of the nuclear landscape.

(i) A number of studies[5,14–19] suggest that the quality and quantity of the data has already reached a point at which the statistical models can approach and possibly surpass the theory-thick models in sheer predictive performance. In this contribution, we shall present strong evidence from machine learning experiments with Support Vector Machines that this is indeed the case.

(ii) In spite of their "black-box" origin, the machine-learning predictions can be directly useful to nuclear experimentalists working at radioactive ion-beam facilities, as well as astrophysicists developing refined models of nucleosynthesis.
(iii) Although not straightforward, it will in fact be possible to gain some insights into the inner workings of nuclear physics through statistical learning experiments, by applying techniques analogous to gene knock-out studies in molecular biology.
(iv) It is fundamental interest to answer, for the field of nuclear physics, the Jaynesian question that was posed above.

2. Key Features of Support Vector Machines

In technical jargon, the Support Vector Machine is a *kernel method*,[2] which, in terms of a classification problem, means that it implicitly performs a nonlinear mapping of the input data into a higher-dimensional *feature space* in which the problem becomes separable (at least approximately). Architecturally, the Support Vector Machine and Multilayer Perceptron are close relatives; in fact the SVM includes the MLP with one hidden layer as a special case. However, SVMs in general offer important advantages over MLPs, including avoidance of the curse of dimensionality through extraction of the feature space from the training set, once the kernel and error function have been specified.

Support Vector Machines may be developed for function approximation (i.e., nonlinear regression) as well as classification. In either case, the output of the machine (approximation to the function or location of the decision hyperplane, respectively) is expressed in terms of a representative subset of the examples of the mapping $\mathbf{x} \to y$ contained in the training set. These special examples are the *support vectors.*

The basic ideas underlying SVMs are most readily grasped by first considering the case of a classification problem involving linearly separable patterns. Suppose some of the patterns are green and the others are red, depending on some input variables defining the n-dimensional input space. To find a decision surface that separates red from green patterns, one *seeks the hyperplane that provides the maximum margin between the red and green examples.* The training examples that define the margin are just the support vectors. In this simple case, an exact solution is possible, but in general errors are unavoidable. When faced with a problem involving nonseparable patterns, the objective then is to locate a decision hyperplane such that the misclassification error, averaged over the training set, is minimized. Guided by the principle of structural-risk minimization,[2] the SVM approach determines an optimal hyperplane by minimizing a cost function that includes a term to reduce the VC dimension[2,6,7] (thereby enhancing generalization capability) and a term that governs the tradeoff between machine complexity and the number of nonseparable patterns.

In practice, the SVM strategy actually involves two steps. The first is to implement a nonlinear mapping $\boldsymbol{\varphi}$: $\mathbf{x} \to \varphi_j(\mathbf{x})$, $j = 1, \ldots, m > n$, from the space of input vectors $\mathbf{x}$ into a higher-dimensional feature space, which is "hidden" from input and output (and corresponds to the hidden layer in MLPs). This is done in terms of an inner-product kernel $K(\mathbf{x}, \mathbf{x}_k) = \boldsymbol{\varphi}^T(\mathbf{x})\boldsymbol{\varphi}(\mathbf{x}_k)$ satisfying certain mathematical conditions, notably Mercer's theorem.[20] The second step is to find a hyperplane that separates (approximately, in general) the features identified in the first step. This is accomplished by the optimization procedure sketched above.

A self-contained introduction to the SVM technique is beyond the scope of the present contribution. Excellent treatments are available in the original work of Vapnik as expounded in Refs. 6,7 and in Haykin's text[2] (see also Ref. 19).

To provide some essential background, let us consider a regression problem corresponding to a map $\mathbf{x} \to y(\mathbf{x})$, where $\mathbf{x}$ is an input vector with n components $x^{(i)}$, and suppose that T training examples indexed by k are made available. Then the optimal approximating function takes the form

$$\hat{y}_{\text{opt}}(\mathbf{x}) = \sum_{k=1}^{T} (\alpha_k - \alpha'_k) K(\mathbf{x}, \mathbf{x}_k) \,, \tag{1}$$

Solution of the optimization problem stated above determines the parameters α_k and α'_k, and the support vectors of the machine are defined by those training patterns for which $\alpha_k \neq \alpha'_k$.

Different choices of the inner-product kernel appearing in Eq. (1) yield different versions of the Support Vector Machine. Common choices include

$$K(\mathbf{x}, \mathbf{x}_k) = (\mathbf{x}^T \mathbf{x}_k + 1)^p \,, \tag{2}$$

corresponding to the *polynomial learning machine* with user-selected power p; a Gaussian form

$$K(\mathbf{x}, \mathbf{x}_k) = \exp\left(-\gamma ||\mathbf{x} - \mathbf{x}_k||^2\right) \tag{3}$$

containing a user-selected width parameter γ, which generates a radial-basis-function (RBF) network; and

$$K(\mathbf{x}, \mathbf{x}_k) = \tanh(\beta_1 \mathbf{x}^T \mathbf{x}_k + \beta_2) \,, \tag{4}$$

which realizes a two-layer (one-hidden-layer) perceptron, only one of the parameters β_1, β_2 being independently adjustable. We also draw attention to a generalization of the RBF kernel (3) introduced recently as a simplified version of what is called ANOVA decomposition,[21] having the form

$$K(\mathbf{x}, \mathbf{x}_k) = \left(\sum_{i=1}^{n} \exp\left[-\gamma \left(x_k^{(i)} - x^{(i)} \right)^2 \right] \right)^d \,, \tag{5}$$

The Support Vector Machine may be considered as a feedforward neural network in which the inner-product kernel, through an appropriate set of m elements

$K(\mathbf{x}, \mathbf{x}_k)$, defines a layer of hidden units that embody the mapping from the n-dimensional input space to the m-dimensional feature space. These hidden units process the input patterns nonlinearly and provide outputs that are weighted linearly and summed by an output unit. As already pointed out, the familiar structures of radial-basis-function networks and two-layer perceptrons can be recaptured as special cases by particular choices of kernel. However, the SVM methodology transcends these limiting cases in a very important way: it automatically determines the number of hidden units suitable for the problem at hand, whatever the choice of kernel, by finding an optimally representative set of support vectors and therewith the dimension of the feature space. In essence, the Support Vector Machine offers a generic and principled way to control model complexity. By contrast, approaches to supervised learning based on MLPs trained by backpropagation or conjugate-gradient algorithms depend heavily on rules of thumb, heuristics, and trial and error in arriving at a network architecture that achieves a good compromise between complexity (ability to fit) and flexibility (ability to generalize).

3. Application to Nuclear Systematics

In this section we summarize the findings of recent explorations of the potential of Support Vector Machines for global statistical modeling of nuclear properties. The discussion will focus on the predictive reliability of SVM models relative to that of traditional "theory-thick" models. The properties that are directly modeled in these initial studies, all referring to nuclear ground states, are (i) the nuclear mass excess $\Delta M = M - A$, where M is the atomic mass, measured in amu, (ii) β-decay lifetimes of nuclides that decay 100% via the β^- mode, and (iii) nuclear spins and parities.

The requisite experimental data are taken from the on-line repository of the Brookhaven National Nuclear Data Center (NNDC) at http://www.nndc.bnl.gov/. The experimental mass values are those of the AME03 compilation of Audi et al.[22]

Extensive preliminary studies have been performed to identify inner-product kernels well suited to global nuclear modeling. Earlier work converged on the ANOVA kernel (5) as a favorable choice, and corresponding results have been published in Ref. 19. More recently, we have introduced a new kernel that yields superior results, formed by the sum of polynomial and ANOVA kernels and named the pA kernel. (Satisfaction of Mercer's theorem is conserved under summation.) The new kernel contains three parameters (p, γ, and d) that may be adjusted by the user. Aside from parameters contained in the inner-product kernel, the SVM procedure involves a constant C giving the user control over the tradeoff between complexity and flexibility, plus an additional control constant ε in the regression case, measuring the tolerance permitted in the reproduction of training data. Thus, SVM models developed with the pA kernel contain four or five adjustable parameters (five in all applications reported here).

To allow for a meaningful evaluation of predictive performance (whether interpolation or extrapolation), the existing database for the property being modeled is divided into three subsets, namely the ***training set***, ***validation set***, and ***test set***. These sets are created by random sampling, consistently with approximate realization of chosen numerical proportions among them, e.g. $(100–2R){:}R{:}R$ for training, validation, and test sets, respectively, with $R < 25$. The training set is used to find the support vectors and construct the machine for given values of the adjustable parameters p, γ, d, C, and ε. The validation set is used to guide the optimal determination of these parameters, seeking good performance on both the training and validation examples. The test set remains untouched during this process of model development; accordingly, the overall error (or error rate) of the final model on the members of the test set may be taken as a valid measure of predictive performance. When one considers how SVM models might be applied in nuclear data analysis during the ongoing exploration of the nuclear landscape, it seems reasonable that consistent predictive performance for 80:10:10 or 90:5:5 partitions into training, validation, and test sets would be sufficient for the SVM approach to be useful in practice.

The SVM approach has been applied to generate a variety of global models of nuclear mass excess, beta-decay lifetimes, and spin/parity, corresponding to different kernels, databases, partitions into training/validation/test sets, and tradeoffs between the relative performance on these three sets. Here we will focus on those models considered to be the best achieved to date. Moreover, due to limited space, we will restrict the discussion to the most salient features of those models and to an assessment of their quality relative to favored traditional global models and to the best available MLP models. Further, more detailed information may be found at the web site http://abacus.wustl.edu/Clark/svmpp.php, which generates SVM estimates of the listed nuclear properties for (Z, N) pairs entered by visitors. This web site will be periodically updated as improved SVM models are developed.

3.1. *SVM Models of Atomic Mass Surfaces*

Development and testing of the SVM mass models to be highlighted here are based on the AME03 data for all nuclides with Z, $N > 16$ and experimental masses having error bars below 4%. This set of nuclides is divided into the four classes: even-Z-even-N (EE) even-Z-odd-N (EO), odd-Z-even-N (OE), and odd-Z-odd-N (OO). Separate SVM regression models were constructed for each such "even-oddness" class. This does introduce some minimal knowledge about the problem domain into the modeling process; one might therefore say that the models developed are not absolutely theory-free. However, the data itself gives strong evidence for the existence of different mass surfaces depending on whether Z and N are even or odd. Knowledge of the integral character of Z and N may, quite properly, bias the SVM toward inclusion of associated quantum effects.[23]

Table 1 displays performance measures for models based on an 80:10:10 target partitioning of the full data set among training, validation, and test sets, respectively. Inspection of the actual distributions of these sets in the $Z-N$ plane shows that substantial fractions of the validation and test sets lie on the outer fringes of the known nuclei, significantly distant from the line of stable nuclides. Accordingly, performance on the test set measures the capability of the models in extrapolation as well as interpolation. Performance on a given data set is quantified by the corresponding root-mean-square (rms) error σ of model results relative to experiment (as is standard in global mass modeling). The "optimized" model parameters are included in Table 1. They show enough differences from one even-oddness class to another to justify development of separate models for the four classes.

Table 1. Performance of SVM global models of atomic mass. All rms σ values are in MeV, and # stands for the number of nuclides in the indicated set. Training, validation, and tests sets are selected randomly with target proportions 80:10:10, respectively. For all 531 nuclides in EE class, the SVM error σ is 0.42 MeV, compared to 1.5 MeV obtained by Bertsch et al.[13]

	Training		Validation		Test		Control Param's		pA Param's		
Classes	#	σ	#	σ	#	σ	$C(10^{-2})$	$\varepsilon(10^{-4})$	p	d	γ
EE	425	0.26	53	0.72	53	0.83	1.307	9.976	2	6	25.84
EO	398	0.48	50	0.35	50	0.53	2.240	9.953	2	5	19.98
OE	397	0.19	50	0.31	50	0.35	1.298	9.801	2	6	25.93
OO	377	0.41	47	0.77	47	0.96	2.238	9.160	2	5	18.06
Overall	1597	0.35	200	0.58	200	0.71					

The results in Table 1 attest to a quality of performance, in both fitting and prediction, that is on a par with the best available from traditional modeling[10,11] and from MLP models trained by an enhanced backpropagation algorithm.[16,18,24] To emphasize this point qualitatively, we display in Table 2 some representative rms error figures that have been achieved in recent work with all three approaches. (We must note, however, that the data sets used for the different entries in the table may not be directly comparable, and the division into training, validation, and test sets does not necessarily have the strict meaning assigned here.) The second SVM model listed in the table was developed for a partitioning of the data into training, validation, and test sets of approximately 90:5:5, obtained by random transfer of nuclides from the validation and test sets of the 80:10:10 model to the training set.

The quality of representation that can be realized through the SVM methodology may be highlighted in another way. Employing the nuclear mass excess values generated by the SVM models of Table 1, we have calculated the Q_α values for eight alpha-decay chains of the superheavy elements 110, 111, 112, 114, 115, 116, and 118. (The alpha-decay Q-value is defined as $Q_\alpha(A,Z) = M(\text{parent}) - M(\text{daughter}) - M(\alpha) = \text{BE}(A-4, Z-2) + \text{BE}(4,2) - \text{BE}(A,Z)$, where BE stands for the binding energy of the indicated nuclide.) Results are presented in graphical and tabular form on the web site http://abacus.wustl.edu/Clark/svmpp.php. For the models of Table 1 (based on an 80:10:10 partition of the assumed AME03 data set), the average rms error of the 38 estimates of Q_α is 0.82 MeV, while the average absolute error is 0.64

Table 2. Performance measures of superior global models of the atomic mass table developed through traditional approaches of nuclear theory (FRDM and HFB2), enhanced backpropagation training of neural networks (MLP), and SVM regression methodology. All σ values are in MeV and # is the number of nuclides. SVM results for two different partitions of the full data set into training, validation, and test sets are shown.

	Training Set		Validation Set		Test Set	
	#	σ	#	σ	#	σ
FRDM[10]	1303	0.68	351	0.71	158	0.70
HFB2[11]	1303	0.67	351	0.68	158	0.73
MLP[16]	1303	0.44	351	0.44	158	0.95
MLP[18,24]	1303	0.28	351	0.40	158	0.71
SVM (80:10:10)	1597	0.35	200	0.58	200	0.71
SVM (90:5:5)	1797	0.31	100	0.29	100	0.49

Mev. We emphasize that these estimates are predictions (rather than fits), since none of the nuclei involved belongs to the validation or test set. Moreover, due to the situation of these superheavy nuclides in the $Z-N$ plane, prediction of the associated Q_α values provides a strong test of extrapolation.

The performance of SVM mass models documented in Tables 1 and 2 and in the alpha-chain predictions gives assurance that this approach to global modeling will be useful in guiding further exploration of the nuclear landscape. However, it is important to gain some sense of when and how it begins to fail. The performance figures for the two sets of SVM models involved in Table 2 are consistent with the natural expectation that if one depletes the validation and test sets of the 80:10:10 partition in favor of an enlarged test set, the predictive ability of the model is enhanced. Conversely, one should be able to "break" the SVM modeling approach by random depletion of the training set of the 80:10:10 model in favor of larger validation and test sets. Eventually the training set will become too small for the method to work at all. The results of a quantitative study of this process are shown in Figure 1. Writing the generic partition as $(100-2R){:}R{:}R$, the error measure increases roughly linearly with R for R greater than 10.

In addition to direct statistical modeling using either SVMs or MLPs, a promising hybrid approach is being explored. Recently, the *differences* $\Delta M_{\rm exp} - \Delta M_{FRDM}$ between experimentally measured masses and the corresponding theoretical masses given by the Finite-Range Droplet Model (FRDM) of Möller, Nix, and collaborators[9,10] have been modeled with a feedforward neural network of architecture 4–6–6–6–1 trained by a modified backpropagation learning algorithm.[18] (The integers denote the numbers of neurons in successive layers, from input to output.) The rms errors on training (1276), validation (344), and test (529) sets are respectively 0.40, 0.49, and 0.41 MeV, where the numbers of nuclides in each of these sets is given in parentheses. In a similar experiment, we have constructed SVM models for EE, EO, OE, and EE classes using pA kernels. Overall rms errors of 0.19, 0.26, and 0.34

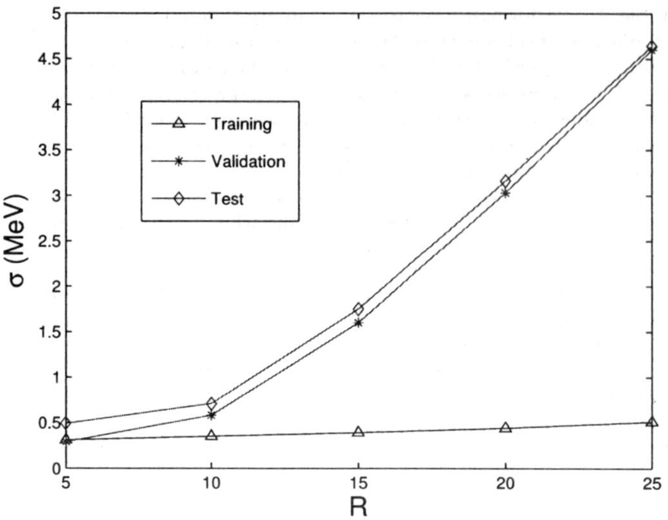

Fig. 1. Overall performance of EE, EO, OE, and EE SVM models as a function of the target percentage R of nuclides assigned to both validation and test sets.

were achieved on the training (1712), validation (213), and test (213) sets, respectively, with little variation over even-oddness classes. Error figures over comparable subsets for the FRDM model in question run around 0.7 MeV, again with relatively little variation from subset to subset. These results suggest that MLPs and SVMs are capable of capturing some 1/2 to 2/3 of the physical regularities missed by the FRDM. It remains to be seen whether the residual error has a systematic component or instead reflects a large number of small effects that will continue to elude global description.

3.2. *SVM models of the Systematics of Beta Halflives*

Another important problem in global modeling involves the prediction of beta-decay halflives of nuclei. As in the case of atomic masses, this is a problem in nonlinear regression. Here we restrict attention to nuclear ground states and to nuclides that decay 100% through the β^- mode. For this presentation, we make the further restriction to nuclides with halflives $T_{1/2}$ below 10^6 s (although we have also included the longer-lived examples in another set of modeling experiments). The Brookhaven NNDC provides 838 examples fitting these criteria. Since the examples still span 9 orders of magnitude in $T_{1/2}$, it is natural to work with $L = \log T_{1/2}(\text{ms})$ and seek an approximation to the mapping $(Z, N) \to L$ in the form of SVMs. Again, we construct separate SVMs for the EE, EO, OE, and OO classes, and again a kernel of type pA is adopted. The full data set is divided by random distribution into training, validation, and test sets in approximately the proportions 80:10:10.

The performance of the favored models is quantified in Table 3. Here we use two measures to assess the accuracy of the SVM results for training, validation, and test nuclides. These are the rms error, again denoted by σ, and the mean absolute error μ of the model estimates of L, relative to experiment.

Detailed studies of beta-decay systematics within the established framework of nuclear theory and phenomenology include those of Staudt et al.[25] (1990), Hirsch et al.[26] (1993), Homma et al.[27] (1996), and Möller et al.[28] (1997). However, comparison of the performance of the SVM models with that of the models resulting from these studies is obscured by the differences in the data sets involved. Most significantly, the data set employed here is considerably larger than those used previously, including as it does many new nuclides far from stability. Analysis of SVM performance on subsets of the data set, now in progress, will yield useful information on the efficacy of SVM models relative to the more traditional ones, as we continue to develop improved global models of beta-decay systematics.

On the other hand, MLP models for the beta-halflife problem have been generated for the same data set as used in our SVM study, allowing a meaningful comparison to be made. The best MLP models created to date[29] show values for the rms error σ over all even-oddness classes of 0.55 in training, 0.61 in validation, and 0.64 in prediction. These values are somewhat larger than those seen in Table 3. However, it must be pointed out that the MLP results were obtained with a smaller training set, so the efficacy of the two statistical methods appears to be about equal at this stage of development. That being the case, it is relevant to note that the recent MLP models represent a distinct advance earlier versions,[15] and that those earlier statistical models already showed better performance over short-lived data sets than the conventional models of Homma et al.[27] and Möller et al.[28]

Table 3. Performance of SVM global models constructed for β-decay halflives below 10^6s. Control parameter $C = 3.861 \times 10^{-3}$ and pA parameters $p = 1$, $d = 8$ for all four classes.

	Training Set			Validation Set			Test Set			Param's	
Classes	#	σ	μ	#	σ	μ	#	σ	μ	$\varepsilon(10^{-4})$	γ
EE	131	0.55	0.18	16	0.57	0.24	16	0.62	0.32	9.155	102
EO	179	0.41	0.15	22	0.42	0.19	22	0.51	0.25	8.401	79
OE	172	0.41	0.15	21	0.47	0.20	21	0.47	0.26	7.862	97
OO	190	0.52	0.18	24	0.40	0.16	24	0.52	0.28	8.896	102
Overall	672	0.47	0.16	83	0.46	0.19	83	0.53	0.27		

3.3. *SVM Models of Ground-State Parity and Spin*

The applications to prediction of atomic masses and beta-decay lifetimes demonstrate the predictive power of SVMs in two important problems of global nuclear modeling that involve function estimation. The final two applications will probe the performance of SVMs in global modeling of the discrete nuclear properties of parity and spin. In essence, these are problems of classification: "Which of a finite number of exclusive possibilities is associated with or implied by a given input pattern?" Support Vector Machines were first developed to solve classification problems, and good SVM classifier software is available on the web.[30] However, for convenience and uniformity we prefer to treat the parity and spin problems with the same SVM regression technique as in the other examples, also using the pA choice of inner-product kernel. In the parity problem, the decision of the regression SVM is interpreted to be positive parity [negative parity] if the machine's output is positive [negative]. In the spin problem, the spin assigned by the machine is taken to be correct if and only if the numerical output (after rescaling) is within ± 0.25 of the correct value (in $\hbar$ units).

As before, all data are taken from the Brookhaven site. For parity and spin, it is especially natural to create separate SVM models for the different even-oddness classes. However, as is well known, all EE nuclei have spin/parity $J^{\pi} = 0^{+}$. Modeling this property is trivial for SVMs, so the EE class may be removed from further consideration. The data in and of itself permits us to do so. Moreover, in the case of spin, the data itself establishes, with a high degree of certainty, that the spin of EO or OE nuclides takes half-odd integral values (in units of $\hbar$), while the spin of OO nuclides is integral. Although this formulation of the parity and spin problems introduces significant domain knowledge into the model-building process, the data alone provides adequate motivation. Nuclei with spin values larger than 23/2 were not considered. The predictive performance that may be achieved with SVM models of parity and spin is illustrated in Tables 4 and 5. Performance is measured by the percentage of correct assignments. Construction of both parity and spin models is based on an 80:10:10 partition of the data into training, validation, and test sets. (As usual, the target distribution is realized only approximately.)

Averaged over even-oddness classes, the overall performance of the parity SVMs is 97% correct on the training set and 95% on the validation set, with a predictive performance on the test set of 94%. Obviously, assigning parity to nuclear ground states is an extremely easy task for Support Vector Machines. One might expect quite a different situation for the spin problem: since there are 12 legitimate spin assignments for the EO or OE nuclides considered (i.e., obeying the rules for addition of angular momenta) and also 12 for the OO class, the chance probability of a correct guess is low. It is then most remarkable that the SVM spin models we have developed perform with very high accuracy in prediction as well as fitting and validation. While some success has been achieved previously in MLP modeling of parity and spin,[31,32] consistent predictive quality within the 80–90 percentile range has been elusive. Within main-stream nuclear theory and phenomenology, the problem of global modeling of ground-state spins has received little attention, and the few attempts have not been very successful. As a baseline, global nuclear structure calculations within the macroscopic/microscopic approach[33] reproduce the ground-state spins of odd-A nuclei with an accuracy of 60% (agreement being found in 428 examples out of 713).

It should be mentioned that in the preliminary investigations described in Ref. 19, the tasks of global modeling of parity and spin with SVMs were in fact treated as classification rather than function-estimation problems. Corresponding SVM classifiers were created using established procedures.[2,30] Based on an RBF kernel, results were obtained that surpass the available MLP models in quality, but are inferior to those reported here in Tables 4 and 5.

Table 4. Performance of SVM global models of nuclear ground-state parity. pA parameter $p = 1$ for all three classes.

	Training		Validation		Test		Control Param's		pA Param's	
Classes	#	Score	#	Score	#	Score	$C\ (10^{-6})$	$\varepsilon\ (10^{-4})$	d	γ
EO	468	99%	58	98%	58	97%	7.629	9.876	17	19.57
OE	462	94%	56	96%	56	93%	7.629	9.921	17	19.41
OO	429	96%	53	94%	53	92%	1.907	9.652	19	19.45
Overall	1357	97%	168	95%	168	94%				

Table 5. Performance of SVM global models of nuclear ground-state spin. pA parameter $p = 1$ for all three classes.

	Training		Validation		Test		Control Param's		pA Param's	
Classes	#	Score	#	Score	#	Score	$C\ (10^{-3})$	$\varepsilon\ (10^{-4})$	d	γ
EO	523	93%	58	90%	58	86%	7.634	9.675	7	202
OE	460	92%	57	88%	57	84%	7.634	8.804	7	239
OO	469	85%	52	88%	52	81%	7.634	9.820	7	192
Overall	1452	90%	167	89%	167	84%				

4. Conclusions

Global statistical models of atomic masses, beta-decay lifetimes, and nuclear spins and parities have been constructed using the methodology of Support Vector Machines. The predictive power of these "theory-thin" models, which in essence are derived from the data and only the data, is shown to be competitive with, or superior to, that of conventional "theory-thick" models based on nuclear theory and phenomenology. Conservative many-body theorists may be troubled by the "black-box" nature of the SVM predictors, i.e., the impenetrability of their computational machinery. However, this alternative, highly pragmatic approach may represent a wave of the future in many fields of science – already visible in the proliferation of density-functional computational packages for materials physics and eventually molecular biology, which, for the user, are effectively black boxes. While it is true that the statistical models produced by advances in machine learning do not as yet yield the physical insights of traditional modeling approaches, their prospects for revealing new regularities of nature are by no means sterile.

Acknowledgements

This research has received support from the U.S. National Science Foundation under Grant No. PHY-0140316. We acknowledge helpful discussions and communications with S. Athanassopoulos, M. Binder, E. Mavrommatis, T. Papenbrock, S. C. Pieper, and R. B. Wiringa. In the regression studies reported herein, we have found the mySVM software and instruction manual created by S. Rüping[34] (Dortmund) to be very useful.

References

1. E. T. Jaynes, *Probability Theory: The Logic of Science* (Cambridge University Press, Cambridge, 2003).
2. S. Haykin, *Neural Networks: A Comprehensive Foundation*, Second Edition (McMillan, New York, 1999).
3. D. E. Rumelhart, G. E. Hinton, and R. J. Williams, in *Parallel Distributed Processing: Explorations in the Microstructure of Cognition*, Vol. 1, edited by D. E. Rumelhart *et al.* (MIT Press, Cambridge, MA, 1986).
4. J. Hertz, A. Krogh, and R. G. Palmer, *Introduction to the Theory of Neural Computation* (Addison-Wesley, Redwood City, CA, 1991).
5. J. W. Clark, T. Lindenau, and M. L. Ristig, *Scientific Applications of Neural Nets* (Springer-Verlag, Berlin, 1999).
6. V. N. Vapnik, *The Nature of Statistical Learning Theory* (Springer-Verlag, New York, 1995).
7. V. N. Vapnik, *Statistical Learning Theory* (Wiley, New York, 1998).
8. S. C. Pieper and R. B. Wiringa, *Annu. Rev. Part. Nucl. Sci.* **51**, 53 (2001).
9. P. Möller and J. R. Nix, *J. Phys.* **G20**, 1681 (1994).
10. P. Möller, J. R. Nix, W. D. Myers, and W. J. Swiatecki, *At. Data Nucl. Data Tables* **59**, 185 (1995).
11. M. Samyn, S. Goriely, P.-H. Heenen, J. M. Pearson, and F. Tondeur, *Nucl. Phys.* **A700**, 142 (2002); S. Goriely, M. Samyn, P.-H. Heenen, J. M. Pearson, and F. Tondeur, *Phys. Rev. C* **66**, 024326 (2002).
12. M. Bender, P.-H. Heenen, and P.-G. Reinhard, *Rev. Mod. Phys.* **75**, 121 (2003); M. Bender, G. F. Bertsch, and P.-H. Heenen, *Phys. Rev. Lett.* **94**, 102503 (2005).
13. G. F. Bertsch, B. Sabbey, and M. Uusnäkki, *Phys. Rev. C* **71**, 054311 (2005).
14. K. A. Gernoth and J. W. Clark, *Neural Networks* **8**, 291 (1995).
15. J. W. Clark, E. Mavrommatis, S. Athanassopoulos, A. Dakos, and K. A. Gernoth, *Fission Dynamics of Atomic Clusters and Nuclei*, edited by D. M. Brink, F. F. Karpechine, F. B. Malik, and J. da Providencia (World Scientific, Singapore, 2001), p. 76. [nucl-th/0109081]
16. S. Athanassopoulos, E. Mavrommatis, K. A. Gernoth, and J. W. Clark, *Nucl. Phys.* **A743**, 222 (2004), and references therein.
17. S. Athanassopoulos, E. Mavrommatis, K. A. Gernoth, and J. W. Clark, in *Advances in Nuclear Physics, Proceedings of the Hellenic Symposium on Nuclear Physics*, in press (2005). [nucl-th/0509075]
18. S. Athanassopoulos, E. Mavrommatis, K. A. Gernoth, and J. W. Clark, in *Advances in Nuclear Physics*, Proceedings of the Hellenic Symposium on Nuclear Physics, in press (2006). [nucl-th/0511088]
19. H. Li, J. W. Clark, E. Mavrommatis, S. Athanassopoulos, and K. A. Gernoth, in

Condensed Matter Theories, Vol. 20, edited by J. W. Clark, R. M. Panoff, and H. Li (Nova Science Publishers, Hauppauge, NY, 2006), p. 505. [nucl-th/0506080]

20. J. Mercer, *Transactions of the London Philosophical Society (A)* **209**, 415 (1909).
21. M. O. Stitson, A. Gammerman, V. Vapnik, V. Vovk, C. Watkins, and J. Weston, in *Advances in Kernel Methods – Support Vector Learning*, edited by B. Schükopf, C. Burges, and A. J. Smola (MIT Press, Cambridge, MA, 1999), p. 285.
22. G. Audi, A. H. Wapstra, C. Thibault, J. Blachot, and O. Bersillon *Nucl. Phys.* **A729** (2003).
23. S. Gazula, J. W. Clark, and H. Bohr, *Nucl. Phys.* **A540**, 1 (1992).
24. S. Athanassopoulos, E. Mavrommatis, K. A. Gernoth, and J. W. Clark, to be published.
25. A. Staudt, E. Bender, K. Muto, and H. V. Klapdor-Kleingrothaus, *At. Data Nucl. Data Tables* **44**, 132 (1990).
26. M. Hirsch, A. Staudt, K. Muto, and H. V. Klapdor-Kleingrothaus, *At. Data Nucl. Data Tables* **53**, 165 (1993).
27. H. Homma, E. Bender, M. Hirsch, K. Muto, and H. V. Klapdor-Kleingrothaus, *Phys. Rev. C* **54**, 2972 (1996).
28. P. Möller, J. R. Nix, and K. L. Kratz, *At. Data Nucl. Data Tables* **66**, 131 (1997).
29. N. Costiris, A. Dakos, E. Mavrommatis, K. A. Gernoth, and J. W. Clark, to be published.
30. T. Joachims (2004), Multi-Class Support Vector Machine, http://www.cs.cornell.edu/People/tj/svm_light/svm_multiclass.html (2004).
31. J. W. Clark, S. Gazula, K. A. Gernoth, J. Hasenbein, J. S. Prater, and H. Bohr, in *Recent Progress in Many-Body Theories*, Vol. 3, edited by T. L. Ainsworth, C. E. Campbell, B. E. Clements, and E. Krotscheck (Plenum, New York, 1992), p. 371.
32. K. A. Gernoth, J. W. Clark, J. S. Prater, and H. Bohr, *Phys. Lett.* **B300**, 1 (1993).
33. P. Möller and J. R. Nix, *Nucl. Phys.* **A520**, 369c (1990).
34. S. Rüping, mySVM, http://www-ai.cs.uni-dortmund.de/SOFTWARE/MYSVM/ (2004).

A TRIBUTE TO MANFRED L. RISTIG ON HIS 70th BIRTHDAY

LESZEK SZYBISZ*

Departmento de Física, Facultad de Ciencias Exactas y Naturales, Universidad de Buenos Aires, Ciudad Universitaria, RA-1428 Buenos Aires, Argentina

and

Departmento de Física, Centro Atómico Constituyentes, CNEA, Av. del Libertador 8250, RA-1429 Buenos Aires, Argentina

szybisz@tandar.cnea.gov.ar

The participants of the 13th International Conference on Recent Progress in Many-Body Theories have had the oportunity of celebrating the 70th birthday of two distinguished colleagues, John W. Clark and Manfred L. Ristig, who made outstanding contributions to theoretical many-body physics.

It is a pleasure and, of course, a privilege to have the possibility of presenting a biographical sketch of M.L. Ristig where I also included a summary of his work. I met him twenty five years ago when I was working at the Universität zu Köln (Cologne).

Fred was born on January 27, 1935 in Erfurt, Germany, and started to study Physics in 1953 at the Friedrich Schiller Universität in Jena. However, in 1958 he interrupted the studies because was taken as "Political prisoner" and transferred to Bautzen, a town in the former East Germany, to work in a factory. Only after three years, just before the building of the Berlin's wall in 1961, he could move to Munich and continue the studies at the Ludwig Maximilian Universität, where he got the Physics Diploma in 1964. In 1966 he presented his Doctoral Thesis on nuclear physics, the advisor was P. Mittelsteadt. In this way he initiated his work in theoretical physics.

Prior to getting the "Habilitation in theoretical physics and venia legendi" at the Universität zu Köln in 1972, he was Assistant at the same university and spent a couple of years, 1969-71, as Research Associate at the Physics Department of the Washington University in St. Louis, Missouri, USA. There he met Eugene Feenberg and John Clark. In passing, it is interesting to mention that when Fred was in St. Louis, Feenberg visited Cologne in order to present his own book on "Theory of Quantum Fluids" published in 1969. At that time, John and Fred began their fruitful collaboration, which extends up to today.

After returning to the Institut für Teoretische Physik at the Universität zu

*Also at the Researcher Career of the Consejo Nacional de Investigaciones Científicas y Técnicas.

Manfred L. Ristig

Köln, Fred was promoted to Docent in 1974, to Professor (C3) in 1980, and finally he got an appointment as Universitätsprofessor in 1987. On the other hand, he was Vertrauensdozent der Studienstiftung des deutschen Volkes (German Scholarship Foundation) (1988-2004).

In the meantime, he visited several times the Physics Department of the Washington University. Furthermore, in 1977 he was working at the Niels Bohr Institute in Copenhagen, Denmark, in 1984 he was a Gordon-Godfrey visiting fellow at the Department of Physics of the University of New South Wales, Australia, and in 1989 he was a visiting professor at the Department of Physics of the Universidad de Granada in Spain.

Fred contributed to the communication among members of the international physical community: he was a member of the international advisory board of the "International Conference on Recent Progress in Many-Body Theories" (1983-97); Committee member and chairman for the Eugene-Feenberg-Award since 1988; he organized several international conferences on different topics, and co-edited proceedings of conferences.

Since 1997 he is Adjoint Professor of Physics at the Washington University in St. Louis. This is a consequence of the productive collaboration with John which along three decates led to about 50 joint papers. They have - together with their students - included systematically state-dependent correlations in the theory of quantum fluids leading to correlated one- and two-body density matrices formalism.

He has supervised a large number of students (Diplom and Doctoral) and visitors awarded with Humboldt and DAAD Fellowships. He has published more than 100 papers, a selection of them is given in a list below.

Fred Ristig's first doctoral student was Eckhard Krotscheck, currently Eckhard is a very well known physicist in the field of microscopic many-body theories. In his Doctoral Thesis it was developed a Fermi Hypernetted Chain (FHNC) resummation method (it should be mentioned that a similar procedure has been independently also proposed by Fantoni and Rosati) to evaluate the equation of state for dense neutron matter and liquid helium. An extension of that formalism in connection with correlated basis functions (CBF) allowed to calculate matrix elements including off-diagonal terms needed in the case of strongly correlated quantum systems.

Let me now refer to some further applications performed by Fred and collaborators. Together with his student Karl E. Kürten he applied the paired-phonon analysis to determine the optimal ground-state wave function of Jastrow type for a binary boson mixture. An extention of the theory of quantum fluids to finite temperature was performed in the doctoral work of G. Senger by employing the minimum principle for the Helmholz free energy to develope a self-consistent variational formalism of liquid ^{4}He. The temperature dependent theory was improved in the Thesis of Thomas Lindenau where calculations on the superfluid phase of liquid helium were done and the lambda-transition was analized. The problem of stability and excitations of helium films have been also tackled. In order to test many-body theories ab-initio calculations of the Ising model in a transverse field were carried out in collaboration with Ray F. Bishop.

It is worth of notice that Fred is now trying with high entusiasm to apply physical theories for understanding biologic processes. In recent years, 2003/4, he was invited as a Guest Lecturer at the Department of Physics of the University of Pretoria, South Africa, to give lectures on subjects related with biology. To illustrate his contribution in this field I include an article at the end the list of selected papers.

It was to me a real pleasure to learn from Fred the basic tools of the CFB theory and the HNC expansion during my stay at the Institut für Teoretische Physik at the Universität zu Köln in the 80's.

Selected publications of Manfred L. Ristig

- J.W. Clark and M.L. Ristig, Embeddeding of the Brueckner approximation in the extended Jastrow scheme, *Phys. Rev. C* **7**, 1792 (1973).
- E. Krotscheck and M.L. Ristig, Long-range Jastrow correlations, *Nucl. Phys.* **A242**, 389 (1975).
- M.L. Ristig and J.W. Clark, Density matrix of quantum fluids, *Phys. Rev. B* **14**, 2875 (1976).
- K.E. Kürten and M.L. Ristig, Atomic impurities in liquid helium, *Phys. Rev. B* **27**, 5479 (1983).
- C.E. Campbell, K.E. Kürten, M.L. Ristig, and G. Senger, Variational density-matrix heory of liquid ^{4}He at nonzero temperature, *Phys. Rev. B* **30**, 728 (1984).
- L. Szybisz and M.L. Ristig, New method of solving the optimized paired-phonon analysis equations and stability of thin films of liquid ^{4}He at $T = 0$ K, *Phys. Rev. B* **40**, 4391 (1989).
- K.A. Gernoth and M.L. Ristig, Excitation spectrum of the vapour-liquid ^{4}He interface, *Phys. Rev. B* **45**, 2969 (1992).
- R.F. Bishop, D.J.J. Farnell, and M.L. Ristig, Ab-initio treatment of the Ising model in a transverse field, *J. Mod. Phys. B* **14**, 1517 (2000).
- T. Lindenau, M.L. Ristig, J.W. Clark, and K.A. Gernoth, Bose-Einstein condensation and the lambda-transiion in liquid helium, *J. Low Temp. Phys.* **129**, 143 (2002).
- B.K. Dellen, M.J. Barber, M.L. Ristig, J. Hescheler, H. Sauer, and M. Wartenberg, [Ca2+] oscillations in a model of energy-dependent Ca2+ uptake by the endoplasmic reticulum, *J. Theor. Biology* (2005).

Review by Manfred L. Ristig

- M.L. Ristig, Momentum distribution in quantum fluids and nuclear matter, in: *From nuclei to particles, International School of Physics "Enrico Fermi", Course LXXIX*, ed. A. Molinari (North-Holland, Amsterdam, 1982), p. 340.

Edited books

- *Recent Progress in Many-Body Theories, Lecture Notes in Physics* **198**, eds. H. Kuemmel and M.L. Ristig (Springer-Verlag, Berlin, 1984).
- *Theory of Spin Lattices and Lattice Gauge Models, Lecture Notes in Physics* **494**, eds. J.W. Clark and M.L. Ristig (Springer-Verlag, Berlin, 1997).
- *Scientific Applications of Neural Nets, Lecture Notes in Physics* **522**, eds. J.W. Clark, T. Lindenau, and M.L. Ristig (Springer-Verlag, Berlin, 1998).
- *Particle Scattering, X-Ray Diffraction, and Microstructure of Solids and Liquids, Lecture Notes in Physics* **610**, eds. M.L. Ristig and K.A. Gernoth (Springer-Verlag, Berlin, 1997).

THE PHYSICS OF LIQUID PARA-HYDROGEN

THOMAS LINDENAU, MANFRED L. RISTIG
Institut für Theoretische Physik, Universität zu Köln
D–50937 Köln, Germany
ristig@thp.uni-koeln.de

KLAUS A. GERNOTH
School of Physics and Astronomy, The University of Manchester
Manchester, United Kingdom
k.a.gernoth@man.ac.uk

JAVIER DAWIDOWSKI
Consejo Nacional de Investigationes Cientificas y Tecnicas,
Centro Atomico Bariloche and Instituto Balseiro,
Comision Nacional de Energia Atomica,
Universidad Nacional de Cuyo, (8400) Bariloche, Argentina
javier@cab.cnea.gov.ar

FRANCISCO J. BERMEJO
C.S.I.C. Dept. of Electricity and Electronics,
University of the Basque Country, P.O. 644, E-48080 Bilbao, Spain
javier@langran.iem.csic.es

Macroscopic systems of hydrogen molecules exhibit a rich thermodynamic phase behavior. Due to the simplicity of the molecular constituents a detailed exploration of the thermal properties of these boson systems at low temperatures is of fundamental interest. Here,we report theoretical and experimental results on various spatial correlation functions and corresponding distributions in momentum space of liquid para-hydrogen close to the triple point. They characterize the structure of the correlated liquid and provide information on quantum effects present in this Bose fluid. Numerical calculations employ Correlated Density-Matrix (CDM) theory and Path-Integral Monte-Carlo(PIMC)simulations. A comparison of these theoretical results demonstrates the accuracy of CDM theory. This algorithm therefore permits a fast and efficient quantitative analysis of the normal phase of liquid para-hydrogen. We compare and discuss the theoretical results with available experimental data.

Keywords: Liquid hydrogen; Monte Carlo; correlated density matrix theory.

1. Introduction

Microscopic, mesoscopic, and macroscopic systems of hydrogen atoms or molecules are of fundamental importance in quantum many-body theory. The properties of a single hydrogen atom led the basis for the atomic shell model. A single hydrogen molecule is the simplest compound and has been an early laboratory for studying chemical binding. Its spectrum exhibits the effects of internal degrees of freedom leading to different levels of excitations and the existence of ortho- and para-hydrogen. The molecule therefore shares common features with nuclei where internal degrees are important and cause violation of the so-called Y-symmetry.[1,2] Atomic hydrogen gas in a strong magnetic field remains a Bose fluid even at zero temperature and can condense into a Bose-Einstein phase under certain conditions[3] like the alkali gases.[4]

Liquid and solid phases of molecular hydrogen and deuterium have been extensively studied, experimentally as well as theoretically. These many-body systems are of continuing interest for a variety of reasons. Hydrogen molecules are the dominant constituents of giant planets.[5] The physics of metallic hydrogen is explored by many researchers to achieve the insulator-metal cross-over.[6]

The hydrogen liquid, close to the triple point, is the object of current research. In this paper we concentrate on a theoretical and experimental analysis of its quantum properties close to the triple point. There are many other open and interesting questions, such as properties of mesoscopic clusters and films, molecular hydrogen in confined geometries, Bose- Einstein condensation in solid and supercooled liquid hydrogen, etc.

We investigate the properties of the one- and two-body reduced density-matrix elements of liquid para-hydrogen in its normal boson phase at low temperatures.The associated Fourier transverse of these quantities reveal the spatial structure of the correlated system. Detailed numerical calculations are performed at the temperature $T = 16\,\mathrm{K}$ and a particle number density $\rho = 0.021\,\text{Å}^{-3}$. The experimental measurements for the liquid structure function[7] and the dispersion law of the collective excitations[8] have been done at the temperature $T = 15.2\,\mathrm{K}$. Quantitative information on single-particle properties have been gained by precision neutron-scattering experiments at $T = 15.7\,\mathrm{K}$ and various pressures that permits to extract the momentum distribution of a molecule in the hydrogen liquid.[9]

The present theoretical analysis is based on the parameter-free microscopic CDM theory[10–12] and PIMC calculations[13] with the central Silvera-Goldman potential as input.[14] Section 2 begins with a quantitative study of the spatial distribution function $g(r)$ and the static structure function $S(k)$ comparing the CDM results with the corresponding PIMC results. Theoretical results on the excitation energies and quasiparticle energies within CDM theory are discussed and compared with the measured dispersion laws in Section 3. The theoretical momentum distribution $n(k)$ is analysed in Sec. 4. Its study is based on a structural factorization of the one-body reduced density-matrix elements $n(r)$ reported earlier[15] that permits a clean separation of particle exchange properties from spatial phase-phase correlations caused by the intermolecular interactions. A short summary is given in Section 5.

2. Spatial Structure

The spatial structure of a homogeneous quantum fluid in thermal equilibrium may be characterized by a set of correlation functions and concomitant Fourier transforms or structure functions.The radial distribution function $g(r)$ describes the spatial correlations between two particles in coordinate space and depends on the relative distance between them, $r = |\mathbf{r}_1 - \mathbf{r}_2|$.The associated static structure function $S(k)$ is the dimensionless Fourier transform of $g(r)$.Its dependence on the relative momentum or wavenumber k can be extracted from measured neutron scattering cross sections.The quantum-mechanical correlations between identical particles become apparent in the cyclic or particle-exchange correlation function $G_{cc}(r)$ and its Fourier inverse $S_{cc}(k)$. Information on the spatial correlations between the phase factors associated with the single-particle wave components of the full N-body density matrix of the thermodynamic state of the quantum fluid is embodied in the phase-phase correlation function $Q(r)$. This quantity is contained in the off-diagonal elements $n(r)$, i.e., the one-body reduced density matrix.

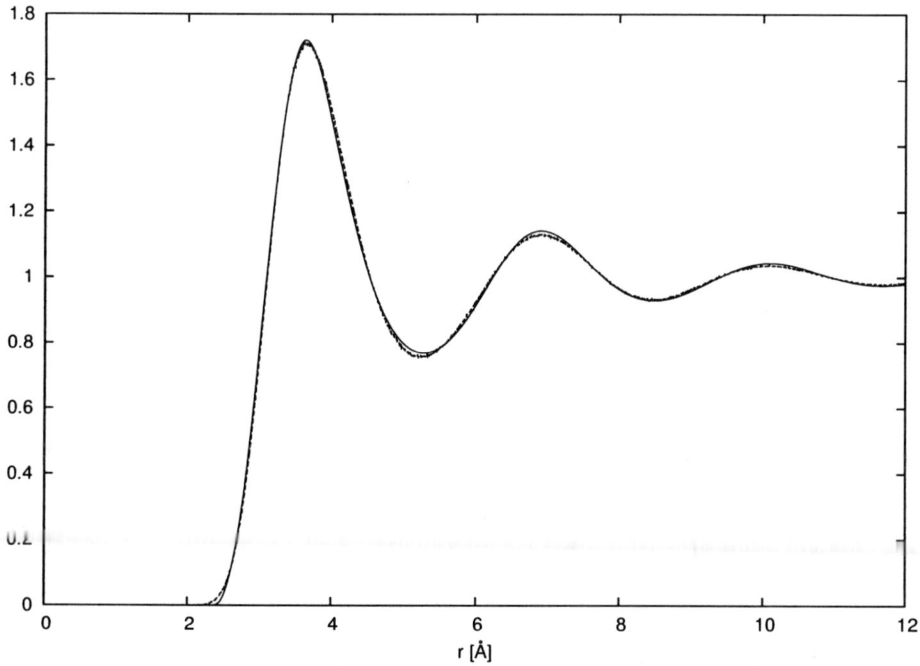

Fig. 1. Radial distribution function $g(r)$ of liquid para-hydrogen (T = 16 K, ρ = 0.021 Å^{-3}). Full line: results of CDM theory, broken line: PIMC simulation results.

This Section deals with a thorough analysis and comparison of theoretical and experimental results on functions $g(r)$ and $S(k)$. CDM theory provides a renormalized Schrödinger equation with zero eigenvalue for the square root of the radial distribution function,

$$\left[-\frac{\hbar^2}{m}\Delta + v(r) + w(r) + v_{coll}(r) + v_{qp}(r)\right]\sqrt{g(r)} = 0\,. \tag{1}$$

The potential energy terms appearing in Eq. (1) are the Silvera potential[14] $v(r)$, the induced potential[12] $w(r)$, the coupling term $v_{coll}(r)$ to the collective excitations,[10] and the quasiparticle coupling term[12] $v_{qp}(r)$. For liquid para-hydrogen close to the triple point the last coupling term is very small and may be ignored. Without the term $v_{qp}(r)$ it is straight-forward to solve the Schrödinger equation by following the elementary calculus designed in Ref. 10. The solution is displayed in Fig. 1. We have checked its numerical accuracy by comparing the CDM results with those of a PIMC calculation for the same temperature and density,[16] finding excellent agreement (see Fig. 1).

The results on the distribution $g(r)$ show the enormous correlation strength that exists in the hydrogen liquid. The first maximum at $r = 3.7\,\text{Å}$ attains a value that is significantly larger than the maximal strength one finds at the first peak of

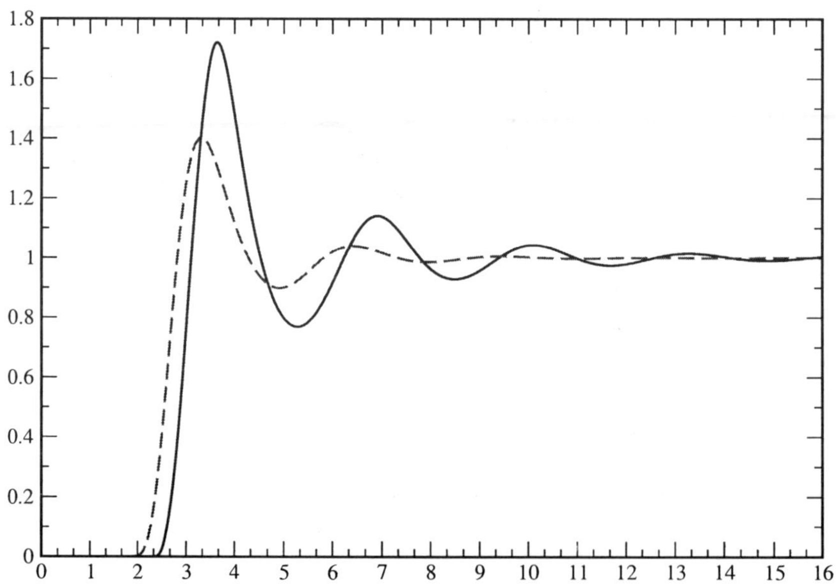

Fig. 2. Comparison of radial distribution functions of liquid para-hydrogen and of normal liquid helium (broken line) under similar thermodynamic conditions (abscissa: relative distance r in units [Å]).

the radial distribution function of liquid helium under similar thermal conditions assuming the realistic Aziz potential[17] as input for the theoretical analysis (Fig. 2).

Next, we turn to a detailed study of the associated static structure function $S(k)$. Within CDM theory as well as in the stochastic PIMC approach this quantity is calculated as the dimensionless Fourier transform

$$S(k) = 1 + \rho \int [g(r) - 1]\, e^{i\mathbf{kr}} d\mathbf{r} \,. \tag{2}$$

Numerical results on the structure function with CDM and PIMC data for $g(r)$ as input are displayed in Fig. 3. Once more, the Monte-Carlo results confirm the high accuracy of the results from CDM theory for liquid para-hydrogen close to the triple point. We emphasize that the integral (2) has been calculated without applying any fitting prescription for the distribution $g(r)$ at large relative distances. The oscillations seen in the stochastic results at wavenumbers $k < 1.5\,\text{Å}^{-1}$ are artifacts of the inherently unavoidable small box size used in the PIMC calculations. In contrast, CDM theory does not suffer from this deficiency since we can easily solve the Schrödinger equation in a sufficiently large interval of relative distances. CDM theory provides therefore an efficient tool for accurate calculations of the isothermal compressibility and isothermal velocity of sound in the limit of vanishing wavenumber.

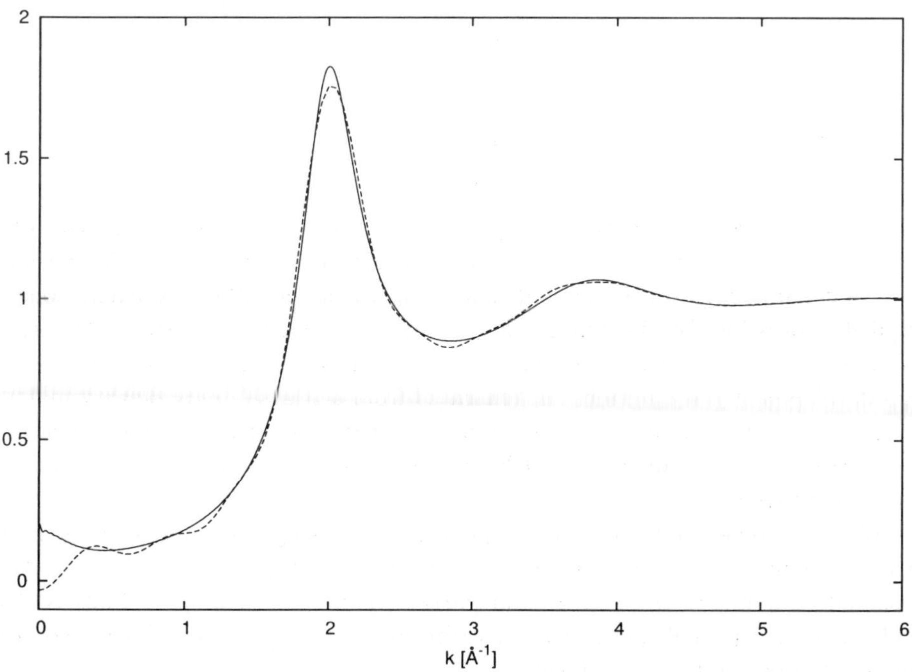

Fig. 3. Numerical results on the structure function $S(k)$ of liquid para-hydrogen by CDM theory and from PIMC calculations (dashed line), see text.

The theoretical results reported may be compared with experimental data from neutron scattering experiments.[7] The data processing yields an experimental structure function with a much higher first peak than found in the theoretical result on $S(k)$. This discrepancy between peak values of about 2.8 experimentally and about 1.8 for the theoretical result is not acceptable. There are theoretical data from molecular dynamics calculations[7] which seem to support the experimental findings (see Figs. 3 and 5 of Ref. 7). However, they are evidently in disagreement with the present results within CDM theory and of PIMC calculations. This problem -at least on the theoretical side- awaits a future careful examination.

3. Excitations

A correlated normal Bose fluid permits at least two different branches of excitations.There is a collective branch of phonons at small wavenumbers k and possibly roton-like excitations at atomic wavelengths. A second branch consists of excited quasiparticles that conserves the total number of constituents.CDM theory enables us to determine the excitation energies of both branches by solving two Euler-Lagrange equations,[12]

$$\frac{\delta F_\lambda}{\delta n_{cc}(k)} = 0\,, \tag{3}$$

with subcondition

$$\sum_{\mathbf{k}} n_{cc}(k) = N \tag{4}$$

and

$$\frac{\delta F_\lambda}{\delta n_{coll}(k)} = 0\,. \tag{5}$$

The thermodynamic potential $F_\lambda[S(k), n_{cc}(k), n_{coll}(k); T, \rho]$ is a functional of the static structure function $S(k)$, the momentum distribution $n_{cc}(k)$ of N quasiparticles, and the occupation-number density $n_{coll}(k)$ of the collective excitations. An explicit expression for the functional F_λ can be constructed within CDM theory by following a systematic approximation procedure. At the present level of formal development the functional is generated from a trial N-body density matrix of Jastrow- type. Future improvements may be derived by a suitable generalization of the Correlated Basis Functions (CBF) formalism for the ground state of a quantum many-body system.[18,19]

We have solved the Euler-Lagrange equations using the approximate thermodynamic potential derived from the Jastrow trial N-body density matrix.[12] At the assumed temperature $T = 16$ K and particle-number density $\rho = 0.021$ Å^{-3} one finds that the optimal momentum distribution of the quasiparticles is excellently represented by the Gaussian form

$$n_{cc}(k) = \exp \beta \left[\mu_0 - \epsilon_0(k)\right] \tag{6}$$

with $\epsilon_0(k) = \hbar^2/2m$ (molecular mass m of hydrogen) and inverse temperature $\beta = (k_B T)^{-1}$. The chemical potential is $\mu \simeq -7.8$ K due to the subcondition (4). The coresponding one-body reduced density matrix is therefore given by

$$n_{cc}(r) = \exp\left[-\pi\left(\frac{r}{\lambda}\right)^2\right] \tag{7}$$

with a thermal wavelength $\lambda \simeq 3.09$ Å. A consequence of this result is the absence of particle-exchange correlations in the hydrogen state considered, explicitly expressed by the result[11,12] $S_{cc}(k) = G_{cc}(r) \simeq 0$. The quasiparticles are therefore distinguishable free hydrogen molecules obeying classical Boltzmann statistics.

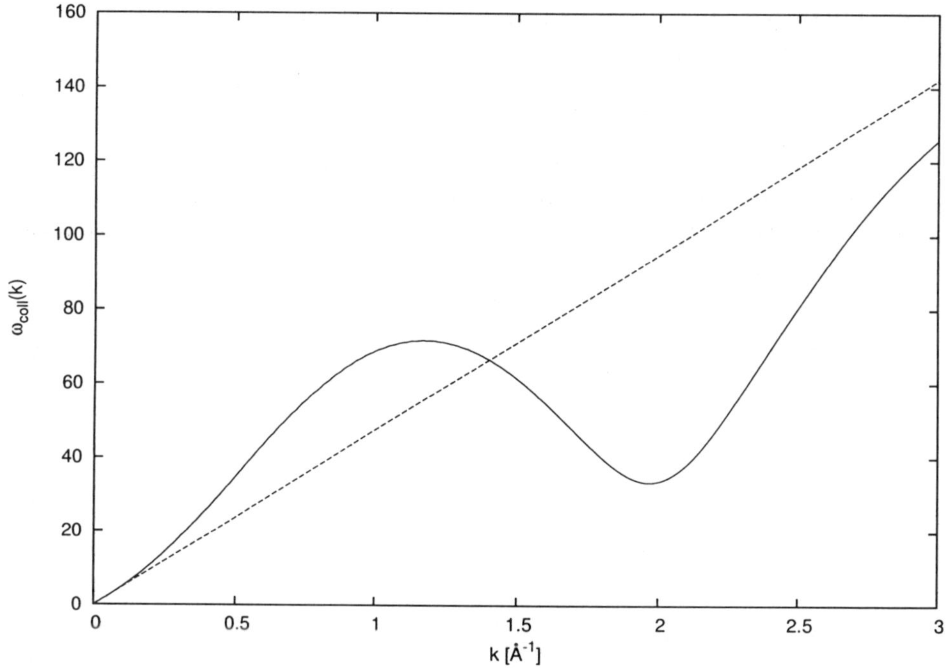

Fig. 4. Energy spectrum of collective excitations in the hydrogen system, in generalized Feynman approximation. The dashed line indicates the linear phonon spectrum in the limit of vanishing wavenumber.

Eq. (5) is a generalized Feynman equation[12,20] that can be easily solved. The optimal excitation energies $\omega_{coll}(k)$ are displayed in Figure 4. This energy branch is of the familiar phonon/roton form. The broken line indicates the phonon limit $\omega_{coll}(k) \simeq \hbar ck$ as k approaches zero. Its slope yields the theoretical isothermal sound velocity $c \simeq 680$ m/s. We may compare these results with the available experimental data[8] (cf. Figures 5 and 8 therein). There is good agreement between the theoretical and the experimental results in the roton region. However, the roton excitations are strongly damped.In contrast,the excitations in the wavenumber region $k < 0.8$ Å^{-1}

are stable, yet the theoretical excitation energies become significantly smaller than the experimental energies by increasing the wavenumber k. The discrepancy requires further investigation.

4. Momentum Distribution

The momentum distribution of a single hydrogen molecule in the liquid is given by the integral

$$n(k) = \rho \int n(r) e^{i\mathbf{kr}} d\mathbf{r} \tag{8}$$

where function $n(r)$ is the (unit-normalized) one-body reduced density matrix. This quantity has been analyzed within CDM theory.[2,15] The formalism yields the structural factor decomposition

$$n(r) = n_c N_0(r) \exp\left[-Q(r)\right] \tag{9}$$

with the strength factor $n_c = \exp Q(0)$. The factor $N_0(r) = n_{cc}(r) + N_c(r)$ embodies the quantum-mechanical effects due to the exchange of identical particles. The function $P(r) = Q(r)/Q(0)$ can be interpreted as the (unit-normalized) phase-phase correlation function.

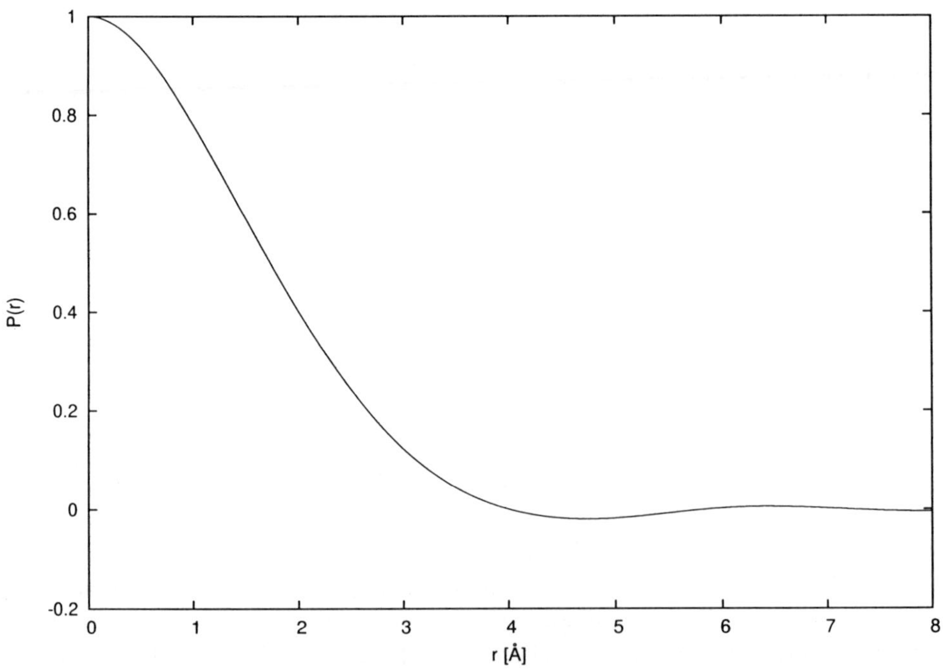

Fig. 5. Numerical results for the phase-phase correlation function $P(r)$ by CDM theory, in HNC/0 approximation.

Employing the hypernetted-chain (HNC) technique for evaluating the functions[15] we calculated $N_0(r)$ and $P(r)$ for liquid hydrogen in HNC/0 approximation. As expected, the exchange correlations $N_c(r)$ are very small and thus

$$N_0(r) \simeq n_{cc}(r)\,. \tag{10}$$

Figure 5 displays theoretical results on function P(r) in HNC/0 approximation. We see that this distribution is well represented by a Gaussian form

$$P(r) \simeq P_G(r) = \exp\left[-\pi\left(\frac{r}{\lambda_P}\right)^2\right], \tag{11}$$

with $\lambda_P \simeq 3.73\,\text{Å}$ at $T = 16$ K and $\rho = 0.021\,\text{Å}^{-3}$. The strength factor n_c is related to the curvature of function $P(r)$ and the total kinetic energy per molecule of the Bose liquid. A PIMC calculation of this energy portion yields 58.6 K at $T = 16$ K and $\rho = 0.021\,\text{Å}^{-3}$. With these input data the strength factor has the value $n_c = 0.118$.

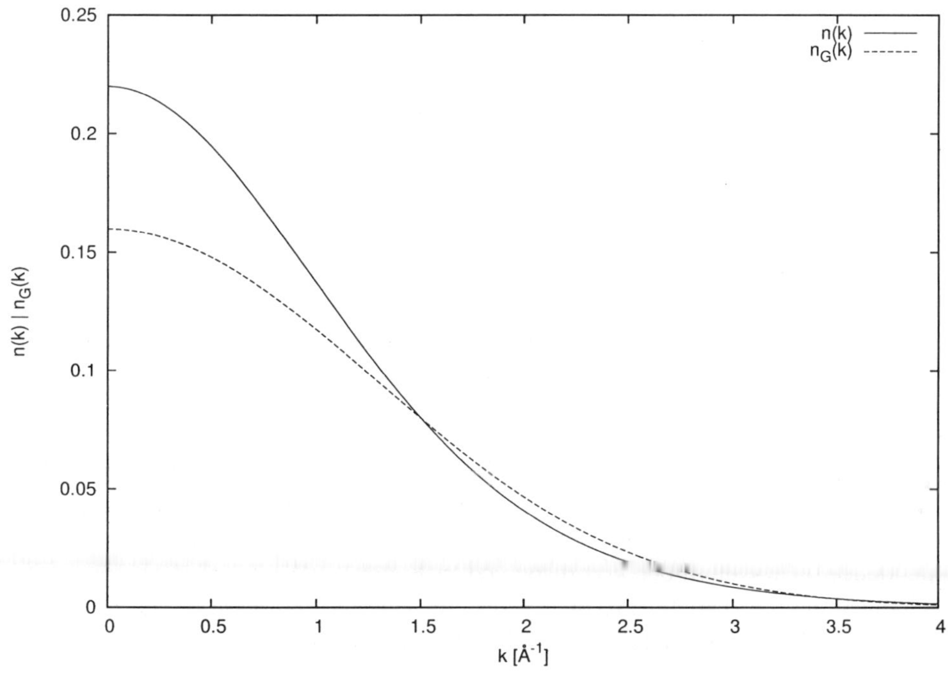

Fig. 6. Theoretical momentum distribution $n(k)$ of a single hydrogen molecule in the liquid. It is compared to the classical Gauss distribution $n_G(k)$ that leads to the same total kinetic energy per molecule as the correlated hydrogen liquid (broken line).

Straightforward multiplication and integration via Eq. (9) yield the one-body elements $n(r)$ and the momentum distribution $n(k)$. Figure 6 represents numerical

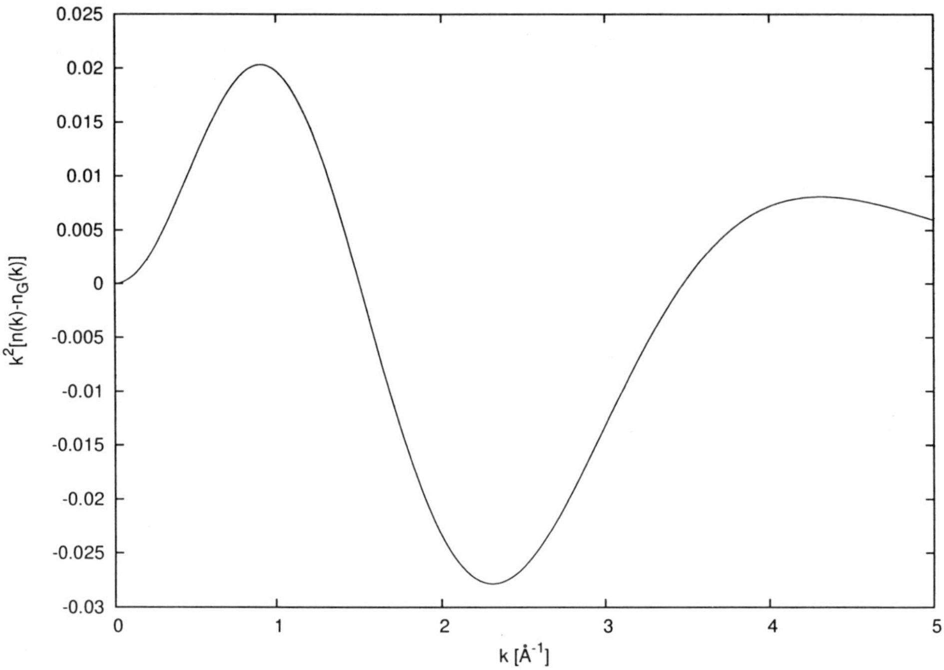

Fig. 7. Deviation of the kinetic energy distribution $k^2n(k)$ of liquid para-hydrogen from the corresponding classical Maxwell energy distribution (cf. Eq. (12)).

results on the dependence of the latter function on wavenumber k. We may compare it with the Gaussian momentum distribution that would hold by assuming that the molecules are free and distinguishable but have the same total kinetic energy per particle as the quantum liquid. This distribution would be characterized by an effective thermal wavelength $\lambda_0 \simeq 1.97\,\text{Å}$ (Figure 6, broken line). Obviously, the difference

$$\Delta N(k) = k^2\left[n(k) - n_G(k)\right] \tag{12}$$

measures the deviation of the kinetic energy distribution of the quantum liquid from the classical Maxwell distribution of distinguishable hydrogen molecules. Its dependence on wavenumber k is displayed in Figure 7. We see that the quantum-mechanical energy distribution of the system favors an increase of the number of molecules with low momenta compared to the classical distribution.

5. Summary

We have presented a theoretical and an experimental analysis of the structure of liquid para-hydrogen close to the triple point.The system is characterized by very strong dynamical spatial correlations induced by the intermolecular forces. However, the repulsion at short relative distances suppresses particle-exchange correlations between and among the hydrogen molecules. For the same reason, the exchange (or cyclic) correlation function and the associated exchange structure function are almost zero everywhere. Furthermore, the quasiparticle momentum distribution is very well approximated by the classical Gaussian distribution of free and distinguishable particles. Similarly,the short-ranged phase-phase correlation function and corresponding structure function are to a very good approximation of Gaussian form. The interplay of exchange and phase-phase coupling, however, generates a significant departure of the single-particle momentum distribution in liquid parahydrogen from the classical Maxwell-Gauss distribution.

The present study employed the formalism of CDM theory for a normal Bose fluid. The results have been compared with PIMC simulation data. The numerical comparison demonstrates the high accuracy of CDM theory when applied to liquid hydrogen at low temperatures. CDM theory is therefore expected to be a very fast, efficient, and reliable tool for a detailed quantitative analysis of normal Bose fluids under similar thermodynamic conditions.

References

1. K. A. Gernoth and M. L. Ristig, in *Particle Scattering, X-Ray Diffraction, and Microstructure of Solids and Liquids*, Lecture Notes in Physics **610**, eds. M. L. Ristig and K. A. Gernoth, (Springer-Verlag, Heidelberg, 2003).
2. P. E. Sokol, R. N. Silver, and J. W. Clark, in *Momentum Distributions*, eds. R. N. Silver and P. E. Sokol, (Plenum, New York, 1989).
3. D. G. Fried, T. C. Killian, L. Willmann, D. Landhuis, S. C. Moss, D. Kleppner, and T. J. Greytak, *Phys. Rev. Lett.* **81**, 3811 (1998).
4. E. A. Cornell and C. E. Wieman, *Rev. Mod. Phys.* **74**, 875 (2002).
5. M. Ross, *Rep. Prog. Phys.* **48**, 1 (1985).
6. S. Bonev, E. Schwegler, T. Ogitsu, G. Galli, *Nature* **431**, 669 (2004).
7. J. Dawidowski, F. J. Bermejo, M. L. Ristig, B. Fak, C. Cabrillo, R. Fernandez-Perea, K. Kinugawa, and J. Campo, *Phys. Rev.* **B69**, 014207 (2004).
8. F. J. Bermejo, B. Fak, S. M. Bennington, R. Fernandez-Perea, C. Cabrillo, J. Dawidowski, R. Fernandez-Diaz, P. Verkerk, *Phys. Rev.* **B60** 15154 (1999).
9. J. Dawidowski, F. J. Bermejo, M. L. Ristig, C. Cabrillo, and S. M. Bennington, *Phys. Rev.* **B** (submitted).
10. G. Senger, M. L. Ristig, K. E. Kuerten, and C. E. Campbell, *Phys. Rev.* **B33**, 7562 (1986).
11. G. Senger, M. L. Ristig, C. E. Campbell, and J. W. Clark, *Ann. Phys. (N.Y.)* **218**, 160 (1992).
12. T. Lindenau, M. L. Ristig, J. W. Clark, and K. A. Gernoth, *J. Low Temp. Phys.* **129**, 143 (2002).
13. K. A. Gernoth, *Ann. Phys. (N.Y.)* **291**, 202 (2001).
14. I. Silvera and V. Goldman, *J. Chem. Phys.* **69**, 4209 (1978).
15. R. Pantfoerder, T. Lindenau, and M. L. Ristig, *J. Low Temp. Phys.* **108**, 245 (1997).
16. K. A. Gernoth, M. J. Harrison, and M. L. Ristig, these Proceedings.
17. R. A. Aziz, V. P. S. Nain, J. S. Carley, W. L. Taylor, and G. T. McConville, *J. Chem. Phys.* **70** 4330 (1979).
18. M. Saarela, *Phys. Rev.* **B33**, 4596 (1986).
19. E. Krotscheck, in *Microscopic Quantum Many-Body Theories and Their Applications*,

Lecture Notes in Physics **510**, eds. J. Navarro and A. Polls (Springer-Verlag, Berlin, Heidelberg 1998).
20. R. P. Feynman, *Phys. Rev.* **94**, 262 (1954).

QUANTUM FLUIDS AND SOLIDS

QUANTUM REFLECTION, EVAPORATION, AND TRANSPORT CURRENTS AT ^{4}He SURFACES

V. APAJA, E. KROTSCHECK, A. RIMNAC

Institut für Theoretische Physik, Johannes-Kepler-Universität, A-4040 Linz, Austria

R. E. ZILLICH

Fraunhofer ITWM, D-67663 Kaiserslautern, Germany

In this work, we study transport currents in excited states. This requires the calculation of particle currents $\langle n|\hat{\mathbf{j}}|n\rangle$ to second order in the excitation amplitudes. For that purpose, we take a well-tested microscopic theory of inhomogeneous quantum liquids and extend it to find the mass currents created when atoms scatter off a surface or when excitations evaporate atoms. This is the first theoretical study of transport phenomena in a quantum liquid based on a quantitative microscopic theory.

Keywords: Quantum reflection; transport currents; atom scattering

1. Introduction

The response of quantum fluids to low-energy perturbations like neutrons or X-rays is nowadays well understood. The situation is different when atoms are scattered off a quantum liquid, either a surface or a cluster. Depending on the species of the atom, Bose symmetry must be properly taken into account. Inside the liquid, it is appropriate to talk about quasiparticle excitations which can, at a surface, either evaporate physical particles, be scattered back, or decay into two or more quasiparticles. In all of these cases *inelastic* processes, where an excitation can decay into one or more excitations of lower energy, become relevant.

We have dealt with the microscopic description of vacuum to vacuum transitions in Refs. 1, 2 and 3; the present work is devoted to the more complicated case of processes that happen inside the liquid. Phonon-decay occurs, for example, in the long-wavelength regime where the phonon has an anomalous dispersion relation. Other important problems are quantum condensation and evaporation. Common to these processes is that they are *non-periodic*.

Our work have been motivated by the experiments of Edwards, Wyatt and collaborators[4–13] on the reflection, condensation, evaporation, and transmission of ^{4}He atoms at a ^{4}He surface or through ^{4}He filaments. Quasiparticle scattering at helium surfaces has been studied theoretically using microscopic many-body theory,[1] by Sobnack and Inkson[14] within a real–space formulation of Beliaev's formalism, and by Dalfovo *et al.*[15] within a pseudo-phenomenological density functional theory.

Both approaches ignore inelastic processes which account for the essential physics[1] and use a questionable definition of the quasiparticle current.

2. Equations of motion and continuity equations

2.1. *Densities and currents*

We write the wave function of an excited state of a Bose liquid as

$$\Psi(\mathbf{r}_1,\ldots,\mathbf{r}_N;t) = \frac{1}{\mathcal{N}(t)^{\frac{1}{2}}} e^{-iH_{00}t/\hbar} e^{\frac{1}{2}\delta U(t)} \Psi_0(\mathbf{r}_1,\ldots,\mathbf{r}_N) \tag{1}$$

$$\delta U(t) = \sum_{i=1}^{N} \delta u_1(\mathbf{r}_i;t) + \sum_{1=i<j}^{N} \delta u_2(\mathbf{r}_i,\mathbf{r}_j;t) + \ldots . \tag{2}$$

where $\Psi_0(\mathbf{r}_1,\ldots,\mathbf{r}_N)$ is the exact ground state or an optimized Jastrow-Feenberg wave function. H_{00} is the ground state energy expectation value, and $\mathcal{N}(t)$ is the normalization integral. Without loss of generality, we can write the excitation operator as a series of multiparticle fluctuations (2).

We are especially interested in the one-body density and the one-body current

$$\rho_1(\mathbf{r}_1;t) = N \int d^3r_2 \ldots d^3r_N \left|\Psi(\mathbf{r}_1,\ldots,\mathbf{r}_N;t)\right|^2 , \tag{3}$$

$$\mathbf{j}_1(\mathbf{r}_1;t) = \frac{\hbar N}{2m} \Im m \int d^3r_2 \ldots d^3r_N \Psi^*(\mathbf{r}_1,\ldots,\mathbf{r}_N;t) \nabla_1 \Psi(\mathbf{r}_1,\ldots,\mathbf{r}_N;t) . \tag{4}$$

These quantities can be expanded in powers of the excitation operator $U(t)$. The *first* and *second order* densities and currents are

$$\delta\rho_1^{(1)}(\mathbf{r}_1;t) = N \int d^3r_2 \ldots d^3r_N \Psi_0^2 \delta U(t), \tag{5}$$

$$\mathbf{j}_1^{(1)}(\mathbf{r}_1;t) = \frac{\hbar N}{2mi} \int d^3r_2 \ldots d^3r_N \Psi_0^2 \nabla_1 \delta U(t) , \tag{6}$$

$$\delta\rho_1^{(2)}(\mathbf{r}_1;t) = \frac{N}{4} \int d^3r_2 \ldots d^3r_N \Psi_0^2 \left[|\delta U(t)|^2 - \langle \Psi_0 | \, |\delta U(t)|^2 \, |\Psi_0\rangle \right] \tag{7}$$

$$\mathbf{j}_1^{(2)}(\mathbf{r}_1;t) = \frac{\hbar N}{8mi} \int d^3r_2 \ldots d^3r_N \Psi_0^2 \left[\delta U^*(t) \nabla_1 \delta U(t) - \text{c.c.} \right] . \tag{8}$$

Note that we have defined the first order quantities (5) and (6) as complex; the physical observables are their real parts. Since these are periodic in time and, hence, average to zero, one has to calculate the second-order quantities to describe net (quasi-)particle transport. Generally, one can write the second-order one-body current in terms of the complex first-order densities as

$$\mathbf{j}_1^{(2)}(\mathbf{r}_1;t) = \frac{\hbar}{8mi} \left[\delta\rho_1^{*(1)}(\mathbf{r}_1;t) \nabla_1 \delta u_1(\mathbf{r}_1;t) + \int d^3r_2 \delta\rho_2^{*(1)}(\mathbf{r}_1,\mathbf{r}_2;t) \nabla_1 \delta u_2(\mathbf{r}_1,\mathbf{r}_2;t) + \ldots - \text{c.c.} \right] . \tag{9}$$

where the $\delta\rho_M^{*(1)}(\mathbf{r}_1,\ldots,\mathbf{r}_M;t)$ are the first-order fluctuating parts of the M-body densities. Truncation of the expansion (9) at the level $M=2$ leads to presently practical working formulas.

2.2. *Equations of Motion*

Whether the above quantities satisfy a continuity equation depends on the equations of motion relating the first-order fluctuations to a weak external field $U_{\rm ext}(\mathbf{r};\omega)e^{-i(\omega+i\epsilon)t}$ + c.c., which is adiabatically turned on with the rate $e^{\epsilon t}$, and finally one sets $\epsilon\to 0^+$.

The excitation operator $\delta U(t)$ is determined by stationarity of the action

$$\mathcal{S}=\int dt\mathcal{L}(t)=\int dt\,\langle\Psi(t)|\,H+U_{\rm ext}(t)-i\hbar\frac{\partial}{\partial t}\,|\Psi(t)\rangle\;. \tag{10}$$

For weak perturbations we may linearize the equations of motion.[16,17] Truncating the series (2) at the level of 2-body fluctuations, the variations of $\mathcal{S}$ with respect to $\delta u_1(\mathbf{r}_1;t)$ and $\delta u_2(\mathbf{r}_1,\mathbf{r}_2;t)$ lead to two linear equations of motion (we omit the driving terms with $U_{\rm ext}(t)$)

$$\begin{aligned}&\frac{\hbar^2}{2mi}\nabla_1\cdot\left\{\rho_1(\mathbf{r}_1)\nabla_1\delta u_1(\mathbf{r}_1;t)+\int d^3r_2\rho_2(\mathbf{r}_1,\mathbf{r}_2)\nabla_1\delta u_2(\mathbf{r}_1,\mathbf{r}_2;t)\right\}\\&+\hbar\delta\dot\rho_1^{(1)}(\mathbf{r}_1;t)=0\,,\end{aligned}\tag{11}$$

$$\begin{aligned}&\frac{\hbar^2}{2mi}\nabla_1\cdot\left\{\big[\rho_2(\mathbf{r}_1,\mathbf{r}_2)-\rho_1(\mathbf{r}_1)\rho_1(\mathbf{r}_2)\big]\nabla_1\delta u_1(\mathbf{r}_1;t)+\rho_2(\mathbf{r}_1,\mathbf{r}_2)\nabla_1\delta u_2(\mathbf{r}_1,\mathbf{r}_2;t)\right.\\&\left.+\int d^3r_3\big[\rho_3(\mathbf{r}_1,\mathbf{r}_2,\mathbf{r}_3)-\rho_2(\mathbf{r}_1,\mathbf{r}_3)\rho_1(\mathbf{r}_2)\big]\nabla_1\delta u_2(\mathbf{r}_1,\mathbf{r}_3;t)\right\}+\text{ same for }(1\leftrightarrow 2)\\&+\hbar\big[\delta\dot\rho_2^{(1)}(\mathbf{r}_1,\mathbf{r}_2;t)-\rho_1(\mathbf{r}_1)\delta\dot\rho_1^{(1)}(\mathbf{r}_2;t)-\rho_1(\mathbf{r}_2)\delta\dot\rho_1^{(1)}(\mathbf{r}_1;t)\big]=0\,.\end{aligned}\tag{12}$$

The most advanced implementation[18] in the bulk liquid has led to very accurate predictions of the phonon-roton spectrum. In non-uniform geometries more compromises have to be made, but with minimal phenomenological input a rather good agreement with experimental data was achieved.[16]

Setting $\delta u_2=0$ in Eq. (11) and defining $\psi(\mathbf{r})e^{-i\omega t}\equiv\delta u_1(\mathbf{r};t)/\sqrt{\rho_1(\mathbf{r})}$, the first equation of motion (11) reduces to a generalized eigenvalue problem

$$H_1(\mathbf{r})\psi_n(\mathbf{r})=\hbar\omega_n\int d^3r'S(\mathbf{r},\mathbf{r}')\psi_n(\mathbf{r}')\,, \tag{13}$$

which is evidently the inhomogeneous generalization of the Feynman dispersion relation $\hbar\omega(k)=\hbar^2k^2/(2mS(k))$. Here

$$S(\mathbf{r},\mathbf{r}')=\delta(\mathbf{r}-\mathbf{r}')+\sqrt{\rho_1(\mathbf{r})}\,[g(\mathbf{r},\mathbf{r}')-1]\,\sqrt{\rho_1(\mathbf{r}')} \tag{14}$$

is the coordinate space *static structure function*, and $H_1(\mathbf{r})$ is a *one-body Hamiltonian*

$$H_1(\mathbf{r})=-\frac{\hbar^2}{2m}\frac{1}{\sqrt{\rho_1(\mathbf{r})}}\nabla\cdot\rho_1(\mathbf{r})\nabla\frac{1}{\sqrt{\rho_1(\mathbf{r})}}=-\frac{\hbar^2}{2m}\nabla^2+V_H(\mathbf{r})-\mu \tag{15}$$

where $V_H(\mathbf{r})$ is a local, self-consistent "Hartree-potential", and μ is the chemical potential. The latter form is valid if the ground state density has been optimized using the condition

$$\left[-\frac{\hbar^2}{2m}\nabla^2 + V_H(\mathbf{r}) - \mu\right]\sqrt{\rho_1(\mathbf{r})} = 0\,. \tag{16}$$

We shall use, as reference states for the further developments, the Feynman states $\psi_n(\mathbf{r})$ and their adjoint states $\phi_m(\mathbf{r}) = \frac{1}{\hbar\omega_m}H_1(\mathbf{r})\psi_m(\mathbf{r})$, with $\langle\phi_m|\psi_n\rangle = \delta_{mn}$.

The approximations that lead to working formulas for the dynamic structure function that both capture the right physics and are practical in the inhomogeneous geometry have been discussed in Ref. 16. These approximations allow us to solve the second equation of motion analytically, and lead to the so-called *convolution approximation* for the density–density response function,

$$\chi(\mathbf{r},\mathbf{r}',\omega) = \sum_{st}\sqrt{\rho(\mathbf{r})}\phi_s(\mathbf{r})\left[G_{st}(\omega) + G_{st}(-\omega)\right]\phi_t(\mathbf{r}')\sqrt{\rho(\mathbf{r}')}\,. \tag{17}$$

Here

$$G_{st}(\omega) = \left[\hbar[\omega - \omega_s + i\epsilon]\delta_{st} + \Sigma_{st}(\omega)\right]^{-1} \tag{18}$$

is the Green's function (phonon propagator), with the self energy

$$\Sigma_{st}(\omega) = \frac{1}{2}\sum_{mn}\frac{V_{mn}^{(s)}V_{mn}^{(t)}}{\hbar(\omega_m + \omega_n - \omega - i\epsilon)}\,. \tag{19}$$

The three-phonon matrix elements $V_{mn}^{(s)}$ are overlap integrals of three Feynman states, see Ref. 16. By setting $\Sigma_{st}(\omega) = 0$ one recovers the random phase approximation (RPA) - which is, equivalently, obtained from time-dependent Hartree theory or time-dependent density functional theory. The three-phonon corrections have the important consequence that the density-density response function can have a non-trivial imaginary part due to phonon decay processes.

2.3. *Transport current*

For calculating scattering probabilities as measured in experiments, we need to keep only those contributions that persist in the homogeneous parts of the system (in the vacuum or deep in the bulk liquid) and that have a non-zero spatial average. We shall call this portion the *transport current.* By a lengthy diagrammatic analysis,[19] the second order current (9) can be written in the suggestive form

$$\mathbf{j}_1^{(2)}(\mathbf{r};t) = \mathbf{j}_{\rm el}^{(2)}(\mathbf{r};t) + \mathbf{j}_{\rm inel}^{(2)}(\mathbf{r};t) = \frac{\hbar}{8mi}\Big[\delta\rho_1^{*(1F)}(\mathbf{r};t)\mathbf{j}_1^{(1F)}(\mathbf{r};t) \tag{20}$$
$$+ \rho_1(\mathbf{r})\int d^3r'\rho_1(\mathbf{r}')\delta_u g^{*(1)}(\mathbf{r},\mathbf{r}';t)\nabla_{\mathbf{r}}\delta u_2(\mathbf{r},\mathbf{r}';t)\Big] + \text{c.c.}\,,$$

where $\delta_u g^{(1)}(\mathbf{r},\mathbf{r}';t)$ is the first-order variation of the pair distribution function $g(\mathbf{r},\mathbf{r}')$ due to the variation of $u_2(\mathbf{r},\mathbf{r}';t)$, keeping the density fixed. $\delta\rho_1^{*(1F)}(\mathbf{r};t)$ and $\mathbf{j}_1^{(1F)}(\mathbf{r};t)$ are the first order density and current in Feynman approximation.

Let us consider the first, elastic current term in Eq. (20). Inserting the normal mode expansions $\delta\rho_1^{(1F)}(\mathbf{r};t) = \sqrt{\rho(\mathbf{r})}\sum_m r_m(t)\phi_m(\mathbf{r}) =$ and $\mathbf{j}_1^{(1F)}(\mathbf{r};t) = \sum_m r_m(t)\nabla[\psi_m(\mathbf{r})/\sqrt{\rho(\mathbf{r})}]$, where, for stationary states, $r_m(t) = r_m e^{i\omega t}$, gives

$$\mathbf{j}_{\rm el}^{(2)}(\mathbf{r};t) = \frac{\hbar}{8mi}\sqrt{\rho_1(\mathbf{r})}\sum_{m,n} r_m(r)r_n^*(t)\phi_m^*(\mathbf{r})\nabla\frac{\psi_n(\mathbf{r})}{\sqrt{\rho_1(\mathbf{r})}} + \text{c.c.}\,.$$

In any homogeneous region both $\phi_m(\mathbf{r})$ and $\psi_n(\mathbf{r})$ are plane waves, hence the above current is also a plane wave unless we have $m = n$. If we average the transport current over length scales that are long compared with the wave lengths of the Feynman phonons, the interference terms cancel out. We may therefore define real, time-independent "channel currents",

$$\mathbf{j}_n(\mathbf{r}) \equiv \frac{\hbar}{8mi}\sqrt{\rho_1(\mathbf{r})}\phi_n^*(\mathbf{r})\nabla\frac{\psi_n(\mathbf{r})}{\sqrt{\rho_1(\mathbf{r})}}\,. \tag{21}$$

Any current describing transport is a linear combination of these channel currents.

In the numerical applications, we deal with the "inelastic" term in Eq. (20) in the so-called *convolution approximation* $\delta_u g^{(1)}(\mathbf{r},\mathbf{r}';t) \approx [S * \delta u_2 * S](\mathbf{r},\mathbf{r}';t)$. The inelastic current becomes in this approximation

$$\mathbf{j}_{\rm inel}^{(2)}(\mathbf{r};t) = \sum_m \mathbf{j}_m(\mathbf{r})\sum_{st}\left|\frac{V_{ms}^{(t)}r_t(t)}{\hbar(\omega-\omega_m-\omega_s+i\epsilon)}\right|^2\,. \tag{22}$$

A few words are in order concerning this "inelastic" current: As long as the energy denominator is non-zero – recall that ω is the frequency of the external perturbation – this is just an elastic correction to the total current. If, however, the energy denominator has a node, the resulting double pole has to be reinterpreted. A node in the energy denominator means that phonons can decay. To maintain a stationary system, one would need to supply a constant source of energy. In that case, one should rather talk about the *decay rate of the current*

$$\begin{aligned}\frac{d}{dt}\mathbf{j}^{(2)}(\mathbf{r};t) &= \frac{d}{dt}\mathbf{j}_{\rm inel}^{(2)}(\mathbf{r};t) = \sum_m \mathbf{j}_m(\mathbf{r})\sum_{st}\left|V_{ms}^{(t)}r_t\right|^2\frac{2\epsilon\, e^{2\epsilon t}}{|\hbar(\omega-\omega_m-\omega_s+i\epsilon)|^2}\\ &\to \sum_m \mathbf{j}_m(\mathbf{r})\sum_{st}\left|V_{ms}^{(t)}r_t\right|^2 2\pi\delta(\hbar\omega-\hbar\omega_m-\hbar\omega_s)\,,\end{aligned} \tag{23}$$

where we have taken in the last line the limit $\epsilon \to 0^+$. This analysis justifies the identification of the second term in the expression (20) with an *inelastic* current.

In the homogeneous limit, the total transport current has, by symmetry, the direction $\mathbf{k}$ of the elastic current, and thus contains no information about the direction of the decay products. The *angular* dependence of the current is obtained as the 2nd-order expectation value of the operator $\delta(\hat{\mathbf{n}} - \hat{\mathbf{n}}_j)\mathbf{j}(\mathbf{r})$, where the unit vector $\hat{\mathbf{n}}$ denotes the scattering direction. For the magnitude of the inelastic current in direction $\hat{\mathbf{n}}$, we get in the homogeneous limit

$$j_{\rm inel}^{(2)}(\hat{\mathbf{n}},t) = \sum_{\mathbf{q}}\delta(\hat{\mathbf{n}}-\hat{\mathbf{q}})\; j_{\mathbf{q}}\; S(q)S(|\mathbf{k}-\mathbf{q}|)\,|r_{\mathbf{q},\mathbf{k}-\mathbf{q}}|^2\,. \tag{24}$$

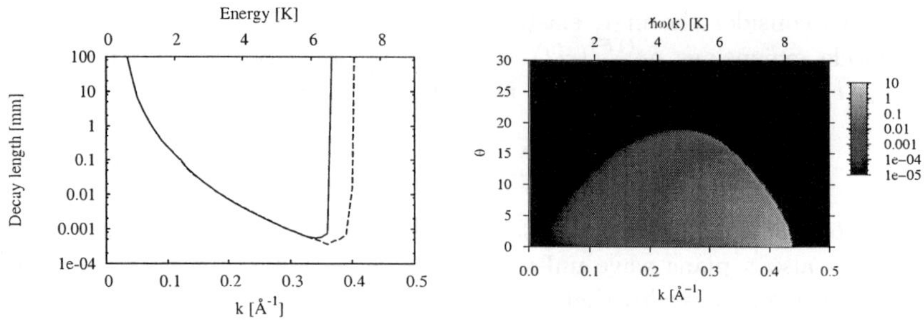

Fig. 1. The left figure shows the phonon decay length in bulk ^{4}He in the present HNC-EL theory (solid line) or using the experimental spectrum (dashed line). Evidently, phonons with intermediate energy decay rapidly into low-energy phonons with $\hbar\omega < 1$ K; only these and the very high-energy phonons traverse appreciable distances of order 1 mm. The right figure shows a logarithmic contour plot of the angular dependence of the decay rate $\frac{d}{dt}j^{(2)}$, shown as a function of k and θ.

3. Applications

3.1. *Phonon decay in bulk ^{4}He*

At the equilibrium density 0.022 Å^{-3} the calculated phonon–roton branch reproduces especially well the experimental spectrum.[18] The imaginary part of the self energy $\Im m\Sigma(\mathbf{k},\omega)$ determines the decay rate of the transport current. To obtain a consistent picture of the phonon decay, we insert the calculated or the experimental dispersion relation into the decay rate (23). This is legitimate at low momenta, where the dynamic structure function is exhausted by the collective mode. The left part of Fig. 1 shows the phonon lifetime due to three phonon processes, converted to decay length using the group velocity of the excitations. Phonons with wave number below $k \sim 0.1$ Å^{-1} and above $k \sim 0.36$ Å^{-1} (above $k \sim 0.42$ Å^{-1} for the experimental spectrum) are fairly stable against decay to two phonons. The shortest calculated decay length is about 4000 Å . Taking the experimental dispersion curve we find very slow decay due to three phonon processes up to about $k \sim 0.5$ Å^{-1}, also seen in recent experiments.[20]

For the angular dependence (24) of the decay rate into direction $\mathbf{e}$ we obtain

$$\frac{d}{dt}j^{(2)}(\mathbf{e}) = \frac{\pi}{2mS(k)}\left|r_{\mathbf{k}}\right|^2 \Im m \sum_{\mathbf{q}} q\,\delta(\mathbf{e}-\hat{\mathbf{q}})\,\frac{\left|\tilde{V}(\mathbf{q},\mathbf{k}-\mathbf{q};\mathbf{k})\right|^2}{\hbar(\omega-\omega_{\mathbf{q}}-\omega_{\mathbf{k}-\mathbf{q}}+i\epsilon)}\,. \tag{25}$$

The right part of Fig. 1 shows $\frac{d}{dt}j^{(2)}(\mathbf{e}) = \frac{d}{dt}j^{(2)}(\theta)$ where $\theta = \arccos(\mathbf{e}\cdot\hat{\mathbf{k}})$ is the scattering angle with respect to the wave vector $\mathbf{k}$. The largest decay rate is obtained for high energy phonons close to the edge of the unstable regime. These decay into lower energy phonons which lie in a narrow cone (small θ), while lower energy phonons decay at a smaller rate, generating phonons in a larger cone.

3.2. *Quantum Reflection from a free ^4He surface*

The effect of "universal quantum reflection"[21] is that the reflectivity of a surface goes to 1 as $\sqrt{E}$ for $E \to 0^+$, where E is the energy of the incoming particle. The regime where universal quantum reflection holds is determined by *low-energy resonances* in the liquid surface; such resonances can exist, for example, if the one-body potential seen by the impinging particle falls off like $1/z^3$ far from the surface.[21,22] Universal quantum reflection has indeed been observed in ^{4}He.[23]

Theoretically, atom scattering is described by the equation of motion

$$-\frac{\hbar^2}{2m}\nabla^2\xi(\mathbf{r}) + V_H^{(4)}(\mathbf{r})\xi(\mathbf{r}) + \int d^3r'\Sigma(\mathbf{r},\mathbf{r}';\omega)\xi(\mathbf{r}') = \hbar\omega\int d^3r' S(\mathbf{r},\mathbf{r}')\xi(\mathbf{r}') \quad (26)$$

To demonstrate the key features we use three sets of input data:

(1) HNC-EL data, with and without the self-energy $\Sigma(\mathbf{r},\mathbf{r}';\omega)$

(2) DMC density profile of Ref. 24 and the one-body potential generated from that profile through Eq. (16). We have fitted the DMC data to the form[24]

$$\rho(z) = \frac{\rho_0}{[1+e^{\beta(z-z_0)}]^\delta}\,. \quad (27)$$

The density $\rho(z)$ and the one–body potential $V_H(z)$ are connected through Eq. (16), which imposes the constraint $\beta^2\delta^2 = -8m\mu/\hbar^2$. The static structure function $S(\mathbf{r},\mathbf{r}')$ and the Feynman phonon wave functions $\psi_m(\mathbf{r})$ and $\phi_m(\mathbf{r})$ are very time consuming to obtain with sufficient accuracy from a Monte Carlo calculation, we therefore generate them using HNC-EL.

(3) A modified HNC-EL density profile, solved from Eq. (16) using the effective potential with an attractive $1/z^3$ tail:[22]

$$V_H^{\mathrm{mod}}(z) = \begin{cases} V_H(z) & \text{when } z < z_0 \\ -\dfrac{C_3}{\left[z - z_0 + \left(\frac{C_3}{V_H(z_0)}\right)^{1/3}\right]^3} & \text{elsewhere,} \end{cases} \quad (28)$$

where C_3 is the Van der Waals constant of a half-space of ^{4}He atoms interacting via the Aziz potential, and z_0 an adjustable matching location. With this modified potential, which is also shown in Fig. 2, we have re-calculated the ground state density, the static structure function and the Feynman phonon wave functions.

Fig. 2 shows the density profiles and the corresponding $V_H(z)$ for the three cases described above. Although the DMC surface region is only slightly wider than the HNC one, the difference is much more pronounced in the effective one-body potential $V_H(z)$, although part of this may be due to stochastic errors in the numerical data.

The left part of Fig. 3 shows our calculated reflection coefficient at low energies for the three cases described above. The density profile obtained from the HNC-EL leads to an energy dependence that is much weaker than what is found experimentally. The DMC density profile is slightly broader, which has some effect

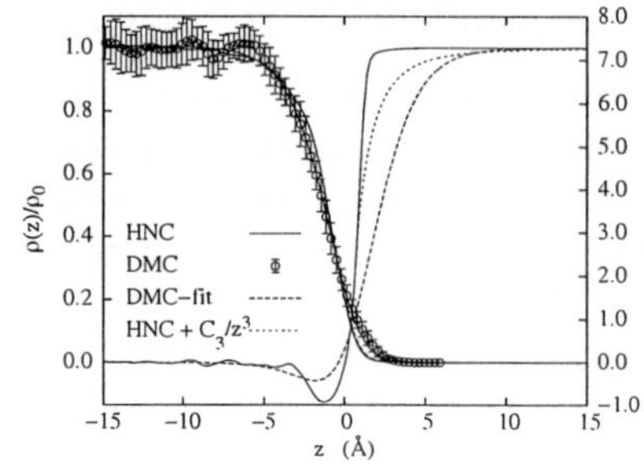

Fig. 2. The figure shows the density profile and the corresponding effective one-body potential for the three cases described in the main text. The original DMC data is also shown to give an idea of how close all the three density profiles really are.

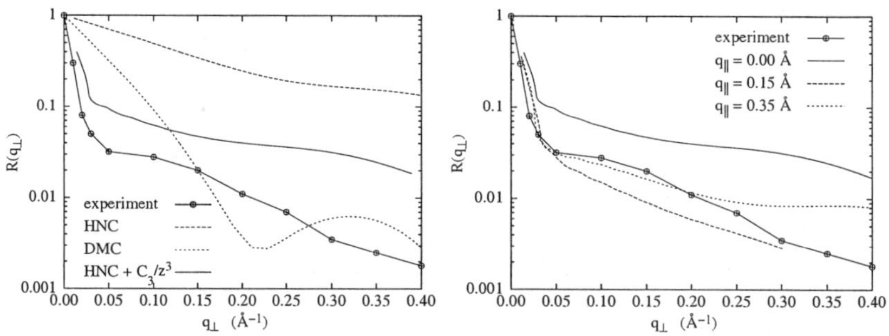

Fig. 3. The left figure shows the energy dependence of the reflectivity obtained from HNC-EL, HNC-EL with the added $1/z^3$ extrapolation, and the DMC data as described in the text. The right figure shows the reflectivity as a function of perpendicular momentum $q_\perp = q\cos(\vartheta)$ for the free ^{4}He surface and different values of the parallel momentum $q_\parallel$ starting from normal incidence $q_\parallel = 0.0/\text{Å}$ to $q_\parallel = 0.6/\text{Å}$ in steps of $0.1/\text{Å}$. In both figures the solid line with markers shows the experimental values displayed in Fig. 2 of Ref. 23.

on the reflectivity. But this is not the full story, in particular the "kink" seen in experiments at $q_\perp \approx 0.05\,\text{Å}^{-1}$ is missing. The modified potential of the HNC-EL calculation leads to a steeper slope of the reflectivity and comes close to the experiment, although the potential is narrower than what the DMC data suggests. We emphasize that the density profiles of the HNC-EL and the modified HNC-EL results are practically indistinguishable: the modification of the potential only affects the outer surface region where the ground–state density falls off exponentially. We conclude from this analysis that the very–low–energy behavior of quantum reflection is such an extremely sensitive function of small details of the density profile that a clear microscopic statement is impossible.

The right part of Fig. 3 shows the reflection probability R for the free liquid surface as a function of the perpendicular component of the wave vector $q_\perp$. Calculations were done for different parallel momenta $q_\parallel$ ranging from 0 to $0.35/\text{Å}$.

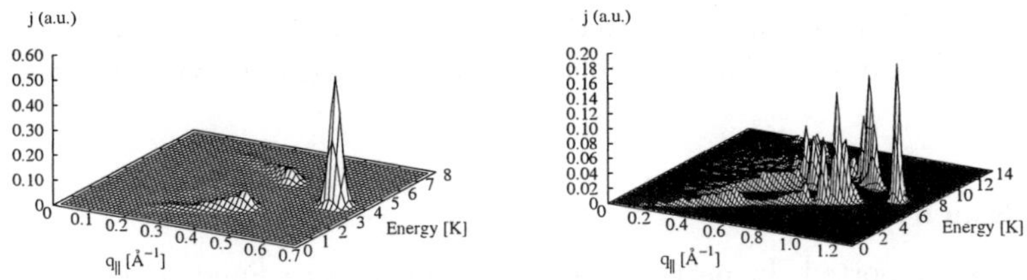

Fig. 4. The rate of inelastic current produced by an impinging particle (normal incidence) characterized by energy and parallel momentum of the decay products. Left panel: Impinging particle energy $\omega = 7.5$K, condensation of an unstable phonon. Right panel: $\omega = 14.8$K, condensation of an R^+ roton.

The experimental findings on the angle dependence of the reflection coefficient is not clear. On one hand, it is reported that R is a function of $q_\perp$ only..[8,23] On the other hand, it has been reported[23] that the static structure function $S(\mathbf{r}, \mathbf{r}')$ in Eq. (26) has a visible effect; our calculations support this finding. Now, scattering at grazing angles amounts to a large parallel momentum, in which case $S(z, z', q_\parallel) \rightarrow \delta(z, z')$. Our calculations are completely consistent with this view, suggesting a re-analysis of the experimental data.

3.3. *Inelastic decay in the surface*

Since both the reflection and the transmission rate are very low, it is, of course, of interest where the particle current goes. We have computed the inelastic current production rate in the free liquid ^{4}He surface for the case of condensation in normal incidence geometry. Figs. 4 show the rates at different energies.

A characterization in terms of energy $\omega_\parallel$ and parallel momentum $q_\parallel$ of the decay products for the case of normal incidence reveals that the majority of the production rate is found at roughly half of the energy of the impinging particle and the maximum parallel momentum allowed by energy conservation. In accordance with momentum conservation in the surface direction we find that two quasiparticles of roughly the same energy are created which travel in opposite direction along the surface. This was also found for ^{4}He scattering off ^{4}He droplets.[3] Comparing energy and parallel momenta of a single quasiparticle created with the evaporation boundary given by $-\mu + \hbar^2 q_\parallel^2 / 2m$ reveals that most of the created quasiparticles remain in the liquid and thus cannot be detected in an evaporation experiment unless a suitable detector is placed in the liquid surface. This conclusion is in agreement with the experiments of Brown and Wyatt:[25] In their experiment on phonon-induced evaporation, the parallel momentum carried away by a ripplon in an inelastic process in the surface is less than $(0.05 \pm 0.05)\text{Å}^{-1}$. For an inelastic scattering process with two ripplons emitted in opposite direction, Brown and Wyatt determine a maximum value of 0.1Å^{-1} for their respective momentum. Repeating the same experiment,

Williams[26] reduced this boundary to $(0.01 \pm 0.01)\text{Å}^{-1}$ and for two ripplons being scattered off an evaporated phonon Williams derives a maximum parallel momentum of 0.07Å^{-1}.

Acknowledgments

This work was supported, in part, by the Austrian Science Fund under project P11098-PHY. R. E. Z wants to thank the Fraunhofer Gesellschaft for financial support. We also would like to thank J. Boronat for the data of Ref. 24.

References

1. C. E. Campbell, E. Krotscheck, and M. Saarela, *Phys. Rev. Lett.* **80**, 2169 (1998).
2. E. Krotscheck and R. Zillich, *Phys. Rev.* **B58**, 5707 (1998).
3. E. Krotscheck and R. Zillich, *J. Chem. Phys.* **22**, 10161 (2001).
4. D. O. Edwards *et al.*, *Phys. Rev. Lett.* **34**, 1153 (1975).
5. D. O. Edwards and P. P. Fatouros, *Phys. Rev.* **B17**, 2147 (1978).
6. H. Baddar and D. O. Edwards, *J. Low Temp. Phys.* **104**, 317 (1996).
7. M. A. H. Tucker and A. F. G. Wyatt, *J. Low Temp. Phys.* **100**, 105 (1995).
8. A. F. G. Wyatt, M. A. H. Tucker, and R. F. Cregan, *Phys. Rev. Lett.* **74**, 5236 (1995).
9. M. Brown and A. F. G. Wyatt, *J. Phys.: Condensed Matter* **15**, 4717 (2003).
10. F. R. Hope and A. F. G. Wyatt, *Phys. Rev. Lett.* **52**, 1528 (1983).
11. G. M. Wyborn and A. F. G. Wyatt, *Phys. Rev. Lett.* **65**, 345 (1990).
12. M. A. H. Tucker and A. F. G. Wyatt, *J. Low Temp. Phys.* **113**, 615 (1998).
13. C. D. H. Williams and A. F. G. Wyatt, *Phys. Rev. Lett.* **91**, 085301 (2003).
14. M. B. Sobnack and J. C. Inkson, *Phys. Rev. Lett.* **82**, 3657 (1999).
15. F. Dalfovo *et al.*, *Phys. Rev. Lett.* **75**, 2510 (1995).
16. B. E. Clements, E. Krotscheck, and C. J. Tymczak, *Phys. Rev.* **B53**, 12253 (1996).
17. V. Apaja and E. Krotscheck, in *Microscopic Approaches to Quantum Liquids in Confined Geometries*, edited by E. Krotscheck and J. Navarro (World Scientific, Singapore, 2002), pp. 205–268.
18. V. Apaja and M. Saarela, *Phys. Rev.* **B57**, 5358 (1998).
19. V. Apaja, E. Krotscheck, A. Rimnac, and R. Zillich, *Quantum Reflection, Evaporation, and Transport Currents in ^{4}He*, to be published, 2006.
20. T. Keller, K. Habicht, R. Golub, and F. Mezei, *Europhys. Lett.* **67**, 773 (2004).
21. C. Carraro and M. W. Cole, *Phys. Rev.* **B45**, 12930 (1992).
22. J. Bölheim, W. Brenig, and J. Stuzki, *Z. Physik* **B48**, 43 (1982).
23. V. U. Nayak, D. O. Edwards, and N. Masuhara, *Phys. Rev. Lett.* **50**, 990 (1983).
24. J. M. Marín, J. Boronat, and J. Casulleras, *Phys. Rev.* **B71**, 144518 (2005).
25. M. Brown and A. F. G. Wyatt, *J. Phys.: Condensed Matter* **2**, 5025 (1990).
26. C. D. H. Williams, *J. Low Temp. Phys.* **113**, 11 (1998).

RADIAL DISTRIBUTION AND LIQUID STRUCTURE FUNCTION FOR LIQUID PARA-HYDROGEN AT LOW TEMPERATURES

KLAUS A. GERNOTH, MATTHEW J. HARRISON

School of Physics and Astronomy, The University of Manchester
Manchester, United Kingdom
k.a.gernoth@man.ac.uk, m.j.harrison@man.ac.uk

MANFRED L. RISTIG

Institut für Theoretische Physik, Universität zu Köln
D–50937 Köln, Germany
ristig@thp.uni-koeln.de

We present theoretical results for the radial distribution function $g(r)$ and the static liquid structure function $S(k)$ of liquid para-hydrogen at low temperatures. The results have been obtained via quantum Monte Carlo Path Integral simulations, classical Monte Carlo calculations, and correlated density matrix theory.

Keywords: Liquid hydrogen; radial distribution function; liquid structure function.

1. Theoretical Concepts

In this paper we present result for the angle-averaged radial distribution function (rdf) $\bar{g}(r)$ and the static structure factor $S(k)$ for liquid molecular para-hydrogen H_2 at low temperatures. For liquids $\bar{g}(r)$ is more commonly referred to as just the radial distribution function and denoted simply by $g(r)$, where $r = |\mathbf{x}_1 - \mathbf{x}_2|$ is the distance between a particle at $\mathbf{x}_1$ and another one at $\mathbf{x}_2$. For liquids $S(k)$ is also known as the liquid structure function, where $k = |\mathbf{k}|$ symbolizes the modulus of the reciprocal vectors $\mathbf{k}$. In this paper we may use $\bar{g}(r)$ and $g(r)$, angle-averaged rdf and rdf, and static structure factor and liquid structure function in, respectively, synonymous meanings.

The full two-body density $\varrho_2(\mathbf{S}, \mathbf{r})$ is defined as the thermodynamic expectation

$$\varrho_2(\mathbf{S}, \mathbf{r}) = \mathrm{Tr}\left\{\hat{\varrho}_2(\mathbf{S}, \mathbf{r})\hat{W}(\beta)\right\} = \langle\hat{\varrho}_2(\mathbf{S}, \mathbf{r})\rangle \tag{1}$$

of the two-body density operator

$$\hat{\varrho}_2(\mathbf{S}, \mathbf{r}) = \sum_{i \neq j = 1}^{N} \delta\left(\mathbf{S} - \mathbf{S}_{ij}\right) \delta\left(\mathbf{r} - \mathbf{r}_{ij}\right) \tag{2}$$

w.r.t. the canonical density operator $\hat{W}(\beta) = Z^{-1}(\beta)\, \mathrm{e}^{-\beta \hat{H}}$, wherein $\hat{H}$ is the Hamiltonian of the system of N particles, $\beta = (k_B T)^{-1}$ with T the temperature, and $Z(\beta) = \mathrm{Tr}\left\{\mathrm{e}^{-\beta \hat{H}}\right\}$ the partition function. In this paper we use Dirac brackets $\langle \cdots \rangle$ to denote thermodynamic averages $\mathrm{Tr}\left\{\cdots \hat{W}(\beta)\right\}$. The symbol $\sum_{i \neq j=1}^{N}$ denotes the double sum over all pairs of particles, i and j, excluding the diagonal terms, for which $i = j$. In Eq. (2) $\mathbf{S} = (\mathbf{x}_1 + \mathbf{x}_2)/2$ is the center of mass and $\mathbf{r} = \mathbf{x}_1 - \mathbf{x}_2$ the relative position of a particle at $\mathbf{x}_1$ and another one at $\mathbf{x}_2$. Similarly $\mathbf{S}_{ij} = (\mathbf{r}_i + \mathbf{r}_j)/2$ and $\mathbf{r}_{ij} = \mathbf{r}_i - \mathbf{r}_j$ denote the analoguous quantities for particles i and j of the N-body system. The two-body density $\varrho_2(\mathbf{S}, \mathbf{r})$ is the probability density for a particle to be at $\mathbf{x}_1$ and another one simultaneously at $\mathbf{x}_2$.

The angle-averaged rdf $g(r)$ is obtained from $\varrho_2(\mathbf{S}, \mathbf{r})$ by means of

$$\bar{g}(r) = \frac{1}{4\pi \varrho_0^2 V} \int_0^{\pi} \int_0^{2\pi} \int_V \varrho\left(\mathbf{S}, r, \theta, \phi\right)\, d\mathbf{S}\, d\phi\, \sin\theta\, d\theta\,, \tag{3}$$

wherein the three-dimensional volume integral $\int_V \cdots d\mathbf{S}$ extends over the volume V occupied by the many-body system. Here θ and ϕ are the polar and azimuthal angles of $\mathbf{r}$ and $\varrho_0 = N/V$ the bulk particle number density. Inserting Eq. (1) and employing

$$\delta\left(\mathbf{r} - \mathbf{r}_{ij}\right) = \frac{\delta\left(r - r_{ij}\right)}{r_{ij}^2} \frac{\delta\left(\theta - \theta_{ij}\right)}{\sin\theta_{ij}}\, \delta\left(\phi - \phi_{ij}\right) \tag{4}$$

yields

$$g(r) = \frac{1}{4\pi \varrho_0 N} \left\langle \sum_{i \neq j=1}^{N} \frac{\delta\left(r - r_{ij}\right)}{r_{ij}^2} \right\rangle. \tag{5}$$

Here θ_{ij} and ϕ_{ij} are the polar and azimuthal angles of $\mathbf{r}_{ij}$ and $r_{ij} = |\mathbf{r}_{ij}|$. The static structure factor $S(k)$ is related to $g(r)$ via

$$S(k) = 1 + \frac{4\pi \varrho_0}{k} \int_0^{\infty} r\left[g(r) - 1\right] \sin(kr)\, dr \quad \text{for} \quad k \neq 0 \tag{6}$$

and

$$S(0) = 1 + 4\pi \varrho_0 \int_0^{\infty} r^2 \left[g(r) - 1\right] dr\,. \tag{7}$$

The form (5) is conducive for computing $g(r)$ in computer simulations. Here we report numerical results for $g(r)$ from quantum Fourier Path Integral Monte Carlo (FPIMC) and classical MC simulations,[1] in which latter case the expectation values $\langle \cdots \rangle$ are the thermodynamic averages w.r.t. the classical Boltzmann distribution. We also calculated $g(r)$ with correlated density matrix (CDM) theory.[2] In all cases the corresponding $S(k)$ is obtained from Eqs. (6) and (7) by means of numerical integration from zero up to the maximum interparticle distance r_m for which $g(r)$ is computed. We point out that we have not extended the integration range beyond r_m by appending an extrapolated long-range behavior of $g(r)$ as is done in other works.[3]

2. Numerical Results

Figure 1 displays the radial distribution function $g(r)$ vs. $r\,[\AA]$ of liquid para-hydrogen at a temperature $T = 20\,\mathrm{K}$ and a bulk particle number density $\varrho_0 = 0.02124\,\AA^{-3}$ from CDM theory (solid line), FPIMC (dotted line), and classical MC (long-dashed line) simulations. The Silvera-Goldman potential is used in our calculations for the interaction of the hydrogen molecules.[4]

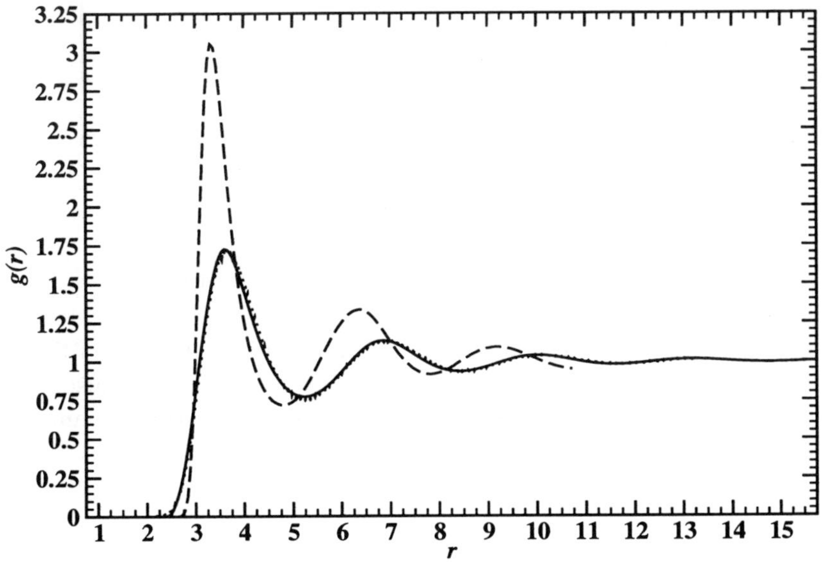

Fig. 1. The radial distribution function $g(r)$ (cf. text for details).

Figure 2 shows the corresponding liquid structure function $S(k)$ vs. $k\,[\AA^{-1}]$ of liquid para-hydrogen, also at $T = 20\,\mathrm{K}$ and $\varrho_0 = 0.02124\,\AA^{-3}$, from CDM (solid line) and FPIMC (dotted line) and classical MC (long-dashed line) simulations. The unphysical small-k behavior of $S(k)$ from the FPIMC and MC simulations is an artefact of the comparatively small r_m achieved in these calculations, $r_m = 13.24\,\AA$ (with 384 molecules in a hexagonal simulation box) and $r_m = 10.71\,\AA$ (with 216 molecules in a cubic simulation box), respectively. In contrast, in the CDM calculations we could choose $r_m = 330\,\AA$, large enough for our CDM calculations to yield the correct small-k behavior, the well-known Ornstein-Zernike rise at finite temperatures, of $S(k)$.

Apart from the small-k behavior of $S(k)$ the CDM and the FPIMC results for

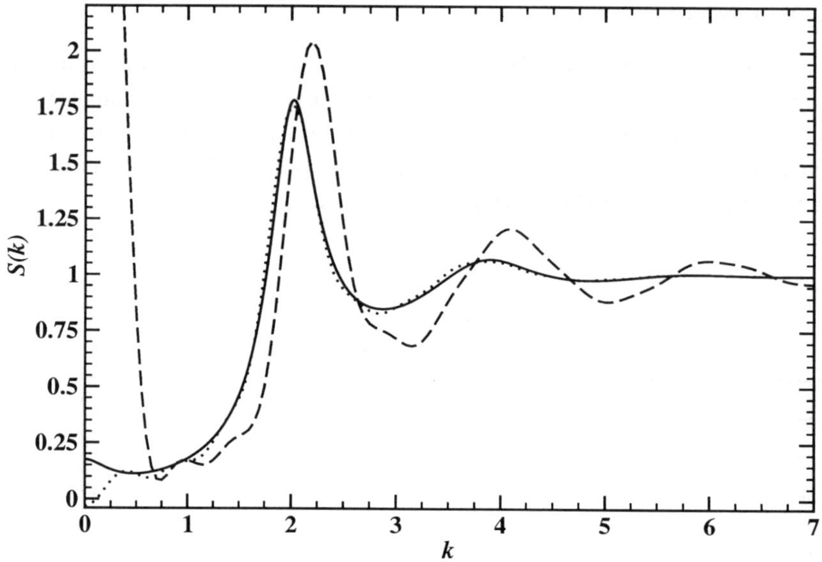

Fig. 2. The liquid structure function $S(k)$ (cf. text for details).

$g(r)$ and $S(k)$ agree remarkably well and can hardly be distinguished on the scale of the plots shown here. Both our FPIMC and classical MC results are also in good agreement with the $g(r)$ and $S(k)$ results reported in Ref. 3 for these quantities at the same T and ϱ_0 as used here.

References

1. K. A. Gernoth, *Ann. Phys. (N.Y.)* **285**, 61 (2000); *Ann. Phys. (N.Y.)* **291**, 202 (2001); *Z. Kristallogr.* **218**, 651 (2003).
2. T. Lindenau *et al.*, *J. Low Temp. Phys.* **129**, 143 (2002); and references contained therein.
3. A. Cunsolo *et al.*, *J. Low Temp. Phys.* **129**, 117 (2002).
4. I. Silvera and V. Goldman, *J. Chem. Phys.* **69**, 4209 (1978).

VARIATIONAL DESCRIPTION OF WEAKLY INTERACTING BOSE GASES IN 3 DIMENSIONS

F. MAZZANTI

Departament de Fısica i Enginyeria Nuclear, Universitat Politècnica de Catalunya
Comte Urgell 187, E-08036 Barcelona, Spain

M. SAARELA*

Institute for Theoretical Physics, Johannes Kepler Universität Linz
Altenberger Str. 69, A-4040 Linz, Austria

V. APAJA

Institute for Theoretical Physics, Johannes Kepler Universität Linz
Altenberger Str. 69, A-4040 Linz, Austria

Static and dynamic properties of a weakly interacting Bose gas of Hard Spheres in three dimensions are studied in the framework of the Correlated Basis Functions (CBF) approximation. Results are compared with explicit expressions for the same quantities derived within the Bogoliubov model. Despite the good agreement in the energy of the groundstate and the excited states, other quantities such as the dynamic structure function present important differences that become more significant when the density is raised.

Keywords: Bose condensate, CBF, Bogoliubov

1. Introduction

The study of dilute systems has achieved considerable interest in the last years, following the experimental realization of Bose-Einstein condensates in low density atomic gases confined in harmonic traps.[1] Systems are termed as *dilute* when the average interparticle distance is much larger than the range of the interaction. The main parameter characterizing the interaction in the dilute regime is the s–wave scattering length a, and the diluteness condition can be expressed in terms of the so called *diluteness parameter* $x = \rho a^3 \ll 1$.

Positive values of the scattering length correspond to interactions that are essentially repulsive at short distances. A model of Hard Spheres (HS), characterized by an infinitely repulsive well at distances smaller than a certain value identified with the diameter of the Hard Sphere, has often been adopted to study the behaviour of

*Permanent address: Department of Physical Sciences, P.O.Box 3000, FIN-90014 University of Oulu, Finland.

low density gases with positive scattering lengths. In this case, a coincides with the radius of the barrier. At low values of x, the scattering length is known to be the only parameter of the potential that affects the groundstate energy of the gas. When x is raised, other quantities such as the *effective range* become also relevant. Typically, differences in the groundstate energy start to be noticeable when $x \geq 10^{-3}$ (Refs. 2,3).

In this work we study the groundstate energy of a gas of Hard Spheres in the CBF approximation[4–6] in comparison with the rather simple prediction of the Bogoliubov model.[7,8] The linear response of the system to density fluctuations induced by an external, time–dependent potential is also analyzed, and the dynamic structure function is compared with the response obtained in the Bogoliubov approximation.

2. Groundstate Properties

Let us consider a gas of $N \gg 1$ spinless bosons of mass m confined in a box of volume Ω and at a constant density $\rho = N/\Omega$. The system is described by the Hamiltonian

$$H = -\frac{\hbar^2}{2m}\sum_{j=1}^{N}\nabla_j^2 + \sum_{1=i<j}^{N} V(r_{ij}) \,, \tag{1}$$

where $V(r)$ is the two–body Hard Sphere potential

$$V(r) = \begin{cases} \infty, & r < a \\ 0 & r > a \end{cases} \tag{2}$$

with the scattering length equal to the core radius a.

A variational description of the gas is possible once a suitable form of the wavefunction has been chosen. In the low x regime, a simple form captivating the leading features induced by particle correlations is enough to attain an accurate description of the groundstate. A suitable choice is a *Jastrow* wavefunction

$$\Psi_0(\mathbf{r}_1, \mathbf{r}_2, \ldots, \mathbf{r}_N) = \prod_{1=i<j}^{N} f(r_{ij}) \tag{3}$$

with a two–body correlation factor $f(r)$ that is the solution of the corresponding Euler–Lagrange problem

$$\frac{\delta E[f]}{\delta f(r)} = 0 \,. \tag{4}$$

It is easy to check, using the Jackson–Feenberg identity, that $f(r)$ and the energy are related by the two–body radial distribution function

$$g(r_{12}) = \frac{N(N-1)}{\rho^2}\frac{\int d\mathbf{r}_3 d\mathbf{r}_4 \cdots d\mathbf{r}_N \Psi_0^2(\mathbf{r}_1, \mathbf{r}_2, \ldots, \mathbf{r}_N)}{\int d\mathbf{r}_1 d\mathbf{r}_2, \cdots d\mathbf{r}_N \Psi_0^2(\mathbf{r}_1, \mathbf{r}_2, \ldots, \mathbf{r}_N)} \,. \tag{5}$$

The hypernetted chain equations (HNC) provide an explicit link between these quantities

$$g(r_{12}) = f(r_{12})\, e^{N(r_{12})+E(r_{12})}$$
$$N(r_{12}) = \rho \int d\mathbf{r}_3\, [g(r_{13}) - 1]\, [g(r_{23}) - N(r_{23}) - 1] \ . \tag{6}$$

where $N(r)$ and $E(r)$ are the sum of nodal and elementary diagrams, respectively. The function $E(r)$ is an input to the theory as there is no closed equation for it, and therefore the solution of the HNC equations depends on the particular approximation used. In the HNC/0 scheme adopted in this work the sum of elementary diagrams is set to zero, a reliable approximation at low densities due to the high connectivity of the diagrams contributing to $E(r)$.

By inverting Eqs. (6), the energy per particle can be written as a function of $g(r)$ alone. In this way, the optimization principle can be cast in the alternative but equivalent form

$$\frac{\delta E[g]}{\delta g(r)} = 0 \ . \tag{7}$$

In terms of the static structure factor

$$S(k) = 1 + \rho \int d\mathbf{r}\, e^{i\mathbf{k}\cdot\mathbf{r}}\, [g(r) - 1] \ , \tag{8}$$

the solution to the optimization problem reads

$$S(k) = \frac{t(k)}{\sqrt{t^2(k) + 2t(k)V_{ph}(k)}} \ , \tag{9}$$

where $t(k) = k^2/2m$ and $V_{ph}(k)$ is the particle–hole potential in momentum space. In the HNC/0 approximation,

$$V_{ph}(r) = g(r)V(r) + \frac{\hbar^2}{m}\, |\nabla\sqrt{g(r)}|^2 + [g(r) - 1]\, \omega_I(r) \ , \tag{10}$$

expressed in terms of the k–space representation of the induced interaction

$$\omega_I(k) = -\frac{1}{2}\, t(k)\, \frac{[2S(k) + 1]\, [S(k) - 1]^2}{S^2(k)} \ . \tag{11}$$

This last quantity depends on $S(k)$, so the solution to the Euler Lagrange problem has to be obtained iteratively, starting from a reasonably good guess for $S(k)$ or $g(r)$.

At very low densities, the groundstate of a dilute gas of weakly interacting bosons can also be described within the Bogoliubov model, where the actual interaction is replaced by a *Fermi pseudopotential* of the delta type

$$V(r) = \frac{4\pi a\hbar^2}{m}\, \delta(\mathbf{r}) \ . \tag{12}$$

The Hamiltonian of the system is written in second quantization and, assuming the number of particles N_0 in the Bose condensate is *macroscopically* large such that the particle creation and annihilation operators can be approximated by numbers $a_0 \approx a_0^\dagger \approx \sqrt{N_0}$, only the leading orders in N_0 are retained. The resulting Hamiltonian can then be diagonalized with the help of a canonical transformation

$$a_{\mathbf{p}} = \mathcal{N}_p \left[b_{\mathbf{p}} - a_p b^\dagger_{-\mathbf{p}} \right] \quad , \quad a^\dagger_{\mathbf{p}} = \mathcal{N}_p \left[b^\dagger_{\mathbf{p}} - a_p b_{-\mathbf{p}} \right] , \tag{13}$$

with

$$\mathcal{N}_p = \sqrt{1 - A_p^2} \quad , \quad A_p = 1 + \frac{m}{4\pi a \hbar^2 \rho} \left[\frac{p^2}{2m} - \epsilon(p) \right] . \tag{14}$$

The new creation and annihilation operators b_α and $b^\dagger_\alpha$ satisfy the commutation relation $[b_\alpha, b^\dagger_\beta] = \delta_{\alpha\beta}$ and therefore create and annihilate bosonic quasiparticles. In units of $\hbar^2/2ma^2$, the energy of the system becomes

$$E = E_0 + \sum_p n_p \epsilon(p) , \tag{15}$$

where

$$E_0 = 4\pi x \left[1 + \frac{128}{15} \sqrt{\frac{x}{\pi}} \right] \tag{16}$$

is the groundstate energy, while

$$\epsilon(p) = \sqrt{16\pi x p^2 + p^4} \quad , \quad n_p = \frac{1}{e^{\epsilon(p)/T} - 1} \tag{17}$$

are the energy and momentum distribution of quasiparticles.

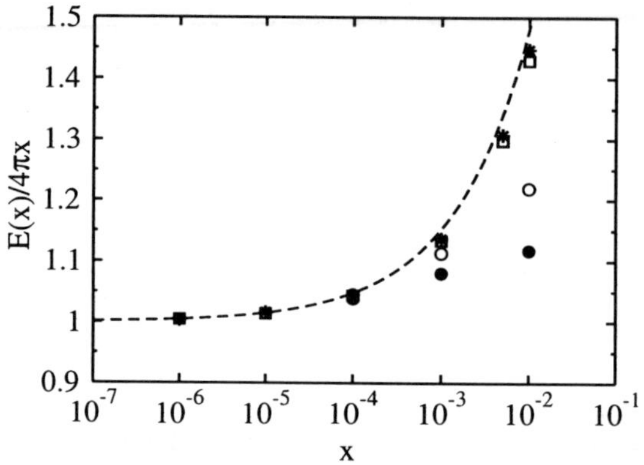

Fig. 1. Groundstate energy per particle of Hard Spheres in the EL approximation (open squares), compared with DMC results (stars), the Bogoliubov prediction (dashed line) and two Soft-Spheres potentials with $R = 5a$ and $R = 10a$ (open and solid circles, respectively).

Results for the groundstate energy of the HS gas compared with Diffusion Monte Carlo (DMC) results from Ref. 2 and the Bogoliubov approximation are reported in Fig. 1. A few points corresponding to the gas of Soft Spheres with different values of the core radius R and the scattering length a have also been included. As it can be seen, the energy per particle of the HS gas is closer to the Bogoliubov prediction, while all energies are approximately equal for $x \leq 10^{-4}$. Universality for the groundstate energy per particle breaks down somewhere below $x = 10^{-3}$. Above this value, parameters other than the scattering length of the potential start to be relevant, while this has little impact on the energy of the HS gas.

The radial distribution function and the static structure factor of the HS gas is reported in Fig. 2 at several values of the diluteness parameter x. Both functions show little structure unless high x's are considered, developing a peak at some x larger than 0.01. For the sake of comparison, the radial distribution function resulting from the Bogoliubov approximation at $x = 0.001$ is also plotted. As it can be seen, the later reproduces qualitatively well the large–r behaviour of the HNC $g(r)$, while it becomes unphysically negative at distances close and smaller than the core radius. This is not surprising considering the form of the pseudopotential in Eq. (12).

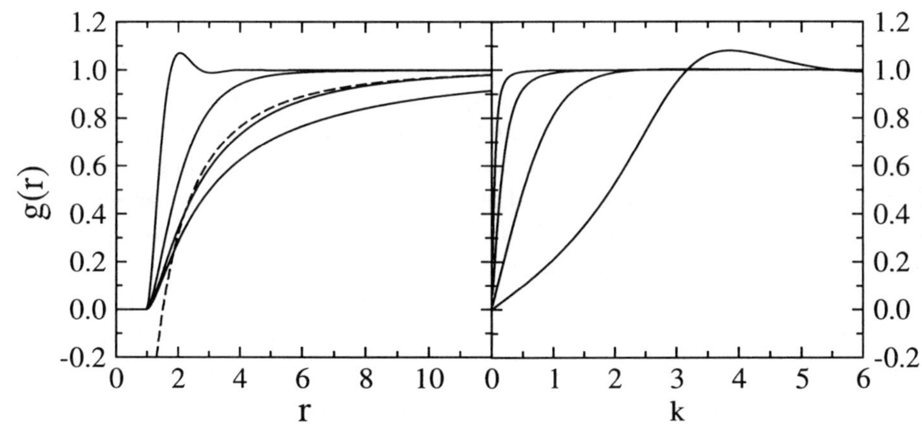

Fig. 2. Two–body radial distribution function and static structure factor of the HS gas. Left panel: $g(r)$ at $x = 0.1$, $x = 0.01$, $x = 0.001$ and $x = 0.0001$, from left to right. Right panel: $S(k)$ at the same values of x, from right to left. The dashed line corresponds to the Bogoliubov prediction at $x = 0.001$.

3. Excited States

The spectrum of excited states of a quantum system can be analyzed by looking at its linear response function to density fluctuations induced by a weak, time–dependent external potential. In the above variational ground state calculation we optimized the total energy with respect to the two-particle correlation function. We can follow a similar approach for the dynamic, time dependent system and use the least–action principle, where one optimizes the action integral[9]

$$\delta S = \delta \int_{t_0}^{t} dt \mathcal{L}(t) \equiv \delta \int_{t_0}^{t} dt \left\langle \Psi(t) \middle| H_0 + \hat{U}_{\rm ext}(t) - i\hbar\frac{\partial}{\partial t} \middle| \Psi(t) \right\rangle = 0 \tag{18}$$

The Hamiltonian contains besides the ground–state Hamiltonian H_0, also the time dependent operator $\hat{U}_{\rm ext}(t) = \sum_{i=1}^{N} U_{\rm ext}(\mathbf{r}_i;t)$, which induces an external, infinitesimal disturbance into the system. The disturbance introduces time dependence in the ground state wave function[10,11,14]

$$\Psi(\mathbf{r}_1,\ldots,\mathbf{r}_N;t) = e^{-iE_0t/\hbar}\frac{1}{\sqrt{\mathcal{N}(t)}}e^{\frac{1}{2}\delta U(\mathbf{r}_1,\ldots,\mathbf{r}_N;t)}\Psi_0(\mathbf{r}_1,\ldots,\mathbf{r}_N)\,. \tag{19}$$

We have separated out the phase with the ground state energy E_0 and the normalization factor $\mathcal{N}(t)$, which also must depend on time. One can think about $\delta U(\mathbf{r}_1,...,\mathbf{r}_N;t)$ as a general complex–valued excitation operator. In the Feynman approximation for δU one allows the time dependence to appear only in the one-particle correlation function, while in the correlated basis function approach (CBF) this dependence is also extended to two-particle correlations

$$\delta U(\mathbf{r}_1,...,\mathbf{r}_N;t) = \sum_i \delta u_1(\mathbf{r}_i;t) + \sum_{i<j}\delta u_2(\mathbf{r}_i,\mathbf{r}_j;t)\;. \tag{20}$$

The new, unknown functions $\delta u_1(\mathbf{r}_i;t)$ and $\delta u_2(\mathbf{r}_i,\mathbf{r}_j;t)$ are then determined by the least-action principle. That defines the continuity equations for the one– and two–particle currents and densities, which are the equations of motion of the system. In this approach all particles are treated on the same footing, including those which are in the Bose condensed state. Special attention to the condensate fraction can be payed by calculating the asymptotic behavior of the time dependent density matrix after the equations of motion have been solved.

Let us now specifically assumer that the system is driven out of the ground state by an infinitesimal external interaction $\tilde{U}_{\rm ext}(k,\omega)$, with a given frequency ω and wave number k. Let us also assume that it induces a change in the one-particle density, $\delta\tilde{\rho}_1(k,\omega)$, of the originally homogeneous system. The response of the system is supposed to be linear and the information on its dynamic properties, excitations and decay modes, is contained in the density–density linear–response function defined as

$$\chi(k,\omega) = \frac{\delta\tilde{\rho}_1(k,\omega)}{\rho_0\tilde{U}_{\rm ext}(k,\omega)}\,. \tag{21}$$

This is a complex function and its imaginary part is proportional to the dynamic structure function by the fluctuation–dissipation theorem,

$$S(k,\omega) = -\frac{1}{\pi}\Im m\left[\chi(k,\omega)\right]. \tag{22}$$

Hence, $S(k,\omega)$ is known once the ratio between the one–body density fluctuations and the external disturbance is established. One can compare the theoretical results directly with experiments because the dynamic response function is accessible through measurements of double differential cross section in inelastic scattering experiments.

In momentum space we can calculate the ratio $\delta\tilde{\rho}_1(k,\omega)/\tilde{U}_{\rm ext}(k,\omega)$ from the one-particle continuity equation and write the linear response function (21) in a general form

$$\chi(k,\omega) = S(k)\left[\frac{1}{\hbar\omega - \varepsilon_F(k) - \Sigma(k,\omega)} - \frac{1}{\hbar\omega + \varepsilon_F(k) + \Sigma^*(k,-\omega)}\right] , \tag{23}$$

where we have collected all the contributions from the two–particle correlation function into the self–energy $\Sigma(k,\omega)$. For real values of the self–energy the response function can have poles which define the collective, elementary excitations. When the decay of the excited modes becomes possible, the self–energy acquires imaginary part and the sharp δ–function in the imaginary part of the response function spreads into a broader peak.

The two-particle continuity equation defines then the self energy. In the uniform limit approximation it becomes.[12–15]

$$\Sigma(k,\omega) = \frac{1}{2}\int \frac{d^3p\, d^3q}{(2\pi)^3\rho_0}\delta(\mathbf{k}+\mathbf{p}+\mathbf{q})\frac{|V_3(\mathbf{k};\mathbf{p},\mathbf{q})|^2}{\hbar\omega - \varepsilon_{\rm F}(p) - \varepsilon_{\rm F}(q)} \tag{24}$$

with the three–phonon coupling matrix element,

$$V_3(\mathbf{k};\mathbf{p},\mathbf{q}) = \frac{\hbar^2}{2m}\sqrt{\frac{S(p)S(q)}{S(k)}}[\mathbf{k}\cdot\mathbf{p}\tilde{X}(p) + \mathbf{k}\cdot\mathbf{q}\tilde{X}(q)] \tag{25}$$

given in terms of the ground–state structure function $S(k)$ and the direct correlation function $\tilde{X}(k) = 1 - S(k)^{-1}$.

The dynamic structure function can also be written as a sum over all states where the Fourier transform of the density fluctuation operator $\rho_{\mathbf{q}} = \sum_{j=1}^{N} e^{i\mathbf{q}\cdot\mathbf{r}_j}$ can excite the system

$$S(q,\omega) = \frac{1}{N}\sum_{\{n\}} |\langle n \mid \rho_{\mathbf{q}} \mid 0\rangle|^2\, \delta\left(E_n - E_0 - \omega\right) . \tag{26}$$

For a weakly interacting Bose gas, the matrix elements can also be evaluated within the Bogoliubov approximation. In second quantization, the density fluctuation operator reads

$$\rho_{\mathbf{q}} = \sum_{\mathbf{k}} a^\dagger_{\mathbf{k}+\mathbf{q}} a_{\mathbf{k}} = a^\dagger_0 a_{-\mathbf{q}} + a^\dagger_{\mathbf{q}} a_0 + {\sum_{\mathbf{k}}}' a^\dagger_{\mathbf{k}+\mathbf{q}} a_{\mathbf{k}} , \tag{27}$$

where a prime in the summation indicates that the condensate terms corresponding to $\mathbf{k} = 0$ and $\mathbf{k}+\mathbf{q} = 0$ have been explicitly separated from the rest. Recalling that $a_0 \approx a_0^\dagger = \sqrt{N_0}$ and the transformation given in Eqs. (13) and (14), one gets

$$\begin{aligned}\rho_{\mathbf{q}} &= \sqrt{N_0}\mathcal{N}_{\mathbf{q}}\left(b^\dagger_{\mathbf{q}} - A_{\mathbf{q}} b_{-\mathbf{q}} + b_{-\mathbf{q}} - A_{\mathbf{q}} b_{\mathbf{q}}\right) \\ &\quad + {\sum_{\mathbf{k}}}' \mathcal{N}_{\mathbf{k}+\mathbf{q}}\mathcal{N}_{\mathbf{k}}\left(b^\dagger_{\mathbf{k}+\mathbf{q}} - A_{\mathbf{k}+\mathbf{q}} b_{\mathbf{k}+\mathbf{q}}\right)\left(b_{\mathbf{k}} - A_{\mathbf{k}} b^\dagger_{-\mathbf{k}}\right) \\ &\equiv \rho^c_{\mathbf{q}} + \rho^{nc}_{\mathbf{q}} ,\end{aligned} \tag{28}$$

where $\rho^c_{\mathbf{q}}$ and $\rho^{nc}_{\mathbf{q}}$ stand for the condensate and non–condensate contributions to the density fluctuation operator, respectively.

The Bogoliubov Hamiltonian is diagonal in a basis characterized by the number of non–interacting quasiparticles in each state of definite momentum. Therefore the action of $b^\dagger_\alpha$ and b_α on the different states of the system can be readily evaluated and leads to a response function that is the sum of a condensate and a non–condensate contribution, resulting from the action of $\rho^c_{\mathbf{q}}$ and $\rho^{nc}_{\mathbf{q}}$, respectively

$$S(q,\omega) = S_c(q,\omega) + S_{nc}(q,\omega) \ . \tag{29}$$

The condensate and non–condensate responses read

$$S_c(q,\omega) = n_0 \frac{q^2}{\sqrt{q^4 + 16\pi x q^2}} \delta\left(\epsilon(q) - \omega\right) \tag{30}$$

and

$$S_{nc}(q,\omega) = \frac{1}{16\pi x} \int d\mathbf{k} n(\mathbf{k}) \left[n(\mathbf{k}+\mathbf{q}) + 1\right] \delta\left(\epsilon(\mathbf{k}+\mathbf{q}) + \epsilon(\mathbf{k}) - \omega\right) \ , \tag{31}$$

where $n(k)$ is the $T = 0$ momentum distribution of particles

$$n(k) = \frac{A_k^2}{1 - A_k^2} \ . \tag{32}$$

The dynamic structure function predicted by the Bogoliubov theory is therefore the sum of two different terms. The condensate part yields a delta peak centered at the quasiparticle energy and is due to the creation of single quasiparticles out of the condensate. The non–condensate contribution comes from the simultaneous creation of pairs of non–interacting quasiparticles with a total energy and momentum equal to the energy and momentum transferred by the external perturbation to the system. The response predicted by the CBF and Bogoliubov theories in logarithmic scale are compared in Fig. 3 for different values of x, where darker regions indicate more strength. In all cases the black curve shows the position of the main peak, corresponding to the condensate contribution in the Bogoliubov response. Both theories agree in the position of this peak at low x, although differences become significant at higher concentrations. In particular, the CBF model predicts a splitting of the peak and the presence of a phonon–roton branch at the highest values considered. On the other hand, the non–condensate part of the Bogoliubov response, corresponding to the background contribution, is quite different from the CBF prediction even at the lowest concentration considered. This is not surprising considering the non–condensate contribution to the Bogoliubov response is of order zero in N_0, and we know the Bogoliubov Hamiltonian is not correct to this order. Furthermore, the non–condensate contribution in the Bogoliubov model is an integral expression involving the momentum distribution of particles (see Eqs. (31) and (32)), while the later is known to be wrongly described in the Bogoliubov model as it decays as $1/k^4$ at large k, thus producing an infinite kinetic energy per particle. This problem is also evidenced when the lowest order energy–weighted sum

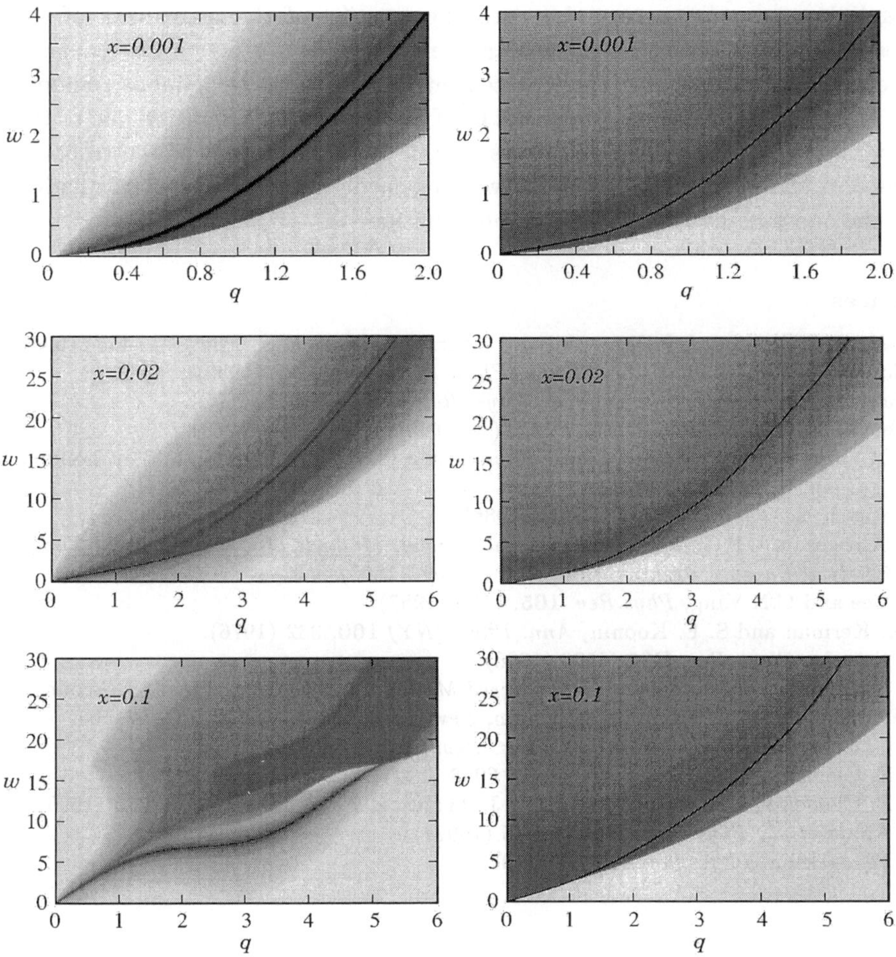

Fig. 3. CBF (solid line) and Bogoliubov (dashed line) response for $x = 0.02$ and $q = 2.25$. The short-dashed vertical line indicates the delta peak corresponding to the condensate contribution to the Bogoliubov response.

rules $m_k(q) = \int w^k S(q,\omega)\, d\omega$ are considered. While the CBF response reproduces exactly the first and second sum rules $m_0(q) = S(q)$ and $m_1(q) = q^2/2m$,[16] this is not the case in the Bogoliubov approximation as in this model the total number of particles is not conserved.

In summary, we have shown that the groundstate energy of a gas of Hard Spheres in 3D is well reproduced by the Bogoliubov prediction at low concentrations where the universality scaling law holds, while deviations indicating the relevance of the detailed shape of the potential start to show up at concentrations close to $x \approx 10^{-3}$. The position of the peak corresponding to the condensate contribution of the Bogoliubov response agrees also with the position of the main peak of the CBF response up to the same value of x, while deviations start to be visible at larger concentrations. The non–condensate contribution to the Bogoliubov response is always different from the more realistic CBF background contribution due to the intrinsic limitations of the model. In this way, a detailed analysis of the response function of a gas of interacting bosons would benefit from a realistic calculation such as the one obtained in the CBF approximation.

References

1. F.Dalfovo, S.Giorgini, L.P.Pitaevskii and S.Stringari, *Rev. Mod. Phys.* **71**, 463 (1999).
2. S.Giorgini, J.Boronat and J.Casulleras, *Phys. Rev.* **A60**, 5129 (1999).
3. F.Mazzanti, A.Polls and A.Fabrocini, *Phys. Rev.* **A67**, 063615 (2003).
4. E. Feenberg, *Theory of Quantum Fluids* (Academic, New York, 1969).
5. C. E. Campbell, in *Progress in Liquid Physics*, edited by C. A. Croxton (Wiley, London, 1978), Chap. 6, pp. 213–308.
6. E. Krotscheck, *Phys. Rev.* **B33**, 3158 (1986).
7. A.A.Abrikosov, L.P.Gorkov and I.E.Dzyaloshinski, *Methods of Quantum Field Theory in Statistical Physics*, Prentice-Hall, New Jersey, (1963).
8. T.D.Lee and C.N.Yang, *Phys.Rev.* **105**, 1119 (1957).
9. A. K. Kerman and S. E. Koonin, *Ann. Phys. (NY)* **100**, 332 (1976).
10. M. Saarela, *Phys. Rev.* **B33**, 4596 (1986).
11. M. Saarela and J. Suominen, in *Condensed Matter Theories*, edited by J. S. Arponen, R. F. Bishop, and M. Manninen (Plenum, New York, 1988), Vol. 3, pp. 157–165.
12. C. C. Chang and C. E. Campbell, *Phys. Rev.* **B13**, 3779 (1976).
13. B. E. Clements *et al.*, *Phys. Rev.* **B50**, 6958 (1994).
14. B. E. Clements, E. Krotscheck, and C. J. Tymczak, *Phys. Rev.* **B53**, 12253 (1996).
15. V. Apaja *et al.*, *Phys. Rev.* **B55**, 12925 (1997).
16. H. W. Jackson, *Phys. Rev.* **A9**, 964 (1974).

STABILITY AND SPECTRA OF SMALL ^{3}He-^{4}He CLUSTERS

J. NAVARRO

IFIC (CSIC-Universidad de Valencia)
Apartado Postal 22085, E-46071-Valencia, Spain
navarro@ific.uv.es

S. FANTONI

SISSA and INFM DEMOCRITOS National Simulation Center
Via Beirut 2-4, I-34014 Trieste, Italy

R. GUARDIOLA

Departamento de Física Atómica y Nuclear
Facultad de Física, E-46100-Burjassot, Spain

A. ZUKER

Institute de Recherches Subatomiques
IN2P3-CNRS/Université Louis Pasteur, F-67037 Strasbourg, France

Diffusion Monte Carlo calculations have been systematically performed to analyze the stability of small mixed ^{3}He-^{4}He clusters, as well as their excitation spectra. The picture that emerges is that of systems with strong shell effects whose binding and excitation energies are essentially determined by the monopole properties of an effective Hamiltonian.

Keywords: Mixed Helium Clusters

1. Introduction

Helium clusters are weakly bound quantum systems, as a consequence of both the large zero-point motion energy and the weak van der Waals interaction between helium atoms. We refer the reader to Refs. 1–3 for recent reviews on the topic. Experimentally, helium clusters are produced by free jet expansion of the gas. The mass of small helium clusters is measured by diffraction through a transmission grating followed by a mass spectrometer detector,[4] which has allowed the unambigous detection[5] of the very weakly bound dimer ^{4}He$_2$.The experimental setup has been improved to detect droplets with masses up to a few hundred amu,[6] and small mixed systems have been identified in this way.[7] The theoretical calculations predict that ^{4}He clusters are bound for any number of atoms. However, a minimum number of about 30 ^{3}He atoms[8,9] is required to form a bound system, due to the larger zero-point motion and the Pauli principle. The case of mixed ^{3}He-^{4}He clusters is very interesting, since they are made of particles with different statistics and masses interacting through the same potential. Theoretical calculations[10–12] predict that few ^{4}He atoms are able to bind any number of ^{3}He atoms, but there exist instability islands for a sufficiently small number of ^{4}He atoms.

This work reports on diffusion Monte Carlo (DMC) calculations performed to analyze the ground state of mixed helium clusters and the low-lying excited states of the ^{3}He atoms. Our calculations enlarge the combinations of bosons and fermions previously considered.[12,13] All calculations have been made using the Aziz HFD-(B) potential[14] for the He-He interaction.

The employed DMC method[15,16] is based on an importance sampling wave function, with the double role of controlling the variance of the ground state energy and to incorporate both the required statistics and the desired spin and angular momentum quantum numbers. Following our previous calculations[12,13] the trial wave function used for the importance sampling is written as the product of three Jastrow factors for the boson-boson, boson-fermion, and fermion-fermion parts, and two Slater determinants for the spin-up and -down fermions. Each of the Jastrow factors is translationally invariant and symmetric under any permutation of particles. The description of the trial wave function is completed by the inclusion of the Feynman-Cohen backflow[17] in the fermion radial coordinates. Due to the presence of the Slater determinants, the importance sampling function is not definite positive. In our calculations, the random process has been constrained using the so-called fixed node approximation: any walker attempting to cross a nodal surface is neglected. As it has been shown,[18,19] the use of this approximation leads to an upper bound to the lowest energies.

2. Clusters with a single ^{3}He atom

The filling of the Slater determinants relies on the single-particle orbital orderings obtained previously[20] from the study of a single ^{3}He atom attached to a cluster of ^{4}He atoms. The resulting spacings resemble those of the rotational spectrum of a diatomic molecule, where the ^{4}He core plays the role of one atom and the ^{3}He is the other atom. This result is in agreement with previous theoretical studies, either based on a density functional approach[21,22] or employing a microscopic atom-atom interaction.[23–25] At low energies each level is uniquely classified by its angular momentum, but vibrational-like excitations appear at higher energies for heavy enough clusters.

For the sake of completeness we summarize here the results of the single-fermion case of relevance for the analysis of systems with more fermions. As there is only one fermion, the product of two Slater determinants reduces to a single-particle wave function $\phi_{n\ell m}(\mathbf{r})$ of radial quantum number n and orbital angular momentum ℓ with projection m. Translational invariance is ensured by referring the fermion coordinate to the bosonic center of mass. Different values of ℓ select a specific angular momentum subspace in which the DMC procedure will drive the wave function so as to minimize the energy. Thus, the radial part of the fermionic wave function can be taken to have the form $\phi_{1\ell m}(r) = r^\ell Y_{\ell m}(\hat{r})$, which ensures that there are no radial nodes ($n = 1$). These single-particle wave functions do not contain a radial confining term, because it is already included in the Jastrow part of the importance sampling wave function. Notice that within the DMC procedure it is not possible to obtain excited levels with $n > 1$, because this will require to impose a strict orthogonality on the physical $n = 0$ ground state. Nevertheless, one may use an indirect procedure based in the moment method[26] to get an upper bound to the energy of the radial excited levels.

Table 1. Energies (in K) of ^{3}He,-^{4}HeN_B clusters with $N_B = 8$ and 20 for several values of the angular momentum. The columns labelled $\epsilon_{n\ell}$ are the ^{3}He separation energies. Values in italics correspond to unbound levels. The last row labelled $N_F = 0$ correspond to the pure ^{4}He cluster.

	$N_B = 8$		$N_B = 20$	
Config	E	$\epsilon_{n\ell}$	E	$\epsilon_{n\ell}$
$1s$	-6.08(1)	-0.94	-35.55(1)	-1.79
$1p$	-5.60(1)	-0.46	-35.15(2)	-1.39
$1d$	*-4.98(1)*	0.16	-34.55(2)	-0.89
$2s$	*-5.10*	0.06	-33.80	-0.04
$N_F = 0$	-5.14(1)		-33.76(2)	

Table 1 displays the energies of the ^{3}He atom as a function of its (n, ℓ) quantum numbers, for the clusters with 8 and 20 ^{4}He atoms. The ground state energies of the pure bosonic droplets are also given, so as to define the separation energies $\epsilon_{n\ell} = E_{n\ell}(N_B, N_F = 1) - E(N_B, N_F = 0)$, also quoted in the Table. Positive values refer to unbound levels. The energy obtained for the $n = 2, \ell = 0$ excited level is above the dissociation limit for $N_B = 8$ or very close to it for $N_B = 20$. In conclusion, for our subsequent analysis of clusters containing up to 18 ^{3}He atoms, the adopted filling order is $1s, 1p, 1d$, although in some cases some attention has been paid to a possible $2s$ intruder.

3. The stability map

One appealing property of mixed He clusters is the existence of instability islands, namely regions around selected values of (N_B, N_F) in which the system is not bound. The present DMC approach, based upon a different level ordering leads to a noticeable energy gain in the $1d$-shell with respect to previous works.[10,12] This fact suggests that some of the clusters previously qualified as metastable, *i.e.* systems with negative energy but less bound than clusters with a smaller number of fermions, could be in fact stable. In Table 2 are compared our predictions with the experimental findings of Ref. 7. In some cases, different (N_B, N_F) clusters can have the same mass, which is the only signal in the experiment. There is an overall agreement between theory and experiment, with the striking exception of cluster (4,4), for which there is no experimental signal but theory predicts it to be bound. It is worth noticing that varying the source conditions can in some cases enhance an otherwise non-visible signal.

Table 2. Comparison between experiment and theory. The mass of the detected signal is given in amu, and it may correspond in some cases to several (N_B, N_F) clusters. The theoretical predictions are indicated with the symbols B, MS and U, which correspond to bound, metastable and unbound systems, respectively.

Mass	Exp.	Theory	Mass	Exp.	Theory	Mass	Exp.	Theory
11	strong	(2,1) B	22	strong	(4,2) B	28	no	(4,4) B
13	no	(1,3) U			(1,6) U			(1,8) U
14	weak	(2,2) B	23	strong	(5,1) B	29	may be	(5,3) B
15	strong	(3,1) B			(2,5) U			(2,7) U
16	no	(1,4) U	24	may be	(3,4) MS	30	may be	(6,2) B
17	no	(2,3) MS	25	weak	(4,3) B			(3,6) B
18	weak	(3,2) B			(1,7) U	31	strong	(7,1) B
19	strong	(4,1) B	26	weak	(5,2) B			(4,5) B
		(1,5) U			(2,6) U			(1,9) U
20	no	(2,4) U	27	strong	(6,1) B			
21	no	(3,3) U			(3,5) B			

The stability map of mixed clusters is displayed in Fig. 1. Our results show that one single ^{4}He atom is able to bind the otherwise unbound cluster with 20 ^{3}He atoms. Notice the stabilities related to the completion of fermionic shells.

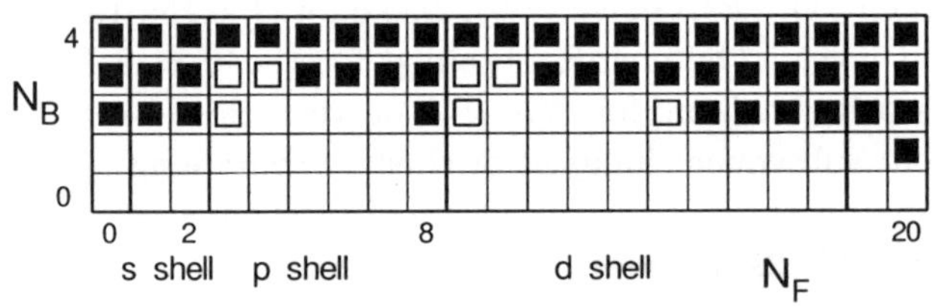

Fig. 1. Stability map of mixed clusters. Solid squares represent truly bound states, and open squares represent metastable states.

4. Energetics of clusters with several fermions

We have computed the binding energies of clusters with N_B from 4 to 20 in steps of 4, and for N_F from 0 to 18. Among the possible trial functions related to a given configuration, we have chosen the two simplest couplings: (a) S_{max}, where in each shell particles are aligned to maximum spin S and then to maximum orbital angular momentum L (Hund's rule); and (b) L_{max}, where particles are aligned to maximum L. In this way, the choice of the two Slater determinants is unique. Other couplings demand linear combination of Slater determinant, except for fully polarized clusters, where all ^{3}He atoms have the same spin orientation.

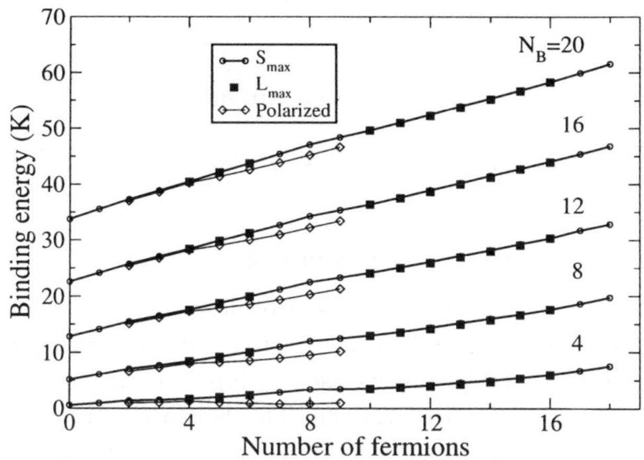

Fig. 2. Binding energies of (N_B, N_F) mixed helium clusters for S_{max} coupling (particles aligned to maximum S and then to maximum L), L_{max} coupling (maximum L), and fully polarized clusters.

In Fig. 2 are displayed the obtained binding energies for the three considered couplings. The present binding energies for $N_F > 8$ are slightly higher that the previous results, the gain being of 0.16 K for $N_F = 9$ and monotonically increasing up to 1.40 K for $N_F = 18$, in the case of $N_B = 8$. This apparently modest increase (up to 6%) is nevertheless relevant for the boundaries of the stability chart. In any case, it gives support to the level ordering used in the present work, as the associated importance sampling function provides a better variational bound than the previous ones. Our results indicate that maximum spin is quite uniformly favored, but the splitting between the two computed S_{max} and L_{max} levels is always smaller than 0.3-0.4 K, too small to be appreciated in the figure.

Basically, the energies grow linearly with the number of fermions but finer details emerge when looking at the separation energies $E(N_B, N_F - 1) - E(N_B, N_F)$, or chemical potentials, which are plotted in Fig. 3. A sudden drop of the separation energies is visible after $N_F = 2$ and 8, corresponding to the closure of the $1s$ and $1p$ shells. The relative minimum appearing at $N_F = 14$ is at first somewhat puzzling. Its origin will become clear in the next section.

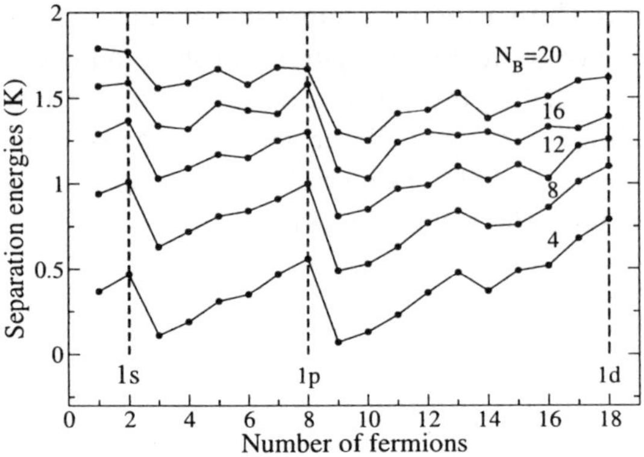

Fig. 3. Separation energies (in K) for normal S_{max} clusters as a function of the number of fermions N_F.

5. Effective monopole interaction analysis

We have adapted the method currently used to describe shell properties in nuclear physics[27] with the purpose of finding an effective one- plus two-body Hamiltonian that could reproduce our DMC results used as input data. In the present case the two-body part will only refer to fermions. The boson cluster is assumed to be a fixed core that simply originates the fermion single-particle energies, leading to a one body potential

$$U = \sum_s n_s \varepsilon_s ,$$

where n_s is the number of particles in shell s, and the sum extends to the occupied shells. The two-body part is defined by matrix elements

$$V^{LS}_{rs,tu} = \langle rs : (LS)|V|tu : (LS)\rangle ,$$

where $|tu : (LS)\rangle$ is a two-particle state in shells t and s, coupled to orbital angular momentum L and spin S, properly antisymmetrized and normalized.

As shown in Ref. 27 the two-body part of the full Hamiltonian can be separated into monopole H_{mS} and multipole H_M contributions. The former is defined by the property of giving the average energy of configurations at fixed number of particles n_r and spin $\mathbf{S_r}$. The closed shells and the one-particle and one-hole states built on them are configurations with a single state. Hence, their energies are entirely given by H_{mS}. We have assumed (and checked in some cases) that the influence of H_M is small, and simply neglected it.

The average matrix elements are defined as

$$V_{rs} = \frac{\sum_{LS} V^{LS}_{rsrs}(2L+1)(2S+1)(1+(-1)^{L+S}\delta_{rs})}{\sum_{LS}(2L+1)(2S+1)(1+(-1)^{L+S}\delta_{rs})} , \quad (1)$$

$$V^S_{rs} = \frac{\sum_L V^{LS}_{rsrs}(2L+1)(1+(-1)^{L+S}\delta_{rs})}{\sum_L (2L+1)(1+(-1)^{L+S}\delta_{rs})} , \quad (2)$$

where V_{rs} is the full (scalar) average of two-body matrix elements, whereas V_{rs}^S are (vector) averages at fixed S. It is convenient to introduce the following combinations

$$a_{rs} = \frac{1}{4}(3V_{rs}^1 + V_{rs}^0) \quad , \quad b_{rs} = V_{rs}^1 - V_{rs}^0 \, , \tag{3}$$

so that

$$V_{rs} = a_{rs} - \frac{3}{4}\frac{\delta_{rs}}{D_r - 1} b_{rs} \, , \tag{4}$$

where $D_r = 2(2l_r + 1)$ is the maximum number of particles in the shell. With these definitions the H_{mS} part of the effective interaction can be written as

$$H_{mS} = U + \sum_{r \leq s} \frac{1}{(1+\delta_{rs})} \left[V_{rs} \, n_r(n_s - \delta_{rs}) + b_{rs} \left(\mathbf{S}_r \cdot \mathbf{S}_s - \frac{3n_r(D_r - n_r)}{4(D_r - 1)} \delta_{rs} \right) \right] \tag{5}$$

The b_{rs} terms give no contribution at the closed shell neither at one-particle and one-hole states. Their energies are fully given by the V_{rs} term, which we refer to as H_m from now on. The advantage of this operation is that it decouples the determination of the V_{rs} and b_{rs} centroids, so that we can proceed with the former first, as they are the ones that give the global features.

In principle, the six necessary centroids —V_{ss}, V_{sp}, V_{sd}, V_{pp}, V_{pd} and V_{dd}— could be extracted from the energies of clusters with two fermions. However, this parameter-free choice has large uncertainties and it is better to reserve it as a consistency check with the results of a more precise fit to the DMC energies. A careful exploration of the two fermion case shows that only the combination $V_{sd} + 3V_{pd}$ appears in the present case, so that the number of independent parameters is in fact reduced to five.

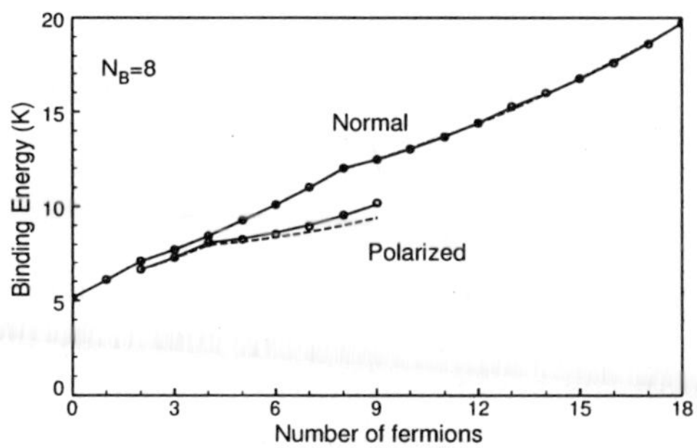

Fig. 4. Comparison of DMC binding energies with the monopole Hamiltonian with spin (continuous line) and without spin terms (dashed line). Circles correspond to the computed DMC values for normal S_{max} clusters. The lower line corresponds to the fully polarized case, and the upper group to the normal clusters. In both cases is $N_B = 8$.

The results for the binding energies in the particular $N_B = 8$ case are given in Fig. 4 for both the S_{max} normal and the fully polarized cases. The agreement for the normal case is quite excellent, but there are significant discrepancies for the polarized case. These can be cured by introducing the full H_{mS} through a single parameter $b_{rs} = b$ so that the contribution of the second line in Eq. (5) becomes

$$b[\,S(S+1)/2 - \sum_r 3n_r(D_r - n_r)/4(D_r - 1)]$$

Once this term is included, the fitted and calculated curves become nearly undistinguishable also for the polarized case.

The full energies, displayed in Figs. 2 and 4 have rather smooth patterns that tell us little about details. The truly sensitive quantities are the fermion separation energies, since a smooth linear approximation to the binding energies would result in a constant. The true test of the monopole description comes in Fig. 5. It can be seen that by introducing H_m there is an enormous improvement in that the shell effects at closures are well reproduced (dashed line) in both S_{max} and L_{max} cases, represented in the left and right panels, respectively.

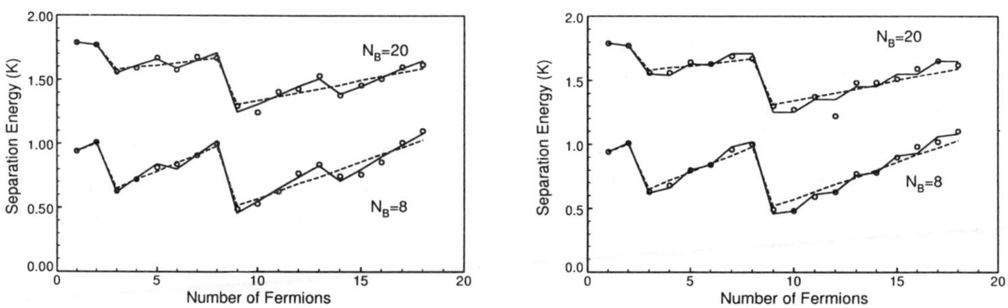

Fig. 5. Comparison of calculated separation energies with the monopole Hamiltonian with spin (H_{mS}, continuous line) and without spin terms (H_m, dashed line). Circles correspond to the computed DMC values. The lower group corresponds to $N_B = 8$, and the upper group to $N_B = 20$. Left and right panels correspond to clusters with $S_{\max}$ and $L_{\max}$, respectively.

However, the more detailed pattern between closures demands the $S(S+1)$ term in the full H_{mS}: the agreement with DMC becomes truly quantitative (full line). Specifically, the drop of the separation energies at the middle of the active shell (*e.g.*, at $N_F = 14$) is very well described by the spin-dependent part of this simple effective interaction, which reflects the role of Pauli's Principle.

In the L_{max} case the numerical fit beautifully reproduces the staggering pattern between $S = 0$ and $S = 1/2$ states. Here we are faced with some interesting physics: the numerical fit was chosen to find good compromise values for V_{pp} and V_{dd}, but it does slightly better than a fit restricted to those over-determined parameters. The most likely reason is to be found in size effects: as fermions are added, the overall radius evolves, and for a self-bound system it should go asymptotically as $N_F^{1/3}$. Therefore, the effective matrix elements should also evolve in a way our simplified

H_{mS} ignores. The numerical fit then emerges as the sound and natural way to define a best compromise value, not only for V_{pp} and V_{dd} but for all the monopole parameters.

Finally, let us mention that the centroids V_{rs}^0 and V_{rs}^1 obtained with the fit to the 19 normal S_{max} clusters agrees with the values obtained directly from the DMC calculation of two-fermion states. The nice agreement (within the large errors of about 0.04 K associated to the latter), thus confirms the consistency of the effective interaction interpretation.

6. Conclusions

A simplified pattern stems from previous density functional calculations, namely, the ^{4}He atoms arrange themselves as a quite rigid core whereas the ^{3}He atoms are distributed in the surface of that core. In the present work this arrangement has been confirmed and pushed down to systems with a small number of constituents. The calculated stability map is in agreement with experiment, except for the (4,4) cluster, predicted to be bound but for which there is no experimental signal.

The analysis in terms of an effective interaction model suggests that the residual interactions between the fermions in the cluster is very weak, of the order of the computational precision achieved (near 0.1 K). This fact is also reflected in the insensitivity of the energies to the (L, S) quantum numbers, the spectrum being essentially determined by the configuration. It appears that very hard DMC calculations lead to results amenable to a very simple and cogent interpretation in terms of the monopole Hamiltonians H_{mS}. Such an analysis points to the basic simplicity of the DMC results and invites to take on the challenge of unearthing the deep reasons of such simplicity.

The ground state of these clusters corresponds to the S_{max} case, where the spin has the maximum possible value in the open shell configuration. We have also analyzed the fully polarized case, with all ^{3}He spins aligned. The normal phase is energetically favored but the polarized one gives rise to bound states, even for a moderately large number of ^{3}He atoms. Fig. 4 suggests the idea that for large enough number of fermions the polarized clusters could become ground states. Though this is only a speculation, it may be also be taken as a strong invitation to push the study of mixed clusters much further.

The mixed systems may have a very rich excitation spectrum, because of the gap between the normal and the polarized phase. The analysis of the intermediate filled configurations, such as $1s^2 1p^3 1d^1$ or $1s^2 1p^2 1d^2$ puts heavy demands on the DMC algorithm but it would be relatively simple in the density-functional method. Though it does not seem possible with the present experimental techniques to measure the spectrum, it is worth remembering that the excitation spectrum plays a relevant role in the production abundances of clusters.

Acknowledgements

This work has been supported by MCyT/FEDER (Spain), grant FIS2004-00912, Generalitat Valenciana (Spain), grant GV2003-002, and MIUR (Italy), grant cofin-2001025498.

References

1. K.B. Whaley (editor), Special issue of *J. Chem. Phys.* **115**, 10065 (2001).
2. E. Krotscheck and J. Navarro (editors), *Advances in Quantum Many-Body Theory*, Vol. 4 (2002), World Scientific (Singapore).
3. M. Barranco, R. Guardiola, E.S. Hernández, R. Mayol, J. Navarro, and M. Pi, *J. Low Temp. Phys.*, **142**, 1 (2006).
4. W. Schöllkopf and J.P. Toennies, *Science* **266**, 1345 (1994).
5. R.E. Grisenti, W. Schöllkopf, J.P. Toennies, G.C. Hegerfeldt, T. Köhler, and M. Stoll, *Phys. Rev. Lett.* **85**, 2284 (2000).
6. R. Brühl, R. Guardiola, A. Kalinin, O. Kornilov, J. Navarro, T. Savas, and J.P. Toennies, *Phys. Rev. Lett.* **92**, 185301 (2004).
7. A. Kalinin, O. Kornilov, W. Schöllkopf, and J.P. Toennies, *Phys. Rev. Lett.* **95**, 113402 (2005).
8. M. Barranco, J. Navarro and A. Poves, *Phys. Rev. Lett.* **78**, 4729 (1997).
9. R. Guardiola and J. Navarro, *Phys. Rev.* **A71**, 035201 (2005).
10. R. Guardiola and J. Navarro, *Phys. Rev. Lett.* **89**, 193401 (2002).
11. D. Bressanini and G. Morosi, *Phys. Rev. Lett.* **90**, 133401 (2003).
12. R. Guardiola and J. Navarro, *Phys. Rev.* **A68**, 055201 (2003).
13. S. Fantoni, R. Guardiola, J. Navarro, and A. Zuker, *J. Chem. Phys.* **123**, 054503 (2005).
14. R.A. Aziz, F.R. McCourt and C.C.K. Wong, *Mol. Phys.* **61**, 1487 (1987).
15. P.J. Reynolds, D.M. Ceperley, B.J. Alder and W.A. Lester Jr., *J. Chem. Phys.* **77**, 5593 (1982).
16. J.W. Moskowitz, K.E. Schmidt, M.A. Lee and H.M. Kalos, *J. Chem. Phys.* **77**, 349 (1982).
17. R.P. Feynman and M. Cohen, *Phys. Rev.* **102**, 1189 (1956).
18. J.W. Moskowitz, K.E. Schimdt, M.A. Lee, and H.M. Kalos, *J. Chem. Phys.* **77**, 349 (1982).
19. P.J. Reynolds, D.M. Ceperley, B.J. Alder, and W.A. Lester Jr., *J. Chem. Phys.* **77**, 5593 (1982).
20. S. Fantoni, R. Guardiola, and J. Navarro, *Phys. Rev.* **A70**, 023206 (2004).
21. F. Dalfovo, *Z. Phys.* **D14**, 263 (1989).
22. M. Barranco, M. Pi, S.M. Gatica, E.S. Hernández, and J. Navarro, *Phys. Rev.* **B56**, 8997 (1997).
23. A. Belić, F. Dalfovo, S. Fantoni, and S. Stringari, *Phys. Rev.* **B49**, 15253 (1994).
24. D. Bressanini, M. Zavaglia, M. Mella, and G. Morosi, *J. Chem. Phys.* **112**, 717 (2000).
25. E. Krotscheck and R. Zillich, *J. Chem. Phys.* **115**, 10161 (2001).
26. O. Bohigas, A.M. Lane and J. Martorell, *Phys. Rep.* **51**, 267 (1979).
27. E. Caurier, G. Martínez-Pinedo, F. Nowacki, A. Poves, and A. Zuker, *Rev. Mod. Phys.* **77**, 427 (2005).

BOSE–EINSTEIN CONDENSATION IN BULK AND CONFINED SOLID HELIUM

L. REATTO, M. ROSSI and D. E. GALLI

Dipartimento di Fisica, Università degli Studi di Milano, Via Celoria 16, 20133 Milano, Italy
luciano.reatto@unimi.it

We address the question if the ground state of solid ^{4}He has the number of lattice sites equal to the number of atoms (commensurate state) or if it is different (incommensurate state). We point out that energy computation from simulation as performed by now cannot be used to decide this question and that the presently best variational wave function, a shadow wave function, gives an incommensurate state. We have extended the calculation of the one–body density matrix ρ_1 to the exact Shadow Path Integral Ground State method. Calculation of ρ_1 at $\rho = 0.031$ Å^{-3} shows that Vacancy–Interstitial pair processes are present also in the exact computation but the simulated system size is is too small to infer the presence of off–diagonal long range order. Variational simulations of ^{4}He confined in a narrow cylindrical pore are also discussed.

Keywords: solid helium, Bose Einstein condensate, porous media

1. Introduction

Since the first pioneering papers in the sixties, the prospect that a solid could have superfluid–like properties has fascinated and attracted a great number of physicists. Thus solid ^{4}He, the best candidate for such supersolid state, has received a considerable theoretical and experimental attention.[1] The underlaying idea, as for the superfluid phase, is the possibility of a Bose–Einstein condensation (BEC) in solids. Penrose and Onsager[2] proved that a solid in which the particles are localized around their lattice positions cannot have a BEC transition. Then a classical solid cannot be a supersolid. This situation is different for a quantum solid, i.e a solid in which the atoms, due to their zero–point motion, execute very large amplitude vibrations around their lattice sites and these vibrations occupy a sizable fraction of the cell size. Andreev and Lifshitz,[3] and independently Chester,[4] were the first to discuss the possible implications of the large zero–point vibrations of the particles in establishing a supersolid state. They argued that when these quantum effects are large enough the ground state might contain lattice defects, like vacancies, which are delocalized and acting as a diluite Bose gas, undergo a BEC transition resulting in superfluid–like properties of the crystal. But this is only one of the possible source of the supersolid behavior. Leggett[5] showed how a very large zero–point motion could be enough, even in absence of ground state defects, for the solid to possess a non classical rotational inertia (NCRI), i.e. a rotational inertia displayed by the solid which is lower than its classical value of rigid body rotation.

The experimental search of the supersolid gave no results[1] until 2004, when Kim and Chan[6] reported to have observed a NCRI effect for solid ^{4}He filling the pores of Vycor during a torsional oscillator experiment. Vycor is a porous glass composed by a network of interconnected worm–like pores. In such a confinement solid ^{4}He will be surely rich of defects, especially vacancies near the pore wall, and these defects were considered responsible for the observed NCRI effect. But later on, NCRI effects were also observed for the bulk solid ^{4}He.[7] These observations have given a new life to the interest on the supersolid state, whose existence is still an open problem. In fact, up today there is no experimental evidence for the zero–point vacancies in bulk solid ^{4}He and arguments against bulk superfluidity have appeared in literature.[8,9]

Leggett showed that BEC, at least at $T = 0$ K, is a sufficient condition for the observation of NCRI effects.[10] So a central quantity to compute is the one–body density matrix $\rho_1(\vec{r}, \vec{r}\,')$, whose Fourier transform represents the momentum distribution. A non–zero limit at large distance of ρ_1 (off–diagonal long–range order, ODLRO) implies BEC and then NCRI. In Ref. 11 we presented the results for ρ_1 computed with the shadow wave function (SWF) in a perfect hcp and in a bcc ^{4}He crystal. We found ODLRO for a range of densities above the melting one. For example at $\rho = 0.031\text{Å}^{-3}$, which corresponds to a pressure of about 54 bars, we found a condensate fraction of $n_c = (2.0 \pm 0.4) \times 10^{-6}$. These results represent the first microscopic indication that BEC can be present in commensurate solid ^{4}He. By commensurate solid, we mean a solid where the number of maxima in the local density $\rho(\vec{r}) = \rho_1(\vec{r}, \vec{r})$ is equal to the number of ^{4}He atoms.[12]

Here we present some new results for the calculation of ρ_1 for a commensurate solid ^{4}He at $T = 0$ K. In the first part we present results obtained with the SWF variational technique. With the SWF, the full Bose symmetry is always preserved, and the crystalline order is an effect of the spontaneously broken translational symmetry.[13] In this way, all the microscopic processes that might be present due to the large zero–point motion of the ^{4}He atoms are in principle allowed. The SWF technique has obtained a number of important successes not only in the description of the ground state properties of both the liquid and the solid phase of ^{4}He, but, for example, it is able to give also an accurate description of longitudinal acoustic phonons[14] and the excitation spectrum of a vacancy.[14] We present also some preliminary results of a computation of ρ_1 with the exact projector method,[15] the shadow path integral ground state (SPIGS). The question of the presence of ground state defects, i.e. if the ground state is incommensurate, is addressed in Sec. 3. Finally we present some new results for the SWF variational description of ^{4}He filling a porous media. In Ref. 16 we presented the phase diagram of ^{4}He confined in a smooth cylindrical pore of radius $R = 13$ Å. There we found that ^{4}He forms a layered structure with no 3D bulk–like crystal in the pore; increasing the density a layer by layer solidification takes place, starting from the outermost layer, and very high pressure (about 200 bar) are needed in order tho have the whole system in the solid phase. In order to confirm our conclusion we have considered a larger pore and also a different pore–He potential. We find that the qualitative phase diagram is only dependent on the confinement, while the quantitative one is sensitive to the details of the potential employed to model the confinement.

2. Off–diagonal long range order in commensurate solid ^{4}He

Recently[11] we have used the exact SPIGS method to test the accuracy of the SWF on diagonal properties like the static structure factor $S(\vec{k})$ and the local density $\rho(\vec{r})$, finding that within the SWF description the localization of ^{4}He atoms is higher than with SPIGS. Fore instance, at density $\rho = 0.031$ Å^{-3}, the Bragg peaks in $S(\vec{k})$ are 7% higher respect to the SPIGS values and the oscillations in $\rho(\vec{r})$ are more marked than in the exact case (see Fig. 1). In our previous computation of

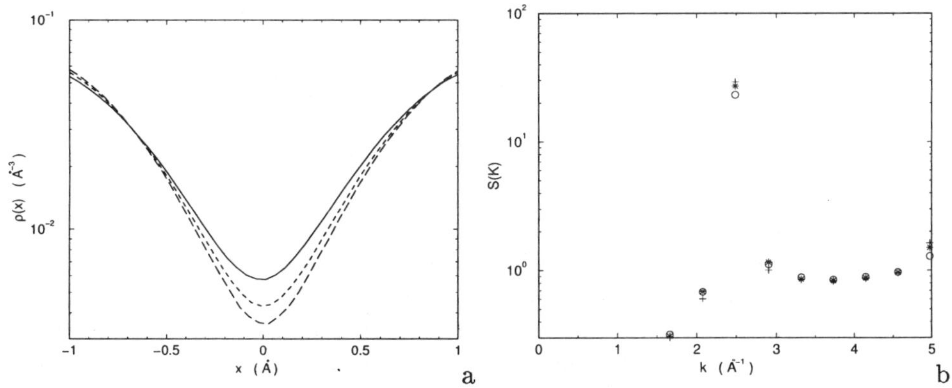

Fig. 1. (a) Local density $\rho(x)$ and (b) static structure factor $S(k)$ (x and k lying along the [1,0,0] direction of the lattice) for a perfect fcc lattice at density $\rho = 0.031$ Å^{-3} computed with the SPIGS method (solid line and circles), with the new SWF with $N_{\text{extra}} = 15$ (dashed line and stars) and with the standard SWF (long dashed line and plus).

the solid with one vacancy,[14] we found an improved variational description if one extra–shadow variable was introduced. Then, following this idea also for the perfect solid, we introduce additional extra–shadows which are integrated over. The new SWF now reads

$$\psi_{\text{new}}^{\text{SWF}}(R) = \int dSdS_{\text{v}}\, F(R,S)L(S,S_{\text{v}}) \quad . \tag{1}$$

The kernel $F(R,S)$ has the usual form[14] which contains direct two–body particle correlations, correlations between the ^{4}He positions and the respective shadow variables and two–body correlations between shadow variables. In the present computation we have used the fully optimized form[17] of these correlation functions. The new factor $L(S,S_{\text{v}})$ contains two–body correlations between the extra–shadow and the usual shadow variables. More precisely, if N is the number of ^{4}He atoms, we introduce $N + N_{\text{extra}}$ subsidiary variables. N of these are coupled to the real variables in the standard way, and $N_{\rm extra}$ are only coupled to the other N subsidiary variables via a pseudopotential which is separately optimized. The two–body correlation factor between the extra–shadow variables and the usual shadow has been chosen of the Jastrow–McMillan type: $f(\mid \vec{s}_{ex} - \vec{s}\mid) = \exp\{-1/2(b_{ex}/\mid \vec{s}_{ex} - \vec{s}\mid)^9\}$. Having chosen the fully optimized form for the standard correlations in Eq.1, the parameter b_{ex} is the only one which has been optimized in this calculation. The inclusion of these extra–shadow variables allows for the introduction of a perturbation in the crystal lattice in the attempt to enhance exchange processes an to obtain a better description of the structure of the lattice and of the ground state energy. We find that these extra–shadows indeed lower the energy (see Table 1) and, in

Table 1. Ground state energy per particle e_0 computed in a perfect fcc lattice at the density $\rho = 0.031$ Å^{-3}.

technique	e_0 (K)
SPIGS	-5.270(5)
new SWF ($N_{\rm extra} = 15$)	-5.066(1)
old SWF	-4.968(2)

addition, bring $S(\vec{k})$ and $\rho(\vec{r})$ in better agreement with SPIGS results (see Fig. 1). We find that the expectation value of the Hamiltonian has a well defined minimum as a function of b^{ex} at fixed $N_{\rm extra}$, but the optimal b^{ex} changes with $N_{\rm extra}$: higher values of b^{ex} tends to destroy the crystal structure and the optimal value of b^{ex}, which always refers to a stable solid order, decreases as $N_{\rm extra}$ increases. By changing $N_{\rm extra}$ and optimizing the energy as function of b^{ex}, one obtains substantially equivalent variational descriptions of the ground state of solid ^{4}He in the range $N_{\rm extra} > N/20$. We have choose to fix $N_{\rm extra} = 15$ in the fcc lattice with $N = 108$ particles in order to not increase too much the number of degrees of freedom in the computation. At the density of the computation, $\rho = 0.031$Å^{-3}, with $N_{\rm extra} = 15$ we obtain a lowering of the energy of about 2% with $b^{ex} = 2.3$ Å. This new SWF also improves the structure as can be seen in Fig. 1.

With this improved SWF, we have computed ρ_1 in a commensurate fcc solid ^{4}He; the results are plotted in Fig. 2 together with an old result for ρ_1 computed with the standard SWF at the same density but in the hcp lattice.[11] Previous computations with standard SWF have shown that there is a very small difference between fcc and hcp lattice results when $\vec{r} - \vec{r}'$ is in the direction of first neighbors. The lower degree of order in the structure of the crystal is also observable in the oscillations of the plateau of ρ_1 which reflect the crystalline order in the system; they are still present but very much damped if compared with the ρ_1 computed with the standard SWF. The plateau starts at shorter distance and the value of ρ_1 at distance of 10 Å obtained with the new SWF at $\rho = 0.031$Å^{-3} is larger nearly by a factor of 40 compared to the previous SWF; however the size of the box are too small to deduce an accurate estimation of the condensate fraction n_c. The

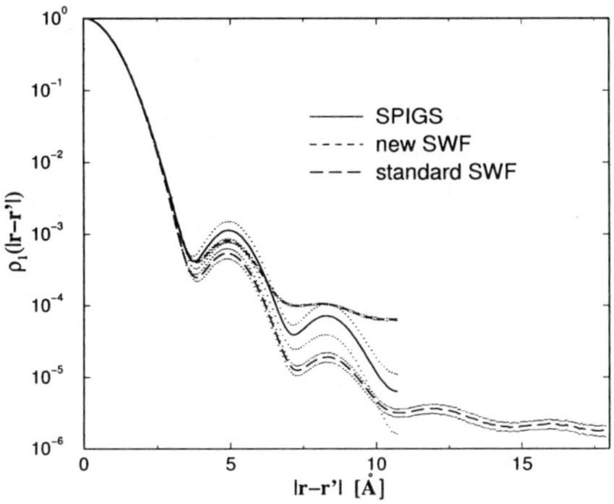

Fig. 2. ρ_1 computed in a perfect solid ^{4}He at $\rho = 0.031\text{Å}^{-3}$ along the nearest neighbor direction in a fcc lattice with the SPIGS method (solid line), in a fcc lattice with the new SWF (dashed line) and in an hcp lattice with a standard SWF (long dashed line). Dotted lines represent statistical uncertainty.

sensitivity of the off–diagonal properties of ρ_1 on the form of the SWF indicates the relevance of computing ρ_1 also by an exact Quantum Monte Carlo method. We have extended the calculation of ρ_1 by projecting in imaginary time this new SWF optimized at $\rho = 0.031\text{Å}^{-3}$ by the SPIGS method.[15] This method is based on the application of the imaginary time evolution operator $\exp\{-\tau\hat{H}\}$ to the SWF and on the splitting of this operator as $\exp\{-\tau\hat{H}\} = [\exp(-\frac{\tau}{P}\hat{H})]^P$ which give rise to a path integral of *linear* polymers. We present preliminary results for $\tau = 0.075$ K^{-1} with a time step $\delta\tau = \frac{\tau}{P} = 0.025$ K^{-1} and the pair–product approximation[18] has been used for the $\delta\tau$ imaginary time propagator. One can see in Fig. 2 that the oscillations in ρ_1 are marked even more than those obtained with the standard SWF, moreover at the computed largest distance the value of ρ_1 seems more in agreement with the one obtained with the standard SWF. This indicates that even if the new SWF improves the variational description of solid ^{4}He, it overestimates off–diagonal properties. In the tail region, the SPIGS ρ_1 is similar but somewhat larger than the result given by the standard SWF, and indicates that the source of the processes which are responsible for the oscillations of ρ_1 is the same as analyzed with SWF: the presence of a vacancy–interstitial pairs as spontaneous fluctuations in the ground state. However the size of the simulated system with SPIGS is too small to see a well defined plateau and we also need to project at larger value of τ, so that we cannot conclude if ODLRO is present or not in the exact ground state of commensurate solid ^{4}He.[12]

3. Is the ground state of solid ^{4}He commensurate or incommensurate?

The presence of zero–point defects is a possible mechanism for the superfluid behavior of solid ^{4}He. Arguments against this possibility and supporting instead the commensurate (C) ground state, i.e. a solid with $N = M$ (where M is the number of lattice sites and N the number of particles), have appeared in the current literature.[8,12,19] They rely upon statements that microscopic computations[15,20] have shown that a vacancy increases the energy of the system by at least 15 K and by an even higher value in the case of an interstitial. On the other hand, even if the experiments give no evidence of zero–point vacancies, these experiments[21] cannot exclude their presence at low temperature at a level below of order of 1%. Moreover, very recently, a phenomenological theory[22] has shown how the low temperature properties of crystalline ^{4}He would be modified if the ground state would be incommensurate finding results which are quite consistent with the experiment. Thus it is significant to critically discuss if the ground state of solid ^{4}He is C or if $N \neq M$, i.e. the ground state is incommensurate (I) in the sense that the lattice parameter inferred from bulk density measurement differs from the one deduced from Bragg scattering. Moreover, very recently, a phenomenological theory[22] has shown how the low temperature properties of crystalline ^{4}He would be modified if the ground state were incommensurate finding results which are quite consistent with the experiment.

It is important to understand that the estimate of the formation energy ΔE_v of a vacancy in solid ^{4}He given either by the variational[20] or by the exact SPIGS[15] computations, is actually based on the computation of the *ground state energy of two different systems*. Assuming these results as evidence that the ground state of solid ^{4}He does not contain zero–point vacancies, fails to recognize that those computations were specifically designed to compute excitation energy of one extra vacancy and not to study the presence of ground state defects. In addition it fails to appreciate rules and limitations that are intrinsically introduced in the simulation of a limited number of particles in a box with periodic boundary conditions (pbc). It is worthy to note that the equilibrium sites of the solid are not introduced at any stage of the computation, either in the SWF or in the exact projection procedure (SPIGS), but the crystalline order, if stable, arises as spontaneous broken symmetry. In order to compute the formation energy of a vacancy two computations are performed: one in which the number N of particles is equal to the number M of lattice sites which fit in the simulation box and satisfy the pbc, and one in which $N = M - 1$. In the second case one finds that the local density continues to have M maxima with essentially the same degree of crystalline order, as measured by the height of the Bragg peaks, as in the case $N = M$. This means that the crystalline order is stable and that there is one mobile vacancy, then that state is I. Both wave functions (wfs), that of the C state and that of the I state, are non negative and give two slightly different energies, but both values represent a ground state energy of two periodically repeated small systems with the nature C or I being imposed by the choice of M,N and by the pbc. Difference between these two energies has been used to estimate the formation energy of an extra vacancy in bulk under the hypothesis of non interacting vacancies. Such kind of computation cannot be used to determine if the ground state of bulk solid ^{4}He is C or I unless one is able to extrapolate these finite size results to the bulk limit also allowing for the effects of the pbc.

This is not surprising, also for classical solids[23] there is no direct computation of the equilibrium concentration $\bar{X}_v$ of vacancies at finite temperature exactly for the same reasons as in the quantum case, i.e. finite size of any system that can be simulated and commensurability effect between crystal lattice and simulation box. In the classical case, $\bar{X}_v$ has been obtained only indirectly by a statistical thermodynamical analysis of an extended system.[23] In a similar way we can expect to get light on the nature of the ground state of bulk crystalline ^{4}He only by considering the wf of an extended system, not of the one which is simulated. Let us consider wfs that are translationally invariant so that the crystalline state arises as spontaneous broken symmetry. An example is given by the Jastrow wf. Another one is SWF which presently gives the best representation of the ground state of solid ^{4}He.[17] It is interesting to note that both these translationally invariant wfs give rise to a ground state with a finite concentration x_v of vacancies . For the Jastrow wf this was the Chester[4] argument, which we briefly summarize here: consider a very large system described by a Jastrow wf of $\mathcal{N}$ particles in volume $\mathcal{V}$

$$\Psi_J(R) = \prod_{i<j}^{\mathcal{N}} e^{-\frac{1}{2}u(r_{ij})} / Q_{\mathcal{N}}^{1/2}, \tag{2}$$

where $r_{ij} = \mid \vec{r}_i - \vec{r}_j \mid$, $R = \{\vec{r}_1, .., \vec{r}_{\mathcal{N}}\}$ and $Q_{\mathcal{N}}$ is the normalization constant

$$Q_{\mathcal{N}} = \int_{\mathcal{V}} dR \prod_{i<j}^{\mathcal{N}} e^{-u(r_{ij})}. \tag{3}$$

The straightforward isomorphism between quantum averages with Ψ_J^2 and averages in classical statistical mechanics was already recognized long ago.[24] In fact, Ψ_J^2 coincides with the normalized probability distribution in configurational space of $\mathcal{N}$ classical particles at an inverse temperature $\beta^* = 1/k_B T^*$ which interact with a pair potential $v^*(r)$ such that $\beta^* v^*(r) = u(r)$. In addition, $Q_{\mathcal{N}}$ is equal to the canonical configurational partition function of this classical system and its logarithm is proportional to the excess Helmholtz free energy. It has been proved[25] that wf (2) has a finite BEC fraction, but it is also known that the equivalent classical system corresponding to Ψ_J^2 is a crystalline solid, when the density is large enough. In addition this solid has a finite concentration of vacancies. In fact, for this equivalent classical solid, the equilibrium concentration of vacancies $\bar{X}_v = (\mathcal{M} - \mathcal{N})/\mathcal{N}$ (where $\mathcal{M}$ is the number of lattice sites) is non–zero even if a single vacancy has a finite cost of local free energy because of the gain in configurational entropy.[26] Another way of expressing this is $Q_{\mathcal{N}}$ of this equivalent classical system has contributions from different pockets in configurational space corresponding to the commensurate state

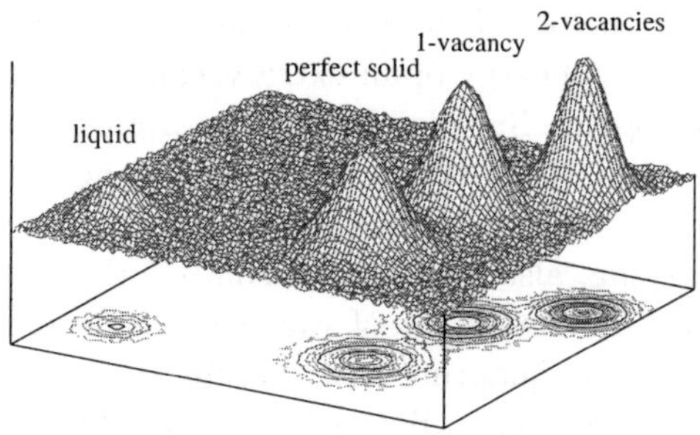

Fig. 3. Schematic landscape of the probability in configurational space of Ψ_J of a bulk system at a density where the solid is stable.

but also to states with a different number of vacancies (see Fig. 3) and the pockets with vacancies give the main contribution to $Q_{\mathcal{N}}$. This translates in the quantum case to say that the wf (2) of a bulk system is describing at the same time states with no vacancies but also with vacancies, and that the overwhelming contribution to the normalization constant $Q_{\mathcal{N}}$ derives from pockets corresponding to a finite concentration of vacancies. The simulation of a small system of N particles with pbc is mimicking expectation values of the quantum Hamiltonian in a restricted pocket in configurational space, the pocket that has been implicitly chosen at the start of the computation by taking $N = M$ or $N = M - 1$. Notice also that the normalization constant (3) is never computed explicitly, but averages are implicitly normalized to the set of configurations that are explored in the MC simulation, i.e. to the chosen pocket.

The previous discussion can be applied also to the SWF. In fact, the normalization of Ψ_{SWF} coincides with the configurational partition function of a classical system of suitable flexible triatomic molecules[13] and for the previous argument it is inessential that the "particles" are monoatomic or are molecular species. Then we can conclude that also this equivalent classical molecular system displays a finite $\bar{X}_v$. In addition one can extend the proof in Ref. 25 of BEC for a Jastrow wf also to the case of SWF. This means that Ψ_{SWF} describes a quantum solid with BEC and vacancies. We can estimate the ground state energy per particle e_G of a bulk system as

$$e_G = e_0 + \bar{X}_v \Delta e_v \tag{4}$$

where e_0 is the energy per particle from a simulation of the C state and Δe_v is the energy for one vacancy obtained as discussed above. So a very relevant question is how big is $\bar{X}_v$. The only estimate of $\bar{X}_v$ has been performed some years ago for a Jastrow wf,[23] but Ψ_J is not a good wf for solid ^{4}He because it gives a solid phase with too much localized atoms. For a SWF $\bar{X}_v$ is not known at the moment.

One should ask if wfs exist which are translationally invariant and describe a commensurate solid, i.e. a wf which is non zero only in the pocket, that was call Ω_0, of the prefect solid in Fig. 3. One might think to take as wf the product $\Psi_{SWF}(R) \times E_{\Omega_0}(R)$ where $E_{\Omega_0}(R)$ is a function which is unity in the domain Ω_0 and it goes smoothly to zero outside it. This wf describes a commensurate solid but the corresponding ground state energy e_0 will be higher of the one given by Ψ_{SWF}. In fact Ψ_{SWF} is small in the transition region between Ω_0 and the 1–vacancy region, for instance, but it is not vanishing small because particles as described by Ψ_{SWF} are rather mobile. As a consequence the factor $E_{\Omega_0}(R)$ will rise the kinetic energy and therefore one should compare this increased kinetic energy with the extra term $X_v \Delta e_v$ in (4) so that only an explicit computation can say if the commensurate state has a lower energy of Eq. (4). Other wfs which describes a commensurate solid are obtained by multiplying the Jastrow wf by one body terms[17] which explicitly break the translational symmetry and contain information on an assumed set of equilibrium positions. However such wfs give a worst energy[17] than the SWF so that such wfs cannot be used to rule out ground state vacancies.

4. ^{4}He filling porous materials

We have recently given a variational description of the behavior of ^{4}He filling a porous medium within the SWF technique.[16] We have modeled the confining medium with a smooth cylindrical pore[27] and the radius, $R = 13$ Å, was chosen to resemble the nominal pore size of Gelsil glasses employed in Ref. 28. The He–pore was chosen to reproduce the He–Si interaction employed in Ref. 29 and gives rise to a potential with an attractive well of -152 K placed at 2.2 Å from the pore wall. We find that ^{4}He forms a distinct layered structure with the atoms arranging themselves in four concentric annuli. The atoms in the layer adjacent to the pore wall are very localized under all conditions of a filled pore making the properties of this layer (0th layer) to be almost independent on the total density of the system. Since this 0th layer is always solid in the considered density range, in the study of the inner layers we have chosen to fix the positions of the atoms in this layer treating them as a part of the substrate. Thus the dynamical system is composed by the three internal layers and we give our results in terms of effective density ρ_{eff} in which we have excluded the high density 0th layer. As we increase the effective density a layer by layer solidification takes place starting from the outermost layer; by solid layer we mean a defected triangular planar lattice wrapped on a cylindrical surface. One of the major features that emerge from these simulations is that the solidification takes place starting from the pore wall, while the widely accepted mechanism for solidification in larger pores predicts that the solid nucleates within the liquid region at the center of the pore.[30] Moreover our system does not completely solidify up to a pressure of about 200 bars, which is very high compared to the experimental solidification pressures. We proposed in Ref. 16 that these features are due to the reduced size of the considered pore. In fact, in the real system, there must be a transition region, between the amorphous layers adsorbed at the pore wall and the crystalline solid in the center.[31] This transition region results insensitive to the pressure[31] and its size for solid ^{4}He has been estimated[31] of order of 10 Å. Then the $R = 13$ Å pore accommodates only this incompressible "transition region".

Another important feature that we find is that the condensate fraction is finite even at a pressure of order 250 bar, i.e. when the system is in the solid phase. We have studied the presence of BEC by computing the single layer contribution to the one–body density matrix $\rho_1(z - z')$ along the pore axis. By studying the long distance tail oscillations of ρ_1, whose maxima are registered with the layer crystalline lattice, it is possible to argue that the dominant contribution to ODLRO is given by the presence of defects such as vacancies rather than of the vacancy–interstitial pairs due to the zero point motion of the ^{4}He atoms, as is the case for bulk ^{4}He.[11]

In order to verify our conclusions on the phase diagram we have performed new SWF simulations with a larger pore of radius $R = 22.5$ Å, which is comparable with the nominal pore size of the Gelsil glass employed in Ref. 32. The structure of the system near the pore wall is still layered and, since the number of particles involved in these simulations is very large (more than 900 atoms), we have chosen to lock the two outermost layers. At the effective density $\rho_{\text{eff}} = 0.05722$ Å^{-3} ($N = 699$ dynamical particles in a $L = 15.57$ Å pore) we have found a 3D crystal in the central region of the pore. In Fig. 4 we show a snapshot of the ^{4}He atom coordinates taken during the simulation run. This crystal shows a coordination number very close to

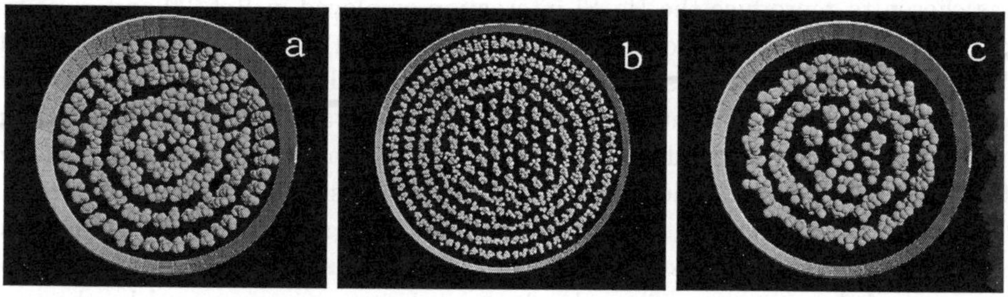

Fig. 4. Snapshot along the pore axis of the ^{4}He atom and shadow particle coordinates during a simulation run (a) of the $R = 13$ Å pore ($\rho_{\text{eff}} = 0.04070$ Å^{-3}), and (b) the $R = 22.5$ Å pore ($\rho_{\text{eff}} = 0.05722$ Å^{-3}) with the old[16] potential, and (c) the $R = 13$ Å pore ($\rho_{\text{eff}} = 0.03320$ Å^{-3}) with the new[33] He–pore potential.

the closed packed solid value. Computing the angular probability density $P(\theta, \phi)$ as described in Ref. 16 we were able to conclude that the crystal in the central region of the pore is a defected polycrystalline hcp lattice (see Fig. 5). The density of this crystal, estimated from the mean distance between the atoms, is about $\rho_{\text{hcp}} = 0.04182$ Å^{-3}, which corresponds[34] to a bulk pressure of about 500 bar, which is still too high compared with the experimental findings. At lower pressure we find no evidence for an hcp crystal and we have also tried to induce the growth of a 3D crystal at much lower pressures inserting a solid nucleus in the central liquid

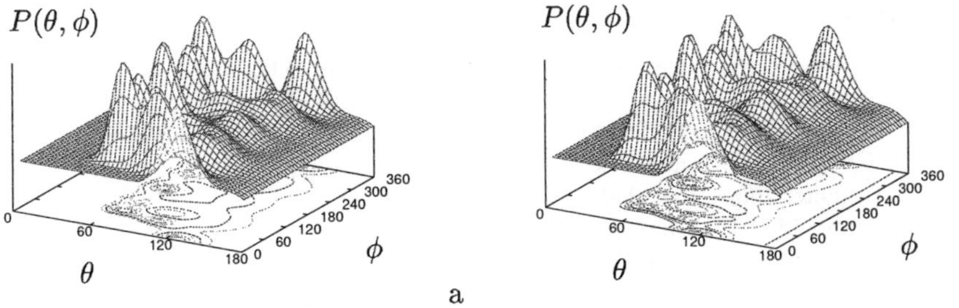

a b

Fig. 5. $P(\theta,\phi)$ computed (a) in a bulk hcp lattice and in the central region of (b) the the $R = 22.5$ Å pore ($\rho_{\text{eff}} = 0.05722$ Å^{-3}).

region. Also in this case no solidification takes place and the system maintains a layered structure with liquid central layers. Unfortunately this system is too large and too computational consuming to be systematically explored with a suitable and reasonable statistics employing the available computational techniques.

In order to study the influence of the pore potential on the transition pressures we have considered a less attractive potential that recently has been proposed for He–silica glass.[33] This potential displays an attractive well of -116 K at 3.6 Å from the pore wall and we have studied with this potential a $R = 13$ Å pore. We find that the ^{4}He atoms arrange themselves in a layered structure composed by the adsorbed locked layer, two cylindrical layers and a central rod. By studying the single layer static structure factor we find that at $\rho_{\text{eff}} = 0.03320$ Å^{-3}, which correspond[34] to a pressure of order 95 bar, we find that the whole system is in the solid phase, the central rod behaving as an 1D solid, even if the the effective density is considerably smaller than the one required for the complete solidification with the more attractive potential of Ref. 16. The coordination number is very close to 12 but the computed angular probability density $P(\theta,\phi)$ in the central region of the pore suggests also in this case that there is no bulk–like 3D solid inside the pore. Instead, as in the previous case,[16] the layers organize themselves as successive triangular planar lattices, but the strong curvature induced by the confinement does not allow the growth of a 3D hcp crystal in the central region.

We can conclude that the qualitative features of the phase diagram, such as the layer by layer transition and the absence of a bulk–like solid inside the pore of 13 Å radius, are only due to the geometry of the system and are insensitive to the employed potential. On the other hand the quantitative results, as the transition pressure and the freezing or melting effective densities are strongly dependent on the potential details. We plan to extend the study with this new He–pore potential to characterize the confined system in a quantitative way and comparing our results with the experimental findings.

References

1. for a review see M. W. Meisel, *Physica* **B178**, 121 (1992).
2. O. Penrose and L. Onsager, *Phys. Rev.* **104**, 576 (1956).
3. A. F. Andreev and I. M. Lifshitz, *Soviet Phys. JETP* **29**, 1107 (1969).
4. G. V. Chester, *Phys. Rev.* **A2**, 256 (1970).
5. A. J. Leggett, *Phys. Rev. Lett.* **25**, 1543 (1970).
6. E. Kim and M.H.W. Chan, *Nature* **427**, 225 (2004).
7. E. Kim and M.H.W. Chan, *Science* **305**, 1941 (2004).
8. D. M. Ceperley and B. Bernu, *Phys. Rev. Lett.* **93**, 155303 (2004).
9. N. Prokof'ev and B. Svistunov, *Phys. Rev. Lett.* **94**, 155302 (2005).
10. A. J. Leggett, *Physica Fennica* **8**, 125 (1973).
11. D. E. Galli *et al.*, *Phys. Rev.* **B71**, 14056 (2005).
12. While this paper was written two papers have been posted (B. K. Clark and D. M. Ceperley, cond-mat/0512547; M. Boninsegni *et al.*, cond-mat/0512103) in which ρ_1 has been computed by path integral Monte Carlo at finite temperature of 0.5 and 0.2 K. Both these computation gives no ODLRO. It is not known if these temperature are small enough to allow an extrapolation to the $T = 0$ K.
13. S. A. Vitiello *et al.*, *Phys. Rev. Lett.* **60**, 1970 (1988).
14. D. E. Galli *et al.*, *Phys. Rev. Lett.* **90**, 175301 (2003).
15. D. E. Galli *et al.*, *J. Low Temp. Phys.* **134**, 121 (2004).
16. M. Rossi *et al.*, *Phys. Rev.* **B72**, 64516 (2005).
17. S. Moroni *et al.*, *Phys. Rev.* **B58**, 909 (1998).
18. D. M. Ceperley, *Rev. Mod. Phys.* **67**, 279 (1995).
19. J. Day and J. Beamish, cond-mat/0601149.
20. F. Pederiva *et al.*, *Phys. Rev.* **B56**, 5909 (1997).
21. R. O. Simmons, *J. Phys. Chem. Solids* **55**, 895 (1994).
22. P. W. Anderson *et al.*, *Science* **310**, 1164 (2005).
23. J. A. Hodgon and F. H. Stillinger, *J. Stat. Phys.* **79**, 117 (1995).
24. W. L. McMillan, *Phys. Rev.* **138**, A442 (1965).
25. L. Reatto, *Phys. Rev.* **183**, 334 (1969).
26. C. Kittel, *Introduction to solid state physics* (Wiley, New York, 1976).
27. G. Stan *et al.*, *Phys. Rev.* **B62**, 2173 (2000).
28. K. Yamamoto *et al.*, *Phys. Rev. Lett.* **93**, 075302 (2004).
29. V. Apaja and E. Krostcheck, *Phys. Rev.* **B67**, 184304 (2003).
30. C. Lie-Zhao *et al.*, *Phys. Rev.* **B33**, 106 (1986).
31. M. Schindler *et al.*, *Phys. Rev.* **B53**, 11451 (1996).
32. J. V. Pearce *et al.*, *Phys. Rev. Lett.* **93**, 145303 (2004).
33. S. A. Khairallah and D.M. Ceperley, *Phys. Rev. Lett.* **95**, 185301 (2005).
34. A. Driessen *et al.*, *Phys. Rev.* **B33**, 3269 (1986).

CONTINUUM THEORY OF SUPERFLOW IN SUPERSOLID ^{4}He: A REVIEW

WAYNE M. SASLOW

Department of Physics, Texas A&M University
College Station, TX 77843-4242, USA
wsaslow@tamu.edu

The recent observation of non-classical rotational inertia in solid ^{4}He for temperatures $T < 200$ mK, and the continued non-observation of unusual flow associated with "zero-point defectons", favors the Leggett picture of Non-Classical Rotational Inertia over the Andreev-Lifshitz zero-point defecton picture of possible superflow in this system. We discuss the nature of the wavefunction, the Leggett one-body phase function upper bound for the superfluid fraction, and more complex forms for the phase-function and how this affects the net momentum density of this many-body system. The general properties of quantum vortices in such a system are briefly discussed.

Keywords: Superfluidity; Supersolid; Solid 4He.

1. Introduction

The present work discusses the microscopic continuum theory of supersolid ^{4}He. We argue that many-body contributions to the mass current that are unimportant for bulk superfluid become essential for a correct description of a supersolid.

Particle flow in a classical crystal, when classical atomic diffusion or hopping is neglected, occurs by lattice displacements. On the other hand, for an electron of mass m_e in the 2p, l=1 state of H, with wavefunction $\psi = \sqrt{\rho}\exp(i\phi)$, the number current $\vec{j} = \rho\vec{v}$ (with $\vec{v} = (\hbar/m_e)\vec{\nabla}\phi$) involves the product of the electron number density and the gradient of the quantum-mechanical (QM) phase of the wavefunction; we call this "phase flow". Mass flow in ordinary superfluids is due to phase flow. In fact, any particle is described by quantum mechanics, so the motion even of a (nearly classical) uranium atom can be described as phase flow. Thus, on including fundamental many-body phase-flow terms that I will later argue can be quantitatively important corrections to this one-body result, all particle motion (even for particles in a crystal) can be attributed to phase flow.

In 1969, Andreev and Lifshitz proposed that solid ^{4}He might display unusual particle flow if lattice defects (impurities, vacancies, interstitials, etc.) become mobile enough for quantum diffusion to dominate thermal diffusion, thus enabling one to think of these defects as being in mobile quantum-mechanical states called "defecton" states.[1] Specifically, they note the possibility of a new type of mass transport "by means of the motion of the zero-point defectons while the lattice sites remain essentially fixed... Thanks to this, a crystal is able to flow through a capillary in a gravitational field". (We previously called this flow of zero-point defectons "unusual lattice flow", although a more accurate phrase would be "zero-point defecton flow".)[a] Ref. 1 proposed that Bose-condensed zero-point defectons will support superfluid flow, and developed a two-fluid hydrodynamics for the system. They did not explicitly invoke the mechanism whereby the zero-point defectons, in a condensed state for which a wavefunction phase can be defined, could be the means for phase flow. However, they develop a superfluid hydrodynamics and from it note the possibility of quantum vortex lines, so they had a type of QM phase in mind. In 1970, Chester cited a result of Reatto on Bose-Einstein Condensation (BEC) for a wide class of Monte Carlo wavefunctions,[3] and noting an analogy between quantum and classical systems, the latter of which can condense into solids, proposed that solid ^{4}He might exist in a periodic state that also undergoes BEC. He also suggested that BEC might occur for ground state vacancies.[4]

Leggett shortly proposed that solid ^{4}He might display Non-Classical Rotational Inertia (NCRI), where under rotation the angular momentum would be less than for a classical system.[5] This occurs when in the rotating frame the lattice is at rest, but there is phase flow with the density fixed. Ref. 5 developed an expression for the flow energy E caused by rotation, and indicated how to obtain an upper limit for the superfluid fraction $f_s = \rho_s/\rho_0$ at temperature $T = 0$, where ρ_s is the superfluid mass density and ρ_0 is the total mass density. Estimates based on analogies to tunnelling in solid ^{3}He led to values of ρ_s/ρ_0 from 0.03%[5] to 0.0001%,[6] but actual calculations based on Ref. 5 led to values from 5% to 30%.[7,8]

Since these proposals there have been many searches for "supersolid" behavior.[9] To date, no experiment has found evidence for crystal flow around a macroscopic object or through a capillary.[10–12] However, using a torsion oscillator with a frequency of about 1000 s^{-1}, Kim and Chan recently have observed NCRI for solid ^{4}He in Vycor glass, with characteristic dimension 7 nm.[13] More recently they have found NCRI for solid ^{4}He both in porous gold, with characteristic dimension on the order of 500 nm;[14] and in bulk, with characteristic dimension on the order of mm.[15] In each case the zero temperature value for ρ_s/ρ_0 is on the order of 1-2%, and the supersolid-to-ordinary solid transition is around 0.2 K. The similar NCRI properties of these systems, with presumably very different defect densities, argues against defecton-based superfluidity. More recently, results confirming the observation of NCRI in bulk solid ^{4}He have been obtained using torsion oscillators of both higher and lower frequencies than 1000 s^{-1}.[16,17]

Recently two theoretical works, both based on the Monte Carlo method, have

[a]Note the apparent contradiction between having the lattice sites fixed yet having the crystal flow, since the lattice sites define the crystal. Such motion must correspond to a density that varies with time. The Gross-Pitaevskii-based work of Pomeau and Rica on a 2d model quantum solid, which studies steady-state flow around a disk, on the other hand finds that the lattice remains constant, rather than flowing.[2]

come down on opposite sides of the issue of whether or not solid ^{4}He can support superfluidity. On the one hand, Ceperley and Bernu,[18] after performing winding number calculations of the superfluid density that indicate superfluidity only for small systems, go on to study individual exchange processes to argue that solid ^{4}He cannot support bulk superflow in solid ^{4}He, and therefore that the results of Refs. 13–15 require another interpretation. On the other hand, Galli, Rossi, and Reatto find a wide range of values of the density where solid ^{4}He can support off-diagonal long-range order, which implies that phase is a well-defined quantity.[19] Given the unsettled status of the microscopic theory, we will assume that Refs. 13–15 indicate bulk superflow in solid ^{4}He, and work within that framework. For a more extended discussion, see the paper by Reatto in these Proceedings.[20]

2. General Arguments

Ref. 5 showed that rotation of an annulus leads, in the rotating frame, to an eigenvalue problem subject to the boundary condition that, on going around the annulus, the phase of an atomic wavefunction changes, in proportion to the rotation rate. Superflow – directed opposite the rotation – may occur if the zero-rotation wavefunction is Bose-condensed into a common periodic state, which then develops a phase whose gradient is, up to a constant, the superfluid velocity. More generally, NCRI shows up as an energy-sensitivity of the energy to the rotation-induced boundary condition. On the other hand, Ref. 5 also noted that the system can adjust to this new boundary condition in at least two other ways, both of which would defeat superflow:[21] (1) it can go to an orthogonal wavefunction with essentially the same energy as the zero-rotation state (e.g. scattering or decay, a violation of a generalized version of the Landau criterion); (2) its zero-rotation wavefunction can be based on individually localized states, each of which can adjust to the new boundary condition by developing its rotation-induced phase change in regions where its amplitude is essentially zero. (In this latter case, the wavefunction is, in a sense, topologically disconnected on trying to trace a path around the sample in the direction of rotation.) If neither of these possibilities occurs, then the following considerations apply.

Consider an annulus of radius R and radial thickness $d \ll R$, containing N identical particles of mass m. This system is periodic with period $L = 2\pi R$ in any tangential coordinate x_i, $i = 1, N$. In the laboratory frame the many-body wavefunction Ψ satisfies the boundary condition that $\Psi(x_i + L, t) = \Psi(x_i, t)$. If the walls move at velocity $v_0 = \omega R$, then (suppressing particle index i) the Hamiltonian is

$$H_0(x,t) = -\frac{\hbar^2}{2m}\frac{\partial^2}{\partial x^2} + V(x - v_0 t), \tag{1}$$

so that the potential due to the moving walls is time-dependent. The system satisfies the Schrodinger equation,

$$H_0\Psi = i\hbar\frac{\partial\Psi}{\partial t}. \tag{2}$$

In the moving frame, where $x' = x - v_0 t$ and $t' = t$, the potential is time-independent. That is

$$H_0(x,t) = H_0(x') = -\frac{\hbar^2}{2m}\frac{\partial^2}{\partial x'^2} + V(x'). \tag{3}$$

Now define

$$\Psi(x,t) = \exp(ikx)\Psi'(x,t), \tag{4}$$

corresponding to a boost by velocity $\hbar k/m$, with k not yet specified. Then $\Psi(x + L, t) = \exp(-ikL)\Psi'(x,t)$ is the boundary condition satisfied by Ψ'. Substituting for Ψ' in the Schrodinger equation and setting $k = mv_0/\hbar = m\omega R/\hbar$ gives

$$[H_0 + \frac{m}{2}(v_0)^2]\Psi' = i\hbar\frac{\partial\Psi'}{\partial t}, \tag{5}$$

where now H_0 has no explicit time-dependence and Ψ' is subject to the boundary condition that on translation by L in the transverse direction, there is a phase change $\Delta\phi = -kL = -mv_0L/\hbar = -m\omega(2\pi R^2)/\hbar$.

This can be turned into a static eigenvalue problem with eigenvalue

$$E^* = E_0(\omega) + \frac{m}{2}\omega^2 R^2. \tag{6}$$

If $E_0(\omega)$ depends upon ω, then the system is superfluid.

3. One-Body Phase

With Ψ_0 the ground state wavefunction, to obtain an upper bound on f_s Leggett assumes that in the rotating frame the wavefunction is given by

$$\Psi = \Psi_0 \exp(i\phi), \tag{7}$$

and that

$$\phi = \Sigma\phi_1(x_i); \tag{8}$$

a one-body phase. Then all particles i have the same superflow flow pattern, given by

$$v_s = (\hbar/m)\vec{\nabla}\phi_1, \tag{9}$$

and this is determined by minimizing the energy

$$E = (1/2)\int \rho(\vec{r})v_s^2(\vec{r})d\vec{r}, \tag{10}$$

subject to the rotation-induced boundary condition $\Delta\phi = -kL = -mv_0L/\hbar$; the negative sign means that the superflow is opposite to the rotation. The minimum value of E provides an upper bound on what Leggett has called the Non-Classical Rotational Inertia (NCRI).

Minimization of E with respect to ϕ yields $\vec{\nabla} \cdot \left(\rho(\vec{r})\vec{v}_s(\vec{r})\right) = 0$:[8] the steady-state continuity equation. From a knowledge of $\rho(\vec{r})$ and the average superfluid velocity $\vec{v}_s = \overline{\vec{v}_s(\vec{r})}$ (with $|\vec{v}_s| = v_0$) one can determine the full superfluid velocity profile $\vec{v}_s(\vec{r})$, and from that, E. The superfluid density ρ_s then follows from $E/V = (1/2)\rho_s v_s^2$, where V is the system volume. The average momentum density satisfies $\vec{g} = \overline{\rho(\vec{r})\vec{v}_s(\vec{r})} = \rho_s \vec{v}_s$. Physically, the minimization condition causes phase flow (from $\vec{v}_s(\vec{r})$) to occur preferentially in regions of low density. The more localized the $\rho(\vec{r})$, the more suppressed is ρ_s/ρ_0 from unity. Note that $\rho(\vec{r})$ is defined on a microscopic level, whereas the average mass density $\rho_0 = \overline{\rho(\vec{r})}$ and ρ_s are defined on the macroscopic level.

Ref. 8 computed ρ_s/ρ_0 as a function of b/a, for a $\rho(\vec{r})$ given as a sum over Gaussians of width b placed on the sites of a fictitious face-centered-cubic lattice of solid ^{4}He, with lattice constant a. In the experiments[13–15] the system was likely a hexagonal close-packed crystal or hcp polycrystal. Since the hcp and fcc lattices have the same local packing, use of an fcc lattice is a reasonable approximation.

For values of $b/a \leq 0.15$ the results of Ref. 8 were only an upper bound, because the computer then available did not have the capability to solve for the superflow pattern accurately.[8] Present-day computers permit convergent calculations. Figure 1 gives values for ρ_s/ρ_0 for the full range of b/a (although for $b/a < 0.10$ the results are only an upper bound). It gives the expected result that ρ_s/ρ_0 approaches unity for a delocalized system (large b/a); for $b/a = 0.30$, $\rho_s/\rho_0 \approx 0.9860$. It also gives the expected result that ρ_s/ρ_0 approaches zero for a localized system (small b/a); for $b/a = 0.10$, $\rho_s/\rho_0 \approx 0.0009$. For $0.10 < b/a \leq 0.20$, ρ_s/ρ_0 is very sensitive to b/a, as shown by the inset. Over 6000 plane waves were needed to obtain ρ_s/ρ_0 at the smaller values of b/a.[22] Although ρ_s is, in general, a tensor, it is diagonal for a cubic lattice; for the fcc lattice considered here, on convergence ρ_s was independent of the direction of superflow.

The exchange considerations of Ref. 6 give b/a near 0.125. The inset yields that $\rho_s/\rho_0 = 0.022$ for $b/a = 0.125$. Considering the sensitivity of ρ_s/ρ_0 to b/a, this is in good agreement with the experimental value $\rho_s/\rho \approx 0.015$. Because ρ_s/ρ_0 is so sensitive to b/a, it would be desirable to have a more accurate $\rho(\vec{r})$ from calculations on solid ^{4}He using wavefunctions possessing BEC.[23,24] Alternative ways exist to calculate ρ_s.[25,26]

4. Nearly Isotropic Superflow for the Hcp Phase of Solid ^{4}He

We have more recently performed calculations for a mass distribution that is a sum of gaussians on an hcp and a bcc lattice.[27] On the scale of the figure for f_s, the results for the hcp and fcc lattices are indistinguishable, for corresponding localization ratios (gaussian width divided by nearest-neighbor distance). Specifically, the hcp lattice has a superfluid mass density tensor that is isotropic to about a part in a

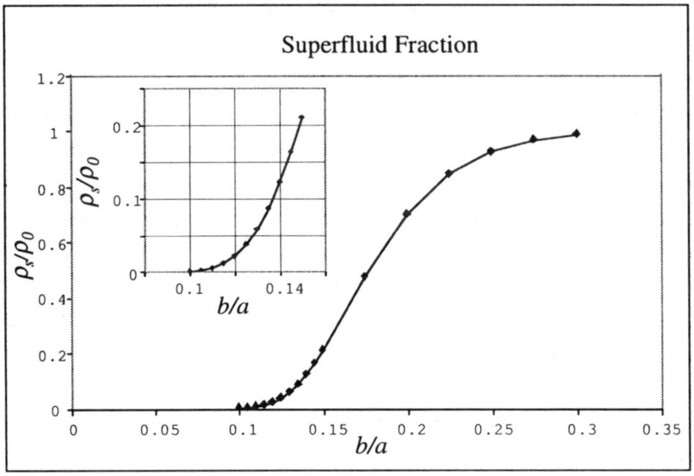

Fig. 1. Superfluid fraction ρ_s/ρ_0 *vs* b/a.

thousand, but it is definitely not completely isotropic, to that same accuracy.

From this we conclude that any spread in the experimental data on f_s likely is not due to the expected run-to-run variations in crystal orientation; f_s should be nearly isotropic.

The bcc lattice, which is more open than the close-packed fcc and hcp lattices, is not expected to be relevant because such a phase does not occur until temperatures on the order of 1K. Nevertheless we have calculated f_s for the bcc lattice.[27] As expected, for a given ratio of gaussian width to nearest-neighbor distance, for the bcc lattice we find values of f_s that are lower than for the fcc and hcp lattices.

5. Two-Body Phase

We have also recently performed calculations for Monte Carlo density profiles provided by Galli and Reatto.[28] For an average particle number density 0.029 Å^3, which is in the solid phase, we find f_s on the order of 0.20. This model of the solid, which is much better than a sum of gaussians, is too superfluid! This calculation, however, is only an upper limit.[5]

One possible resolution is to make the phase function ϕ more complex.[b] Thus consider a ϕ that depends on all coordinates. Let us expand it in the form[28]

$$\phi = \sum_{i}^{N} \phi_1(x_i) + \frac{1}{2(N-1)} \sum_{i \neq j}^{N} \phi_2(x_i, x_j) + \dots \tag{11}$$

[b]Another possibility is that the system adjusts in such a way that the internal structure of the wavefunction is changed, so that the effect of rotation is not described solely by employing a phase function. This means that the original ground state "scatters" into a new ground state.

The two body contributions to the phase represent the physical effect of backflow, i.e. correlated motion of atoms, and dots represent more complex backflow effects.[29] By symmetry, $\phi_2(x_i, x_j) = \phi_2(x_j, x_i)$. There is no constant term in either coordinate, since that would be part of ϕ_1.

In terms of ϕ, which must be symmetric in all coordinates, we have

$$\vec{v}_{s1}(x_1, \ldots) \equiv \frac{\hbar}{m}\vec{\nabla}_1\phi = \vec{v}_s(1) + \frac{1}{N-1}\sum_{i\neq 1}^{N}\vec{v}_s(1,i) + \ldots, \tag{12}$$

where $\vec{v}_s(1) \equiv \frac{\hbar}{m}\vec{\nabla}_1\phi_1(x_1)$ and $\vec{v}_s(1,i) \equiv \frac{\hbar}{m}\vec{\nabla}_1\phi_2(x_1, x_i)$. If the $N-1$ two-body velocities are to yield a net velocity that is comparable to those from the one-body velocity, then the two-body velocity should be on the order of the one-body velocity.

The results that follow require the following definitions: the one-body density

$$\rho_1(1) \equiv \rho_1(x_1) \equiv N\int |\Psi_0|^2 dx_2 \ldots dx_N, \tag{13}$$

where $\int dx_1\rho_1(x_1) = N$; the two-body density

$$\rho_2(1,2) \equiv \rho_2(x_1, x_2) \equiv N(N-1)\int |\Psi_0|^2 dx_3 \ldots dx_N, \tag{14}$$

where $\int \rho_2(x_1, x_2)dx_2 = (N-1)\rho_1(x_1)$; and the three-body density

$$\rho_3(1,2,3) \equiv \rho_3(x_1, x_2, x_3) \equiv N(N-1)(N-2)\int |\Psi_0|^2 dx_4 \ldots dx_N, \tag{15}$$

where $\int \rho_3(x_1, x_2, x_3)dx_3 = (N-2)\rho_2(x_1, x_2)$. Note that $\rho_n \sim (N/V)^n$.

In terms of these densities, and in the limit $N \to \infty$, if the one-body and two-body velocities are of the same order of magnitude, the flow energy is given by

$$\Delta E = \frac{m}{2}\int \rho_1(x_1)(\vec{v}_s(1))^2 dx_1 + \frac{m}{N}\int \rho_2(x_1, x_2)\vec{v}_s(1)\cdot\vec{v}_{s,2}dx_1 dx_2$$
$$+\frac{m}{2N^2}\int \rho_3(x_1, x_2, x_3)\vec{v}_{s,2}\cdot\vec{v}_{s,3}dx_1 dx_2 dx_3. \tag{16}$$

The second and the third terms of this equation represent the contribution of backflow to the flow energy having truncated backflow to the pair level.

The condition that E be a minimum relative to variations in $\phi_1(x_1)$ leads to

$$0 = \frac{1}{\hbar}\frac{\delta\Delta E}{\delta\phi_1} = \vec{\nabla}_1\cdot[\rho_1(x_1)\vec{v}_s(1) + \frac{1}{N}\int dx_2\rho_2(x_1, x_2)\vec{v}_s(1,2)] \equiv \vec{\nabla}_1\cdot\vec{j}_1, \tag{17}$$

thus defining the net one-body current density $\vec{j}_1$. The condition that E be a minimum relative to variations in $\phi_2(x_1, x_2)$ leads to (in the limit $N \to \infty$)

$$0 = \frac{1}{\hbar}\frac{\delta\Delta E}{\delta\phi_2} = \vec{\nabla}_1\cdot[\rho_2(x_1, x_2)\vec{v}_s(1) + \frac{1}{N}\int dx_3\rho_3(x_1, x_2, x_3)\vec{v}_s(1,3)] \equiv \vec{\nabla}_1\cdot\vec{j}_{1,2}, \tag{18}$$

thus defining the net two-body current density $\vec{j}_{1,2}$.

From a knowledge of ρ_1, ρ_2, and ρ_3, and the average flow velocity $\vec{v}_0 = \overline{\vec{v}_s(1)}$ as a source the, the above two conditions are a set of linear equations that determine the full velocity profiles $\vec{v}_s(1)$ and $\vec{v}_s(1,2)$.

6. Superfluid Density

To obtain an expression for the superfluid density, we rewrite (16) as

$$\Delta E = \frac{m}{2}\int dx_1 \vec{v}_s(1)\cdot\vec{j}(1) + \frac{m}{2N}\int dx_1 dx_2 \vec{v}_s(1,2)\cdot\vec{j}(1,2). \tag{19}$$

Because $\vec{v}_s(1)$ and $\vec{v}_s(1,2)$ involve $\vec{\nabla}_1$, in both of these terms we can integrate by parts, where we take the integration direction along the flow. Each term becomes a surface term and a product of a phase with a divergence. However, by (17) and (18), these divergences are zero. Moreover, the quantity $\vec{v}_s(1,2)$ is strictly periodic, so its phase is periodic, and therefore its two surface terms cancel. All that remains is the surface term associated with the change in phase of $\vec{v}_s(1)$ as a particle goes around the system. Defining Σ_1 this surface of area A and $\hat{v}_0$ the unit vector normal to Σ_1, i.e. the direction of the imposed velocity $\vec{v}_0$, then we have

$$\begin{aligned}\Delta E &= \frac{\hbar}{2}\int_{\Sigma_1} dA_1(\Delta\phi_0)\hat{v}_0\cdot[\rho_1(x_1)\vec{v}_s(1) + \frac{1}{N}\int dx_2\rho_2(x_1,x_2)\vec{v}_s(1,2)]\\ &= \frac{\hbar}{2}A(\Delta\phi_0)j_0 = \frac{m}{2}ALv_0j_0 \equiv \frac{m}{2}V\rho_s v_0^2,\end{aligned} \tag{20}$$

where $V = AL$ and $j_0 = \rho_s v_0$. There is no contribution to j_0 from the part of $\rho_2(x_1,x_2)$ that is independent of x_1 because the integral of x_1 is over a strictly periodic function of x_1. Hence, once we know the flow pattern in terms of v_0, we can compute j_0 and determine $\rho_s = j_0/v_0$.

So far the theory has been very general. Let us therefore apply it to a homogeneous bose liquid. Here the previous equations simplify drastically. Then the two-body density $\rho_2(x_1,x_2)$ is a function only of $|\vec{x}_1 - \vec{x}_2|$ and we expect the two-body phase $\phi_2(x_1,x_2)$ likewise to be a function only of $|\vec{x}_1 - \vec{x}_2|$. As a consequence, $\rho_2(x_1,x_2)$ is even in $\vec{x}_1 - \vec{x}_2$, but $\vec{v}_s(1,2)$, which involves $\vec{\nabla}_1|\vec{x}_1 - \vec{x}_2|$, is odd in $\vec{x}_1 - \vec{x}_2$. Hence the second term in (20) goes to zero. Therefore the two body terms do not enter explicitly in determining ρ_s. Moreover, in (17), the two-body terms similarly are zero, so that they do not even affect the one-body terms. Hence the two-body terms, i.e. backflow, are irrelevant to determining the one-body flow pattern and the total energy of a uniform flow of a translationally invariant system. Therefore we can neglect two-body terms in determining the superfluid density in a fluid, as is done conventionally. We believe this is also true of the higher-order terms when applied to a superfluid. The one-body phase alone then gives a $T = 0$ K superfluid fraction of unity. This answers (in the negative) a question posed by Leggett, of whether a uniform and time-reversal invariant system can have a superfluid fraction that is less than unity.[30]

Additional complexity in the phase function has appeared previously in the context of quantum fluids, both for excitations (where the approach is very close to that taken above),[31] and for vortex lines.[32,33]

7. Quantum Vortices

One striking difference between superflow in bulk liquid and bulk solid ^{4}He is that the critical velocity v_c is much lower for bulk solid ($\approx$ 5 μm/s, as opposed to $\approx$ 1 m/s).[15] In torsion oscillator experiments, during an oscillation the velocity amplitude ranges from zero to a maximum value v_m. If $v_m < v_c$, there is no degradation of the superflow, but if $v_m > v_c$, there is degradation during that part of the oscillation where $v > v_c$. The observed v_c is on the order of what one would obtain for a single unit of circulation.[15]

In vortex nucleation the major energy cost is to nucleate the core, where the local density tends to go to zero. In a uniform superfluid there are no preferred places for nucleation, but in a nonuniform superfluid the regions between atoms are preferred; this is because the local density is already small, so it costs the solid much less energy than the liquid to further reduce the local density to zero. Without calculating in detail the vortex flow velocity field, it is not possible to say how much less energy would be required to create a vortex line in supersolid ^{4}He relative to superfluid ^{4}He. Nevertheless, the following considerations should apply.

Far from the vortex core, the average velocity should define a local superflow pattern very like that used in obtaining Figure 1. Near the vortex core the flow pattern can be rather different. In an ordinary superfluid, the energy per unit length L of a vortex line at the center of a cylinder of radius R is given by $E/L = \pi\rho_s(\hbar/m_4)^2\ln(R/a)$, where a is a length on the order of the size of the vortex core. In the supersolid this same form will hold, with the reduced value of ρ_s. In addition, because the energy of nucleation (which involves the energy at small distances) is expected to be relatively small (independently of the smaller value of ρ_s), the value of a for the solid should be relatively large.

Since the energy to thermally excite a vortex line is proportional to ρ_s, which at low temperature is suppressed from ρ_0 by about a factor of fifty, and since v_c for thermal generation of a vortex line is proportional to its energy,[34] one would expect a suppression of v_c by about this factor. This is suggestive, but much too small to explain the low observed values of v_c.[15]

Because the superfluid density is so low, the vortices should be very deformable. Driven by the flow fields of other vortices, they will tend to move with their cores in the regions between the lattice sites. This will cause stretching and unstretching of the vortices, as the vortex cores locally move around the repulsive lattice potential.

8. Final Remarks

In closing we note a curious aspect of the *macroscopic* theory. If the normal fluid fraction f_n is due to thermal excitations, then $f_n = 0$ at $T = 0$. Moreover, as seen above, any theory of the supersolid is expected to give $f_s < 1$ at $T = 0$. Therefore a certain mass fraction is missing, and must somehow be attributed to a lattice mass fraction: $1 = f_n + f_s + f_L$. The dynamical consequences of this are not yet clear, although it is expected that at low frequencies Umklapp processes will lock the normal fluid and the lattice, giving an effective lattice mass fraction of $f_n^* = f_n + f_L$.[22]

Acknowledgments

I would like to acknowledge valuable conversations with E. Krotscheck, D. Galli, and L. Reatto. This work was supported in part by DOE grant No. DE-FG03-96ER45598.

References

1. A. F. Andreev and I. M. Lifshitz, *Sov. Phys. JETP* **29**, 1107 (1969).
2. Y. Pomeau and S. Rica, *Phys. Rev. Lett.* **72**, 2426 (1994).
3. L. Reatto, *Phys. Rev.* **A183**, 334 (1969).
4. G. V. Chester, *Phys. Rev.* **A2**, 256 (1970).
5. A. J. Leggett, *Phys. Rev. Lett.* **25**, 1543 (1970).
6. R. L. Guyer, *Phys. Rev. Lett.* **26**, 174 (1971).
7. J. F. Fernandez and M. Puma, *J. Low Temp. Phys.* **17**, 131 (1974).
8. W. M. Saslow, *Phys. Rev. Lett.* **36**, 1151 (1976).
9. For a review, see M. W. Meisel, *Physica* **B178**, 121 (1992).
10. A. Andreev, K. Keshishev, L. Mezkov-Deglin and A. Shal'nikov, *Sov. Phys. JETP Lett.* **9**, 306 (1969).
11. D. S. Greywall, *Phys. Rev.* **B16**, 1291 (1977).
12. J. Day, T. Herman, and J. Beamish, *Phys. Rev. Lett.* **95**, 035301 (2005).
13. E. Kim and M. H. W. Chan, *Nature* **427**, 225 (2004).
14. E. Kim and M. H. W. Chan, *J. Low Temp. Phys.* **138**, 859 (2005).
15. E. Kim and M. H. W. Chan, *Science* **305**, 1941 (2004).
16. K. Shirahama (private communication).
17. J. Reppy (private communication).
18. D. M. Ceperley and B. Bernu, *Phys. Rev. Lett.* **93**, 155303 (2004).
19. D. E. Galli, M. Rossi, and L. Reatto, *Phys. Rev.* **B71**, 140506 (2005).
20. L. Reatto, *Proceedings of QMBT13*.
21. C. N. Yang, *Rev. Mod. Phys.* **34**, 644 (1962); F. Bloch and H. E. Rorshach, *Phys. Rev.* **128**, 1697 (1962); W. Kohn, *Phys. Rev.* **133**, A171 (1964).
22. W. M. Saslow, *Phys. Rev.* **B71**, 092502 (2005).
23. D. N. Lowy and C.-W. Woo, *Phys. Rev.* **B13**, 3790 (1976).
24. D. Ceperley, G. Chester, and M. Kalos, *Phys. Rev.* **B17**, 1070 (1978).
25. E. L. Pollock and D. M. Ceperley, *Phys. Rev.* **B30**, 2555 (1984).
26. D. M. Ceperley, *Rev. Mod. Phys.* **67**, 279 (1995).
27. W. M. Saslow and S. Jolad, cond-mat/0511214, submitted to *Phys. Rev.* **B**.
28. W. M. Saslow, D. Galli, L. Reatto, in preparation.
29. In principle, ϕ can depend on the positions of all of the particles. Including only two-body correlations in ϕ causes three-body correlations to affect ϕ, and hence the superfluid density. See Ref. 28.
30. A. J. Leggett, *J. Stat. Phys.* **93**, 927 (1998).
31. E. Krotscheck and C. J. Tymczak. See C. J. Tymczak, Ph.D dissertation, *Excitations in Liquid ^{4}He Films,* Texas A&M University (1995).
32. S. A. Vitiello, L. Reatto, and M. H. Kalos, in *Condensed Matter Theories,* edited by V. C. Aguilera-Navarro Plenum, New York, 1989.
33. G. Ortiz and D. M. Ceperley, *Phys. Rev. Lett.* **75**, 4642 (1995).

34. J. S. Langer and M. E. Fisher, *Phys. Rev. Lett.* **19**, 560 (1967); S. V. Iordanskii, *Zh. Eksp. Teor. Fiz.* **48**, 708 (1965) [*Sov. Phys. JETP* **21**, 467 (1965)].

ANALYSIS OF AN INTERATOMIC POTENTIAL FOR THE CONDENSED PHASES OF HELIUM

SEBASTIAN UJEVIC[†] and S. A. VITIELLO[‡]

Instituto de Física, Universidade Estadual de Campinas
13083 Campinas–São Paulo, Brazil
[†]*sut@ifi.unicamp.br*, [‡]*vitiello@ifi.unicamp.br*

A recent interatomic potential, that includes two- and three-body interactions, is used to study the liquid and solid equations of state of ^{4}He and other properties of this system. The high-order contributions are explicitly computed by multi-weight diffusion Monte Carlo. It turns out that this is an excellent interatomic potential for the description of condensed phases of helium atoms systems.

Keywords: Liquid and solid helium; Interatomic potential; Equation of state.

1. Introduction

In theoretical investigations of systems of helium atoms where great accuracy is required efforts are still needed to better understand its interatomic potential, in particular its three-body contributions. This work analyzes some of the properties of a system of ^{4}He atoms using a recent interatomic potential $V(R)$ put together[1] to describe the He-He interactions. Its two-body part includes retardation effects in the dipole-dipole dispersion term[2] of the potential proposed by Korona and coworkers[3] that models the intermediate atomic interactions by applying infinite order symmetry adapted perturbation theory.

The three-body potential has a damped Axilrod-Teller Muto (ATM)[4,5] term and a exchange part. By its very nature, the ATM term is not valid at short-range distances, it needs to be damped. We have introduced this damping by a product of three Tang-Toennies functions.[6] As the exchange term we employ the Cohen and Murrell[7] form. Its amplitude and also the repulsive short-range parameter δ of the Tang-Toennies damping function were determined as described in Ref. 1 using experimental data.

The three-body exchange contribution to the interatomic potential is much weaker than the ATM triple-dipole interaction. It is from the intermediate and short-range distances that come its most significant contributions. They are important to explain many properties of systems of helium atoms.[7,8] The analytical expression of Cohen and Murrell was obtained by fitting *ab initio* results of helium trimmers in an isosceles triangle geometry. In the next section we give details of the simulations. The last section is devoted to our results and conclusions.

2. The Simulations

The simulations for the liquid and solid phases started from a *hcp* structure and were performed with 180 atoms in a cubic cell with periodic boundary conditions. The minimum imagine convention is adopted in all the simulations. For calculations of the three-body interactions, we slightly modify this convention[9] in order to correctly compute the size of the third side of the triangles. Tail corrections are considered for the two-body and damped ATM interactions for distances greater than half the side of the simulation cell.

To calculate the ground-state energies of the systems of ^{4}He atoms we use the diffusion Monte Carlo (DMC) method.[10,11] In this method an efficient exploration of the configuration space requires a guiding function for the diffusion process. In the liquid phase a guiding function of the Jastrow form is used: $\Psi_J(R) = \prod_{i<j} f(r_{ij})$, where $f(r_{ij}) = \exp[-u(r_{ij})/2]$. This function correlates pairs of particles with a pseudo-potential of the McMillan form, $u(r_{ij}) = (b/r_{ij})^5$, where b is a parameter. For the solid phase a Nosanov-Jastrow form function $\Psi_{NJ}(R) = \Psi_J(R)\Phi(R)$ is considered, where $\Phi(R) = \prod_i \exp\left[-\frac{C}{2}(\mathbf{r}_i - \mathbf{l}_i)^2\right]$ represents a mean field term that explicitly localize the particles around given lattice sites $\mathbf{l}_i$. The guiding function can explicitly include three-body correlations with the advantages of a fast convergence and small fluctuations, at a price of more evolved programming.

3. Results and Conclusions

The liquid and solid phases of helium atoms systems were investigated by performing independent runs at several densities. In Table 1 we present binding energy results for both phases together with experimental data.[12,13] We see from Table 1 that the results for the liquid and solid phases are in excellent agreement, within statistical fluctuations, with very reliable experimental data. A good description of the condensed phases of the systems of helium atoms requires the inclusion of the damped ATM and exchange terms in the interatomic potential. In Table 1 we also present the contributions of the three-body terms to the total energy of the system. These results are subject only to statistical uncertainties. They were obtained by performing a multi-weight diffusion Monte Carlo calculation as discussed in References 9 and 14. In the liquid phase the contribution of the damped ATM interaction is approximately 2-3 per cent of the modulus of the two-body binding energy. In the solid phase the contribution of the damped ATM term is about 3 per cent of the modulus of the two-body binding energy at the lowest density. It goes up to 13 per cent at the highest density.

We have built analytical equations of state (EOS) for the liquid and solid phases by fitting our binding energy results to a function given by $E(\rho) = E_0 + A\left[(\rho - \rho_0)/\rho_0\right]^2 + B\left[(\rho - \rho_0)/\rho_0\right]^3$, where E_0, A, B and ρ_0 are fitting parameters. In Fig. 1 we compare the computed analytical EOS for the liquid and solid phases with the experimental values of References 12 and 13. As we can see, an excellent agreement between theory and experiment is achieved simultaneously

Table 1. Binding (E), experimental (Exp), damped ATM (E^D) and exchange (E^J) energies for the liquid and solid phases computed at several densities (ρ).

ρ	E	$Exp.$	E^D	E^J	ρ	E	$Exp.$	E^D	E^J
		Liquid					**Solid**		
21.86	-7.171(8)	-7.170	0.121(1)	0.024(2)	30.11	-5.538(10)	-5.56	0.273(1)	-0.002(2)
23.20	-7.120(8)	-7.114	0.140(1)	0.028(2)	30.88	-5.265(10)	-5.28	0.291(1)	-0.006(2)
24.83	-6.897(9)	-6.893	0.167(1)	0.032(2)	32.55	-4.490(10)	-4.50	0.331(1)	-0.018(2)
25.36	-6.778(9)	-6.782	0.175(1)	0.034(2)	34.41	-3.298(10)	-3.32	0.381(1)	-0.041(2)

for the solid and liquid phases.

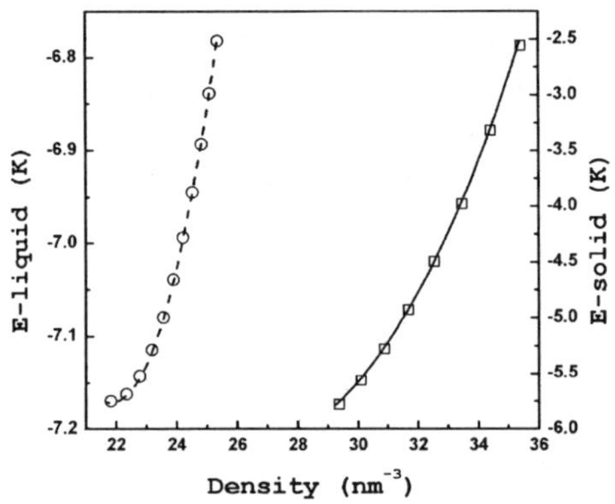

Fig. 1. Analytical equations of state for the liquid (left) and solid (right) phases. The circles and the squares stand for the experimental data of References 12 and 13, respectively.

In order to test the accuracy of our liquid and solid analytical EOS, we have calculated the melting and freezing densities using the Maxwell double tangent construction. We obtain 25.95(31) nm^{-3} as the freezing density and 28.93(31) nm^{-3} as the melting density. These values are in excellent agreement with the experimental data,[15,16] 25.970 and 28.568 nm^{-3}, respectively.

Table 2. Values of the equilibrium density (ρ_0), sound velocity (c) and isothermal compressibility (K) computed at the liquid phase.

	ρ_0 (nm^{-3})	c (m/s)	K (atm^{-1})
EOS	21.82(14)	235.06(80)	0.0126(6)
Exp-EOS	21.820(5)	236.37(29)	0.01250(9)

We have also used the liquid EOS to calculate the equilibrium density, the sound velocity, and the isothermal compressibility. The results are presented in Table 2, the last two quantities were evaluated at the computed equilibrium density ρ_0. At the same table we also display values of these quantities determined by using an analytical equation of state obtained by a fit of experimental data (Exp-EOS).

For both EOS the same equilibrium density is obtained. Moreover this value is in excellent agreement with the accepted value, 21.834 nm^{-3}.[15] The values of the sound velocity in Table 2 are in agreement. However it is well know that calculations of this quantity by an analytical EOS does not give satisfactory results. In fact, both values of the sound velocity do not agree with the experimental datum, 238.30 m/s.[17] The values of the isothermal compressibility in Table 2 are in excellent agreement with the experiment,[17] 0.0123 atm^{-1}.

In summary, our results have shown that the interatomic potential used in this work successfully describes some properties of a system of helium atoms in both the solid and liquid phases with great accuracy. The three-body contributions we have considered are essential to obtain in both phases agreement between theory and experiment. However, further work is still need to understand in more details the nature of the high-order interactions of this system.

Acknowledgments

This work was conducted, in part, using the facilities of the "Centro Nacional de Processamento de Alto Desempenho em São Paulo". It was partially funded by the "Fundação de Amparo à Pesquisa do Estado de São Paulo - FAPESP" and "Fundo de Apoio ao Ensino, à Pesquisa e à Extensão da Unicamp."

References

1. S. Ujevic and S. A. Vitiello, *Phys. Rev.* **B**, accepted for publication (2006).
2. A. R. Janzen and R. A. Aziz, *J. Chem. Phys.* **107**, 914 (1997).
3. T. Korona, H. L. Williams, R. Bukowski, B. Jeziorski, and K. Szalewicz, *J. Chem. Phys.* **106**, 5109 (1997).
4. B. M. Axilrod and E. Teller, *J. Chem. Phys.* **11**, 299 (1943).
5. Y. Muto, *Proc. Phys. Math. Soc. Jpn.* **17**, 629 (1943).
6. K. T. Tang and J. P. Toennies, *J. Chem. Phys.* **80**, 3726 (1984).
7. M. J. Cohen and J. N. Murrell, *Chem. Phys. Lett.* **260**, 371 (1996).
8. V. F. Lotrich and K. Szalewicz, *J. Chem. Phys.* **112**, 112 (2000).
9. S. Ujevic and S. A. Vitiello, *J. Chem. Phys.* **119**, 8482 (2003).
10. J. W. Moskowitz, K. E. Schmidt, M. A. Lee, and M. H. Kalos, *J. Chem. Phys.* **77**, 349 (1982).
11. P. J. Reynolds, D. M. Ceperley, B. J. Alder, and W. A. Lester, *J. Chem. Phys.* **77**, 5593 (1982).
12. R. de Bruyn Ouboter and C. N. Yang, *Physica.* **B144**, 127 (1987).
13. D. O. Edwards and R. C. Pandorf, *Phys. Rev.* **140A**, 816 (1965).
14. S. Ujevic and S. A. Vitiello, *Phys. Rev.* **B71**, 224518 (2005).
15. J. E. Berthold, H. N. Hanson, H. J. Maris, and G. M. Seidel, *Phys. Rev.* **B14**, 1902 (1976).
16. A. Driessen, E. van der Poll, and I. F. Silvera, *Phys. Rev.* **B33**, 3269 (1986).
17. B. M. Abraham *et al.*, *Phys. Rev.* **A1**, 250 (1970).

LIQUID ^{4}He ADSORBED FILMS ON VERY ATTRACTIVE SUBSTRATES

IGNACIO URRUTIA AND LESZEK SZYBISZ

Departamento de Física, Comisión Nacional de Energía Atómica, Av. Gral. Paz 1499 (RA-1650) San Martín, Buenos Aires, Argentina,
Departamento de Física, Facultad de Ciencias Exactas y Naturales, Universidad de Buenos Aires, Ciudad Universitaria Pab. 1 (RA-1428) Buenos Aires, and also at Consejo Nacional de Investigaciones Científicas y Técnicas, Av. Rivadavia 1917 (RA-1033) Buenos Aires.
iurrutia@cnea.gov.ar and szybisz@tandar.cnea.gov.ar

Adsorbed films of liquid ^{4}He are analized, in the framework of Density Functionals Theories (DF). In these systems, when the substrate becomes increasingly attractive, the thin films of ^{4}He approaches the quasi-bidimensional limit. We study this strongly attractive substrate regime with two DF, the Orsay-Trento (OT) and a recent Hybrid proposal (Hyb), focusing in the energy behavior. It is showed that OT does not reproduce the correct limiting energy curve, and it implies that this functional could not provide reliable results for very strongly attractive substrates like Graphite (Gr). In other hand, with the Hyb DF, the correct energy behavior is found for the adsorption energy of ^{4}He on Gr. These results show that OT should not be applied to quasi 2D (confinement) situations, and that Hyb DF provides a much more realistic description.

Keywords: Adsorbed films; liquid ^{4}He; Density Functional.

1. Introduction

The correct description of strongly confined systems is a central task in modern Density Functional theories for quantum and classical fluid systems,[1] and even in other more fundamental or *ab initio* approach as Hypernetted Chain Theory.[2] The quasi two dimensional confinement behavior may be reached througth the adsorption of a thin fluid film on a strongly attractive planar substrate. We are interested in the physical description of such an adsorbed thin films of ^{4}He. Until now, ^{4}He adsorbed on the surface of a strongly attractive subtsrate has not been succesfully traced from any DF theory.[2] However, precise results are known about the strong substrate limit from a variety of Monte Carlo (MC) techniques.[3–5] In this work we have studied adsorbed ^{4}He thin films on planar surfaces with two different DF theories. One of this is the most extended DF,[6] called OT DF. The other one, the Hyb DF[7,8] is a recent proposal due to the authors. The obtained results are compared with essentially exact results of MC.

2. Brief review of the basic equations

In a DF theory the ground-state energy per unit area A of N ^{4}He atoms embedded by $U_{\rm sub}(z)$ is

$$\frac{E_{\rm gs}}{A} = \frac{E_{\rm kin}+E_{\rm sc}+E_{\rm sub}}{A} = \frac{\hbar^2}{2m}\int_0^\infty dz \left(\frac{d}{dz}\sqrt{\rho(z)}\right)^2 + \int_0^\infty dz\,\rho(z)\,e_{\rm sc}(z) + \int_0^\infty dz\,\rho(z)\,U_{\rm sub}(z)\,, \tag{1}$$

where $e_{\rm sc}(z)$ is the self-correlation energy per particle, its expression depends on which version of DF is applied.[9] The Hartree's equation, obtained from the functional derivative of Eq. (1) is

$$\left[-\frac{\hbar^2}{2\,m}\frac{d^2}{dz^2} + V_H(z) + U_{\rm sub}(z)\right]\sqrt{\rho(z)} = \mu\,\sqrt{\rho(z)}\,, \tag{2}$$

where μ is the chemical potential and $V_H(z)$ the Hartree mean-field potential.[9] Solving this equation, we can find the $\rho(z)$ density distribution and μ. In practice, Eq. (2) is solved for a fixed coverage n_c defined as the number of particles N per unit area A

$$n_c = \frac{N}{A} = \int_0^\infty dz\,\rho(z)\,. \tag{3}$$

In this paper we shall focus the attention on the adsorption of ^{4}He films on strongly attractive substrates. The substrate potential adopted was CCZ[10] for Li and Mg substrates, and a (6-9) standard potential for Al and Gr.[11]

3. 2D energy limit

In the literature one may often find that, either for classical fluids[1] or for quantum fluids,[2] the behavior in the confined 2D limit is determined by assuming

$$\rho(z) = \rho_{\rm 2D}\,\delta(z)\,, \tag{4}$$

and introducing this form into the $E_{\rm sc}$ term of Eq. (1).

However, we propose that such a limit should be studied by shrinking the liquid by means of very strongly attractive external potentials and looking for the solution of the Hartree's equation for $\rho(z)$ given by (2).

In the limit $n_c = N/A \to 0$ the self-correlation energy $E_{\rm sc}$ goes to zero, and the total ground-state energy per particle becomes the "binding energy" of a single ^{4}He atom on the solid substrate

$$\begin{aligned} e_{\rm sa} &= \frac{E_{\rm gs}(n_c\to 0)}{N} = \frac{E_{\rm kin}(n_c\to 0)}{N} + \frac{E_{\rm sub}(n_c\to 0)}{N} \\ &= e_{\rm kin}(n_c\to 0) + e_{\rm sub}(n_c\to 0)\,. \end{aligned} \tag{5}$$

When by increasing n_c the first layer is formed on a strongly attractive surface, the sum of the kinetic and substrate contributions remains constant

$$e_{\rm sa} = e_{\rm kin}(n_c) + e_{\rm sub}(n_c)\,. \tag{6}$$

This feature is due to the fact that the profile of the films increases without changing its shape. Hence, in the case of a thin helium film the self-correlation energy per particle can be evaluated as a function of coverage in the following way

$$\varepsilon_{sc} = e_{gs} - (e_{kin} + e_{sub}) = e_{gs} - e_{sa} \ . \qquad (7)$$

4. Results

We perform a systematic study of ε_{sc} with OT DF, in the regime of strongly attractive substrates. Figure 1 indicates how the limiting behavior of the energy is reached when increasingly attractive substrates are analyzed. Results for Li, Mg, Al, and a family of hypothetic substrates (which does not correspond to any substance) with (9-6) potentials of depth: D=160K, 640K, 2560K, are shown. Clearly, at more deeper potential substrate, the energy plot becomes closer to the limiting OT results. The limiting behavior of the OT energy, does not reproduce the essentially exact result obtained with Green Functional Montecarlo[3] (GFMC) for the 2D ^{4}He.

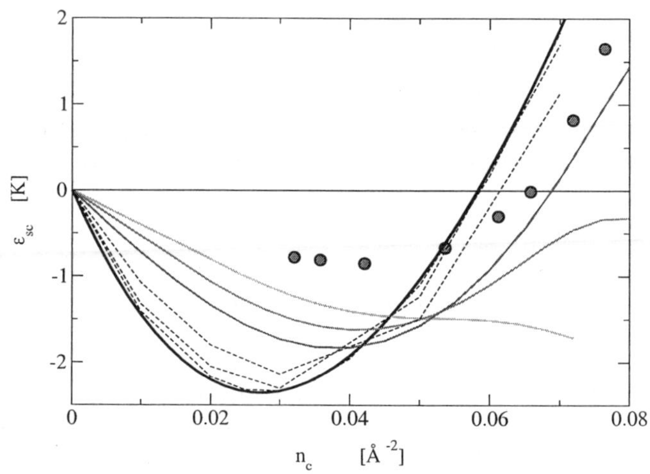

Fig. 1. Energy of adsorbed ^{4}He film with OT. Approach of ε_{sc} to the limiting very attractive substrate regime. From top to bottom, the three continuous lines show results of adsorbed ^{4}He in Li, Mg, Al, and with dashed lines are ploted the hypothetical substrates with a depth of D=160K, 640K, 2560K. In continuous bold line is ploted limiting curve energy obtained with OT DF. Circles correspond to the GFMC results for that limit.

In Fig. 2 in continuous trace it is ploted the energy results for the adsorption of ^{4}He in Gr with Hyb DF. Results on the same substrate from GFMC are showed with squares.[4] For comparison, on dashed line can be viewed the OT limiting curve. May be noticed that Gr substrate is more stronger attractive than Al (see Fig. 1).

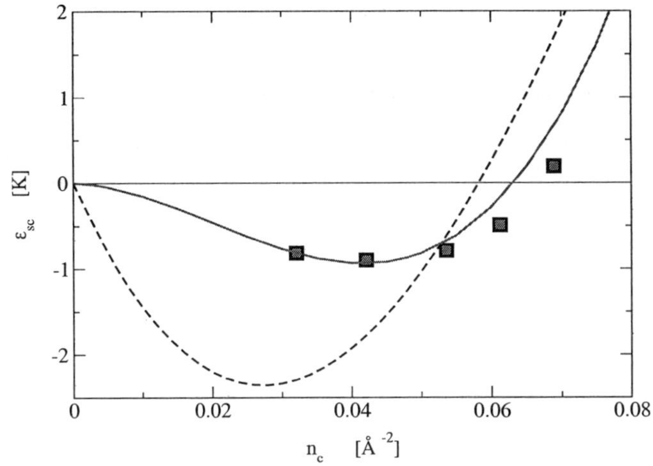

Fig. 2. Continuous line plots the Hyb DF results for the ^{4}He adsorbed in Gr, while squares show GFMC calculations for the same substrate. Dashed line are the results of OT for the limiting behavior.

5. Conclusion

The problem of the correct description of a ^{4}He liquid strongly confined by a substrate potential to a nearly 2D geometry was addressed in the framework of DF theories. Using OT, it was determined that the energy obtained thougth the calculations of the adsorption on strongly attractive substrates does not reproduce the correct limiting behavior. And thus, this DF is not appropiate for treating of high confinement ^{4}He situations, like the adsorption on Gr substrates. On the contrary, using Hyb, the adsorption energy on Gr reproduces the GFMC results for the same substrate. It is the first time that such a nearly 2D confinement of ^{4}He system is correctly reproduced by a DF.

Acknowledgements

This work was supported in part by the Ministry of Culture and Education of Argentina through Grants ANPCyT PICT No. 03-08450, CONICET PIP No. 5138/05 and UBACyT No. X298.

References

1. Y. Rosenfeld, M. Schmidt, H. Löwen, and P. Tarazona, *Phys. Rev.* **E55**, 4245 (1997).
2. B. E. Clements, H. Forbert, E. Krotscheck, and M. Saarela, *J. Low Temp. Phys.* **95**, 849 (1994).
3. P. A. Whitlock, G. V. Chester, and M. H. Kalos, *Phys. Rev.* **B38**, 2418 (1988).
4. P. A. Whitlock, G. V. Chester, and B. Krishnamachari, *Phys. Rev.* **B58**, 8704 (1998).
5. M. C. Gordillo, D. M. Ceperley, *Phys. Rev.* **B58**, 6447 (1998).

6. F. Dalfovo, A. Lastri, L. Pricaupenko, S. Stringari, J. Treiner, *Phys. Rev.* **B52**, 1193 (1995).
7. L. Szybisz, I. Urrutia, *J. Low Temp. Phys.* **138**, 337 (2005).
8. L. Szybisz, I. Urrutia, *Phys. Lett.* **A338**, 155 (2005).
9. L. Szybisz, *Eur. Phys. J.* **B14**, 733 (2000).
10. A. Chizmeshya, M. W. Cole, and E. Zaremba, *J. Low Temp. Phys.* **110**, 677 (1998).
11. G. Vidali, G. Ihm, H. Y. Kim, and M. W. Cole, *Surf. Sci. Repts.* **12**, 133 (1991).

QUANTUM MONTE CARLO STUDIES OF MANY–BODY SYSTEMS AND QUANTUM COMPUTATION

MONTE CARLO SIMULATION OF BOSON LATTICES

VESA APAJA

Institut für Theoretische Physik, Johannes-Kepler Universität, A-4040 Linz, Austria

OLAV F. SYLJUÅSEN

Nordita, Blegdamsvej 17, DK-2100, Copenhagen Ø, Denmark

Boson lattices are theoretically well described by the Hubbard model. The basic model and its variants can be effectively simulated using Monte Carlo techniques. We describe two newly developed approaches, the Stochastic Series Expansion (SSE) with directed loop updates and continuous–time Diffusion Monte Carlo (CTDMC). SSE is a formulation of the finite temperature partition function as a stochastic sampling over product terms. Directed loops is a general framework to implement this stochastic sampling in a non–local fashion while maintaining detailed balance. CTDMC is well suited to finding exact ground–state properties, applicable to any lattice model not suffering from the sign problem; for a lattice model the evolution of the wave function can be performed in continuous time without any time discretization error. Both the directed loop algorithm and the CTDMC are important recent advances in development of computational methods. Here we present results for a Hubbard model for anti–ferromagnetic spin–1 bosons in one dimensions, and show evidence for a dimerized ground state in the lowest Mott lobe.

Keywords: Optical lattices; Antiferromagnetic Boson systems; Dimerization

1. Introduction

An optical lattice can be made by applying orthogonal laser beams to an ultracold gas of atoms. As a result ^{87}Rb or ^{23}Na atoms can be trapped to form a perfect lattice. These atoms have a hyperfine spin 1, with either a ferromagnetic (^{87}Rb) or antiferromagnetic (^{23}Na) interaction. Unpolarized ^{23}Na atoms have spin–correlated Mott insulating states,[1] and it has been suggested[2] that the ground state is a dimer phase in one, two and three dimensions. An effective Hamiltonian of spin–1 bosons in an optical lattice has the Bose–Hubbard form, supplemented with a term that describes the spin interaction of atoms on the same lattice site,[3]

$$H = -t \sum_{\langle ij\rangle,\alpha} \left(a^\dagger_{i\alpha} a_{j\alpha} + cc. \right) - \mu \sum_i \hat{n}_i + \frac{U}{2} \sum_i \hat{n}_i(\hat{n}_i - 1) + \frac{U_2}{2} \sum_i \left(\vec{S}_i^2 - 2\hat{n}_i \right), \quad (1)$$

where the spin components $\alpha = x, y, z$ form a basis where the spin operator at site i is written in terms of boson operators $a^\dagger_\alpha$ as $S_i^\alpha = -i\epsilon^{\alpha\beta\gamma} a^\dagger_\beta a_\gamma$. ($\epsilon^{\alpha\beta\gamma}$ is the totally antisymmetric Levi-Civita tensor). In this basis the antiferromagnetic spin interaction has no sign problem. The value of U_2/U can be determined from experimental scattering lengths.

SSE[4] is an exact scheme, where the quantum partition function is expanded as a power series in a given basis $\{|s\rangle\}$,

$$Z = \mathrm{Tr}\left\{e^{-\beta H}\right\} = \sum_s \sum_{n=0}^{\infty} \frac{-\beta^n}{n!} \langle s|H^n|s\rangle \ , \tag{2}$$

where $\beta = 1/(k_B T)$ is the inverse temperature. Typically the low–temperature phases we are simulating have at most 1–3 atoms per site, so even with the spin degrees of freedom the required number of single–site states is rather small. The Hamiltonian (1) couples at most two neighboring sites, so SSE splits the partition function to a sum of products of bond and site operations.

To generate the terms in SSE we employ directed loop updates.[5] The basic idea is to pick a few well chosen elementary update operations, that change the bond or site operators to each other. Each update affects only a single site (and operator) at a time and is applied in a loop among the operators in the product that is the current term of the SSE sum. The outcome is a new product of operators, and, thanks to the looping, this new product is also a term in the partition function. In addition one controls the number of operators in the product by adding or removing diagonal operators; they don't change the state so their addition or removal won't disrupt the continuity of the sequence. The rules for how the loop travels follow from the detailed balance condition, which is used to ascertain that the operators appear in the products with proper weights.

If the Hilbert space is finite one can formulate Diffusion Monte Carlo (DMC) in continuous time.[6,7] DMC is a stochastic simulation of the imaginary-time evolution operator $e^{-H\tau}$. For an infinitesimal time step $d\tau$ this operator describes one out of three possible actions: 1) No action, the current state is unchanged, 2) Transition to a new state, and 3) The weight of the current state changes. Action 3) is needed due to the in general non-Markovian nature of the evolution operator. The probabilities for the different actions $p(i)$, $i = 1, 2, 3$ can be read directly off the Hamiltonian. The probability $p(2)$ is determined by off-diagonal elements of the Hamiltonian and is therefore of order $d\tau$. Also $p(3)$ is of order $d\tau$ because it describes the deviation from Markovian evolution, thus only $p(1)$ is of order unity. It follows that the DMC can be treated as a continuous–time simulation of a radioactive (multi-channel) decay problem, where decay–times are generated according to the exponential distribution: $\tau_{\mathrm{decay}} = -\ln(r)d\tau/(1 - p(1))$ where r is a random number uniformly distributed between 0 and 1. Having generated a decay–time the system is moved directly to the time of decay and the appropriate decay process is chosen dependent on the ratio $p(2)/p(3)$. CTDMC can be combined with other standard improvements of DMC such as importance sampling, reweighting and forward-walking.[8]

2. Results

Fig. 1 shows the phase boundaries of the two lowest Mott insulating phases in the cases $U_2/U = 0.2$ and $U_2/U = 0.4$. Scattering between spin states stabilizes the second lobe, while the first lobe is slightly reduced in size.

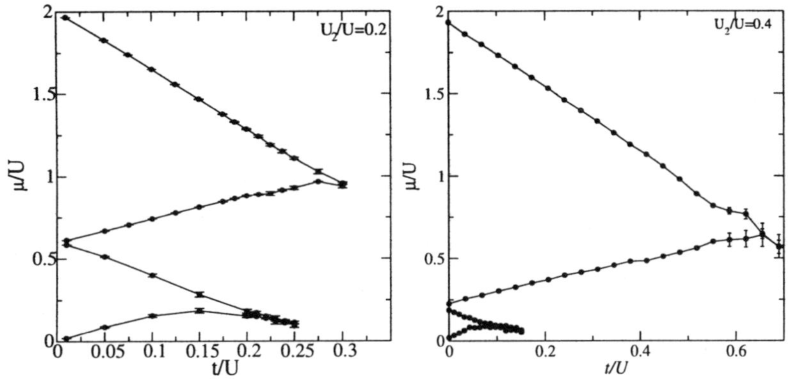

Fig. 1. The phase boundaries between Mott insulating and superfluid regions of a 1D chain for $U_2/U = 0.2$ (left panel) and $U_2/U = 0.4$ (right panel) at zero temperature. The results were obtained using CTDMC to compute ground state energies $E_L(N)$ for different particle numbers N and system sizes L. The upper(lower) boundary μ_{p+} (μ_{p-}) of the Mott lobe with p bosons per site was obtained from the finite size values for $\mu_{p\pm} = \pm(E_L(pL \pm 1) - E_L(pL))$ extrapolated to $L \to \infty$ using $L = 8, 16, 20$ and 24.

To study the structure of the Mott insulating phase with one atom per site we have measured the bond hopping or dimer susceptibility

$$\mathcal{X}(q) = \frac{1}{t^2 \beta N}(\langle N_q N_{-q} \rangle - \langle N_q \rangle \langle N_{-q} \rangle) , \tag{3}$$

where N_q is the Fourier transform of N_i, the number of hopping operators on bond i. Because of the on–site spin scattering, the hopping operators tend to couple pairwise adjacent sites, indicated by a peak at wave number $q = \pi$. For a dimer state in an infinite system this peak height should diverge, and we demonstrate this in Fig. 2. For larger than 32 sites the computation of the susceptibility becomes exceedingly slow, as autocorrelation times increase rapidly. This is due to our non-optimal treatment of the terms of type $a_x^\dagger a_x^\dagger a_y a_y$ which changes two spin indices on the same site simultaneously. In order to build such terms out of loop changes where only one spin index changes we have included intermediate, auxiliary terms $a_x^\dagger a_y$ in the Hamiltonian. These auxiliary terms are generally present when the loop is being built, but we do not allow a loop to close if any of them are present.

The left panel of Fig. 3 show how the dimer state is suppressed as one moves from the insulating to the superfluid state. The hopping parameter t is kept fixed and we move out of the insulating phase by increasing μ. The corresponding density, as it increases from unity, is shown in the right panel of Fig. 3.

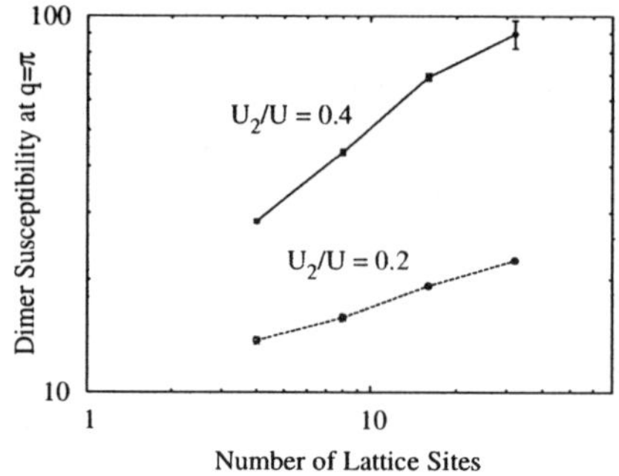

Fig. 2. Finite size scaling of the dimer susceptibility $\mathcal{X}(q)$ at $q = \pi$ in the Mott insulating phase of a 1D chain at $\beta = 150$ for two values of U_2/U indicated in the figure. We show the results for $L = 4$, 8, 16 and 32 sites, those for $U_2/U = 0.4$ were computed at $t = 0.1$, $\mu = 0.1$, and those for $U_2/U = 0.2$ at $t = 0.15$, $\mu = 0.25$.

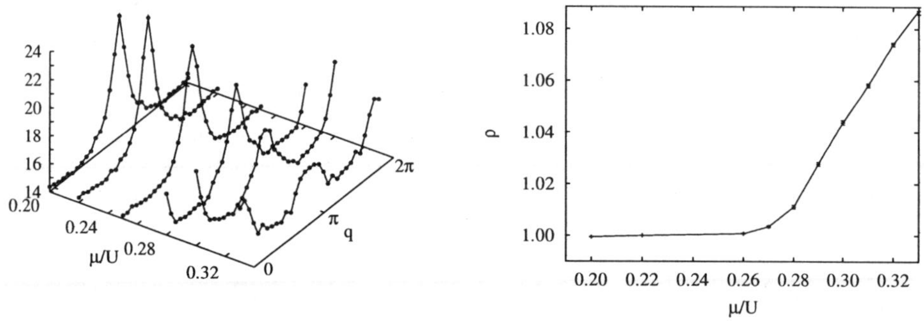

Fig. 3. The left panel shows the dimer susceptibility $\mathcal{X}(q)$ at the upper edge of the first Mott lobe (see the left panel of Fig. 1) of a 1D chain with 32 sites. The susceptibility is plotted as a function of the chemical potential μ/U and the wave number q. The data was computed at $t = 0.15$, $\beta = 150$ and $U_2/U = 0.2$. The statistical error is less than 0.2 in the vertical scale. The right panel shows the corresponding density.

References

1. E. Demler and F. Zhou, *Phys. Rev. Lett.* **88**, 163001 (2002).
2. S. K. Yip, *Phys. Rev. Lett.* **90**, 250402 (2003).
3. A. Imambekov, M. Lukin, and E. Demler, *Phys. Rev. Lett.* **93**, 120405 (2004).
4. A. W. Sandvik and J. Kurkijärvi, *Phys. Rev.* **B43**, 5950 (1991).
5. O. F. Syljuåsen and A. W. Sandvik, *Phys. Rev.* **E66**, 046701 (2002).
6. E. Farhi and S. Gutmann, *Ann. Phys.* **213**, 182 (1992).
7. O. F. Syljuåsen, *Phys. Rev.* **B71**, 020401(R) (2005).
8. M. P. Nightingale and C. J. Umrigar, in *Advances in Chemical Physics* **105**, eds. D. M. Ferguson *et al.* (John Wiley, NY, 1998), chapter 4.

THERMAL ENTANGLEMENT IN SPIN SYSTEMS

N. CANOSA[†] and R. ROSSIGNOLI[‡]

Departamento de Física, Facultad de Ciencias Exactas, Universidad Nacional de La Plata, C.C.67, La Plata (1900), Argentina

canosa@fisica.unlp.edu.ar[†], *rossigno@fisica.unlp.edu.ar*[‡]

We examine the entanglement of thermal states of n spins interacting through an XYZ type Heisenberg coupling in the presence of a uniform magnetic field, by evaluating the negativities of bipartite partitions of the whole system and subsystems. The corresponding limit temperatures for entanglement are also examined. Results indicate that limit temperatures for global entanglement depend on the type of partition and are higher than those limiting pairwise entanglement, and that their behavior with anisotropy and applied magnetic field may differ significantly from that of the corresponding mean field critical temperature.

Keywords: Quantum Entanglement; Spin Chains, Finite Temperature

1. Introduction

In the last few years, the concept of quantum entanglement has shed new light on the study and analysis of correlations in many-body quantum systems. Entanglement refers to the non-local correlations without classical analogue that can be exhibited by composite quantum systems, constituting one of the most fundamental and intriguing features of quantum mechanics. Recognized already by Schrödinger,[1] interest on entanglement has been triggered in recent years by its potential for enabling totally new forms of information transmission and processing.[2] It plays a key role in the field of quantum information,[3] where it is considered a resource.

A mixed state ρ of a two component quantum system $A+B$ (assumed distinguishable) is said to be *entangled* or *inseparable* if it cannot be written as a convex combination of product densities, i.e., if a representation of the form $\rho = \sum_\alpha q_\alpha \rho_A^\alpha \otimes \rho_B^\alpha$, with $q_\alpha \geq 0$, $\sum_\alpha q_\alpha = 1$ and $\rho_A^\alpha, \rho_B^\alpha$ density matrices for each component, is not feasible.[4] When such an expansion exists, ρ is termed *separable* or *classically correlated*, since it can be constructed by local operations and classical communication.[3,4] Separable states fulfill, therefore, Bell inequalities and other classical properties[5,6] which can be violated by entangled states. Pure states ($\rho^2 = \rho$) are separable if and only if they are product states ($\rho = \rho_A \otimes \rho_B$), but in the case of mixed states the determination of separability is in general a highly non-trivial problem,[7] in the sense that no easily computable necessary and sufficient separability condition exists, except for simple systems such as a pair of qubits (systems with Hilbert space dimension 2). Accordingly, while bipartite pure state entanglement can be measured through the entropy of a subsystem,[8] no rigorous computable entanglement measure is available for mixed states, except for two qubits.[9] Nonetheless, several simple necessary separability conditions exist, in particular the Peres-Horodecki criterion of positive partial transpose (PPT),[10,11] which is sufficient for two-qubit or qubit+qutrit systems. Moreover, the ***negativity***,[12] which is a measure of the degree of violation of the PPT criterion, provides an approximate entanglement measure for mixed states, being an entanglement monotone.[12] It can be also proved[13] that any state is separable if it is sufficiently close to the fully mixed state I_d/d (with d the full system dimension and I_d the ensuing identity).

For many component systems, the characterization of entanglement is more complex.[14] The are several possible partitions of the whole system and of subsystems, which may or may not be entangled. Entanglement of global partitions does not imply, however, entanglement between corresponding subsystems. Moreover, separability of all global bipartitions does not necessarily entail full separability, which occurs when ρ can be expressed as a convex combination of product densities for each component. Nevertheless, any state becomes again fully separable if its deviation from the full random state is sufficiently small.[13]

Thermal entanglement[15–19] refers to the entanglement of mixed states of the form $\rho(T) \propto \exp[-H/T]$, with $T > 0$ the temperature and H the system Hamiltonian, which are, for instance, the natural initial states in NMR based realizations of quantum computers.[3] The previous statement implies that for any finite system, no matter how small, there always exists an upper *limit temperature* for entanglement T_L above which $\rho(T)$ becomes fully separable, i.e., *classically* correlated. Similarly, lower limit temperatures exist for the entanglement of any bipartition of the full system or subsystem. An important question that immediately arises is what is the nature and behavior of these limit temperatures in interacting systems and how do they compare with the corresponding mean field critical temperature.[18,19] Another interesting feature is the possible emergence of entanglement at *finite* temperature when the ground state is separable, due to excited entangled states.[15,18] In this work we will discuss some of these issues in a quantum spin chain with a Heisenberg XYZ type coupling placed in a uniform magnetic field, a system useful for modelling different physical realizations of quantum gates, investigating in particular the limits for entanglement of global bipartitions.[19] The latter will be shown to be higher than those of the more frequently studied (and easier to evaluate) pairwise entanglement between two spins.[15–17]

2. Calculations and results

We will consider n qubits or spins described by a Hamiltonian of the form

$$H = bS_z - \sum_{i<j}(v_x^{ij} s_x^i s_x^j + v_y^{ij} s_y^i s_y^j + v_z^{ij} s_z^i s_z^j)\,, \tag{1}$$

where $s^i_{\alpha=x,y,z}$ is the spin component at site i, $S_z = \sum_i s^i_z$ the total spin z-component and b accounts for the Zeeman coupling to a uniform magnetic field. We will here examine a one dimensional cyclic chain with nearest neighbor coupling ($v^{ij}_\alpha = v_\alpha \delta_{i+1,j}$, with $n+1 \equiv 1$). The global thermal entanglement between p and $m = n - p$ selected qubits will be analyzed through the associated negativity[12]

$$N_{p-m}[\rho] = \tfrac{1}{2}(\mathrm{Tr}|\rho^{t_P}| - 1)\,, \tag{2}$$

where t_p denotes the corresponding partial transposition. Eq. (2) is just the absolute value of the sum of the negative eigenvalues of ρ^{t_P}, vanishing for separable ρ. Similarly, $N_{p-m}[\rho_k]$ will measure the entanglement between p and $m = k - p$ selected qubits of a subsystem of $k < n$ qubits, described by the reduced density matrix $\rho_k = \mathrm{Tr}_{n-k}\,\rho$. The hierarchy[12] $N_{A-B} \geq N_{A-C_B}$ holds if C_B is a subsystem of B, implying the relation $T_{A-B} \geq T_{A-C_B}$ between the ensuing limit temperatures.[19]

Figure 1 (left) depicts the thermal negativities N_{1-m} between one qubit and the rest for a chain with $n = 8$ qubits and $v_y = v_x$, $v_z = 0$. The maximum attainable value[19] for N_{1-m} is 1/2. For $b = 0$, the ground state is entangled and all depicted negativities decrease monotonously as T increases, vanishing at *different* temperatures. The global negativity N_{1-7} is the largest (reaching saturation at $T = 0$) and most persistent, in agreement with the previous hierarchy. Thus, there is a temperature interval $[T_{1-6} - T_{1-7})$ where the qubit is still entangled with the rest of the chain, but not with any adjacent subset of $m \leq 6$ contiguous qubits, and a larger interval $[T_{1-1} - T_{1-7})$ where entanglement between adjacent qubits is absent even though the qubit remains entangled with larger subsets. The most persistent negativity in this case is actually that of the partition in 4+4 *non-contiguous* qubits.[19]

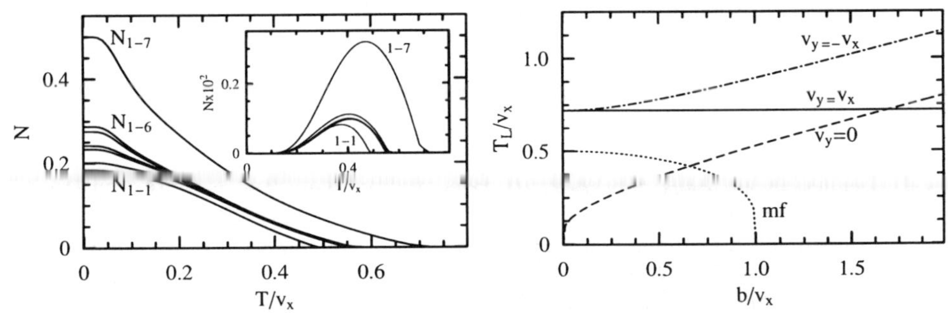

Fig. 1. Left: Thermal behavior of negativities N_{1-m} in a $n = 8$ spin chain with $v_y = v_x$ at zero field ($b = 0$), measuring the entanglement between one and m spins in the whole system ($m = 7$) and in subsystems of $m+1$ contiguous spins ($m = 1, \ldots, 6$). N_{1-1} corresponds to pairwise entanglement. The inset depicts the same quantities for $b = 1.5v_x$, in which case the ground state is fully aligned (separable). Right: Limit temperatures for nonzero negativity between one qubit and the rest (N_{1-7}) for different anisotropies v_y/v_x, as a function of the magnetic field. It is strictly constant in the XX case $v_y = v_x$. The mean field critical temperature (mf) is also depicted.

On the other hand, for $b = 1.5v_x$ the ground state is separable (the aligned state with $|S_z| = n/2$) but entanglement *arises for* $T > 0$, exhibiting a non-monotonous behavior (inset). A remarkable feature is that the limit temperature for all global negativities, including N_{1-7}, can be shown to be strictly *independent* of the uniform field b in all XX or XXZ type models,[19] as seen in the right panel. This entails the emergence of global entanglement for $T > 0$ even when the ground state is aligned (large $|b|$). Limit temperatures T_{1-m} of depicted subsystem negativities ($m \leq 6$) are nearly (but not exactly) independent of b as well, although this does not hold for other non-contiguous subsystem partitions.[19] In contrast, in XYZ type models ($v_x \neq v_y$) limit temperatures for global negativities *depend on* b, *increasing* for large fields,[19] as seen in the right panel for $v_z = 0$. Although large fields favor alignment, T_L/v_x increases due to the largest average energy spacing. Note that the behavior of global limit temperatures is quite different from that of the mean field critical temperature (limiting the symmetry-breaking solution), which always vanishes for large fields[19] ($b > v_x$ in Fig. 1). Other aspects of the thermal entanglement of these type of models, including fully-connected systems, can be found in Ref. 19.

Entanglement provides thus a deeper understanding of quantum correlations in interacting many-body systems. The present analysis shows that as the temperature increases, quantum correlations are lost through a cascade of limit temperatures associated with the vanishing of different types of entanglement, with pairwise and subsystem entanglement vanishing before global entanglement. Small finite systems regain in this sense a critical-like behavior in relation with quantum correlations, becoming classically correlated (although not uncorrelated) above T_L. The present results also indicate that the behavior of these limit temperatures may differ substantially from that of the corresponding mean field critical temperature.

Acknowledgments

The authors acknowledge support from CONICET (NC) and CIC (RR) of Argentina.

References

1. E. Schrödinger, *Naturwissenschaften* **23**, 807 (1935).
2. C.H. Bennett et al., *Phys. Rev. Lett.* **70**, 1895 (1993); *Phys. Rev. Lett.* **76**, 722 (1996); C.H. Bennett and D.P. DiVincenzo, *Nature* (London) **404**, 247 (2000).
3. M.A. Nielsen and I. Chuang, *Quantum Computation and Quantum Information* (Cambridge Univ. Press, UK, 2000).
4. R.F. Werner, *Phys. Rev.* **A40**, 4277 (1989).
5. M.A. Nielsen and J. Kempe, *Phys. Rev. Lett.* **86**, 5184 (2001).
6. R. Rossignoli, N. Canosa, *Phys. Rev.* **A66**, 042306 (2002).
7. A.C. Dogherty, P.A. Parrilo and F.M. Spedalieri, *Phys. Rev.* **A69**, 022308 (2004).
8. C.H. Bennett et al., *Phys. Rev.* **A54**, 3824 (1996).
9. W.K. Wootters, *Phys. Rev. Lett.* **80**, 2245 (1998).
10. A. Peres, *Phys. Rev. Lett.* **77**, 1413 (1996).
11. M. Horodecki, P. Horodecki, and R. Horodecki, *Phys. Lett.* **A223**, 1 (1996).

12. G. Vidal and R.F. Werner, *Phys. Rev.* **A65**, 032314 (2002).
13. L. Gurvits, H. Barnum, *Phys. Rev.* **A66**, 062311 (2002); *ibid* **A68**, 042312 (2003).
14. W. Dür, J.I. Cirac, and R. Tarrach, *Phys. Rev. Lett.* **83**, 3562 (1999); W. Dür and J.I. Cirac, *Phys. Rev.* **A61**, 042314 (2000).
15. M.C. Arnesen, S. Bose and V. Vedral, *Phys. Rev. Lett.* **87**, 017901 (2001).
16. T.J. Osborne and M.A. Nielsen, *Phys. Rev.* **A66**, 032110 (2002).
17. X. Wang, *Phys. Rev.* **A64**, 012313 (2001); *ibid* **A66**, 034302 (2002).
18. N. Canosa, R. Rossignoli, *Phys. Rev.* **A69**, 052306 (2004).
19. R. Rossignoli, N. Canosa, *Phys. Rev.* **A72**, 012335 (2005); *Phys. Rev.* **A** (in press).

LIMITS ON THE POWER OF SOME MODELS OF QUANTUM COMPUTATION

GERARDO ORTIZ[1], ROLANDO SOMMA[1], HOWARD BARNUM[1], and EMANUEL KNILL[2]

[1] *Los Alamos National Laboratory, Los Alamos, NM 87545*

[2] *Mathematical and Computational Sciences Division, NIST, Boulder CO 80305*

We consider quantum computational models defined via a Lie-algebraic theory. In these models, specified initial states are acted on by Lie-algebraic quantum gates and the expectation values of Lie algebra elements are measured at the end. We show that these models can be efficiently simulated on a classical computer in time polynomial in the dimension of the algebra, regardless of the dimension of the Hilbert space where the algebra acts. Similar results hold for the computation of the expectation value of operators implemented by a gate-sequence. We introduce a Lie-algebraic notion of generalized mean-field Hamiltonians and show that they are efficiently (*exactly*) solvable by means of a Jacobi-like diagonalization method. Our results generalize earlier ones on fermionic linear optics computation and provide insight into the source of the power of the conventional model of quantum computation.

Keywords: Quantum Computation; Lie algebraic Quantum Models; Computational Complexity; Generalized Entanglement.

1. Introduction

Quantum models of computation are widely believed to be more powerful than classical ones. Although this has been shown to be true in a few cases, it is still important to determine when a quantum algorithm for a given problem is more resource efficient than any classical one, or, conversely, when a classical algorithm is just as efficient as any quantum counterpart. In general, one needs to know whether it is worth investing in building a quantum computer (QC) and what is required for success. In this paper, we show close connections between these issues and the efficient (or *exact*) solvability of Hamiltonians. In particular, we show that a class of quantum models we call generalized mean-field Hamiltonians (GMFHs)[1] is efficiently solvable and furthermore does not provide a stronger-than-classical model of computation: A quantum device engineered to have dynamical gates generated by Hamiltonians from such a set cannot directly simulate universal efficient quantum computation and can be efficiently simulated by a classical computer (CC).

An algorithm is a sequence of elementary instructions that solves instances of a problem. It is said to be efficient if the resources required to solve problem instances of size N are polynomial in N (poly(N)) resources. Typically, the size of a problem instance is the number of bits required to represent it, and the relevant resources are time and space. In the last few years it has been shown that many pure-state quantum algorithms can be efficiently simulated on a CC when the extent of entanglement is limited (see, e.g., Refs. 2, 3) or when the quantum gates available are far from allowing us to build a set of universal gates.[4–6] Here, we focus on a Lie algebraic analysis to obtain other situations where quantum algorithms can be efficiently simulated by CCs. The so-called generalized coherent states (GCSs)[7] and the notion of generalized entanglement [8,9] play a decisive role in our analysis.

2. Lie-algebraic model of Quantum Computation (LQC)

The algorithms considered here make use of the Lie-algebraic model of quantum computing (LQC). An LQC algorithm begins with the specification of a semisimple, compact M-dimensional real Lie algebra $\hat{\mathfrak{h}}$ of skew-Hermitian operators acting on a finite-dimensional Hilbert space $\mathcal{H}$, with Lie bracket $[\hat{X}, \hat{Y}] := \hat{X}\hat{Y} - \hat{Y}\hat{X}$. Without loss of generality, the action is irreducible. The algorithm begins with a maximum-weight state $|\mathsf{hw}\rangle$ in $\mathcal{H}$ and applies gates expressed as exponentials $e^{\hat{X}}$ for certain $\hat{X}$ in $\hat{\mathfrak{h}}$. The output of the algorithm is a noisy expectation of an operator in $\sqrt{-1}\hat{\mathfrak{h}}$ or in $e^{\hat{\mathfrak{h}}}$. LQC algorithms cannot trivially be classically simulated because of the possibility that the dimension of $\mathcal{H}$ is exponential in the specification complexity of $\hat{\mathfrak{h}}$ and $|\mathsf{hw}\rangle$. In order to precisely define the model of LQC we require some results from the theory of Lie algebras. See Ref. 10 for a textbook covering the basic theory of Lie algebras.

Our intention is to restrict observables and Hamiltonians to operators in $\sqrt{-1}\hat{\mathfrak{h}}$. The dimension of $\mathcal{H}$ may be exponential in M. Since we wish to implement computations with resources that are polynomial in M, our knowledge of $\hat{\mathfrak{h}}$ cannot involve explicit matrix representations of its operators. We therefore assume that $\hat{\mathfrak{h}}$ is specified as an abstract Lie algebra $\mathfrak{h}$ together with a "maximum weight" w characterizing its action on $\mathcal{H}$. For computational purposes, we also use a small-dimensional faithful representation of $\mathfrak{h}$. To be specific, we use the adjoint representation, but for efficiency, one can choose the first fundamental representation instead. We use the following notational conventions: Objects with a "hat" ($\hat{}$) belong to the representation of $\mathfrak{h}$ on $\mathcal{H}$. Objects with an "overline" ($\bar{}$) belong to the chosen faithful representation. Lie algebraic objects with neither a hat nor an overline are associated with the abstract Lie algebra (representation unspecified). Implicit in these conventions are the representational isomorphisms $\mathfrak{h} \to \hat{\mathfrak{h}}$ and $\mathfrak{h} \to \bar{\mathfrak{h}}$.

For the purpose of efficient representation, it is convenient to work with the complexification $\mathbb{C}\mathfrak{h}$ of $\mathfrak{h}$ and use a Cartan-Weyl (CW) basis (see, for example, Ref. 11) for $\mathbb{C}\mathfrak{h}$. Thus, we assume a decomposition $\mathbb{C}\mathfrak{h} = \mathfrak{h}_D \oplus \mathfrak{h}^+ \oplus \mathfrak{h}^-$, where $\mathfrak{h}_D$ is a Cartan subalgebra (CSA), and $\mathfrak{h}^+$ and $\mathfrak{h}^-$ are algebras of generalized raising and lowering operators, respectively. $\mathfrak{h}_D$ is linearly spanned by named elements $h_1, \ldots h_r$, and $\mathfrak{h}^\pm$ by $e^\pm_{\alpha_1}, \ldots, e^\pm_{\alpha_l}$. Thus, $M = 2l + r$. The α_j are linear functionals on $\mathfrak{h}_D$ called the positive roots of $\mathfrak{h}_D$. The abstract Lie algebra is specified by the identities $[h_k, h_l] = 0$, $[h_k, e^\pm_{\alpha_j}] = \pm a_{jk} e^\pm_{\alpha_j}$, $[e^+_{\alpha_j}, e^-_{\alpha_j}] = \sum_k b_{kj} h_k$, and for $j \neq k$, $[e^+_{\alpha_j}, e^\pm_{\alpha_k}] = c_{jk} e^+_{\alpha_j \pm \alpha_k}$. The bases of $\mathfrak{h}_D$ and $\mathfrak{h}^\pm$ may be chosen so that the "structure constants" a_{kj}, b_{kj} and c_{jk} are ratios of integers with poly(M) digits. The structure constants do not uniquely specify the action of $\hat{\mathfrak{h}}$ on $\mathcal{H}$. According to the representation theory of semisimple complex Lie algebras, this action is uniquely specified by its "maximum weight", which is a linear functional w on $\mathfrak{h}_D$ given by its values $w(h_k)$ on the distinguished basis of $\mathfrak{h}_D$. The $w(h_k)$ are integral and are the eigenvalues of $\hat{h}_k$ on the unique state $|\mathsf{hw}\rangle$ annihilated by $\hat{\mathfrak{h}}^+$: $\hat{h}_k|\mathsf{hw}\rangle = w(h_k)|\mathsf{hw}\rangle$. The state $|\mathsf{hw}\rangle$ is called the "maximum weight state" of the representation and its orbit under $e^{\hat{\mathfrak{h}}}$ generates the family of GCS of $\mathcal{H}$ with respect to $\hat{\mathfrak{h}}$. The Hermitian inner product of $\mathcal{H}$ and the Hermitian transpose of operators on $\mathcal{H}$ induce a corresponding Hermitian transpose operation on $\mathbb{C}\mathfrak{h}$. We assume that the CW basis is chosen so that the Hermitian transpose is given by $h_k^\dagger = h_k$ and $(e_{\alpha_j}^+)^\dagger = e_{\alpha_j}^-$. We also assume that the linear space on which $\bar{\mathfrak{h}}$ acts is endowed with a Hermitian inner product for which the representation $\bar{\mathfrak{h}}$ is skew-Hermitian and the Hermitian transpose matches the one defined for $\mathfrak{h}$.

The formal specification of an LQC algorithm requires the structure constants of an abstract CW decomposition of $\mathfrak{h}$ and the weight coefficients $w(h_k)$ determining $|\mathsf{hw}\rangle$. The specification complexities of $\mathfrak{h}$ and $|\mathsf{hw}\rangle$ are the number of bits required to represent the numerators and denominators of the structure constants and the $w(h_k)$. Thus they are polynomial in the dimension of $\mathfrak{h}$ and $\log\max(w(h_k))$. The *gates* of the algorithm may be unitary exponentials $e^{t\hat{X}}$, with $\hat{X}$ members of the CW basis. The gate's resource requirement is the number of bits required to represent t plus $|t|$. More generally, we can allow as gates any $e^{\hat{H}}$ with $\hat{H} \in \hat{\mathfrak{h}}$, where the resource requirement is given by the specification complexity of $e^{\hat{H}}$ (defined below). There are several alternatives for how the algorithm's output is obtained. We consider two. In the first, the output is obtained by measuring the expectation of an operator $\hat{A} \in \mathbb{C}\hat{\mathfrak{h}}$. In the second, it is obtained from the absolute value of the expectation of an operator $\hat{U} \in e^{\mathbb{C}\hat{\mathfrak{h}}}$. The resource cost of making the measurement is proportional to the sum of number of bits of precision and the specification complexity of $\hat{A}$ or $\hat{U}$. The specification complexity of $\hat{A}$ is that of A (the corresponding operator in the abstract Lie algebra $\mathfrak{h}$) and is given by the number of bits used to represent the coefficients of A when expressed in the CW basis. If $\hat{U}$ is of the form $e^{\hat{H}}$ with $\hat{H} \in \mathbb{C}\hat{\mathfrak{h}}$, its specification complexity is that of H plus $\max(|H_\alpha|)$ where H_α ranges over the coefficients of H expressed in the CW basis. Our assumption about the resource cost of measurement (i.e., the number of bits b of precision) makes the LQC model just defined very powerful but physically unreasonable. In particular, an LQC algorithm gives exponentially better precision than an algorithm of similar resource cost for the standard quantum computational model. The standard quantum algorithm would need to be repeated exponentially many times in b to return an expectation value with b bits of precision. In the standard quantum computational model, the hypothetical ability to determine expectation values with b bits of precision using resources polynomial in b implies the ability to efficiently solve problems in $\#P$, the class of problems associated with the ability to count the number of solutions to NP-complete problems such as satisfiability. This is a consequence of more general results in Ref. 12.

3. Efficient simulatability of LQC

A natural question is when and how LQC can efficiently simulate, or be simulated by, standard quantum or classical computation. The measurement models we introduced for LQC have the same form as many typical problems in physics, which involve the evaluation of correlation functions

$$\langle \hat{W} \rangle = \mathsf{Tr}[\rho \hat{W}], \tag{1}$$

where $\rho = \sum_{s=1}^{L} p_s|\phi_s\rangle\langle\phi_s|$ is the density operator of the system ($p_s > 0; \sum_s p_s = 1$), $|\phi_s\rangle$ are pure states, and $\hat{W}$ is a Hermitian or unitary operator acting on $\mathcal{H}$. In general, the dimension d of $\mathcal{H}$ increases exponentially in the problem size N, where the problem size is determined by quantities such as the volume or number of particles of the system. An algorithm to evaluate $\langle\hat{W}\rangle$ with accuracy ϵ is efficient if the amount of resources required is bounded by $\text{polylog}(d) + \text{poly}(1/\epsilon)$.

An efficient quantum algorithm to evaluate Eq. (1) exists if the state ρ (or a good approximation to it) can be efficiently prepared on a QC and if $\hat{W}$ can be efficiently measured by using, for example, the indirect techniques described in Refs. 13, 14. Unfortunately, known classical algorithms for this purpose typically require resources polynomial in the dimension d, which can be exponential in the problem size N. However, if the problem can be specified Lie algebraically, this classical complexity can be greatly reduced and exponential rather than polynomial accuracy is efficiently achievable.

Theorem 1. With ρ as defined following Eq. (1), if $|\phi_s\rangle = e^{\hat{A}_s}|\text{hw}\rangle$ are GCS's of $\hat{\mathfrak{h}}$ ($\hat{A}_s \in \hat{\mathfrak{h}}$) and $\hat{W} \in \mathbb{C}\hat{\mathfrak{h}}$, then $\langle\hat{W}\rangle$ can be classically computed to accuracy ϵ in time polynomial in $log(1/\epsilon)$ and the sum of the specification complexities of $\mathfrak{h}$, $|\text{hw}\rangle$, W, A_s and p_s.

Proof: We have $\langle\hat{W}\rangle = \sum_{s=1}^{L} p_s\langle\text{hw}|\hat{W}_s|\text{hw}\rangle$, where $\hat{W}_s = e^{-\hat{A}_s}\hat{W}e^{\hat{A}_s}$. In the CW basis,

$$\hat{W}_s = \sum_{k=1}^{r} u_k^s \hat{h}_k + \sum_{j=1}^{l} v_j^{s+}\hat{e}_{\alpha_j}^{+} + v_j^{s-}\hat{e}_{\alpha_j}^{-}, \tag{2}$$

where $u_k^s, v_j^{s\pm} \in \mathbb{C}$. To obtain these coefficients, we can compute $\bar{W}_s$ in the adjoint representation: $\bar{W}_s = e^{-\bar{A}_s}\bar{W}e^{\bar{A}_s} = \sum_{k=1}^{r} u_k^s\bar{h}_k + \sum_{j=1}^{l} v_j^{s+}\bar{e}_{\alpha_j}^{+} + v_j^{s-}\bar{e}_{\alpha_j}^{-}$. To compute the u_k^s and $v_j^{s\pm}$ to accuracy δ requires computing the matrix exponentials $e^{\pm\bar{A}_s}$, and matrix multiplication followed by an expansion of the resulting matrix in terms of the CW basis. The matrix exponentials can be obtained to accuracy δ' (in the 2-norm) in time polynomial in $\log(1/\delta')$ and the maximum of the entries of the $\bar{A}_s$ by direct series expansion or other, more efficient methods.[15] Matrix multiplication and basis expansion increase the 2-norm error by at most a constant factor, so that the u_k^s and v_j^s can be efficiently obtained to the desired accuracy.

Using the property that the $\hat{e}_{\alpha_j}^{\pm}$ either map $|\text{hw}\rangle$ to an orthogonal state or annihilate it we rewrite Eq. (1) as

$$\langle\hat{W}\rangle = \sum_{s=1}^{L} p_s \sum_{k=1}^{r} u_k^s w(h_k) \tag{3}$$

and this sum can be evaluated efficiently with respect to the given specification complexities. ■

The following variant of Thm. 1 holds for $\hat{W} = e^{\hat{H}}$ with $\hat{H} \in \mathbb{C}\hat{\mathfrak{h}}$.

Theorem 2. If $|\phi_s\rangle = e^{\hat{A}_s}|\text{hw}\rangle$ ($\hat{A}_s \in \hat{\mathfrak{h}}$) are GCS's of $\hat{\mathfrak{h}}$ and $\hat{W} = e^{\hat{H}}$ with $\hat{H} \in \mathbb{C}\hat{\mathfrak{h}}$, then $|\langle\hat{W}\rangle|^2$ can be classically computed to accuracy ϵ in time polynomial in $\log(1/\epsilon)$ and the sum of the specification complexities of $\mathfrak{h}$, $|\mathsf{hw}\rangle$, W, A_s and p_s.

Proof: We first define the operators $\hat{\Pi} = |\mathsf{hw}\rangle\langle\mathsf{hw}|$ and $\hat{B}_s = e^{-\hat{A}_s}\hat{W}e^{\hat{A}_s}$. We expand $|\langle\hat{W}\rangle|^2$ as

$$|\langle\hat{W}\rangle|^2 = \sum_{s,s'} p_s p_{s'} \langle\phi_s|\hat{W}|\phi_s\rangle\langle\phi_{s'}|\hat{W}^\dagger|\phi_{s'}\rangle = \sum_{s,s'} p_s p_{s'} \operatorname{tr} \hat{O}_{s,s'},$$
$$\text{with } \hat{O}_{s,s'} := \hat{\Pi}\hat{B}_s\hat{\Pi}\hat{B}^\dagger_{s'}\hat{\Pi}. \tag{4}$$

$\hat{O}_{s,s'}$ is proportional to $\hat{\Pi}$ and its trace is the constant of proportionality. We can express $\hat{\Pi}$ as a limit of operators in $e^{\mathbb{C}\mathfrak{h}}$. Let $L = \sum_{k=1}^{r} w(h_k)h_k$ and define ω by $\hat{L}|\mathsf{hw}\rangle = \omega|\mathsf{hw}\rangle$. Then $\langle\psi|\hat{L}|\psi\rangle < \omega$ for $|\psi\rangle \neq |\mathsf{hw}\rangle$, from which it follows that $\hat{\Pi} = \lim_{t\to\infty} e^{-t\omega}e^{t\hat{L}}$.[8] Because the eigenvalues of $\hat{L}$ are integral, convergence is exponentially fast in t. Let

$$\hat{E}(t) = \sum_{s,s'} p_s p_{s'} e^{-3\omega t} e^{t\hat{L}}\hat{B}_s e^{t\hat{L}}\hat{B}^\dagger_{s'} e^{t\hat{L}}. \tag{5}$$

$\hat{E}(t)$ is positive definite Hermitian and converges to $\sum_{s,s'} p_s p_{s'} \hat{O}_{s,s'}$ as $t \to \infty$. For a given t, we can compute $\bar{E}(t)$ by computing exponentials and multiplying matrices in the adjoint representation. Observe that the maximum eigenvalue $\kappa(t)$ of $\hat{E}(t)$ converges exponentially fast to $|\langle\hat{W}\rangle|^2$. To compute $\kappa(t)$ we first determine $\bar{Q}(t)$ such that $\bar{E}(t) = e^{\bar{Q}(t)}$. With the assumed Hermitian inner product on the adjoint representation, $\bar{E}(t)$ is positive definite. Thus, there is a unique Hermitian $\bar{Q}(t)$ satisfying $\bar{E}(t) = e^{\bar{Q}(t)}$, and $\bar{Q}(t)$ is necessarily in $\sqrt{-1}\bar{\mathfrak{h}}$. The operator $\bar{Q}(t)$ can be obtained via any conventional efficient diagonalization procedure for non-negative definite matrices. We can then use an efficient Jacobi-like diagonalization procedure[16] to obtain unitary operators $\bar{U}(t) \in e^{\bar{\mathfrak{h}}}$ and $\bar{q}(t) \in \bar{h}_D$ such that $\bar{U}(t)\bar{q}(t)\bar{U}(t)^\dagger = \bar{Q}(t)$. The maximum eigenvalue of $\hat{E}(t)$ is given by the exponential of the maximum eigenvalue of $\hat{q}(t)$. At this point we require a number of results from the representation theory of Lie algebras. For example, see Ref. 10. The element $q(t)$ induces an alternative order on the roots, according to which a root α_j is positive if $\alpha_j(q(t))$ is positive. (To remove degeneracies, it may be necessary to slightly perturb $q(t)$.) For this ordering, we determine simple roots β_k and corresponding members $h'_k \in \mathfrak{h}_D$ such that h'_k is isomorphic to h_k via a member of the Weyl group. We can expand $\bar{q}(t) = \sum_j q_j \bar{h}'_j$. Uniqueness of maximal weights in representations of Lie algebras implies that the maximum eigenvalue of $\hat{q}(t)$ is given by $w'(q(t)) = \sum_j w(h_k)q_j$.

We claim that the necessary steps can be implemented with polynomial resources in the dimension of the Lie algebra and the number of digits of precision of $\kappa(t)$. The matrix and root manipulations can be implemented efficiently, but with respect to the precision of entries of the matrix. It is necessary to realize that unless the weight w is sufficiently small, $\bar{E}(t)$ converges to 0 exponentially fast in t. However, because the $|w(h_k)|$ are polynomial in M, the rate of convergence to 0 is bounded by $e^{-\mathrm{poly}(M)}$. To compute $\kappa(t)$ to a desired number P of digits of precision, it suffices to compute in the low dimensional matrix representation with a precision of $\mathrm{poly}(M) + \mathrm{poly}(P)$ digits, which can still be done with polynomial resources. The relevant Weyl group transformations can be done efficiently using one of the constructive proofs of the transitivity of the Weyl group. See, for example, the proof of Thm. 2.63 in Ref. 17. ■

The theorem still holds if *nonunitary* operators in the Lie group are allowed as "gates", an idea implicit (in the context of matchgate circuits) in Ref. 5, and made explicit in Ref. 18. Here, it allows postselection of measurement results corresponding to efficiently Lie-theoretically describable operators.

Important special cases motivating these results are *fermionic linear optics* quantum computation (and equivalent *matchgate* models introduced by Valiant) which is efficiently classically simulatable,[5,18,19] and models which also include linear fermionic operators ($\mathfrak{so}(2N+1)$) for which an extension of the canonical Bogoliubov mapping exists.[20] Natural bosonic analogues of the fermionic results also exist, which shows the simulatability of quantum computational models in which coherent states are acted on by linear optical circuits, and measured via homodyne detection,[6] and of models with initial multimode squeezed states and squeezing gates as well as linear ones.[21] Like LQC with the second measurement strategy, these involve the efficient simulation, in the dimension of a Lie algebra, of a computational model in which coherent states of a Lie group with gates generated by the algebra constitute the initial states and computation. However, in the bosonic case the relevant algebra is not semisimple, and the relevant irreps are infinite-dimensional.

We can now address the important question of the classical simulatability of LQCs.

Theorem 3. For both LQC measurement schemes, the result of an LQC algorithm $\mathcal{A}$ can be obtained using classical computation in time polynomial in the specification complexity of $\mathcal{A}$.

Proof: The action of the gates of the algorithm result in the state $|\phi\rangle = \prod_{m=1}^{t} e^{\hat{A}_m}|\mathsf{hw}\rangle$ where $\hat{A}_m \in \mathfrak{h}$. Let

$$\langle\hat{W}\rangle = \langle\mathsf{hw}|\prod_{m=1}^{t} e^{-\hat{A}_m}\hat{W}\prod_{m=1}^{t} e^{\hat{A}_m}|\mathsf{hw}\rangle. \tag{6}$$

The result of the algorithm is $\langle\hat{W}\rangle$ if $\hat{W}\in\hat{\mathfrak{h}}$ or $|\langle\hat{W}\rangle|$ if $\hat{W}\in e^{\mathbb{C}\hat{\mathfrak{h}}}$. The result can be computed by generalizing the algorithms given in the proofs of Thms. 1 and 2. All that is required is to compute the full product $\prod_{m=1}^{t} e^{\bar{A}_m}$ instead of the single exponential required for Thms. 1 and 2. The complexity of the method is polynomial with respect to the specification complexity of $\mathcal{A}$. ■

The meaning of Thm. 3 can be expressed in terms of generalized entanglement.[1,8,9] A pure state is generalized unentangled (GU) with respect to a preferred set $\mathcal{O}$ of observables if it is extremal among states considered as linear functionals on $\mathcal{O}$, otherwise it is generalized entangled. In a Lie algebraic framework, a GU state is a GCS of a semisimple compact Lie algebra. Thus, Thm. 3 states that if a quantum computation does not create generalized entanglement with respect to a polynomial-dimensional semisimple compact Lie algebra, such a computation can be efficiently simulated to exponential precision on a CC. We also remark that, because it cannot access all pure states during the computation, but only the submanifold of generalized unentangled ones, such a computation cannot directly simulate standard quantum computation.

For applications to physics simulation, the following corollary shows that higher-order correlation functions can also be computed efficiently, provided the order is not too large.

Corollary 4. *Let* $\hat{W}^1, \ldots, \hat{W}^q$ *be operators in* $\mathbb{C}\hat{\mathfrak{h}}$. *For fixed* q, *the expectation value of correlation functions of the form* $\langle \hat{W}^1 \cdots \hat{W}^q \rangle = \sum_{s=1}^{L} p_s \langle \mathsf{hw}| e^{-\hat{A}_s} \hat{W}^1 \cdots \hat{W}^q e^{\hat{A}_s} |\mathsf{hw}\rangle$, *can be computed on a CC in time polynomial in* $\log(1/\epsilon)$, *and the sum of the specification complexities of* $\mathfrak{h}$, $|\mathsf{hw}\rangle$, W^j, A_s *and* p_s.

The complexity of our algorithm for computing the correlation function in the corollary is exponential in q.

Proof: We outline an efficient algorithm for computing the desired correlation function. First we expand each $\hat{W}^j_s = e^{-\hat{A}_s} \hat{W}^j e^{\hat{A}_s}$ in the CW basis as in the proof of Thm. 1. The desired correlation is given by $\sum_s \langle \mathsf{hw}| \prod_j \hat{W}^j_s |\mathsf{hw}\rangle$. We formally multiply the CW basis expressions for the $\hat{W}^j_s$ to obtain sums P_s of formal products of members of CW basis representing the $\prod_j \hat{W}^j_s$. Each product of CW basis members is standardized by using the commutation rules so that each term is a product where all lowering (raising) operators follow (precede) members of $\mathfrak{h}_D$. This is similar to the procedure of Wick's theorem. After this transformation, terms which retain some lowering or raising operators contribute nothing to the correlation functions. The remaining terms' contribution is easily computed using $\langle \mathsf{hw}|\hat{h}_k|\mathsf{hw}\rangle = w(h_k)$. The contribution to the complexity of the procedure of the formal multiplication and standardization procedure grows exponentially in q. The number of terms that arise is bounded by $\text{poly}(M)^q$, so that for fixed q, the complexity remains polynomial in the given specification complexities. Further details are available in Ref. 22. ■

4. Generalized Mean-field Hamiltonians and efficient (exact) solvability

The algorithms given above can also be used to analyze certain interacting physical models. We use the term GMFH[1] for Hamiltonians belonging to $\sqrt{-1}\mathfrak{h}$ for $\mathfrak{h}$ in a sequence of semisimple compact operator Lie algebras of dimension $M \leq \text{polylog}(d)$ acting on d-dimensional Hilbert spaces. A GMFH is necessarily specified in terms of a basis of $\mathfrak{h}$ that can be efficiently transformed to a CW basis. An example of a GMFH is given by the N spin-1/2 Ising model in a transverse magnetic field $\hat{H}_I = \sum_{j=1}^{N} (g\sigma_x^j \sigma_x^{j+1} + \sigma_z^j)$, where $\hat{H}_I$ is an element of the Lie algebra $\mathfrak{so}(2N)$, with dimension $M = 2N^2 - N \equiv \text{polylog}(d)$, where $d = 2^N$. Interestingly, this model can be exactly solved and, as we will show, this result can be extended to any GMFH. We say that a Hamiltonian acting on a d dimensional Hilbert space can be efficiently (*exactly*) solved when any one of its eigenvalues and a description of the corresponding eigenstate can be obtained and represented to precision ϵ in $\text{polylog}(d) + \text{poly}(1/\epsilon)$ computational operations on a CC. In general, this definition makes sense when we focus on Hamiltonians describing the interactions of N-body systems, where d increases exponentially with N.

Theorem 5. GMFHs can be efficiently solved.

Proof: Let $\hat{H}_{MF}$ be a GMFH in $\sqrt{-1}\hat{\mathfrak{h}}$ given in terms of a CW basis of $\mathfrak{h}$ as in Eq. (2). We show that to solve $\hat{H}_{MF}$ it suffices to diagonalize it according to

$$\hat{H}_D = \hat{U}\hat{H}_{MF}\hat{U}^\dagger = \sum_{k=1}^{r} \varepsilon_k \hat{h}_k \,, \tag{7}$$

with $\varepsilon_k \in \mathbb{R}$ and $\hat{U} \in e^{\hat{\mathfrak{h}}}$ unitary. The eigenvalues of $\hat{H}_{MF}$ are shared with those of $\hat{H}_D$. A description of the corresponding eigenspaces consists of an eigenspace of $\hat{H}_D$ transformed by $\hat{U}^\dagger$, where $\hat{U}$ may be described by a sequence of LQC gates. According to the representation theory of Lie algebras, the eigenspaces of $\hat{H}_D$ consist of weight states of $\hat{\mathfrak{h}}$, which can be obtained from the highest weight state by applying lowering operators. They are characterized by linear functionals λ on H_D of the form $\lambda(h_k) = w(h_k) - \sum_l n_l \alpha_l(h_k)$ where the n_l are non-negative integers. Which choices of n_l correspond to weight states is readily determined from the representation theory of Lie algebras. Once we have expanded $\hat{H}_D = \sum_k \varepsilon_k \hat{h}_k$, the eigenvalue corresponding to λ is readily computed as $\lambda(H_D) = \sum_k \varepsilon_k \lambda(h_k)$.

To efficiently diagonalize $\hat{H}_{MF}$ and obtain a specification of $\hat{U}$, we compute in the adjoint representation and apply a generalization of the Jacobi method[16,23] to $\bar{H}_{MF}$. It yields an exponentially converging diagonalization and an expression for $\hat{U}$ in terms of a sequence of exponentials of members of the $\mathfrak{su}(2)$ subalgebras of $\mathbb{C}\hat{\mathfrak{h}}$ generated by the pairs $\hat{e}^{\pm}_{\alpha_j}$. This suffices for our purposes. ∎

Example. The fermionic Hamiltonians $\hat{H}_{MF} = \sum_{i,j=1}^{N} t_{ij}(c_i^\dagger c_j - \delta_{ij}/2) + u_{ij} c_i^\dagger c_j^\dagger +$ h.c., where the operator $c_i^\dagger$ (c_i) creates (annihilates) a spinless fermion at the ith site, belong to a representation of the Lie algebra $\mathfrak{so}(2N)$ of dimension $M = 2N^2 - N \leq \text{polylog}(d)$. A faithful representation of $\mathfrak{so}(2N)$ is given by $c_i^\dagger c_j - \delta_{ij}/2 \leftrightarrow T_{i,j} - T_{N+j,N+i}$, $c_i^\dagger c_j^\dagger \leftrightarrow T_{i,N+j} - T_{j,N+i}$, and $c_i c_j \leftrightarrow T_{N+i,j} - T_{N+j,i}$, where the $2N \times 2N$ matrices $T_{kk'}$ have +1 in the kth row and k'th column, and zeros otherwise. Therefore, we write the matrix of $\hat{H}_{MF}$ in this representation and apply the Jacobi algorithm to diagonalize it. The result is equivalent to the one given by the Bogoliubov transformation,[24] where the Hamiltonian maps as $\hat{H}_{MF} \to \hat{H}_D = \sum_{k=1}^{r} \varepsilon_k(\gamma_k^\dagger \gamma_k - 1/2)$, where the operator $\gamma_k^\dagger$ (γ_k) creates (annihilates) a fermionic quasiparticle in the kth mode.

5. Efficient preparation of Generalized Coherent States

Although LQC algorithms and GMFHs can be efficiently simulated or solved on a CC, it may still be useful to implement the algorithms or simulate GMFHs with QCs. In particular, there may be problems where a key component is expressed in terms of LQC or GMFHs but a more complex quantum computation is required to determine the information of interest. One case of interest is where the LQC or GMFH component requires preparing a GCS.

One way for such a GCS to arise is as the ground state of a GMFH. According to the next theorem, such GCSs are efficiently preparable on a QC that has efficient access to the LQC initial state and gates.

Theorem 6. Let a GCS $|\phi\rangle$ of $\hat{\mathfrak{h}}$ be specified as the ground state of a Hamiltonian $\hat{H} \in \sqrt{-1}\hat{\mathfrak{h}}$. Then $|\phi\rangle$ can be prepared using resources polynomial in the specification complexity of H on a QC with the ability to initialize $|\mathsf{hw}\rangle$ and efficiently apply LQC gates.

Proof: It suffices to determine a $\hat{U} \in e^{\hat{\mathfrak{h}}}$ expressed as a polynomial product of LQC gates such that $\hat{H} = \hat{U}\hat{H}_D\hat{U}^\dagger$ with $\hat{H}_D \in \hat{\mathfrak{h}}_D$ such that D induces the root order associated with the CW basis. (See the proof of Thm. 2 for how an element of $\mathfrak{h}_D$ induces a root order.) The state $|\phi\rangle$ is then obtained as $\hat{U}|\mathsf{hw}\rangle$ and hence is efficiently preparable using LQC operations. To determine $\hat{U}$ we can first use the generalization of the Jacobi method as discussed previously. This yields an element of $\mathfrak{h}_D$ that does not necessarily induce the desired root order. To complete the determination of $\hat{U}$ requires using a sequence of Weyl reflections to obtain the desired root order. The sequence may be obtained using the method mentioned at the end of the proof of Thm 2. ■

6. Conclusions

Our results provide analogues of the Gottesman-Knill theorem[4] (cf. also Refs. 25, 26) concerning the efficient simulatability of Clifford-group computational models, and of results on the simulatability of certain multimode coherent-state and squeezed-state computational models.[6,21] One might hope for a treatment, perhaps based on Lie groups and groups of Lie type, that will unify these results, specifically those based on (1) finite dimensional semisimple Lie algebras, (2) Bosonic linear optics with homodyne detection (tied to an infinite-dimensional irreducible representation of a solvable Lie algebra) and possibly squeezing (involving a nilpotent Lie algebra), and (3) Clifford groups and semigroups. Our results cast additional light on what aspects of quantum computation are responsible for its possibly greater than classical computational power. It is a crucial fact that the generators of its gate-set, though their number can be chosen to grow polynomially, generate an exponentially large Lie algebra acting on an exponentially large Hilbert space. If the growth of the dimension of the generated Lie algebra is polynomial, a computation with this gate set using compatible state preparations and measurements can be simulated with polynomial efficiency on a classical computer by working in a low-dimensional faithful representation of the Lie algebra. What other algebraically constrained models of quantum computation are efficiently classically simulatable? Such structures may underlie the efficient solvability of further classes of Hamiltonians of condensed matter models, which go beyond the GMFHs, such as those solvable via a Bethe-type *Ansatz*.

Acknowledgements

We thank L. Viola for discussions and L. Gurvits for discussions and pointing out Ref. 23, and the US DOE and NSA for support. Contributions to this work by NIST, an agency of the US government, are not subject to copyright laws.

References

1. R. Somma, G. Ortiz, H. Barnum, E. Knill, and L. Viola, *Phys. Rev.* **A70**, 042311 (2004).
2. G. Vidal, *Phys. Rev. Lett.* **91**, 147902 (2003).
3. R. Jozsa and N. Linden, *Proc. Roy. Soc. Lond.* **A459**, 2011 (2003).
4. D. Gottesman, *Stabilizer Codes and Quantum Error Correction* (PhD Thesis, Caltech, Pasadena, CA, 1997).
5. L. Valiant, in *Proc. 33rd Ann. ACM Symposium on the Theory of Computation (STOC'01)* (ACM Press, El Paso, Texas, 2001), pp. 114-123.
6. S. D. Bartlett, B. C. Sanders, S. L. Braunstein, and K. Nemoto, *Phys. Rev. Lett.* **88**, 097904/1 (2002), quant-ph/0109047.
7. A.M. Perelomov, *Commun. Math. Phys.* **26**, 222 (1972). R. Gilmore, *Rev. Mex. de Física* **23**, 143 (1974). W. Zhang, D. H. Feng, and R. Gilmore, *Rev. of Mod. Phys.* **62**, 867 (1990).
8. H. Barnum, E. Knill, G. Ortiz, and L. Viola, *Phys. Rev.* **A68**, 032308/1 (2003), quant-ph/0207149.
9. H. Barnum, E. Knill, G. Ortiz, R. Somma, and L. Viola, *Phys. Rev. Lett.* **92**, 107902/1 (2004).
10. J. Fuchs, *Affine Lie Algebras and Quantum Groups* (Cambridge University Press, Cambridge, 1992).
11. J. F. Cornwell, *Group Theory in Physics* (Academic Press, London, UK, 1989).
12. S. Fenner, F. Green, S. Homer, and R. Pruim, *Proc. R. Soc. Lond.* **A455**, 3953 (1999).
13. G. Ortiz, J. Gubernatis, E. Knill, and R. Laflamme, *Phys. Rev.* **A64**, 022319/1 (2001).
14. R. Somma, G. Ortiz, J. Gubernatis, E. Knill, and R. Laflamme, *Phys. Rev.* **A65**, 042323/1 (2002), quant-ph/0108146.
15. C. Moler and C. Van Loan, *SIAM Review* **45**, 3 (2003).
16. N. J. Wildberger, *Proc. Am. Math. Soc.* **119**, 649 (1993).
17. A. Knapp, *Lie Groups: Beyond an Introduction* (Birkhäuser, Boston, 1996), (second printing, 2001).
18. E. Knill, Tech. Rep. LAUR-01-4472, Los Alamos National Laboratory (2001), quant-ph/0108033.
19. B. Terhal and D. DiVincenzo, *Phys. Rev.* **A65**, 032325/1 (2002).
20. H. Fukutome, M. Yamamura, and S. Nishiyama, *Prog. Theor. Phys.* **57**, 1554 (1977).
21. S. Bartlett and B. Sanders, *Phys. Rev. Lett.* **89**, 207903/1 (2002).
22. R. Somma, *Quantum Computation, Complexity, and Many-Body Physics* (PhD Thesis, Los Alamos National Laboratory and Instituto Balseiro, Los Alamos, NM, 2005).
23. M. Kleinsteuber, U. Helmke, and K. Hüper, *SIAM J. Matrix Anal. Appl.* **26**, 42 (2004).
24. J. Blaizot and G. Ripka, *Quantum Theory of Finite Systems* (MIT Press, London, UK, 1986).
25. M. Nielsen and I. Chuang, *Quantum Information and Computation*, Cambridge University Press (2000), p. 464.
26. S. Aaronson and D. Gottesman, *Phys. Rev.* **A70**, 052328 (2004).

FREE ROTATION OF DOPANTS IN SUPERFLUID HELIUM CLUSTERS

S. PAOLINI

SISSA – Scuola Internazionale Superiore di Studi Avanzati,
via Beirut 2-4, 34014 Trieste, Italy

S. MORONI

INFM DEMOCRITOS National Simulation Center, via Beirut 2-4, 34014 Trieste, Italy

We present quantum Monte Carlo calculations of the effective rotational constant B of several cromophore molecules embedded in He clusters, as a function of the cluster size. The predictive power of the computed B values is demonstrated not only by their agreement with available measurements, but also by their use in the assignment of several lines in both infrared and microwave spectra. The simulation results complement and extend the experimental information, offering insight into the relationship between structural and dynamical properties and the onset of superfluidity. The range of cluster sizes studied in our simulations includes systems of several tens particles, intermediate between the small-cluster and the nanodroplet regimes. In this size range we find unexpected trends for the evolution of B towards its asymptotic nano-droplet value.

Keywords: quantum Monte Carlo; Helium clusters; superfluidity.

1. Introduction

Computer simulation of quantum many-body systems has considerably advanced in recent years, allowing in some cases to determine the low-lying spectrum of excited states. One of the fields in which such progress is being actively pursued is the study of the rotational dynamics of small molecules solvated in He and para-H_2 clusters.

In the last decade, these systems have attracted increasing experimental and theoretical interest,[1–4] because they offer the possiblilty of studying manifestations of superfluidity at a microscopic scale. In an elegant "microscopic Andronikashvili experiment"[5] the vibrational spectrum of an OCS molecule seeded into He droplet was shown to have a broad peak for the fermionic ^{3}He isotope, while exhibiting sharp rotational resolution for the bosonic ^{4}He, which is superfluid at the temperature of the order of 10^{-1} K determined by evaporative cooling of the droplets. Even more strikingly, rotational resolution appears when as few as about 60 ^{4}He atoms coat the OCS molecule within a large ^{3}He droplet, thus justifying the introduction of the concept of "molecular superfluidity".

By now, several molecules solvated in ^{4}He droplets have been studied experimentally, and the spectra within this superfluid environment reveal a free-rotor character with the same symmetry than the molecule in gas phase, but with an increased moment of inertia. Much theoretical effort[2,6] has been devoted to the prediction of the renormalization of the effective rotational constants, and to the understanding of the responsible mechanisms. A key role has been attributed to the concept of "adiabatic following", leading to consider different dynamical regimes (i.e. "light" or "heavy" rotors, whose rotational constant is weakly or strongly modified by the quantum solvent, respectively).

Recent progress in the experimental techniques has made it possible to selectively assign ro-vibrational lines to clusters of given size. In the first such experiment,[7] the effective rotational constant $B(N)$ of OCS@He$_N$ clusters has been measured for N up to 8. It decreases monotonically, undershooting the nanodroplet value B_∞ at $N > 5$, which implies a subsequent turnaround. These findings spawned further experimental and theoretical studies aimed at a detailed characterization of the onset and the evolution of molecular superfluidity in the small and intermediate size regime.

In this paper we review accurate numerical results obtained with the reptation quantum Monte Carlo (RQMC) method, which enlighted the relationship between rotational dynamics, structural properties and incipient superfluidity, and predicted a nontrivial dependence of B on N over a wide size range, with the asymptotic value not yet reached at completion of the first solvation shell.

2. Theory

The RQMC method[8] exploits a discretized path-integral representation of a state $|\Psi_\beta\rangle = e^{-\beta H/2}|\Psi\rangle$, where Ψ is an explicitly known trial wave function, which approaches the exact ground state $|\Phi_0\rangle$ of the Hamiltonian H for $\beta \to \infty$. Using the identity $e^{-\beta H} = (e^{-\epsilon H})^P$, with $\epsilon = \beta/P$, and a short time approximation for the imaginary time propagator, $\mathcal{G}(R, R'; \epsilon) \simeq G(R, R'; \epsilon) = \langle R'|e^{-\epsilon H}|R\rangle$, expectation values of quantum operators on the state $|\Psi_\beta\rangle$ are in a form amenable to Monte Carlo integration via a generalized Metropolis algorithm (R denotes the generalized coordinates of all the particles in the system):

$$Q_\beta = \frac{1}{\langle\Psi_\beta|\Psi_\beta\rangle}\langle\Psi_\beta|Q|\Psi_\beta\rangle \simeq \int dX\,\Psi(R_0)\Psi(R_P)\Pi_{i=1}^P \mathcal{G}(R_{i-1}, R_i; \epsilon)Q(R_{P/2})$$

$$= \int dX P(X)Q(R_{P/2}). \quad (1)$$

Here X is a time-discretized path which visits R_i at time $i\epsilon$, and we assume that Q is diagonal in the coordinate representation.

The simulation amounts to generate a set of M paths $\{X_\alpha\}$, with $\alpha = 1, \ldots, M$, distributed with probability density $P(X)$. Monte Carlo estimates of ground-state properties are then obtained as averages over the paths. Many of the results to be discussed in the following are based on correlation functions in imaginary time. Within the path integral scheme, the evaluation of correlation functions requires an extended time range, but it is otherwise straightforward: $C_{AB;\beta}(\tau) = \langle\Psi_\beta|Q_A(\tau_0)Q_B(\tau_0 + \tau)|\Psi_\beta\rangle$ is estimated as $\frac{1}{MN_i}\sum_{\alpha,i} Q_A(R_i^{(\alpha)})Q_B(R_{i+k}^{(\alpha)})$, where $k = \tau/\epsilon$, $R_i^{(\alpha)}$ refers to the i-th time slice of the α-th sampled path, and the N_i values of i span the range $\beta/2\epsilon < i < P - k - \beta/2\epsilon$.

In comparison with the diffusion Monte Carlo method,[9] which is more commonly used for ground-state simulations, RQMC features a much easier access to pure estimators of diagonal[8] and off-diagonal[10] operators, correlated sampling,[11,12] and imaginary-time correlation functions[13,14] –whereas it is less efficient for the calculation of the ground-state energy.

The reptation or "slithering snake" algorithm,[8] borrowed from the simulation of classical polymers,[15] uses a suitable *a priori* sampling distribution $T(X, X')$ to move from $X = \{R_0, \ldots, R_P\}$ to $X' = \{R_1, \ldots, R_{P+1}\}$ by adding one slice at the head of the "snake" and chopping off one slice from the tail (the reverse move, which adds to the tail and removes from the head with obvious changes in the indices, is also possible). Using the definition of $P(X)$ in Eq. 1, the probability of accepting the move according to the generalized Metropolis algorithm reads

$$A(X, X') = \min\left\{1, \frac{\Psi_{P+1}}{\Psi_P}\frac{\Psi_1}{\Psi_0}\frac{\mathcal{G}(R_P, R_{P+1}; \epsilon)}{\mathcal{G}(R_0, R_1; \epsilon)}\frac{T(X', X)}{T(X, X')}.\right\}, \tag{2}$$

If the move is accepted, the indices of the path X' are relabeled from 0 to P.

The random choice of the direction of the move (either the head or the tail) implies that it takes on the average $\propto P^2$ steps to refresh all the slices. The prefactor can be reduced[8] by a factor $\simeq \Delta^2$ if we can add a number Δ of slices per move without reducing the acceptance too much (Δ would be tuned to maximal efficiency). Taking into account the increased computational cost of such multiple moves, the net gain in CPU time is $\simeq \Delta$. A better strategy is the bounce algorithm,[11] whereby the direction changes only after a move has been rejected. In this way one takes on the average several steps in the same direction if the acceptance probability is high. Furthermore, late-stage rejections of expensive multiple moves are avoided.

Importance sampling[16] can be introduced in RQMC in a natural way, largely independent of the particular choice for the short-time approximation which enters the probability distribution $P(X)$ of the paths. The importance-sampled Green's function is defined as $\tilde{G}(R_i, R_j) = \Psi(R_j)G(R_i, R_j)/\Psi(R_i)$. One readily recognizes in the acceptance rate of Eq. 2 a factor $\tilde{G}(R_P, R_{P+1})/\tilde{G}(R_1, R_0)$ (apart from a finite time step error). If we were able to directly sample $\tilde{G}$, we could achieve acceptance 1. While this is not feasible in general, it can be used as a guide to improve the efficiency via a suitable choice of the sampling distribution T. To this purpose, it proves useful to consider the approximation routinely used in diffusion Monte Carlo[9] for the importance-sampled Green's function, which reads

$$\tilde{G}(R_i, R_j) \simeq e^{-(R_{P+1}-R_P-2\lambda\epsilon\nabla \ln \Psi(R_P))^2} e^{-\epsilon(E_L(R_{P+1})+E_L(R_P))/2}, \tag{3}$$

where $E_L(R) = \frac{1}{\Psi(R)}H\Psi(R)$ is the local energy, and λ is either $\hbar^2/2m$, with appropriate mass, for the translational degrees of freedom, or $\hbar^2/2I$, with I the molecular moment of inertia, for the angular coordinates. By choosing $T(X, X') \propto e^{-(R_{P+1}-R_P-2\lambda\epsilon\nabla \ln \Psi(R_P))^2}$, and using Eq. 3, the acceptance probability, Eq. 2, reduces to

$$A(X, X') \simeq e^{-\epsilon(E_L(R_{P+1})+E_L(R_P))/2}/e^{-\epsilon(E_L(R_1))+E_L(R_0))/2}. \tag{4}$$

This expression is not a strict equality because in general we have different short time approximations for the Green's function G and its importance-sampled version $\tilde{G}$ (the exact

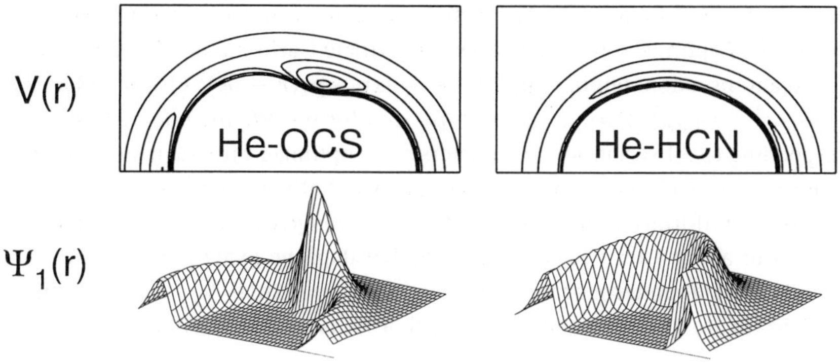

Fig. 1. Top panels: potential energy surfaces of He-OCS (left) and He-HCN (right). The molecule sits on the x asis, with its center of mass at the origin and the Oxygen (Hydrogen) atom in the positive direction. The x range is -6:6, and the y range is 0:6 (in Å). Contour levels start from $V = 0$ to negative values, spaced by 10 K. Lower panels: He-molecule correlation terms optimized for clusters with 20 He atoms.

expression of the acceptance probability, not shown here, can be straightforwardly obtained once the choice of the approximation $\mathcal{G}$ used in Eq. 1 is specified). Equation 4 shows both the strength and the weakness of the reptation algorithm. For good wave functions and systems not too large, such as those studied here, the fluctuations of the local energy are small: therefore importance sampling makes the acceptance probability correspondingly high, and RQMC is extremely efficient. For comparison, Variational Path Integral[17–19] –a related method which uses path integrals for ground state simulations– does not use the trial function for importance sampling. On the other hand, since it allows one-particle moves (instead of the reptation move of a whole time slice), it is expected to have a better scaling with the number of particles.

The results presented here have been obtained using a simple approximation to the quantum propagator, $\mathcal{G}(R, R') = e^{-(R-R')^2/4\lambda\epsilon} e^{-\epsilon(V(R)+V(R'))/2}$. For high accuracy, i.e. small time-step bias, we need time steps of the order of 10^{-3} K^{-1}. With typical values of several tenths of K^{-1} for the projection time β and about 1 K^{-1} for the range τ of the imaginary-time correlation functions, this requires paths of thousands of slices. Since it specifically benefits from low rejection (see above), the RQMC algorithm is not necessarily more efficient with better actions and larger time steps.

We consider a realistic model Hamiltonian with pair interactions. Both the radial He-He potential[20] and the anisotropic He-molecule potential energy surface (PES) are taken from ab-initio quantum chemistry calculations. Contour plots of the PES for OCS,[21] a prototype "heavy rotor", and HCN,[22] a typical "light rotor", are given in the upper panels of Fig. 1. The trial wave function is of the Jastrow form,

$$\Psi(R) = \Pi_{i<j} \exp(-u(r_{ij}))\Pi_i \exp[-\sum_J P_J(\cos(\theta_i))u_J(\rho_i)], \tag{5}$$

where r_{ij} is a He-He pair distance, ρ_i is a He-molecule distance, and θ_i is the angle between ρ_i and the molecular axis. The radial functions u and u_J of Eq. 5 contain variational parameters optimized with the variational Monte Carlo method.[9] We have used a De Michelis-Reatto pseudopotential[23] $p_1/r^{p_2} + p_3 \exp(-p_4(r-p_5)^2)$ for u and a Padé form $(p_{1,J} + p_{2,J}\rho^3 + p_{3,J}\rho^5 + p_{4,J}\rho^7)/(\rho^5 + p_{5,J}\rho^6)$ for u_J. We include up to five values of J, corresponding to 30 variational parameters to be optimized for each system size. While for complexes with 1 He atom one could use an essentially exact wave function, the above trial function is uniformly rather good in the whole range of cluster sizes, giving a variational energy of about 1 or 2 K per particle higher than the exact ground state value. The He-molecule correlation term of the trial function $\Psi_1 = \exp(-\sum_J P_J u_J)$, optimized for clusters of 20 He atoms doped with either OCS or HCN, is displayed in the lower panels of Fig. 1.

3. Results and discussion

The rotational excitation energies are obtained from imaginary-time correlation functions,[13,24]

$$C_J(\tau) = \frac{4\pi}{2J+1}\langle \sum_{M=-J}^{J} Y^*_{JM}(\mathbf{n}(\tau))Y_{JM}(\mathbf{n}(0))\rangle = \langle P_J(\mathbf{n}(\tau)\cdot\mathbf{n}(0))\rangle, \tag{6}$$

where $\mathbf{n}$ is the orientation of the molecular axis and $\langle\rangle$ denotes a ground-state average.

These imaginary-time correlation functions are linear combinations of exponentials, with decay constants and coefficients determined by excitation energies and spectral weights, respectively. Because of the rotational invariance of the Hamiltonian, selection

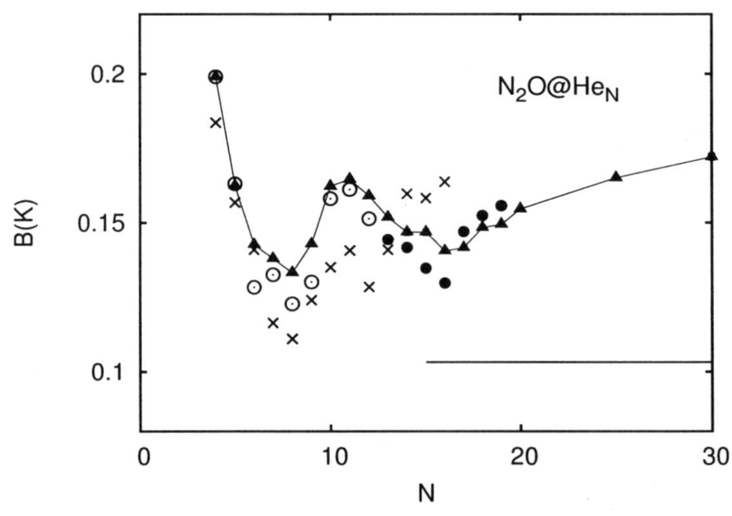

Fig. 2. Evolution of the effective rotational constant B, in K, with the number N of He atoms for $N_2O@He_N$ clusters. Empty[31] and filled[32] circles: experiment. Triangles and thin line: RQMC[26]. Crosses: POITSE[33]. The thick horizontal line indicates the measured nanodroplet limit.[35]

rules imply that C_J involves only states with angular momentum J. The scarcity of low-lying excitations due to the bosonic nature of the quantum solvent further implies that very few states contribute for given J. In these very peculiar conditions,[25] it is possible to extract their energies reliably by a multiexponential fit. In some cases, a single state dominates the $J = 1$ sector of the spectrum, and a rotational analysis including higher values of J yields the rotational constant and the centrifugal distortion of an effective linear rotor;[13,26,27] in other situations, one finds two $J = 1$ states with relevant spectral weight, corresponding to two series of rotational transitions which smoothly correlate with the known a-type and b-type lines of the binary complex.[24,28] All these findings are in quantitative agreement with the available experimental information.[7,27,29–31]

As an example of the high accuracy of the RQMC results, we compare in Fig. 2 the computed[26] and measured[31,32] evolution of the effective rotational constant with the number of He atoms for $N_2O@He_N$ clusters. A Diffusion Monte Carlo simulation[33] using the POITSE[34] method shows a larger discrepancy with experiment, particularly in the size range where the theoretical data predate the experiment: for N between 12 and 16, Ref. 33 predicts an increasing $B(N)$, which is the opposite of the observed behavior.[32]

Further witness of the high predictive power of the RQMC results is given by their successful use in the assignment of infrared transitions in para-Hydrogen clusters doped with CO,[36] as well as in the search and assignment of microwave transitions in He clusters doped with HCCCN.[37]

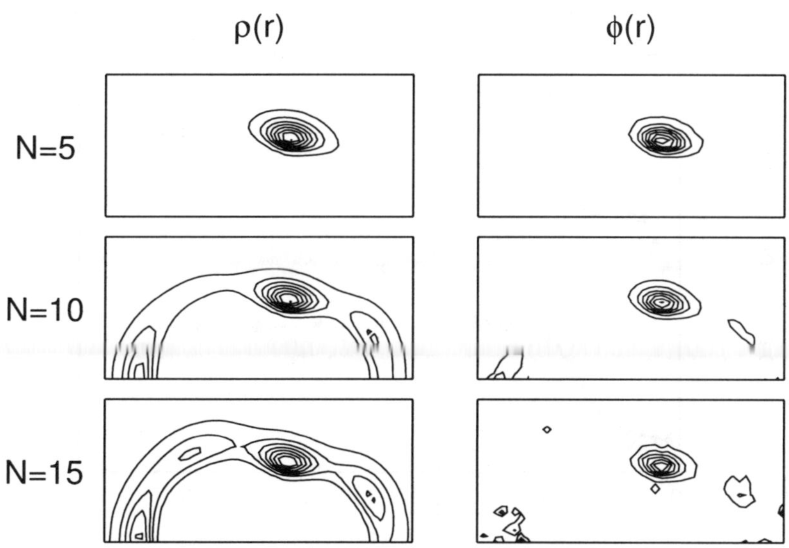

Fig. 3. Contour plots of density profiles, $\rho(\mathbf{r})$ (left panels), and angular correlations, $\phi(\mathbf{r})$ (right panels), for $OCS@He_N$ clusters[13].

In addition to accurate rotational energies, computer simulation provides estimates of other physical properties not directly probed experimentally, which give insight into the onset and evolution of molecular superfluidity. The left panels of Fig. 3 depict the He density, $\rho(\mathbf{r})$, for OCS@He$_N$ clusters of various sizes. For N up to 5, the He density is confined to an equatorial "donut" around the molecular axis, in correspondence to the principal minimum of the He-OCS PES of Fig. 1. For N between 6 and 8 the He density caps the Oxygen pole[38] and for larger clusters the quantum solvent fully coats the molecular impurity. The right panels of Fig. 3 show the correlation, $\phi(\mathbf{r})$, between the angular momentum of the molecule and the density of angular momentum of He atoms.[13] For $N = 5$, $\rho(\mathbf{r})$ and $\phi(\mathbf{r})$ are very similar, indicating that He atoms follow adiabatically the molecular rotation and therefore contribute to the effective moment of inertia. When the size of the cluster increases, the angular momentum correlation remains small in regions where the He density becomes large, implying a decoupling between He and molecular rotation which in turn signals the onset of superfluidity. One can further see that the peak value of $\phi(\mathbf{r})$ decreases in the size range reported in the figure. This explains the turnaround of the rotational constant mentioned in the Introduction.

We note in passing that, if the "morphed" version of the He-OCS PES[21] is adopted, the He density –after filling the donut– distributes more evenly among the two molecular poles,[13] producing a less sharp turnaround of the effective rotational constant and hindering to some extent the agreement with experiment.[38] Comparison between simulation data and measurements can be used to gauge the accuracy of the interparticle potential: in this specific case, the ab-initio PES is better than the recommended morphed version.[21]

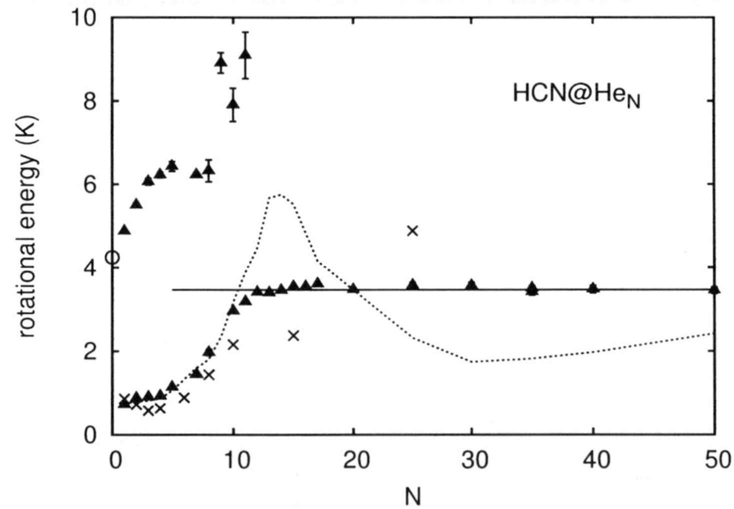

Fig. 4. Rotational energies for HCN@He$_N$ clusters. Triangles, RQMC[28]; crosses, POITSE[40]; cirlce, the gas phase value; solid line, nanodroplet limit[42]. The dashed line indicates the energy of a density fluctuation mode[24,28] whose coupling with molecular rotation is associated with the presence of two $J = 1$ excitations up to $N \approx 10$.

We turn now to discuss the evolution of $B(N)$ for relatively large cluster sizes, namely beyond completion of the first solvation shell (which occurs between $N = 15$ and 20 for the molecules considered here). Prior to RQMC studies, for heavy rotors such as OCS, $B(N)$ was believed to approach the value B_∞ found in large droplets before completion of the first solvation shell.[2,3,39] Such a behavior was rationalized[2] in terms of a so-called local nonsuperfluid part of the solvent "adiabatically following" the molecular rotation. The breakdown of adiabatic following for light rotors such as HCN, and early simulations of HCN@He_N clusters,[40] spurred the view that light rotors would instead attain the asymptotic B_∞ value for much larger sizes. Furthermore, the concept of adiabatic following, which depends on both the He-molecule PES and the gas-phase rotational constant B_0, was invoked[41] as the key physical quantity to explain the observed values of B_∞/B_0, which are small for heavy rotors and large for light rotors:[3] both fast rotation and small anisotropy of the PES would favor decoupling of molecular rotation from the solvent.

The RQMC results present a remarkably different scenario. The evolution of $B(N)$ for N_2O, a moderately heavy rotor, depicted in Fig. 2, indicates that the rotational constant remains significantly higher than the nanodroplet value well beyond completion of the first solvation shell. Apart from details in the turnaround region, a similar behavior is also observed for clusters doped with other heavy rotors such as CO_2[27] and OCS.[13,38] The experimental data, which for CO_2 and N_2O cover the size range up to one full solvation shell, confirm our simulation results.

For the prototype light rotor HCN, instead, the computed rotational energy (see Fig. 4) approaches the measured nanodroplet limit[42] at $N \approx 12$ and stays nearly constant up the

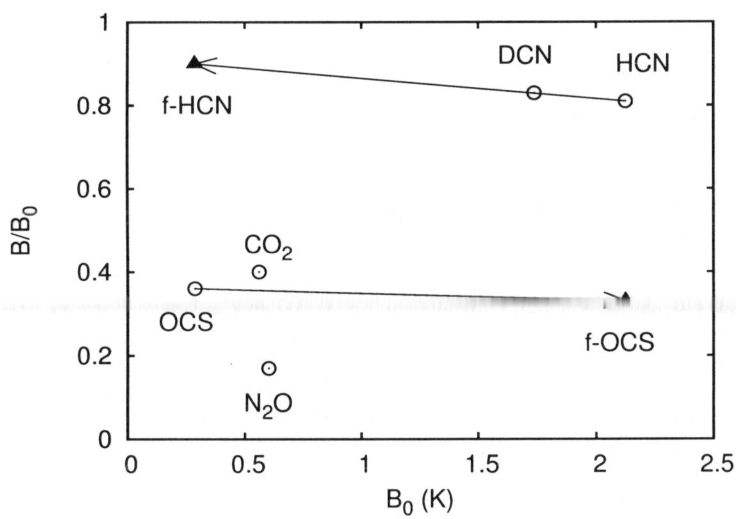

Fig. 5. Effective rotational constant B in the nanodroplet limit relative to its gas-phase value B_0. Circles: experimental data for the heavy rotors OCS, CO_2, N_2O and for the light rotors HCN and DCN. Triangles:[28] estimated RQMC values for fudged OCS and fudged HCN.

largest cluster studied, $N = 50$. A similar behavior found for CO[24] is backed by experimental data up to $N = 20$.[29] A general statement of fast convergence to the nanodroplet limit for light rotors would be in contrast with a recent measurement of B_∞ for CO, which reports a value somewhat smaller than the value observed[29] and computed[24] in medium-size clusters. However, it is clear from the HCN case that a general statement of slow convergence for light rotors is not tenable, either.

We finally consider the amount of renormalization of B upon solvation. Fig. 5 displays B_∞/B_0 vs. B_0 for various molecules: In addition to the upward trend of the measured values (circles), the figure displays the results of computer simulation[28] for two fictitious molecules (triangles): A "fudged OCS", with the same interaction with He as OCS, but with the gas-phase rotational constant of HCN, and a "fudged HCN", with the same interaction as HCN but the gas-phase rotational constant of OCS. The purpose of the calculations for the fictitious systems is to disentangle the effect of the He-molecule interaction from the effect of the rotational speed, i.e. the B_0 value. Despite some uncertainty in the extrapolation of B_∞ from the simulation of relatively small systems (the calculated values are likely to be slightly overestimated), the result clearly shows that –for given interaction– the value of B_∞/B_0 remains almost constant, actually with a weak downward trend (also found in the experimental data for HCN and DCN, neglecting the small difference in the He-molecule PES between the two isotopomers).

Our results suggest that B_∞/B_0 is essentially determined by the strength and anisotropy of the PES, while being largely independent of the gas-phase rotational inertia. According to this view, the small value $B_\infty/B_0 = 0.17$ measured[35] for N_2O, which was considered a somewhat "anomalous" case, simply reflects the stronger modulation of the PES.

Acknowledgements

We acknowledge allocation of computer resources by the INFM Iniziativa Calcolo Parallelo.

References

1. C. Callegari, K. K. Lehmann, R. Schmied and G. Scoles, *J. Chem. Phys.* **115**, 10090 (2001)
2. Y. Kwon, P. Huang, M. V. Patel, D. Blume, and K. B. Whaley, *J. Chem. Phys.* **113**, 6469 (2000).
3. J.P. Toennies and A.F. Vilesov, *Angew. Chem. Int. Ed.* **43**, 2622 (2004).
4. M. Barranco, R. Guardiola, S. Hernández, R. Mayol, J. Navarro, and M. Pi, *J. Low Temp. Phys.* **142**, 1 (2006).
5. S. Grebenev, J.P. Toennies, and A.F. Vilesov, *Science* **279**, 2083 (1998).
6. C. Callegari, A. Conjusteau, I. Reinhard, K. K. Lehmann, G. Scoles, and F. Dalfovo, *Phys. Rev. Lett.* **83**, 4108 (1999).
7. J. Tang, Y.J. Xu, A.R.W. McKellar, and W. Jäger, *Science* **297**, 2030 (2002).
8. S. Baroni and S. Moroni, *Phys. Rev. Lett.* **82**, 4745 (1999); S. Baroni and S. Moroni, in *Quantum Monte Carlo Methods in Physics and Chemistry*, ed. P. Nightingale and C.J. Umrigar. NATO ASI Series, Series C, Mathematical and Physical Sciences, Vol. 525, (Kluwer Academic Publishers, Boston, 1999), p. 313, also available at cond-mat/9808213.

9. B. L. Hammond, W. A. Lester and P. J. Reynolds, *Monte Carlo Methods in Ab Initio Quantum Chemistry* (World Scientific, Singapore, 1994).
10. S. Moroni and M. Boninsegni, *J. Low Temp. Phys.* **136**, 129 (2004).
11. C. Pierleoni and D. M. Ceperley, *ChemPhysChem* **6**, 1 (2005).
12. S. De Palo, S. Moroni and S. Baroni, in *Quantum Monte Carlo: Recent Advances and Common Problems in Condensed Matter and Field Theory*, Eds. M. Campostrini, M. P. Lombardo and F. Pederiva (ETS, Pisa, 2001).
13. S. Moroni, A. Sarsa, S. Fantoni, K.E. Schmidt, and S. Baroni, *Phys. Rev. Lett.* **90**, 143401 (2003).
14. S. De Palo, S. Conti, and S. Moroni, *Phys. Rev.* **B69**, 035109 (2004).
15. A. K. Kron, O. B. Ptitsyn, A. M. Skvortsov and A. K. Fedorov, *Molek. Biol.* **1**, 576 (1967) [*Molec. Biol.* **1**, 487 (1967)].
16. M. H. Kalos and P.A. Whitlock, *Monte Carlo Methods* (Wiley, New York, 1986).
17. D. M. Ceperley, *Rev. Mod. Phys.* **67**, 279 (1995).
18. A. Sarsa, K. E. Schmidt, and W. R. Magro, *J. Chem. Phys.* **113**, 1366 (2000).
19. D. Galli and L. Reatto, *Mol. Phys.* **101**, 1697 (2003).
20. T. Korona, H. L. Williams, R. Bukowski, B. Jeziorski, and K. Szalewicz, *J. Chem. Phys.* **106**, 5109 (1997).
21. J.M.M. Howson and J.M. Hutson, *J. Chem. Phys.* **115**, 5059 (2001).
22. R. R. Toczylowskski, F. Doloresco, and S. M. Cybulski, *J. Chem. Phys.* **114**, 851 (2000).
23. C. De Michelis and L. Reatto, *Phys. Lett.* **50A**, 275 (1974).
24. P. Cazzato, S. Paolini, S. Moroni, and S. Baroni, *J. Chem. Phys.* **120**, 9071 (2004).
25. In general the inverse Laplace transform is a notoriously ill-conditioned problem. See, e.g., J. E. Gubernatis and M. Jarrell, *Phys. Rep.* **269**, 135 (1996).
26. S. Moroni, N. Blinov, and P.-N. Roy, *J. Chem. Phys.* **121**, 3577 (2004).
27. J. Tang, A. R. W. McKellar, F. Mezzacapo, and S. Moroni, *Phys. Rev. Lett.* **92**, 145503 (2004).
28. S. Paolini, S. Fantoni, S. Moroni, and S. Baroni, *J. Chem. Phys.* **123**, 114306 (2005)
29. J. Tang, A.R.W. McKellar, *J. Chem. Phys.* **119**, 754 (2003).
30. J. Tang and A. R. W. McKellar, *J. Chem. Phys.* **121**, 181 (2004).
31. Y. Xu, W. Jäger, J. Tang, and A. R. W. McKellar, *Phys. Rev. Lett.* **91**, 163401 (2003).
32. Y. Xu, N. Blinov, W. Jäger and P.-N. Roy, *J. Chem. Phys.* **124**, 081101 (2006).
33. F. Paesani and K. B. Whaley, *J. Chem. Phys.* **121**, 5293 (2004).
34. D. Blume, M. Lewerenz, P. Niyaz, and K. B. Whaley, *Phys. Rev.* **E55**, 3664 (1997).
35. K. Nauta and R. E. Miller, *J. Chem. Phys.* **115**, 10254 (2001).
36. S. Moroni, M. Botti, S. De Palo, A.R.W. McKellar, *J. Chem. Phys.* **122**, 094314 (2005).
37. W. Topic and W. Jäger, private communication.
38. S. Moroni and S. Baroni, *Comp. Phys. Commun.* **169**, 404 (2005).
39. F. Paesani, A. Viel, F.A. Gianturco, and K.B. Whaley, *Phys. Rev. Lett.* **90**, 73401 (2003).
40. A. Viel and K. B. Whaley, *J. Chem. Phys.* **115**, 10186 (2001).
41. M. V. Patel, A. Viel, F. Paesani, P. Huang, and K. B. Whaley, *J. Chem. Phys.* **118**, 5011 (2003).
42. A. Conjusteau, C. Callegari, I. Reinhard, K. K. Lehmann, and G. Scoles, *J. Chem. Phys.* **113**, 4840 (2000).
43. K. von Haeften, S. Rudolph, I. Simanovski, M. Havenith, R. E. Zillich, and K. B. Whaley, *Phys. Rev.* **B73**, 054502 (2006).

PROJECTED ENTANGLED STATES: PROPERTIES AND APPLICATIONS

F. VERSTRAETE
Institute for Quantum Information, CALTECH, USA
f.verstraete@iqi.caltech.edu

M. WOLF, D. PÉREZ-GARCÍA, J.I. CIRAC
Max-Planck-Institut für Quantenoptik, Hans-Kopfermann-Str. 1, D-85748 Garching, Germany
ignacio.cirac@mpq.mpg.de

We present a new characterization of quantum states, what we call Projected Entangled-Pair States (PEPS). This characterization is based on constructing pairs of maximally entangled states in a Hilbert space of dimension D^2, and then projecting those states in subspaces of dimension d. In one dimension, one recovers the familiar matrix product states, whereas in higher dimensions this procedure gives rise to other interesting states. We have used this new parametrization to construct numerical algorithms to simulate the ground state properties and dynamics of certain quantum-many body systems in two dimensions. .

Keywords: Quantum Information; DMRG; Entangled states.

1. Introduction

The description of many–particle systems in Quantum Mechanics is, in general, an untractable problem.[1] The reason is that the dimension of the Hilbert space corresponding to the whole system scales exponentially with the number of particles. If we call d the dimension of the Hilbert space associated to each particle, the number of coefficients that have to be taken into account in order to write a state of N particles is generically d^N. Thus, even for two–level systems ($d = 2$) and for 50 particles there is no enough computer memory in the world to store those coefficients.

In some particular instances, however, the relevant states can be described (or at least approximated) with a smaller number of parameters. Thus, one may find families of states within which one obtains a good description of the state one is considering. For example, one may perform a variational calculation in order to determine the ground state energy of a particular Hamiltonian system in terms of a set of states which are characterized by a few number of parameters, and still find a very satisfactory result. For most set of states, however, this is not feasible, since even though the family may have very few parameters, when determining expectation values of physical observables one must perform a number of calculations which again scales exponentially with the number of particles. Thus, it is desirable not only to have families of states which represent the states of a particular system, but also with which one may be able to determine physical quantities in an efficient way.

In this paper we consider one such family of states that we introduced before and that is composed of what we call projected entangled–pair states (PEPS).[2–6] This name stems from the fact that the states are built by considering maximally entangled states of pairs of particles which are then projected onto smaller subspaces. Each PEPS is characterized by some parameters, whose number is specified by D. The family of PEPS is complete in the sense that for a sufficiently large value of D all states in the Hilbert space are PEPS (for N finite; for N infinite the PEPS are dense). After defining the PEPS, we will give some examples and enumerate some of their properties. In particular, we will show how to construct local Hamiltonians for which the PEPS are ground states, thus giving them a physical characterization. We will also show how to determine expectation values of tensor products of operators corresponding to each of the particles, i.e. N–particle correlation functions. For the sake of simplicity, we will consider here (what we call) PEPS in one and two dimensions. In one dimension the PEPS coincide with the well known matrix product states (MPS)[7,8] (also known as finitely correlated states). In two dimensions, PEPS also allows us to describe the states of interesting systems. Note that the generalization of PEPS to higher dimensions is straightforward using the ideas presented here for the two–dimensional case. On the other hand, the extension of PEPS to mixed states[5,9] will not be considered here.

The main applications of the PEPS are in the fields of simulating quantum many–body systems and in quantum computation. In the first case, one can reconstruct the ground state of interacting Hamiltonians, the state of a system at finite temperature, as well as to simulate its time evolution. In the second case, by performing local measurements only on some particular PEPS, one can carry out any quantum computation[10] following the measurement–based quantum computing model introduced in Ref. 11.

This paper is organized as follows. In Secs. 2 and 3 we introduce PEPS in one and two dimensions, respectively. We also give some examples, discuss some of their properties, and show how to evaluate N–particle correlation functions. In Sec. 4 we explain some of their applications, including the simulation of quantum systems as well as quantum computing. Most of the ideas and results presented here can be found in Refs. 2, 3, and some of the applications in Refs. 2–6.

2. PEPS in 1D: Matrix Product States

2.1. *Definition*

Let us start by recalling the representation introduced in Ref. 4 of the state Ψ of N d–dimensional systems in terms of Matrix Product States (MPS).[7,8,12,13] For that, we substitute each physical system k by two auxiliary systems a_k and b_k of dimension D (except at the extremes of the chain). Systems b_k and a_{k+1} are in an unnormalized maximally entangled state $|\phi\rangle = \sum_{n=1}^{D} |n,n\rangle$. The state Ψ is obtained by applying a linear operator $Q_k : \mathbb{C}^D \otimes \mathbb{C}^D \to \mathbb{C}^d$ to each pair a_k, b_k that maps the auxiliary systems onto the physical systems, i.e.

$$\begin{aligned} |\Psi\rangle = & \quad Q_1 \otimes Q_2 \otimes \ldots Q_N \; |\phi\rangle \ldots |\phi\rangle \\ = & \sum_{s_1,\ldots,s_N=1}^{d} \mathrm{F}_1(A_1^{s_1}, \ldots, A_N^{s_N}) |s_1, \ldots, s_N\rangle, \end{aligned} \tag{1}$$

where the matrices A_k^s have elements $(A_k^s)_{l,r} = \langle s|Q_k|l,r\rangle$. Note that the indices l and r of each matrix A_k^s are related to the left and right bonds of the auxiliary systems with their neighbors, whereas the index s denotes the state of the physical system. The function F_1 is just the trace of the product of the matrices, i.e. it contracts the indices l, r of the matrices A according to the bonds. Thus, the state Ψ does not uniquely determine the matrices A. In fact, for any invertible matrices X_k, Ψ is invariant under

$$A_k \to X_k A_k X_{k+1}^{-1}. \tag{2}$$

The picture introduced here is basically one dimensional, since each auxiliary system is entangled only to one nearest neighbor. Thus, these states appear to be better suited to describe 1D systems, with short range interactions, since a small local dimension D may give a good approximation to the real state of the whole system.

As shown in Ref. 4, every state can be represented in the form (1) as long as the dimension D is sufficiently large. We consider a_k and b_k as composed of d–level subsystems, $a_k^1, \ldots, a_k^N$ and $b_k^1, \ldots, b_k^N$, respectively, and write $|\phi\rangle$ as a tensor product of maximally entangled states ϕ_d between b_k^l and a_{k+1}^l. For $k = 2, \ldots, N$, we choose the operators $P_k = \mathbf{1}_{a_k^k} \otimes \langle \eta_k|$ where $|\eta_k\rangle$ is a state for all particles but a_k^k, and contains $|\phi_d\rangle$ for each pair a_k^l-b_k^l $(l > k)$ and $|0\rangle$ for the rest. The action of P_k is to teleport[14] the entangled pairs such that at the end one has one entangled pairs between the first system and all the rest, while leaving all the other auxiliary particles in $|0\rangle$. Finally, the operator P_1 is the product of two operators. The first acts on particles a_1 and transforms $|0\rangle^N \to |\Psi\rangle$. The second is $\mathbf{1}_{a_1^1} \otimes \langle \eta_1|$, where $|\eta_1\rangle = |0\rangle_{b_1} \otimes |\phi_d\rangle^{\otimes N-1}$. This operator first prepares the desired state Ψ in particles a_1 and then uses the available entangled pairs to teleport it to the the rest of the particles.

2.2. *Example: AKLT State*

We consider spin one particles interacting according to the AKLT Hamiltonian[12]

$$H_{\mathrm{AKLT}} = \sum_k \mathbf{S}_k \cdot \mathbf{S}_{k+1} + \frac{1}{3} \sum_k (\mathbf{S}_k \cdot \mathbf{S}_{k+1})^2 \tag{3}$$

where $\mathbf{S} = (S_x, S_y, S_z)$ are spin operators. The ground state of this Hamiltonian is a MPS called AKLT state with all Qs equal

$$Q = \sum_{s=1}^{3} |s\rangle\langle B_s|, \tag{4}$$

where $B_{1,2,3}$ are the Bell states $|0,1\rangle \pm |1,0\rangle$ and $|0,0\rangle - |1,1\rangle$. Thus, $A^1_{\mathrm{AKLT}} = \sigma_x$, $A^2_{\mathrm{AKLT}} = i\sigma_y$ and $A^3_{\mathrm{AKLT}} = \sigma_z$, i.e. proportional to the Pauli operators.

2.3. *Example: Cluster State*

The 1-D cluster state, introduced by Briegel and Raussendorf,[11] is the simplest example of a MPS for spin 1/2. It is the ground state of a Hamiltonian involving 3-body interactions given by

$$H_{clus} = \sum_k \sigma^x_{k-1} \sigma^z_k \sigma^x_{k+1}. \tag{5}$$

The corresponding bonds are qubit bonds ($D = 2$) and the projector Q is given by

$$Q = |0\rangle\langle 0,+| + |1\rangle\langle 1,-|, \tag{6}$$

where $|\pm\rangle = |0\rangle \pm |1\rangle$.

2.4. *Matrix Product States as ground states of Local Hamiltonians*

As we have seen, the AKLT state is the ground state of a spin Hamiltonian with local interactions. The reason for that is simply that up to a constant, it can be rewritten as

$$H_{\mathrm{AKLT}} = \sum_k P_k, \tag{7}$$

where P_k projects onto the spin 2 sector of the total spin of particles k and $k+1$, and the AKLT state does not have any component on this sector. Therefore, $P_k \geq 0$ and $P_k|\Psi_{\mathrm{AKLT}}\rangle = 0$. This argument can be extended to show that every MPS is the ground state of some local interaction Hamiltonian. Here by local we mean that it contains terms which involve at most r_D operators corresponding to particles separated by a distance smaller than r_D. The number r_D depends on D and typically grows with it. In order to proceed with this argument, we note that the reduced density operator ρ_r of a block of r particles (say from particle 1 to particle r) has the same range as the operator obtained by replacing the states of particles a_1 and b_r by the identity operator and applying the operators $Q_1 \ldots Q_r$ as before. Thus, the rank of ρ_r is at most D^2, independent of r. For r sufficiently large, it is thus possible to find an operator $P_1 \geq 0$ such that $P_1\rho_r = 0$, which immediately implies that the original MPS is an eigenstate of the (positive semidefinite) Hamiltonian $H = \sum_k P_k$ with zero eigenvalue, and therefore a ground state.

As an application, we show how to build a family of simple Hamiltonians involving two–body nearest–neighbor interactions for which one can determine the ground state. We consider N spin 1 systems in a chain and a family of MPS where A^α are 2×2 matrices. It is very simple to show that for any choice of the A^α, the state Ψ is the ground state of a Hamiltonian involving nearest neighbor interactions only. The idea is that the reduce density operator for two neighboring spins $\rho_{k,k+1}$ has (at most) rank 4, and it is always possible to find a positive operator Q (of rank equal to 5) acting on the subspace orthogonal to that spanned by the range of ρ (i.e. $Q\rho = 0$), so that $H|\Psi\rangle = 0$ with

$$H = \sum_k Q_{k,k+1} \geq 0. \tag{8}$$

2.5. *Expectation values*

Any expectation value of states in the form (1) can be readily evaluated in terms of the matrices A. For example

$$\langle\Psi|O_1 \otimes \ldots O_N|\Psi\rangle = \mathrm{Tr}(E^{O_1} \ldots E^{O_N}), \tag{9}$$

where

$$E^O := \sum_{\alpha,\beta=-1}^{1} \langle\alpha|O|\beta\rangle \left[A^\alpha \otimes \bar{A}^\beta\right], \tag{10}$$

where the bar denotes complex conjugation. This formula implies that any correlation function must generically decay as an exponential function, i.e., $\langle O_k O_{k+\Delta}\rangle \propto e^{-\Delta/\xi}$ as $\Delta \to \infty$.[7] The reason is as follows. We have

$$\langle O_k O_{k+\Delta}\rangle = \frac{\langle\Psi|\mathbf{1} \otimes \ldots \otimes \mathbf{1} \otimes O_k \otimes \mathbf{1} \ldots \otimes \mathbf{1} \otimes O_{k+\Delta} \otimes \mathbf{1} \otimes \ldots \otimes \mathbf{1}|\Psi\rangle}{\langle\Psi|\mathbf{1} \otimes \ldots \otimes \mathbf{1}|\Psi\rangle)}. \tag{11}$$

Both in the numerator and denominator we will have a term $(E^\mathbf{1})^\Delta$ in the trace. Let us denote by λ_n the sorted eigenvalues of $E^\mathbf{1}$, (i.e., $\lambda_n \geq \lambda_{n+1}$). In the denominator and for Δ sufficiently large only its largest eigenvalue λ_1 will survive. In the numerator two things can happen. Either the term that corresponds to the largest eigenvalue may vanish, in which case the second (or higher) largest eigenvalue will contribute. Then we will have the exponential decay. Alternatively, we will have that the correlation function will saturate to some given value (i.e. $\xi \to \infty$), indicating the presence of long range order.

2.6. *Entropy of blocks*

For a MPS, if we consider a block of r neighboring systems, it is very simple to show that the von Neumann entropy of the corresponding reduced density operator is bounded by $2\log D$. The reason is that, as explained above, such density operator has at most rank equal to D^2. This is the reason why, MPS cannot describe well critical systems, where the entropy grows with the logarithm of the block size.[15]

3. Two dimensions

States in the form (1) have also been used to represent 2D systems.[16] For simplicity let us consider a 2D square lattice of $N := N_h \times N_v$ systems. The idea there is to numerate them in such a way that they can be regarded as a long 1D system. In general, this method cannot be extended to larger systems since nearest neighbor interactions in the original 2D system (for example between 11 and 20) give rise to long interactions in the effective 1D description. Moreover, using the arguments presented in the previous section the entropy of some blocks does not scale as the area of the block, as it is expected for 2D configurations.[15] For example, the block formed by systems from 6–15 has at most a constant entropy of $2\log_2 D$.

For 2D systems the natural generalization of MPS can be done again by using projected entangled–pair states (PEPS). Each physical system of coordinates (h, v) is represented by four auxiliary systems $a_{h,v}$, $b_{h,v}$, $c_{h,v}$, and $d_{h,v}$ of dimension D (except at the borders of the lattice). Each of those systems is in a maximally entangled state ϕ with one of its neighbor. The state Ψ is obtained by applying to each site one operator $Q_{h,v}$ that maps the auxiliary systems onto the physical systems:

$$|\Psi\rangle = \sum_{s_{1,1},\ldots,s_{N_h,N_v}=1}^{d} \mathrm{F}_2(\{A_{h,v}^{s_{h,v}}\})|s_{1,1},\ldots,s_{N_h,N_v}\rangle. \tag{12}$$

Here, the A's are four–index tensors with elements

$$(A_{h,v}^{s})_{u,d,l,r} = \langle s|Q_{h,v}|u,d,l,r\rangle. \tag{13}$$

As in the 1D case, we associate each index of such tensors to each direction (up, down, left, and right). Thus, the position with coordinates (h, v) is represented by a tensor $(A_{h,v}^{s})_{u,d,l,r}$ whose index s represents the physical system, and the other four indices are associated with the bonds between the auxiliary systems and the neighboring ones. The function F_2 contracts all these indices u, d, l, r of all tensors A according to those bonds. Note that we can generalize this construction to any lattice shape and dimension, and that, by extending the construction given in the previous section, one can readily show that any state can be written as a PEPS.

3.1. *Example: Cluster State*

The 2-D cluster state has very interesting properties as it can be used as a resource for performing universal quantum computation (see below). It is the ground state of a Hamiltonian involving 5-body interactions given by

$$H_{clus} = \sum_{jk} \sigma^x_{j,k-1}\sigma^x_{j-1,k}\sigma^z_{j,k}\sigma^x_{j+1,k}\sigma^x_{j,k+1}. \tag{14}$$

The corresponding bonds are qubits ($D = 2$) and the projector Q is given by

$$Q = |0\rangle\langle 0,0,+,+| + |1\rangle\langle 1,1,-,-|, \tag{15}$$

where the virtual qubits are ordered clockwise.

3.2. *PEPS as ground states of Local Hamiltonians*

If we consider the reduced density operator ρ_B corresponding to a block of particles B, it will have a range which will be contained in that of an operator in which we would substitute the auxiliary particles that lie in the border of the block by the identity operator and then we would project. Thus, the rank is bounded by D^M, where M is the number of particles that lie at the border of the block. Thus, for a sufficiently large block we can always find a positive semidefinite operator Q fulfilling $Q\rho_B = 0$, and thus construct a local Hamiltonian for which the PEPS is a ground state. Apart from that, the von Neumann entropy of ρ_B will be bounded by $M \log_2 D$, a bound which scales with the "area" of the block.

3.3. *Expectation values*

We show now how to determine expectation values of operators in the state Ψ (12). We consider a general operator $O = \prod_{h,c} O_{h,c}$ and define the four–indices tensor

$$(E_{O_{h,c}})_{\tilde{u},\tilde{d},\tilde{l},\tilde{r}} := \sum_{s,s'=1}^{d} \langle s|O_{h,c}|s'\rangle (A^s)_{u,d,l,r} \bar{A}^{s'}_{u',d',l',r'} \tag{16}$$

where the indices are combined in pairs, i.e., $\tilde{u} = (u,u'), \tilde{d} = (d,d'), \tilde{l} = (l,l')$, and $\tilde{r} = (r,r')$. One can easily show that $\langle\Psi|O|\Psi\rangle = \mathrm{F}_2(E_{O_{h,c}})$. Thus, the evaluation of expectation values consists of contracting indices of the tensors E. In order to show how to do this in practice, we notice that the tensors associated to the first and last rows, once contracted, can be reexpressed in terms of a MPS. In particular, we define [compare (1)]

$$|U_1\rangle := \sum_{d_1\dots d_N=1}^{D^2} \mathrm{F}_1(E^{d_1}_{1,1}\dots E^{d_N}_{1,N})|d_1\dots d_N\rangle, \tag{17}$$

$$\langle U_{N_v}| := \sum_{u_1\dots u_N=1}^{D^2} \mathrm{F}_1(E^{u_1}_{N,1}\dots E^{u_N}_{N,N})\langle u_1\dots u_N|. \tag{18}$$

Here we have used the short–hand notation $E_{h,c} := E_{O_{h,c}}$, and the fact that the tensors in the first and last rows have at most three indices. Thus, the horizontal indices (l,r) of the tensors play the role of the indices of each matrix, whereas the vertical ones (d) plays the role of the indices corresponding to the physical systems in 1D. Analogously, the rows $2, 3, \ldots, N_v - 1$ can be considered as matrix product operators (MPO),[5]

$$U_k := \sum_{d_1, u_1 \ldots = 1}^{D^2} \mathrm{F}_1(E_{1,1}^{d_1,u_1}, \ldots, E_{1,N}^{d_N,u_N}) |d_1 \ldots d_N\rangle\langle u_1 \ldots u_N|.$$

We have $\langle\Psi|O|\Psi\rangle = \langle U_N|U_{N-1}\ldots U_2|U_1\rangle$.

The evaluation of expectation values poses a serious problem since the number of indices proliferate after each contraction. For example, the vector $|U_2\rangle := U_2|U_1\rangle$ can be written as the MPS (17) but with the substitution

$$E_{1,n}^{d_n} \to \sum_{d_n=1}^{D^2} E_{1,n}^{d_n} \otimes E_{2,n}^{d'_n, d_n}. \tag{19}$$

This last tensor has more (right and left) indices than the original one. Thus, every time we apply the MPO U_k to a MPS $|U_{k-1}\rangle$ the number of indices increases, and thus the problem soon becomes intractable. Now we review the numerical algorithm introduced in Ref. 2 to determine $\mathrm{F}_2(E_{O_{h,c}})$ and to overcome this problem.

Given an unnormalized MPS, $|\psi_A\rangle$, parameterized by $D \times D$ matrices, $\{A_k^s\}$, the goal is to find another MPS, $|\psi_B\rangle$, parameterized by $D_f \times D_f$ matrices, $\{B_i^s\}$, where $D_f < D$ is a prescribed number. This has to be done such that $K := \| |\psi_A\rangle - |\psi_B\rangle \|^2$ is minimal, i.e., such that that $|\psi_B\rangle$ gives the best approximation to $|\psi_A\rangle$. We have developed an algorithm that achieves this task in an iterative way. The key insight is that K is quadratic in all components of the matrices $\{B_k^s\}$, and hence if all these matrices are fixed except one of them (say B_j^s) K is quadratic in the components of B_j^s; the optimal choice for B_j^s thus amounts to solving a trivial system of linear equations. Having done that, one moves to the next site $j+1$, fixes all other ones and repeats the same procedure. After a few sweeps back and forth the optimal MPS is found. Note that the error in the approximation is exactly known and if it becomes too large one can always increase D_f; in all relevant situations we encountered the error could be made very small even with moderate D_f. The same reasoning holds for MPS defined with periodic instead of open boundary conditions. In this latter case considered here, one can further simplify the system of linear equations by performing a singular value decomposition of B_j^s and keeping only one of the unitary matrices at each step, analogously as one does in DMRG.[17]

Thus, in order to evaluate an arbitrary expectation value we first determine the MPS $|\tilde{U}_2\rangle$ which is the closest to $U_2|U_1\rangle$ but with a fixed dimensions D_f of the corresponding matrices. Then, we determine $|\tilde{U}_3\rangle$, which is the closest to $U_3|\tilde{U}_2\rangle$, and continue in this vein until we finally determine $\langle\Psi|O|\Psi\rangle \simeq \langle U_N|\tilde{U}_{N-1}\rangle$. Interestingly enough, this method to calculate expectation values and to determine optimal approximations to MPS can be adapted to develop very efficient algorithms to determine the ground states of 2D Hamiltonians and the time evolution of PEPS by extending DMRG and the time evolution schemes to 2D (see next section).

4. Applications

4.1. *Quantum simulations*

PEPS can be used to approximate the states that appear in standard problems in quantum–many body physics. The idea can be summarized as follows. For a fixed dimension D we determine the tensors A (which characterize the PEPS Ψ) such that (in some sense) it is as close as possible to the one wants to simulate. By increasing D one can obtain better approximations until the desired precision is reached. The determination of the tensors A is carried out as follows: first, all the tensors except one, say the one corresponding to the k–th site, are fixed to some "good guess" (or chosen randomly). Then the tensor A_k is determined by minimizing some "cost" function, which makes sure that we approach the desired state. The minimization of such function corresponds to solving a generalized eigenvalue problem $Mx = \lambda Nx$ or to a set of linear equations $Mx = b$, where x is a vector formed by all the elements of A, and the matrices M and N, or the vector b can be determined in terms of the rest of the As. Once this is done, one chooses another location k' and determines $A_{k'}$ by using the same procedure. By continuing in this way, one reaches a (in principle, local) minimum of the cost function, and therefore one may obtain a good approximation to the desired state.

Let us show first of all how to determine the ground state of some given Hamiltonian H using the method outlined above. In this case the function to be minimized is

$$f(A_k) = \langle\Psi|H|\Psi\rangle/\langle\Psi|\Psi\rangle. \tag{20}$$

Given the fact that Ψ depends linearly on the coefficients of the tensor A_k, we can write $f(x) = x^T Mx/x^T Nx$, where M and N can be determined using the methods exposed in previous sections to determine expectation values. Minimizing this function one corresponds to solving the generalized eigenvalue problem $Mx = \lambda Nx$ where λ is minimal. Implementing this scheme numerically is rather simple, although it may be very time consuming. The algorithm can be speeded up by choosing a specific order in which the different locations are selected for the iteration, so that one can store some information in each step ans use it in later steps. Furthermore, one can make use of the conditions (2) in order to simplify further the problem. For the case of 1D Hamiltonians with open boundary conditions, this algorithm coincides with the well known DMRG[17–19] (except for the fact that the latter has a very efficient way of finding a "good guess" to start with). For periodic boundary conditions,[4] it outperforms DMRG and one can gain several orders of magnitude in precision by using our algorithm. Note that, in contrast to Monte Carlo methods,[20] the algorithm is not restricted to bosonic problems or not frustrated ones, and thus in two or higher dimensions may allow to solve problems which have been untractable so far.[2]

One can also determine (real or imaginary) time evolution of a system of particles interacting according to some given Hamiltonian H.[5] The idea here is to decompose the evolution in small time steps, and ea ch of this time steps to use the ideas outlined above. That is, one can easily determine $|\Psi'\rangle = e^{-iH\delta t}|\Psi_0\rangle$, where Ψ_0 is a MPS. It is simple to show that Ψ' is also a PEPS but with a higher value of D. Now one can find a PEPS Ψ_1 with the original D which is as close as possible to Ψ'. The figure of merit is now $||\Psi' - \Psi_1||_2^2$, which is again quadratic on the coefficients of the tensor A_k corresponding to Ψ_1, and thus the minimization can be carried out by solving a simple problem as explained in the previous section. After several rounds of minimizations, one obtains all the A_k and can apply another short time evolution. In one dimension, this approach is reminiscent of the one introduced and developed in Refs. 21–23, although the one presented here is optimal.

In addition, one can also introduce auxiliary particles to purify mixed states and thus develop algorithms which fork for density operators[5] (as opposed to the pure states considered here). In this way, for example, one can determine finite temperature properties or the evolution in the presence of decoherence.[5,9]

4.2. *Quantum Computation*

The existence of entangled states of several particles offers the possibility of performing certain computational tasks in times much shorter than the ones taken by common (classical) computers. By acting on a system entangled to to other systems, one modifies the state of the whole system at the same time, which leads to an important speed up in several computations.[10] In particular, if one could build a quantum computer one could perform several tasks which are not possible now, like for example to decompose very large numbers (of $n \gg 1$ digits) into prime factors in a time that scales polynomially with n.[24]

The standard scenario for quantum computation consists of a set of qubits (two–level systems) which can be manipulated using single– and two–qubit gates, and which can be individually measured.[25–27] In fact, by concatenating these operations it is possible to implement any arbitrary unitary operation to the qubits, and to perform arbitrary measurements. A seemingly different scenario consists of preparing a fixed entangled state and performing local measurement (i.e. in each single qubit) only, without the need of two–qubit gates.[11,28] The entangled state which is used is nothing else but the 2D cluster state which we have introduced in the previous section, and thus it is a PEPS. In fact, it is possible to understand this new way of quantum computation in terms of the standard one[3] by reexpressing it in terms of the artificial particles that build up the PEPS. In that case, certain local measurements effectively teleport[14,29] the state of one artificial qubit from one location to the next (sitting to its right) and, at the same time, apply a desired single-qubit operation. Other local measurements teleport the state of two artificial qubits, which are lying on the same vertical line at two neighboring locations, to the next locations to their right, and, at the same time, a two–qubit gate is applied. Thus, by a sequence of measurements the states of the artificial qubits which are lying on the vertical line all the way to the left are teleported to the right of the cluster state and, at the same time, the desired unitary operation (i.e. the quantum computation) is performed. In a certain sense, the horizontal axis of a cluster state plays the role of "time" whereas the vertical one plays the role of the qubits contained in the quantum register. Once this is recognized, it is simple to determine other PEPS which can act as a universal resource for quantum computation, in as much the same way like the cluster states. Furthermore, it can be shown[3] that all stabilizer states are also PEPS with $D = 2$. Stabilizer states play an essential role quantum error correction codes.[30–32]

4.3. *Further results*

In Ref. 33 we have quantified how well matrix product states approximate exact ground states of 1-D quantum spin systems as a function of the number of spins and the entropy of blocks of spins. We have also investigated the convex set of local reduced density operators of translational invariant systems. The results give a theoretical justification for the high accuracy of renormalization group algorithms, and justifies their use even in the case of critical systems.

The projected entangled pair state (PEPS) representation of quantum states on two-dimensional lattices induces an entanglement based hierarchy in state space. We have shown[34] that the lowest levels of this hierarchy exhibit an enormously rich structure including states with critical and topological properties as well as resonating valence bond states. We have proven, in particular, that coherent versions of thermal states of any local 2D classical spin model correspond to such PEPS, which are in turn ground states of local 2D quantum Hamiltonians. This correspondence maps thermal onto quantum fluctuations, and it allows us to analytically construct critical quantum models exhibiting a strict area law scaling of the entanglement entropy in the face of power law decaying correlations. Moreover, it enables us to show that there exist PEPS within the same class as the cluster state, which can serve as computational resources for the solution of NP-hard problems.

We have also investigated quantum phase transitions (QPTs) in spin chain systems characterized by local Hamiltonians with matrix product ground states.[35] We have shown how to theoretically engineer such QPT points between states with predetermined properties. While some of the characteristics of these transitions are familiar, like the appearance of singularities in the thermodynamic limit, diverging correlation length, and vanishing energy gap, others differ from the standard paradigm: In particular, the ground state energy remains analytic, and the entanglement entropy of a half-chain stays finite. Examples demonstrate that these kinds of transitions can occur at the triple point of 'conventional' QPTs.

References

1. R. P. Feynman, *Int. J. Theor. Phys.* **21**, 467 (1982).
2. F. Verstraete and J. I. Cirac, cond-mat/0407066.
3. F. Verstraete and J.I. Cirac, quant-ph/0311130.

4. F. Verstraete, D. Porras and J.I. Cirac, cond-mat/0404706.
5. F. Verstraete, J.J. Garcia-Ripoll and J. I. Cirac, cond-mat/0406426.
6. F. Verstraete, M.A. Martín-Delgado and J.I. Cirac, *Phys. Rev. Lett.* **92**, 087201 (2004).
7. M. Fannes, B. Nachtergaele and R.F. Werner, *Comm. Math. Phys.* **144**, 443 (1992).
8. S. Ostlund and S. Rommer, *Phys. Rev. Lett.* **75**, 3537 (1995).
9. M. Zwolak and G. Vidal, cond-mat/0406440.
10. A thorough review about classical and quantum information and computation can be found in A. Galindo and M.A. Martin-Delgado, *Rev. Mod. Phys.* **74**, 347 (2002).
11. H. Briegel and R. Raussendorf, *Phys. Rev. Lett.* **86**, 910 (2001).
12. I. Affleck, T. Kennedy, E.H. Lieb and H. Tasaki, *Commun. Math. Phys.* **115**, 477 (1988).
13. J. Dukelsky, M.A. Martin-Delgado, T. Nishino, and G. Sierra, *Europhys. Lett.* **43**, 457 (1997).
14. C.H. Bennett, G. Brassard, c. Crepeau, R. Jozsa, A. Peres, and W. Wootters, *Phys. Rev. Lett.* **70**, 1895 (1993).
15. See G. Vidal, J. I. Latorre, E. Rico, and A. Kitaev, *Phys. Rev. Lett.* **90**, 227902 (2003) and references therein.
16. S. Liang and H. Pang, *Phys. Rev.* **B49**, 9214 (1994); S. R. White, *Phys. Rev. Lett.* **77**, 3633 (1996); T. Xiang, J. Lou, and Z. Su, *Phys. Rev.* **B64**, 104414 (2001).
17. S. R. White, *Phys. Rev. Lett.* **69**, 2863 (1992).
18. S. R. White, *Phys. Rev.* **B48**, 10345 (1992).
19. See, for example, *Density-Matrix Renormalization*, Eds. I. Peschel, *et al.*, (Springer Verlag, Berlin, 1999).
20. D.M. Ceperley and B.J. Alder, *Phys. Rev. Lett.* **45**, 566 (1980).
21. G. Vidal, *Phys. Rev. Lett.* 91, 147902 (2003);
22. A. Daley, C. Kollath, U. Schollwoeck, and G. Vidal, *J.Stat.Mech.: Theor.Exp.* P04005 (2004);
23. S. R. White and A.E. Feiguin, cond-mat/0403310.
24. P.W. Shor P. W., *Proc. of the 3rd Annual Symposium on the Foundations of Computer Science* (IEEE Computer Society Press, Los Alamitos, Ca, 1994) p. 124.
25. M. Nielsen and I. Chuang, *Quantum computation and quantum information*, (Cambridge University Press, Cambridge, U.K, 2000).
26. C. H. Bennett, *Phys. Today*, Oct., 24, (1994).
27. S. Lloyd, *Scientific American*, Oct., 140, (1995).
28. R. Raussendorf and H.J. Briegel, *Phys. Rev. Lett.* **86**, 5188 (2001).
29. D. Gottesman and I. Chuang, *Nature* **402**, 390 (1999).
30. P.W. Shor, *Phys. Rev.* **A52**, 2493 (1995).
31. A. M. Steane, *Phys. Rev. Lett.* **77**, 793 (1996).
32. J. Preskill, *Proc. R. Soc.* **454**, 385 (1998).
33. F. Verstraete and J. I. Cirac, cond-mat/0505140.
34. F. Verstraete, M. Wolf, D. Perez-Garcia, and J.I. Cirac, quant-ph/0601075.
35. M. Wolf, F. Verstraete, G. Ortiz, and J. I. Cirac, cond-mat/0512180.

QUANTUM MONTE CARLO STUDY OF OVERPRESSURIZED LIQUID ^{4}HE AT ZERO TEMPERATURE

LEANDRA VRANJEŠ MARKIĆ

Faculty of Natural Sciences, University of Split, N. Tesle 12, 21 000 Split, Croatia
Institut für Theoretische Physik, Johannes Kepler Universität, A 4040 Linz, Austria
leandra@pmfst.hr

JORDI BORONAT, JOAQUIM CASULLERAS, CLAUDIO CAZORLA

Departament de Física i Enginyeria Nuclear, Campus Nord B4-B5,
Universitat Politècnica de Catalunya, E-08034 Barcelona, Spain
jordi.boronat@upc.es

We present a diffusion Monte Carlo simulation of metastable superfluid ^{4}He at zero temperature and pressures beyond freezing ($\sim$ 25 bar) up to 275 bar. The equation of state of liquid ^{4}He is extended to the overpressurized regime, where the pressure dependence of the static structure factor and the condensate fraction is obtained. Along this large pressure range, excited-state energy corresponding to the roton has been determined using the release-node technique. Our results show that both the roton energies and the condensate fraction decrease with increasing pressure but do not become zero. We compare our calculations to recent experimental data in overpressurized regime.

Keywords: ^{4}He; excitations; Monte Carlo

1. Introduction

Fluids in metastable phases, both below the saturated vapor pressure and above the freezing point, present a research topic of fundamental interest for both experiment and theory.[1] The extreme purity of liquid helium at very low temperatures allows researchers to avoid nucleation on impurities and walls, and thus makes it the best suited system for the study of homogeneous nucleation, an intrinsic property of the liquid.

The negative pressure regime has been extensively studied by Caupin, Balibar and collaborators using an acoustic technique, in which high intensity ultrasound bursts are focused in bulk helium and the possible nucleation of bubbles or crystals is studied by shining laser light through the acoustic focal region.[2,3] In liquid ^{4}He they have measured a negative pressure only 0.2 bar above the spinodal point of -9.6 K predicted by microscopic theory.[4,5] By slightly modifying their experimental setup, so as to avoid nucleation on the walls, the same group has recently succeeded in pressurizing the small quantities of liquid ^{4}He up to 160 bar at temperatures 0.05 K $< T <$ 1 K.[6] This is the highest pressure ever realized in liquid metastable ^{4}He and is much larger than the liquid-solid equilibrium pressure, which at $T = 0$ K is 25.3 bar. All along this large increase of the liquid pressure beyond the freezing point no solidification was observed. It is not clear if a limit exist to how far one can pressurize liquid helium. Namely, there is a significant difference in the nature of the two metastable phases. At negative pressure, there exists an end point (spinodal point) where the speed of sound becomes zero. Since at this point compressibility becomes infinite it is thermodynamically forbidden to cross it maintaining a homogeneous liquid phase. So, below that pressure the liquid brakes into droplets. Such type of an end point does not exist on the overpressurized side. However, as the excitation energy of rotons decreases with pressure (a fact well-known from the stable phase), Schneider and Enz[7] suggested that the pressurized phase has also an end point corresponding to the pressure where the excitation energy of the roton might vanish. On the other hand, Jackson*et al.*[8] and Halinen *et al.*,[9] find another instability that could cause the liquid-solid transition in ^{4}He which involves a soft mode having 6-fold symmetry in the two-body correlations.

It is also interesting to find out what the nature of metastable liquid ^{4}He at these high pressures is; superfluid or liquid. Recent experiments have shown that by immersing liquid ^{4}He in different porous media one can create metastable phases, at both negative pressures and pressures above the stable phase, that have a long enough lifetime to allow study by neutron scattering.[10,11] In this way information on the phonon-roton excitations has been obtained for negative pressures up to -5 bar[10] and overpressures up to ~40 bar.[11] The results of the latter experiment by Pearce *et al.*[11] show the roton excitation falling with pressure in the overpressurized regime, between 25 and 38.5 bar, while no roton is observed in the solid phase. At the liquid-solid transition, the roton energy is still finite, which means that ^{4}He is superfluid when it crystallizes. This result seems to be in disagreement with a recent experimental work by Yamamoto *et al.*[12] who reported superfluid transition temperature approaching $T_c = 0$ at a pressure $P \sim 35$ bar in a porous material, which implies the existence of a quantum phase transition from superfluid to normal liquid at zero temperature. It is possible that the differences observed in the two experiments are a consequence of the different pore diameters used (44 Å in Ref. 11 and 25 Å in Ref. 12), but additional work is needed to confirm this argument. In the direction of quantum phase transition, Nozières[13] has predicted recently that the condensate fraction could vanish at a certain pressure, and therefore a normal liquid before solidification would be possible.

Contrary to the negative pressure region where theoretical knowledge is rather complete,[1] the overpressurized liquid has remained nearly unexplored, especially at the level of microscopic treatment. Recently, we have applied diffusion Monte Carlo (DMC) method to the overpressurized phase up to $P \sim 275$ bar.[14,15] We considered it probably the best suited way to deal with this metastable regime because the physical phase of the system is controlled by the trial wave function used for importance sampling. Our results, in particular those for condensate fraction and the excitation energy of the roton, show that metastable helium remains superfluid all along the large pressure increase from the solidification up to the highest pressures studied. In this paper, we will first present our calculations of the ground state and then proceed to excitations corresponding to the roton.

2. Ground state

The DMC method is nowadays a well-known tool in the of study quantum fluids and solids at zero temperature. For this reason we shall only give here the basic expressions. This fully microscopic approach solves stochastically the imaginary-time Schrödinger equation.

$$-\hbar\frac{\partial\Psi(\boldsymbol{R},t)}{\partial t} = (H - E_t)\Psi(\boldsymbol{R},t) \ , \tag{1}$$

where E_t is a constant acting as a reference energy, $\boldsymbol{R} \equiv (\boldsymbol{r}_1, \ldots, \boldsymbol{r}_N)$ is a *walker* in Monte Carlo terminology and the N-particle Hamiltonian is given by expression

$$H = -\frac{\hbar^2}{2m}\sum_{i=1}^{N}\boldsymbol{\nabla}_i^2 + \sum_{i<j}^{N} V(r_{ij}) \ . \tag{2}$$

The He-He interaction $V(r)$ corresponds to the HFD-B(HE) Aziz potential.[18] The usual practice is to introduce the trial wave function $\psi(\boldsymbol{R})$ for importance sampling and to rewrite the Schrödinger equation in terms of $\Phi(\boldsymbol{R},t) = \Psi(\boldsymbol{R},t)\psi(\boldsymbol{R})$. In the limit $t \to \infty$ only the lowest energy eigenfunction, not orthogonal to $\psi(\boldsymbol{R})$, survives. For the ground state of bosonic system, such as liquid ^{4}He, the DMC gives exact results apart from the statistical errors.

The trial wave function for the simulation of the liquid in its ground state is of Jastrow type. As in the previous calculations in the stable domain,[16] we have used a model proposed by Reatto[17] which includes nearly optimal short and medium range two-body correlations

$$\psi(\mathbf{R}) = \prod_{i<j}\left(-\frac{1}{2}\left(\frac{b}{r_{ij}}\right)^5 - \frac{L}{2}exp\left[-\left(\frac{r_{ij}-\lambda}{\Lambda}\right)^2\right]\right) , \tag{3}$$

This model has variational parameters L, λ, Λ and b, which were optimized in a variational Monte Carlo calculation.[14]

In order to compare our results with the behavior of solid ^{4}He, we have also carried out DMC simulations of the crystalline hcp phase. In this case, the trial wave function is given by a Nosanow-Jastrow model

$$\psi_{NJ}(\boldsymbol{R}) = \psi_J(\boldsymbol{R})\prod_i^N h(r_{iI}), \tag{4}$$

where $h(r)$ is a gaussian function linking every particle i to a fixed lattice point $\boldsymbol{r}_I$.

We have assumed periodic boundary conditions in all the simulations. Since the densities in this study are relatively large it has been necessary to carry out a detailed analysis of the parameters influencing the simulation in order to eliminate possible bias. The most important check concerns residual size effects. This is achieved by summing proper tail corrections to the partial components of the energy (potential and kinetic) and by calculating the energies using different number of particles N. We have found this number increasing with density, from $N = 150$ near freezing to $N = 250$ at the highest density. In addition, the dependence on the mean population of walkers and the time step in the employed second-order algorithm[16] has also been carefully determined to eliminate any possible systematic error.

The complete equation of state, from the spinodal point up to the highest densities is plotted in Fig. 1. We have found that DMC energies are accurately param-

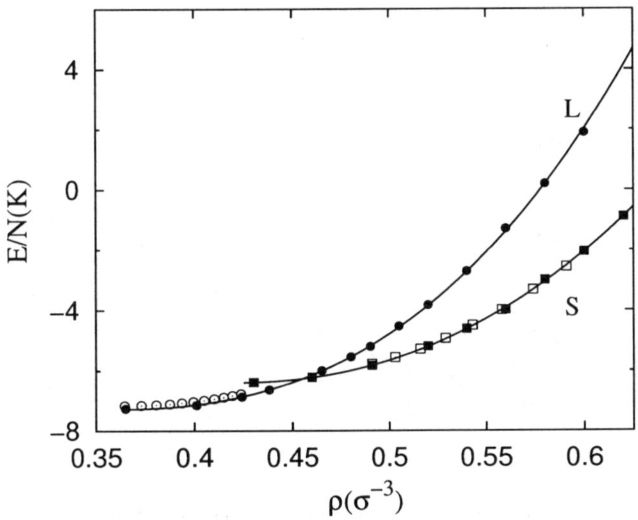

Fig. 1. Energy per particle of liquid ^{4}He from the equilibrium density up to the highest density calculated, 0.6 σ^{-3}, (solid circles). The solid line corresponds to the fit to the DMC energies using Eq. (5), and the open circles are experimental data in the stable regime from Ref. 19. DMC results for the solid phase are shown as solid squares and compared with experimental data from Ref. 20 (open squares). The error bars of our data are smaller than the size of the symbols.

eterized, from the spinodal point up to the highest densities in our calculation, by the analytical form[14]

$$e(\rho) = e_0 + e_1\,(\rho/\rho_c - 1)\,(1 - (\rho/\rho_c - 1)) + b_3(\rho/\rho_c - 1)^3 + b_4(\rho/\rho_c - 1)^4 \;, \quad (5)$$

with $e = E/N$, and $\rho_c = 0.264\ \sigma^{-3}$ ($\sigma = 2.556$ Å) the spinodal density. The rest of parameters in Eq. (5) are $e_0 = -6.3884(40)$ K, $e_1 = -4.274(31)$ K, $b_3 = 1.532(12)$ K, and $b_4 = 1.433(24)$ K, the figures in parenthesis being the statistical errors.[14] Fig. 1 also shows DMC results for the energies of the solid phase, calculated using the Nosanow-Jastrow trial wave function and an hcp lattice. In all the density regime studied, the system is *artificially* maintained in a homogeneous liquid phase. The comparison between the liquid and solid phase simulations shows clearly that DMC is effectively able to study the overpressurized liquid phase in spite of not being the ground-state (minimum energy) configuration, which obviously corresponds to the solid phase beyond the freezing point. This can be achieved by the DMC method because the physical phase is implicitly contained in the importance sampling trial wave function and it is not changed along the simulation. DMC should arrive at the *true* ground state (solid) in the limit of infinite simulation time. Since achievement of this limit requires breaking of symmetry imposed by the importance sampling wave function, it is not observed in the usual time schedules. In particular, in the course of our simulations we have observed no signal of freezing and therefore the results obtained correspond unambiguously to the metastable liquid phase.

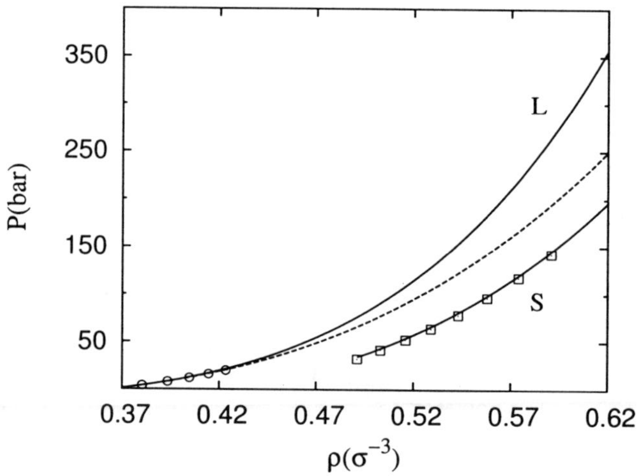

Fig. 2. Pressure as a function of the density. The solid lines stand for the DMC results obtained from the equations of state of the liquid and solid phases shown in Fig. 1. The dashed line is the extrapolation from experimental data;[6,21] the symbols correspond to experimental data for the liquid[19] and solid phases.[20]

Using the equation of state (5), we have obtained the pressure from its thermodynamic definition

$$P(\rho) = \rho^2(\partial e/\partial \rho). \tag{6}$$

The results, shown in Fig. 2, reproduce accurately the experimental data[19] of the pressure as the function of the density in the stable regime and predict a pressure $P \simeq 275$ bar at the highest density evaluated, $\rho = 0.6\ \sigma^{-3}$. They are compared in the same figure with the analytic form suggested in Ref. 6, adjusted to Abraham's experimental data.[21] Below the freezing point, both curves agree but they give significantly different values at higher densities; the difference amounts to ~ 100 bar at $\rho = 0.6\ \sigma^{-3}$. As a matter of comparison, the pressure of the solid phase, derived from the DMC equation of state is also shown in Fig. 1. Similar discrepancies between our results and the extrapolations of the experimental data from the stable region appear at high pressure results of the speed of sound and amount to almost 200 m/s at the highest density.[14]

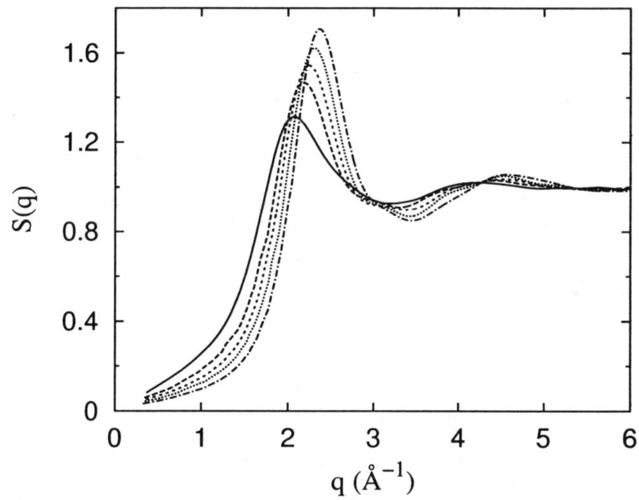

Fig. 3. Static structure function of the liquid phase for different densities. From bottom to top in the height of the main peak, the results correspond to densities 0.365, 0.438, 0.490, 0.540, and 0.6 σ^{-3}.

An important quantity in the study of quantum liquids is the static structure factor $S(q) = \langle \rho_{\boldsymbol{q}} \rho_{-\boldsymbol{q}} \rangle / N$, with $\rho_{\boldsymbol{q}} = \sum_{i=1}^{N} e^{i\boldsymbol{q}\cdot\boldsymbol{r}_i}$. Our results are reported in Fig. 3, for densities ranging from the equilibrium up to the highest densities studied. The results show the expected behavior: when ρ increases, the strength of the main peak increases and moves to higher momenta in a monotonic way. At low momenta, the slope of $S(q)$ decreases with the density, following the limiting behavior $\lim_{q\to 0} S(q) = \hbar q/(2mc)$ driven by the speed of sound c. A characteristic feature of a solid phase is the presence of high-intensity peaks of the static structure function in the reciprocal lattice sites. Following the overpressurized liquid phase, we have not observed this feature which confirms the liquid nature of the system. According to Schneider and Enz, the instability of the liquid against the solid ought to be accompanied by the blowup of the main peak of $S(q)$.[7] Here, we see a rather slow growth of the main peak with pressure which indicates that the predicted instability is located much higher in pressure.

A characteristic signature of bulk superfluid ^{4}He is a finite value of its condensate fraction, i.e., the fraction of particles occupying the zero-momentum state. As usual in a homogeneous system, we have extracted the condensate fraction n_0 from the long range behavior of the one-body density matrix, $\lim_{r\to\infty} \rho(r) = n_0$. To this end,

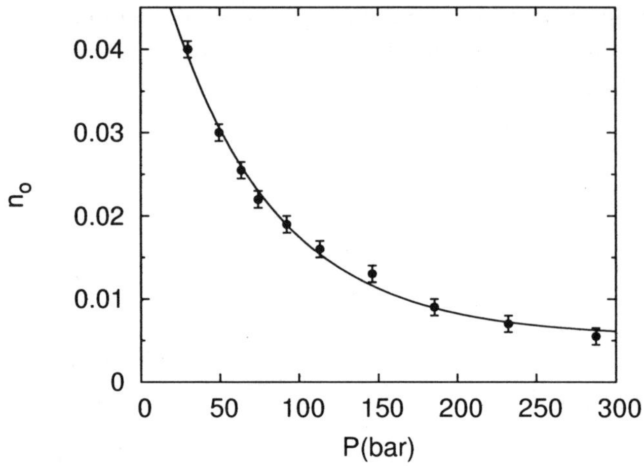

Fig. 4. Condensate fraction of liquid ^{4}He in the overpressurized regime. The line is an exponential fit to the DMC results.

$\rho(r)$ is sampled by means of the quotient

$$\left\langle \frac{\psi(\boldsymbol{r}_1,\ldots,\boldsymbol{r}_i+\boldsymbol{r},\ldots,\boldsymbol{r}_N)}{\psi(\boldsymbol{r}_1,\ldots,\boldsymbol{r}_i,\ldots,\boldsymbol{r}_N)} \right\rangle , \tag{7}$$

evaluated in the configuration space, over a set of random displacements $\boldsymbol{r}$ of particle i. The results obtained for n_0, from the melting pressure up to nearly 300 bar, are plotted in Fig. 4. The line on top of the data corresponds to an exponential fit which reproduces quite accurately our DMC results. As one can see in the figure, n_0 decreases quite fast until $P = 100$ bar and then the slope decreases, approaching a value $n_0 \simeq 0.005$ at the highest density. With the same procedure, we obtained[16] $n_0 = 0.084(1)$ at the equilibrium density $\rho_0 = 0.365\ \sigma^{-3}$, value which is compatible with PIMC estimations at low temperature[22] (0.069(10) at $T = 1.18$ K and 0.087(10) at $T = 1.54$ K). An exponential decay of n_0 with density up to pressures ≈ 80 bar was also obtained using the variational path integral method (VPI) in Ref. 23. VPI method, using a similar estimator as in PIMC, gives n_0=0.069(5) at ρ_0.

3. Excited state

It has been experimentally demonstrated that the energy of the roton excitations reduces with rising pressure.[11,24] The vanishing of the roton energy at some pressure has been proposed as the intrinsic instability limit of the liquid against a solid. This hypothesis led us to carry out a DMC released-node (RN) calculation of the roton energy beyond the freezing point. The same methodology was used in the past in a DMC calculation of the phonon-roton spectrum at equilibrium and freezing densities[26] arriving at an accurate description of the experimental data. The simulation of the roton is more involved than the simulation of the ground state because of the sign problem associated with the excited state wave function. As a trial wave function for importance sampling we have taken an eigenstate of the total momentum operator which incorporates backflow correlations, as originally proposed by Feynman and Cohen,[25] $\psi_{BF}(\boldsymbol{R}) = \psi_e(\boldsymbol{R})\psi_J(\boldsymbol{R})$, where

$$\psi_e(\boldsymbol{R}) = \sum_{i=1}^{N} e^{i\boldsymbol{q}\cdot\tilde{\boldsymbol{r}}_i} \tag{8}$$

with $\tilde{\boldsymbol{r}}_i = \boldsymbol{r}_i + \sum_{j\neq i}^{N} \eta(r_{ij})\boldsymbol{r}_{ij}$, and $\eta(r) = \lambda \exp[-((r-r_b)/\omega_b)^2]$. In the DMC implementation of the program we have used a superposition of the states with momenta $\boldsymbol{q}$ and $-\boldsymbol{q}$ which are degenerate in energy. This enabled us to avoid working with a complex wave function. In this way the calculation of the excited state energy turned into a fermion-like problem since the resulting trial wave function is real but not positive everywhere.[26] In a first step, we have used the fixed-node (FN) approximation, which provides an upper bound to the roton energy. We have verified that the introduction of backflow correlations in the trial wave function produces results quite close to experimental data at the equilibrium density, especially near the roton minimum. The nodal constraint imposed by FN is removed, in a second step, by using the RN technique. In the RN approach, walkers are allowed to cross the nodal surface imposed by ψ and survive for a finite lifetime t. This is achieved by introducing the auxiliary guiding wave function $\psi_g(\mathbf{R})$, positively defined everywhere, which approaches the $|\psi(\mathbf{R})|$ away from the nodal surface and is non-zero in the nodes. The function $\psi_g(\mathbf{R}) = \psi_{\mathbf{J}}(\mathbf{R})(\psi_{\mathbf{e}}^2(\mathbf{R}) + a^2)^{1/2}$ achieves this goal for the proper choice of parameter a. The excited state energy is estimated through an exponential fit $E(t) = E_r + Ae^{-(t/\tau)}$, with t the released time. The uncertainty of this extrapolation is under control since in all cases the difference between considering the last calculated point in released time or E_r is of the same order as the statistical noise. The energy of the roton is then expressed as the difference between the excited and the ground state energy $\epsilon_r = E_r - E_0$. Since both the ground and the excited state energy have statistical errors, the resulting errors for the roton energy are quite large and difficult to reduce.

The results for the roton energy as a function of the pressure are shown in Fig. 5 and compared to experimental data obtained by neutron scattering experiments on superfluid ^{4}He in a porous media and up to 40 bar.[11] From 0 to 40 bar, the measured ϵ_r decreases linearly with the pressure and our data reproduces well this behavior. However, increasing the pressure we find that this slope is reduced. At the highest density studied the roton energy is still different from zero ($\epsilon_r = 2.8 \pm 1.2$ K at $\rho = 0.58\ \sigma^{-3}$). At each density, the number of particles has been adjusted to be as close to the roton momentum as possible. Due to the finite size of the simulation cell only discrete values of $\boldsymbol{q}$ are accessible , so corrections to the energy due to this fact are possible. However, we estimate them to be less than 0.5 K in all cases. In Fig. 6 we show the obtained values of the roton momentum and compare them with the measurements in the stable[24] and overpressurized regime.[11] Our results do not follow linear dependence with pressure, as the data in the stable liquid regime

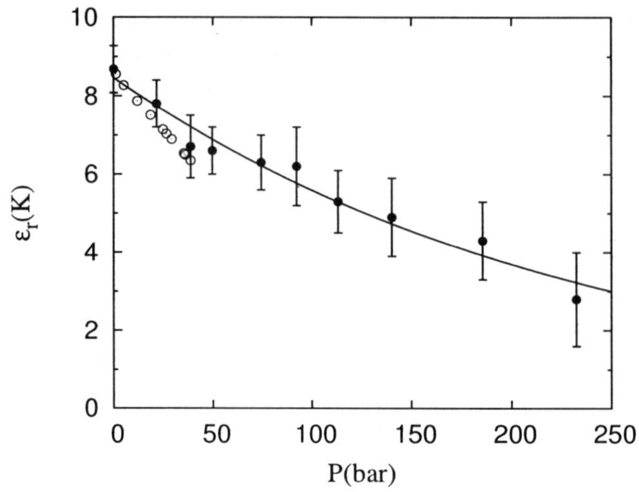

Fig. 5. Roton energy as a function of the pressure (solid circles). Open circles stand for experimental data from Ref. 11. The line is an exponential fit to the DMC data.

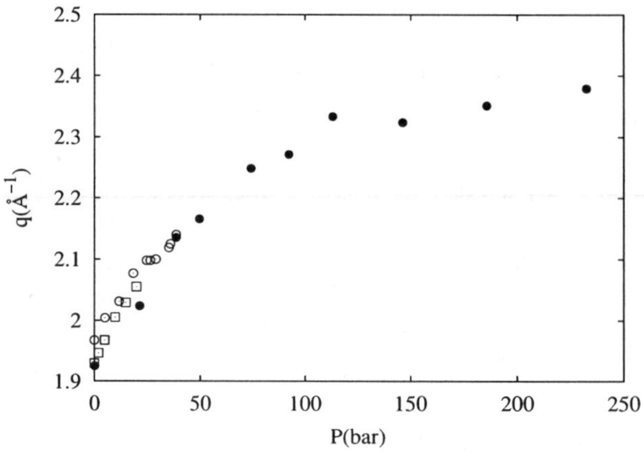

Fig. 6. Roton momentum as a function of the pressure (solid circles). Open circles stand for experimental data from Ref. 11 and open square for experimental data from Ref. 24.

might suggest.[24] As the pressure is increased, the slope of the roton momentum as a function of pressure decreases.

4. Conclusion

It has been shown that overpressurized metastable liquid ^{4}He can be studied with DMC method.[14,15] Along the pressure range from freezing to almost 300 bar, no signature of liquid/solid instability appeared. The finite value of condensate fraction and roton gap imply helium ^{4}He remained superfluid despite the fact that the pressure increased more than 10 times. Static structure factor, condensate fraction and roton energy are however driven by the density (which increased less than 50% from freezing to maximum density studied) and not by the pressure. As can be seen in the Fig. 2, equal density increments in the stable and metastable regime produce clear differences in the pressure increase. This leads, in the density range studied, to an approximated exponential decrease with rising pressure of magnitudes like n_0 and ϵ_r.

Acknowledgements

We thank M. Barranco, F. Caupin, J. Navarro, and M. Saarela for useful discussions. Partial financial support from DGI (Spain) Grant No. BFM2002-00466 and Generalitat de Catalunya Grant No. 2001SGR-00222 is gratefully acknowledged. We acknowledge the support of Central Computing Services at the Johannes Kepler University in Linz, where part of the computations were performed.

References

1. *Liquids Under Negative Pressures*, edited by A. R. Imre *et al.* (Kluwer Ac. Publishers, Dordrecht, 2002).
2. F. Caupin and S. Balibar, *Phys. Rev.* **B64**, 064507 (2001).
3. S. Balibar, *J. Low Temp. Phys.* **129**, 363 (2002).
4. J. Boronat *et al.*, *Phys. Rev.* **B50**, 3427 (1994).
5. G. H. Bauer *et al.*, *Phys. Rev.* **B61**, 9055 (2000).
6. F. Werner, G. Baume, A. Hobeika, S. Nascimbene, C. Herrman, F. Caupin, and S. Balibar, *J. Low. Temp. Phys.* **136**, 93 (2004).
7. T. Schneider and C. P. Enz, *Phys. Rev. Lett.* **125**, 1186 (1971).
8. A. D. Jackson *et al.*, *Phys. Rev.* **B24**, 105 (1981).
9. J. Halinen *et al.*, *J. Low Temp. Phys.* **121**, 531 (2000).
10. F. Albergamo *et al.*, *Phys. Rev. Lett.* **92**, 235301 (2004).
11. J. V. Pearce, J. Bossy, H. Schober, H. R. Glyde, D. R. Daughton, and N. Mulders, *Phys. Rev. Lett.* **93**, 145303 (2004).
12. K. Yamamoto, H. Nakashima, Y. Shibayama, and K. Shirahama, *Phys. Rev. Lett.* **93**, 075302 (2004).
13. P. Nozières, *J. Low Temp. Phys.* **137**, 45 (2004).
14. L. Vranješ, J. Boronat, J. Casulleras, J. *Low Temp. Phys.* **138**, 43 (2005).
15. L. Vranješ, J. Boronat, J. Casulleras, C. Cazorla, *Phys. Rev. Lett.* **95**, 145302 (2005).
16. J. Boronat and J. Casulleras, *Phys. Rev.* **B49**, 8920 (1994).
17. L. Reatto, *Nucl. Phys.* **A328**, 253 (1979).
18. R. A. Aziz *et al.*, *Mol. Phys.* **61**, 1487 (1987).
19. R. De Bruyn Ouboter and C. N. Yang, *Physica* **B44**, 127 (1987).
20. D. O. Edwards and R. C. Pandorff, *Phys. Rev.* **140**, 816 (1965).
21. B. Abraham *et al.*, *Phys. Rev.* **A1**, 250 (1970).
22. D. M. Ceperley and E. L. Pollock, *Can. J. Phys.* **65**, 1416 (1987).
23. S. Moroni and M. Boninsegni, *J. Low Temp. Phys.* **136**, 129 (2004).
24. M. R. Gibbs *et al.*, *J. Phys.: Condens. Matter* **11**, 603 (1999).
25. R. P. Feynman and M. Cohen, *Phys. Rev.* **102**, 1189 (1956).
26. J. Boronat and J. Casulleras, *Europhys. Lett.* **38**, 291 (1997).

COLD ATOMS AND FERMIONS AT THE BEC–BCS CROSSOVER

SPIN 1/2 FERMIONS IN THE UNITARY REGIME AT FINITE TEMPERATURE

AUREL BULGAC, JOAQUIN E. DRUT

Department of Physics, University of Washington, Seattle, WA 98195-1560, USA

PIOTR MAGIERSKI

Faculty of Physics, Warsaw University of Technology, ul. Koszykowa 75, 00-662 Warsaw,POLAND

We have performed a fully non-perturbative calculation of the thermal properties of a system of spin 1/2 fermions in 3D in the unitary regime. We have determined the critical temperature for the superfluid-normal phase transition. The thermodynamic behavior of this system presents a number of unexpected features, and we conclude that spin 1/2 fermions in the BCS-BEC crossover should be classified as a new type of superfluid.

Keywords: Unitary regime; superfluidity; superconductivity.

1. Introduction

The unitary regime is the situation in which the scattering length a is much larger than the average inter-particle distance: $n|a|^3 \gg 1$, where n is the number density.[1,2] At zero temperature, systems in this regime are widely believed to be superfluid, with a coherence length and an inter-particle distance of comparable magnitude. Such zero temperature problem has been considered by several authors.[3–5] It was shown experimentally in 2002 that these systems are (meta)stable, and they have been extensively studied ever since.[6] From the theoretical point of view, the typical treatment is based on an idea put forward by Eagles, Leggett and others.[7] Their approach assumes a BCS-like form for the many-body wave function, which is then used for all values of a. The main problem with this treatment is that, close to the unitary regime, the fraction of non-condensed pairs becomes of order one,[8] and so a mean field description becomes questionable (even including fluctuations).

To determine the thermal properties of fermions in the unitary regime, we have placed the system on a 3D cubic spatial lattice, with periodic boundary conditions. Since the system under consideration is dilute, the interaction that captures the physical situation is a zero-range two-body interaction $V(\mathbf{r}_1 - \mathbf{r}_2) = -g\delta(\mathbf{r}_1 - \mathbf{r}_2)$, with a momentum cut-off $\hbar k_c$. Details of the method can be found in Ref. 9.

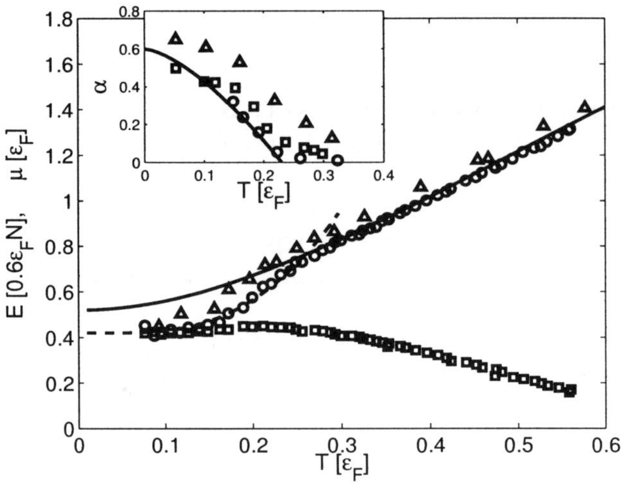

Fig. 1. The total energy $E(T)$ is shown with open circles for a 8^3-lattice and triangles for a 6^3-lattice. The chemical potential $\mu(T)$ shown with squares for the 8^3-lattice. The Bogoliubov-Anderson phonon and fermion quasiparticle contributions $E_{ph+qp}(T)$ are shown as a dashed line. The solid line is $E_{Fg}(T) - 0.6\varepsilon_F N(1-\xi_n)$, where $E_{Fg}(T)$ is the energy of a free Fermi gas. Upper left inset: condensate fraction $\alpha(T)$, with circles for 10^3, squares for 8^3 and triangles the 6^3 lattices respectively. The solid curve is $\alpha(T) = \alpha(0)[1-(T/T_c)^{3/2}]$.

2. Results and Conclusions

The results of our simulations are partially summarized in Fig. 1 (see Ref. 9 for further details). The thermodynamic quantities we computed present a number of features that can be easily identified. First, as $T \to 0$ the energy tends to the $T = 0$ results obtained by other groups.[4,5] This confirms those results, as the algorithms they used suffer from a sign problem, which is not the case in the present approach. Second, there is a low temperature regime and a high temperature regime, separated by what we argue is the critical temperature for the onset of superfluidity, which we estimated to be $T_c = 0.23(2)$.

For $T < T_c$, the T-dependence of the energy can be accounted for by the two types of elementary excitation expected for this system: boson-like Bogoliubov-Anderson phonons and fermion-like gapped Bogoliubov quasiparticles. Their contributions can be estimated assuming that the system is a Fermi superfluid at $T = 0$, with a compressibility and pairing gap as determined from results in Ref. 4.

The chemical potential μ is essentially constant for $T < T_c$, a fact reminiscent of the behavior of an ideal Bose gas in the condensed phase, even though we know the system is strongly interacting and superfluid. This unexpected result implies lack of fermionic degrees of freedom at those temperatures.

Universality of the unitary regime together with $\mu(T) = const.$ for $T < T_c$

implies that

$$E(T) = N\frac{3}{5}\epsilon_F \xi\left(\frac{T}{\epsilon_F}\right), \quad \xi\left(\frac{T}{\epsilon_F}\right) = \xi_s + \zeta\left(\frac{T}{\epsilon_F}\right)^n, \quad n = \frac{5}{2} \tag{1}$$

which is the temperature dependence of an ideal Bose condensed gas. According to our results $n = 5/2$ to about 10%.

Above T_c the system is expected to become normal. The energy behaves like the energy of an ideal Fermi gas, plotted in Fig. 1 with a vertical offset. This is surprising, because the estimated pair-breaking temperature is $T^* \simeq 0.55\epsilon_F$ (see Refs. 10, 7), implying that for $T_c < T < T^*$ there should be a noticeable fraction of non-condensed pairs. Our results show no hint of their presence in this temperature interval.

The condensate fraction $\alpha(T)$, as defined in Ref. 8, and evaluated at $r = L/2$ pair separation, is shown in the inset of Fig. 1. $\alpha(T)$ defines the off-diagonal long range order of the two-body density matrix.[11] The temperature dependence of $\alpha(T)$ is again consistent with $T_c = 0.23(2)$. Moreover, as $T \to 0$ we recover the $T = 0$ results in Ref. 8. The T-dependence of $\alpha(T)$ resembles that of an ideal Bose gas, which comes as a surprise as well.

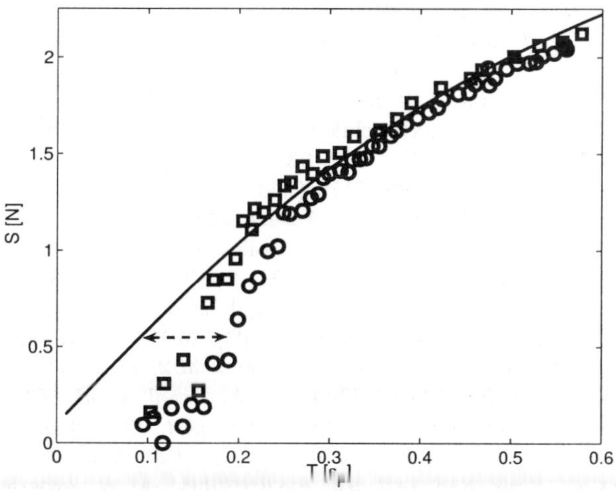

Fig. 2. Entropy per particle as a function of temperature. Our data in circles for the 8^3 lattice and in squares for the 6^3 lattice; free Fermi gas in full line (corresponding to extreme BCS limit). Arrows indicate possible adiabatic cooling or heating processes.

At resonance, universality allows us to estimate the entropy as $S(T) = (\frac{5}{3}E(T) - \mu(T)N(T))/T$, see Fig. 2. The knowledge of $S(T)$ allows to establish a temperature scale at unitarity. Extending the suggestion of Ref. 12, from known T in the BCS limit, one determines the corresponding $S(T_{BCS})$. By adiabatically tuning the system to the unitary regime one can use $S(T_{BCS}) = S(T_{unitary})$ to determine T in the unitary regime. Especially at very low temperature, our results show relatively large errors in S due to both statistical and finite size effects.

Summarizing, we performed a fully non-perturbative calculation of the thermal properties of a system of spin 1/2 fermions at unitarity. We determined the critical temperature for superfluidity to be $T_c = 0.23(2)$. The thermodynamic behavior of this system presents a number of unexpected features, suggesting that spin 1/2 fermions in the BCS-BEC crossover qualify as a new type of superfluid.

Acknowledgements

This work was supported by the Department of Energy under grant DE-FG03-97ER41014, and by the Polish Committee for Scientific Research (KBN) under Contract No. 1 P03B 059 27. Use of computers at the Interdisciplinary Centre for Mathematical and Computational Modelling (ICM) at Warsaw University is gratefully acknowledged.

References

1. *The Many-Body Challenge Problem* formulated by G.F. Bertsch in 1999, see R.A. Bishop, *Int. J. Mod. Phys.* **B 15**, *iii*, (2001).
2. T.-L. Ho, *Phys. Rev. Lett.* **92**, 090402 (2004).
3. G.A. Baker, Jr., *Int. J. Mod. Phys.* **B 15**, 1314 (2001); H. Heiselberg, *Phys. Rev.* **A63**, 043606 (2001).
4. J. Carlson, *et al. Phys. Rev. Lett.* **91**, 050401 (2003) ; S.-Y. Chang *et al. Phys. Rev.* **A70**, 043602 (2004) ; G. Astrakharchik *et al. Phys. Rev. Lett.* **93**, 200404 (2004).
5. G. Astrakharchik *et al. Phys. Rev. Lett.* **93**,
6. K.M. O'Hara, *et al., Science*, **298**, 2179 (2002); T. Bourdel, *et al., Phys. Rev. Lett.* **91**, 020402 (2003); C. A. Regal, *et al., Nature* **424**, 47 (2003); K.E. Strecker, *et al., Phys. Rev. Lett.* **91**, 080406 (2003); J. Cubizolles, *et al., Phys. Rev. Lett.* **91**, 240401 (2003); S. Jochim, *et al., Phys. Rev. Lett.* **91**, 240402 (2003); K. Dieckmann, *et al., Phys. Rev. Lett.* **89**, 203201 (2002); C.A. Regal, *et al., Phys. Rev. Lett.* **92**, 083201 (2004); M. Greiner, *et al., Nature* **426**, 537 (2003); M.W. Zwierlein, *et al., Phys. Rev. Lett.* **91**, 250401 (2003); S. Jochim, *et al., Science* **302**, 2101 (2003); M. Bartenstein, *et al., Phys. Rev. Lett.* **92**, 120401 (2004); C.A. Regal, *et al., Phys. Rev. Lett.* **92**, 040403 (2004); M.W. Zwierlein, *et al., Phys. Rev. Lett.* **92**, 120403 (2004); C. Chin *et al. Science* **305**,1128 (2004); J. Kinast *et al., Phys. Rev. Lett.* **92**, 150402 (2004); M. Bartenstein, *et al., Phys. Rev. Lett.* **92**, 203201(2004); M.W. Zwierlein *et al. Nature*, **435**, 1047 (2005); J. Kinast *et al. Science* **307**, 1296 (2005).
7. D.R. Eagles, *Phys. Rev.* **186**, 456 (1969); A.J. Leggett, in *Modern Trends in the Theory of Condensed Matter*, eds. A. Pekalski and R. Przystawa, Springer–Verlag, Berlin, 1980; *J. Phys. (Paris) Colloq.* **41**, C7–19 (1980); P. Nozières and S. Schmitt–Rink, *J. Low Temp. Phys.* **59**, 195 (1985); C.A.R. Sá de Mello *et al., Phys. Rev. Lett.* **71**, 3202 (1993); M. Randeria, in *Bose–Einstein Condensation*, eds. A. Griffin *et al.* Cambridge University Press (1995), pp 355–392.
8. G.E. Astrakharchik *et al.* cond-mat/0507483.
9. A. Bulgac, J.E. Drut, P. Magierski, cond-mat/0505374
10. A. Perali *et al. Phys. Rev. Lett* **92**, 220404 (2004)
11. C.N. Yang, *Rev. Mod. Phys.* **34**, 694 (1962).

12. L. Carr *et al.* *Phys. Rev. Lett.* **92**, 150404 (2004).

DECONFINEMENT AND COLD ATOMS IN OPTICAL LATTICES

M. A. CAZALILLA

Donostia Int'l Physics Center (DIPC), Manuel de Lardizabal, 4. 20018-Donostia, Spain.
waxcagum@sq.ehu.es

A. F. HO

*School of Physics and Astronomy, The University of Birmingham, Edgbaston, Birmingham B15 2TT, UK.**
andrew.j.ho@imperial.ac.uk

T. GIAMARCHI

University of Geneva, 24 Quai Enerst-Ansermet, CH-1211 Geneva 4, Switzerland.
Thierry.Giamarchi@physics.unige.ch

Despite the fact that by now one dimensional and three dimensional systems of interacting particles are reasonably well understood, very little is known on how to go from the one dimensional physics to the three dimensional one. This is in particular true in a quasi-one dimensional geometry where the hopping of particles between one dimensional chains or tubes can lead to a dimensional crossover between a Luttinger liquid and more conventional high dimensional states. Such a situation is relevant to many physical systems. Recently cold atoms in optical traps have provided a unique and controllable system in which to investigate this physics. We thus analyze a system made of coupled one dimensional tubes of interacting fermions. We explore the observable consequences, such as the phase diagram for isolated tubes, and the possibility to realize unusual superfluid phases in coupled tubes systems.

Keywords: Cold atoms; Superconductivity; Luttinger liquids.

1. Introduction

As is well known, interactions have drastic effects on the behavior of bosonic and fermionic systems, and change drastically their properties compared to those of noninteracting particles. Understanding the properties of such strongly correlated systems is a particularly challenging problem. The effects of interaction are also considerably enhanced when the dimension of the system is reduced. Very naturally this quest for strongly correlated systems has thus led to investigations of low dimensional systems, in particular in condensed matter systems. Among these low dimensional systems, one dimensional ones play a special role. In such systems, interaction effects are particularly strong since there is no way for particles to avoid

*Present address: CMTH Group, Dept. Physics, Blackett Laboratory, Imperial College, Prince Consort Road, London SW7 2BW, UK.

each other. For fermions interactions are known to change drastically the properties compared to the canonical ones of a Fermi liquid. The one dimensional interacting fermionic system is indeed one of the very few solvable case of a non-fermi liquid, known as a Luttinger liquid.[1] For boson systems, here again the interactions lead to properties quite different from the ones of weakly interacting bosons in one dimension, and that are much closer to the ones of fermions: the interacting one dimensional bosonic systems is again a Luttinger liquid.

For fermions the progress in material science and nanotechnology have recently made possible to realize such one dimensional systems. Another class of materials in which such low dimensional systems could be achieved with an unprecedented level of control has been recently provided by cold atoms.[2,3] Indeed cold atoms not only provide the possibility to realize both bosonic and fermionic low dimensional systems, but the interactions and kinetic energy can be controlled at will using optical lattices and Feshbach resonances.[4–6]

Such systems allow also to tackle another question that is of importance for a large class of materials. Since the effects of interactions vary enormously with the dimension, it is crucial to understand how one goes from the one dimensional behavior to a more conventional three dimensional one. Understanding such dimensional crossovers is crucially important for materials such as the organic superconductors,[7] and even more complicated anisotropic materials such as the high Tc superconductors. Cold atomic systems thus provide a unique system for which these questions could be investigated. This can be done by studying the properties of quasi-one dimensional systems made of many one dimensional tubes coupled together. Such systems have been both investigated[8] and realized[3] for the case of bosons, and it is now possible to achieve similar trapping in optical lattices for the fermionic case as well.[9] In these notes we thus focuss on such fermionic systems. We show how the peculiarity of cold atoms leads to interesting properties already at the level of an isolated 1D system. We then examine in details the effect of the coupling between the tubes and shows that such systems can be potential realizations of triplet superconductors. We expand in these notes on Ref. 10 and refer the reader to this paper for technical details and more specialized references.

2. Fermions, one dimension

Let us first consider interacting one dimensional fermions. We use the following extension of the traditional Hubbard model

$$H = -\sum_{\sigma,m} t_\sigma \left(c^\dagger_{\sigma m} c_{\sigma m+1} + \text{h.c.}\right) + U \sum_m n_{\uparrow m} n_{\downarrow m}. \tag{1}$$

This Hamiltonian describes a 1D Fermi gas on a spin dependent periodic potential. t_σ is the hopping from site m to site $m+1$ for the spin species σ and U the local interaction. In the cold atoms context the spin index refers to two hyperfine states, or two different types of atoms (*e.g.* ^{6}Li and ^{40}K). Even though there may be no *true* spin symmetry, we use the spin language to describe this binary mixture. It is important to point out differences between (1) and the standard Hubbard model. First because the system is in an optical lattice it is possible to control separately the hopping of the up and down particles. As we discuss below this allows for a richer phase diagram than the one of the standard Hubbard model. Second, and most importantly, there is no mechanism in the cold atom context that allows to relax the "spin" orientation, since this corresponds to two different isospin states. As a result the number of spin up and spin down is *separately* conserved. This is at strong variance with a standard condensed matter situation where one generally impose a *unique* chemical potential for the spin up and spin down particles. The important difference, as shown in Fig. 1, is that there are, for the case of equal number of spin up and spin down species, only two Fermi points instead of four. For such a system, the various operators corresponding to the different types of

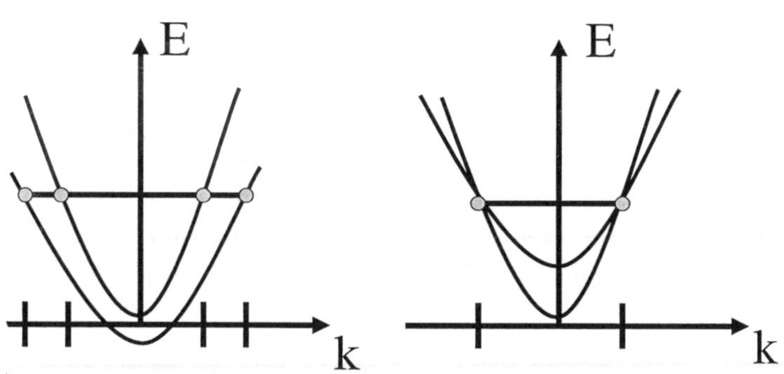

Fig. 1. Relation dispersion $E(k)$ for a system with spin dependent hopping. Left: in the traditional case there are spin relaxation phenomena ensuring the equilibrium of the two spin species. There is thus a unique chemical potential for both species. This implies the presence of four fermi points. Right: for cold atoms, there is no such relaxation and the number of spin up and down are separately conserved. For an equal number of spin up and down, there is thus only two Fermi points since the two fermi momenta $k_{F\uparrow} = k_{F\downarrow}$.

order in the tubes are of two types.[1] There are charge density or spin density order described by

$$O_\rho(x) = \sum_{\sigma,\sigma'} \psi_\sigma^\dagger(x)\delta_{\sigma,\sigma'}\psi_{\sigma'}(x) \tag{2}$$

$$O_\sigma^a(x) = \sum_{\sigma,\sigma'} \psi_\sigma^\dagger(x)\sigma_{\sigma,\sigma'}^a\psi_{\sigma'}(x) \tag{3}$$

where σ^x, σ^y, σ^z are the Pauli matrices The operator O_ρ is simply the total density, whereas $O_\sigma^a(x)$ measures the spin density (along the direction $a = x, y, z$). We have used the notation in the continuum $c_m \to \psi(x)$ as the operator destroying a fermion at point x. The density contains a $q \sim 0$ component and a $q \sim 2k_F$ one. The operators giving the $2k_F$ modulation of the charge or spin density are

$$O_{\rm CDW}(x) = \psi^\dagger_{R\uparrow}\psi_{L\uparrow}(x) + \psi^\dagger_{R\downarrow}\psi_{L\downarrow}(x) = \frac{e^{-2ik_Fx}}{\pi\alpha} e^{i\sqrt{2}\phi_\rho}\cos(\sqrt{2}\phi_\sigma) \tag{4}$$

$$O^x_{\rm SDW}(x) = \psi^\dagger_{R\uparrow}\psi_{L\downarrow}(x) + \psi^\dagger_{R\downarrow}\psi_{L\uparrow}(x) = \frac{e^{-2ik_Fx}}{\pi\alpha} e^{i\sqrt{2}\phi_\rho}\cos(\sqrt{2}\theta_\sigma) \tag{5}$$

$$O^y_{\rm SDW}(x) = -i(\psi^\dagger_{R\uparrow}\psi_{L\downarrow}(x) - \psi^\dagger_{R\downarrow}\psi_{L\uparrow}(x)) = \frac{-e^{-2ik_Fx}}{\pi\alpha} e^{i\sqrt{2}\phi_\rho}\sin(\sqrt{2}\theta_\sigma) \tag{6}$$

$$O^z_{\rm SDW}(x) = \psi^\dagger_{R\uparrow}\psi_{L\uparrow}(x) - \psi^\dagger_{R\downarrow}\psi_{L\downarrow}(x) = \frac{e^{-2ik_Fx}}{\pi\alpha} e^{i\sqrt{2}\phi_\rho} i\sin(\sqrt{2}\phi_\sigma) \tag{7}$$

where we have used the notation $\psi^\dagger_R$ (resp. $\psi^\dagger_L$) to denote the operator creating a fermion with a momentum close to $+k_F$ (resp. $-k_F$). These operators have a simple representation in terms of continuous field $\phi_{\rho,\sigma}$ (resp. $\theta_{\rho,\sigma}$) describing the collective excitations of charge and spin densities (resp. current densities). This representation is the so-called bosonization representation and is very useful to identify the types of order in the system. For more details on this representation see e.g. Ref. 1. Here we will only use that ϕ and $\nabla\theta$ have canonical conjugation commutation relations, so order in one of the field implies exponentially decreasing correlations in the other. Besides the above mentioned types of order, other instability in the system are the superfluid instabilities of the singlet or triplet type

$$O_{\rm SS}(x) = \sum_{\sigma,\sigma'} \sigma\psi^\dagger_{R,\sigma}(x)\delta_{\sigma,\sigma'}\psi^\dagger_{L,-\sigma'}(x) \tag{8}$$

$$O^a_{\rm TS}(x) = \sum_{\sigma,\sigma'} \sigma\psi^\dagger_{R,\sigma}(x)\sigma^a_{\sigma,\sigma'}\psi^\dagger_{L,-\sigma'}(x) \tag{9}$$

where SS denotes singlet pairing whereas TS is triplet pairing. These operators describe paring with zero total momentum. Other pairings are of course possible but are usually less relevant. These operators become

$$O_{\rm SS}(x) = \psi^\dagger_{R\uparrow}\psi^\dagger_{L\downarrow}(x) + \psi^\dagger_{L\uparrow}\psi^\dagger_{R\downarrow}(x) = \frac{1}{\pi\alpha} e^{-i\sqrt{2}\theta_\rho}\cos(\sqrt{2}\phi_\sigma) \tag{10}$$

$$O^x_{\rm TS}(x) = \psi^\dagger_{R\uparrow}\psi^\dagger_{L\uparrow}(x) + \psi^\dagger_{L\downarrow}\psi^\dagger_{R\downarrow}(x) = \frac{1}{\pi\alpha} e^{-i\sqrt{2}\theta_\rho}\cos(\sqrt{2}\theta_\sigma) \tag{11}$$

$$O^y_{\rm TS}(x) = -i(\psi^\dagger_{R\uparrow}\psi^\dagger_{L\uparrow}(x) - \psi^\dagger_{L\downarrow}\psi^\dagger_{R\downarrow}(x)) = \frac{-1}{\pi\alpha} e^{-i\sqrt{2}\theta_\rho}\sin(\sqrt{2}\theta_\sigma) \tag{12}$$

$$O^z_{\rm TS}(x) = \psi^\dagger_{R\uparrow}\psi^\dagger_{L\downarrow}(x) - \psi^\dagger_{L\uparrow}\psi^\dagger_{R\downarrow}(x) = \frac{e^{2ik_Fx}}{\pi\alpha} e^{-i\sqrt{2}\theta_\rho}\sin(\sqrt{2}\phi_\sigma) \tag{13}$$

We will not dwell on the solution[10] of the model here and will only quote the results. Compared to the standard case of the Hubbard model the main effect of the spin dependent hopping is to open a spin gap both for repulsive and attractive interactions (the standard Hubbard model being massless for repulsive interactions). For attractive interactions the spin gap corresponds to the formation of singlets. In the bosonization language this corresponds to the order $\phi_\sigma \to 0$. Correlation functions containing θ_σ or $\sin(\sqrt{2}\phi_\sigma)$ thus decay exponentially to zero. The leading instabilities are thus the charge density wave (CDW) order or a singlet superfluid (SS) instability. Physically it means that fermions of opposite spin pair and that these pairs behave roughly as bosons and can then condense to give a superfluid (SS) or crystallize to give a charge density wave (CDW).

For repulsive interactions the hopping difference induce an order in $\phi_\sigma \to \pi/\sqrt{8}$. Correlation functions containing θ_σ or $\cos(\sqrt{2}\phi_\sigma)$ decay exponentially to zero while $\sin(\sqrt{2}\phi_\sigma)$ tends to a constant. The leading instabilities are thus a spin density wave order along the z direction (SDW$_z$) or a triplet superfluid instability along z (TS_z). Which one of these instabilities dominates is dictated by the decay of the correlations of the charge sector that remains massless. For repulsive interactions, the SDW$_z$ one dominates, while the TS$_z$ one decays with a larger power law and is thus subdominant. A summary of the phase diagram is given in Fig. 2. For

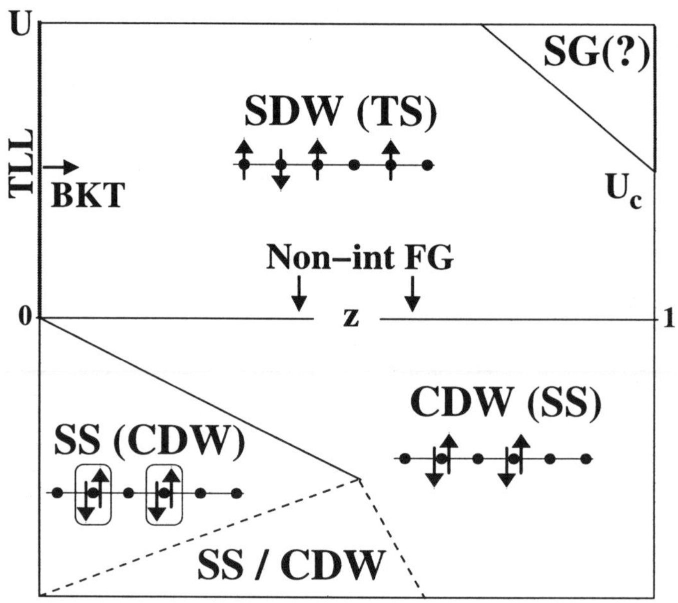

Fig. 2. Schematic phase diagram for the model in Eq. (1) with equal number of spin up and down fermions away from half-filling. The interaction strength is U and $z = |t_\uparrow - t_\downarrow|/(t_\uparrow + t_\downarrow)$. All phases (SDW: spin density wave, CDW: charge density wave, SS singlet superfluid, TS triplet superfluid) exhibit a spin gap Δ_s (however, $\Delta_s = 0$ for $U = 0$ and $z = 0$ with $U > 0$. A cartoon of the type of order characterizing each phase is also shown. In the area between dashed lines the dominant order (either CDW or SS) depends on the lattice filling. In the SG phase, spin up and down fermions are segregated (demixed).

repulsive interactions the spin sector is thus still massive but the spin gap does not correspond to the formation of singlets. On the contrary it is due to a breaking of the spin rotation symmetry with formation of an ising-like antiferromagnetic order along the z direction. Note that even if the spin sector is massive, the spin density wave order is not perfect (except at a commensurate filling of exactly one fermion per site) due to the fluctuations in the charge sector. Contrarily to the case of the Hubbard model where the SDW order in the three directions decays in a similar way (due to the spin rotation symmetry) here the SDW_z order decays as a power law because of charge fluctuations while $\mathrm{SDW}_{x,y}$ order decays exponentially fast because of the presence of the gap in the spin sector. We refer the reader to Ref. 10 for more details on the various correlation functions and ways to probe the existence of such a gap by Raman spectroscopy.

3. Coupled Fermionic tubes

Let us now turn to the case of coupled tubes. The coupling between the tubes can be described by the single particle hopping term

$$H_\perp = -t_\perp \sum_{\langle\alpha,\beta\rangle} \sum_{m,\sigma} c^\dagger_{\sigma,m,\alpha} c_{\sigma,m,\beta} \tag{14}$$

where α and β are the tube indexes and $\langle\alpha,\beta\rangle$ denotes nearest neighbor tubes. Because for cold atomic gases the interactions are short range, there is no need to take into account interactions between different tubes. If the single particle hopping (14) is of the same order of magnitude than the ones in (1) then the system is a three dimensional system, while if $t_\perp = 0$ the tubes are uncoupled. The Hamiltonian (14) is thus able to describe the dimensional crossover between these two situations. For bosons the effects of such a term have been investigated in Ref. 8. For fermions treating such a single particle hopping is an extremely challenging problem. Indeed, contrarily to the case of boson, the average of a single fermion operator does not exist, and thus the term (14) cannot be decoupled simply. Various approximations have thus been used to tackle this term and we refer the reader to Ref. 11, 7, 1 for details and further references.

However for the particular case of the fermionic system investigated in these notes, one is in a much more favorable situation because of the presence of the spin gap. In the case of attractive interactions, the fermions form singlet pairs, that essentially behave as bosons. One is thus essentially led back to the bosonic case of Ref. 8. We thus focus here on the more interesting repulsive case. Because of the presence of the spin gap the situation is now quite different to the one where each tube is described by the simple Hubbard model. In the latter case the spin excitations are massless and the single particle correlations decay as power laws

$$\langle T_\tau c_{\alpha,\sigma,}(x) c^\dagger_{\alpha,\sigma}(0)\rangle \propto \left(\frac{1}{x}\right)^{\frac{1}{4}[K_\rho+1/K_\rho]+\frac{1}{2}} \tag{15}$$

where K_ρ is an interaction dependent parameter (the Luttinger liquid parameter) characterizing the isolated tube. In that case a simple scaling analysis to second order in perturbation in the single particle hopping term (14) shows that it scales as

$$t_\perp^2 L^{4-\frac{1}{2}[K_\rho+1/K_\rho]-1} \tag{16}$$

and is thus in general a relevant perturbation, driving the system away from the isolated tube fixed point. However in the presence of a spin gap the *single particle* excitations now decay exponentially

$$\langle T_\tau c_{\alpha,\sigma,}(x) c^\dagger_{\alpha,\sigma}(0)\rangle \propto e^{-|x|/\xi} \tag{17}$$

where ξ is the correlation length induced by the presence of the spin gap (typically of order $\xi = v_F/\Delta_s$ if v_F is the Fermi velocity). The single particle hopping is now an irrelevant perturbation to (1). Physically it simply means that because of the presence of the spin gap it is now impossible for a single particle to leave a tube since it would take away one spin and thus destroy the spin gap. One would therefore need a critical value of $t_\perp$ for this to happen. However, although the single particle hopping is an irrelevant operator, it can generate relevant perturbations at higher order.[1] Such relevant perturbations corresponds to hopping between different tubes that do not destroy the spin gap.

Quite generally, as shown in Fig. 3, two types of relevant perturbations are possible when looking at the second order term generated by (14). The first one

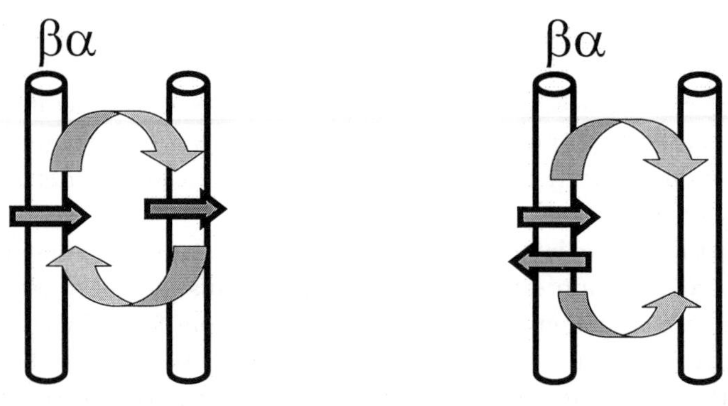

Fig. 3. Relevant coupling generated to second order by the single particle intertube hopping. In order to avoid breaking the spin gap, a correlated hop of two objects must be done simultaneously. Left: particle-hole hopping. Since this term exchanges two particles on two different tubes, it corresponds to a standard density-density or spin-spin coupling term. This is the superexchange term. Right: particle-particle hopping. This term allows a pair to jump from one tube to the next and is thus analogous to the Josephson coupling between superconductors.

corresponds to particle-hole hopping between the tubes, it is thus a density-density or spin-spin term, while the second one is the hopping of two particles between between the tubes and thus corresponds to a Josephson coupling. The simplest way to analyze the terms generated by (14) is to use the bosonization representation.

The intertube tunnelling term becomes

$$H_\perp \propto -t_\perp \sum_{\sigma=\pm1, r=\pm1} \sum_{\langle\alpha,\beta\rangle} \int dx e^{\frac{i}{\sqrt{2}}[r\phi_{\rho,\alpha}(x)-\theta_{\rho,\alpha}(x)+s(r\phi_{\sigma,\alpha}(x)-\theta_{\sigma,\alpha}(x))]} \times$$
$$e^{\frac{-i}{\sqrt{2}}[r\phi_{\rho,\beta}(x)-\theta_{\rho,\beta}(x)+s(r\phi_{\sigma,\beta}(x)-\theta_{\sigma,\beta}(x))]} \quad (18)$$

Going to second order in (14), using the fact that ϕ_σ is now ordered, and the fact that all operators leading to exponential decay must be eliminated to get the leading operator, one sees that a first surviving term is

$$H_{\mathrm{SDW}} = J_{\mathrm{SDW}} \sum_{\alpha,\beta} \int dx O^\dagger_{\mathrm{SDW}_z,\alpha}(x) O_{\mathrm{SDW}_z,\beta}(x) \quad (19)$$

The value of the coupling constant can be obtained by standard second order perturbation theory and is of the order of $J_{\mathrm{SDW}} \sim t_\perp^2/\Delta_s$. This term is thus a standard superexchange term coupling the antiferromagnetic spin modulation on two different tubes. Quite naturally this term would tend to stabilize the SDW_z instability that would develop in an isolated tube and lead to a three dimensional ordered antiferromagnetic phase.

There is however another term that survives. This is a term where two particles can hop from one tube to the other. It is of the form

$$H_{\mathrm{TS}} = J_{\mathrm{TS}} \sum_{\alpha,\beta} \int dx O^\dagger_{\mathrm{TS}_z,\alpha}(x) O_{\mathrm{TS}_z,\beta}(x) \quad (20)$$

This term corresponds to the hopping of a pair of fermions, that are in a triplet paring state from one tube to the other. Singlet hopping is here cancelled because of the presence of the spin gap. Because a superfluid pair has a global momentum of zero, the pair hopping does not contain the oscillating $2k_F$ factor that is present in the superexchange term (see (4)). Here again the coupling constant is of the order of $J_{\mathrm{TS}} \sim t_\perp^2/\Delta_s$.

Both the superexchange and the Josephson term tend to stabilize their corresponding type of order. What phase is realized normally depends crucially on what is the leading one dimensional instability. Usually the Fermi momenta of all tubes are the same and there is no oscillatory factors in (19). Since the SDW order is dominant for the isolated tube, and both J_{SDW} and J_{TS} are of the same order of magnitude a simple RPA treatment of the coupling term shows that the SDW phase is stabilized. This is the situation depicted in Fig. 4(a). This situation is normally the one realized in condensed matter systems. It corresponds to the stabilization of a three dimensional antiferromagnetic order for repulsive interactions. However the situation can be much richer if the Fermi momenta of the different tubes are different. Note that for cold atoms this is the rule rather than the exception because of the presence of the parabolic confining potential that makes each tube different depending from its distance from the center. This situation can also be reinforced artificially by adding an additional modulation of the optical lattice. In that case one has $k_F^\alpha - k_F^\beta \neq 0$ which means that the oscillatory factors remain in (19). Such factors considerably weaken the intertube superexchange. Physically this means that the antiferromagnetic fluctuations on the neighboring tubes are now incommensurate with each other and thus cannot couple very well as depicted in Fig. 4(b). On the other hand the Josephson term which is a $q \sim 0$ transfer between the tube is not affected by such a difference of Fermi momentum and remains unchanged, as schematically shown in Fig. 4(c). One is now in a situation where this term can become dominant[10] *even* if the triplet superfluid instability in the isolated tube is a subdominant instability. One would thus be in a situation to stabilize a three dimensional triplet superfluid phase.

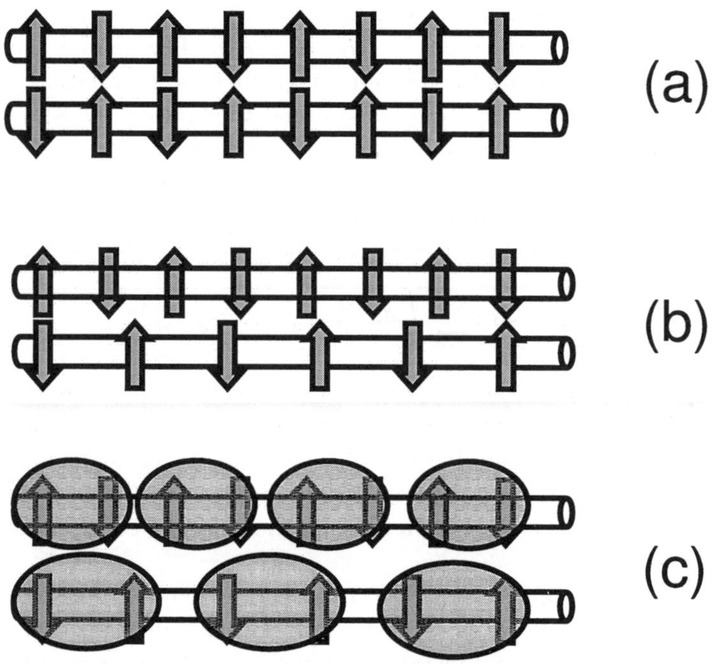

Fig. 4. Schematic diagrams for the true long range order induced by inter-tube particle-pair or particle-hole hopping. a) Antiferromagnetic order for $U > 0$ and different velocities, with equal Fermi momentum (*i.e.* equal number of fermions in each tube). b) Frustration (incommensurability) to the antiferromagnetic order between tubes when the Fermi momentum is different between the tubes. c) No such frustration effects when the order parameter carries zero total momentum, namely for the superfluidity.

This is a rather unique situation. From the theoretical point of view such a stabilization of a subdominant instability by intertube coupling is quite interesting and potentially relevant to other situations as well. In particular, this mechanism is similar, albeit for a triplet superconductor, to the one advocated to stabilize singlet superconductivity for coupled stripes.[12] Such cold atomics system could thus be controlled systems in which to check for the feasibility of such a mechanism. More directly it would be very interesting to have a realization for a triplet superconductor. Triplet superconductivity is indeed quite rare. Besides Helium 3, strontium ruthenate is the only one candidate well identified in condensed matter.[13] Organic superconductors, another system made of coupled chains, is also a potential candidate for such unusual superconductivity.[14] Cold atomic gases of fermions could thus help shed a light on the mechanisms and properties of unusual superconductivity in these very anisotropic systems.

Acknowledgements

M.A.C. is supported by *Gipuzkoako Foru Aldundia* and MEC (Spain) under grant FIS-2004-06490-C03-00, A.F.H. by EPSRC(UK) and DIPC (Spain), and T.G. by the Swiss National Science Foundation under MANEP and Division II.

References

1. T. Giamarchi, *Quantum Physics in One Dimension* (Oxford University Press, Oxford, 2004).
2. M. Greiner *et al.*, *Nature* **415**, 39 (2002).
3. T. Stöferle *et al.*, *Phys. Rev. Lett.* **92**, 130403 (2004).
4. S. Inouye *et al.*, *Nature* **392**, 151 (1998).
5. E. Timmermans, K. Furuya, P. W. Milonni, and A. K. Kerman, *Phys. Lett. A* **258**, 228 (2001).
6. M. Holland, S. J. J. M. F. Kokkelmans, M. L. Chiofalo, and R. Walser, *Phys. Rev. Lett.* **87**, 120406 (2001).
7. T. Giamarchi, *Chem. Rev.* **104**, 5037 (2004).
8. A. F. Ho, M. A. Cazalilla, and T. Giamarchi, *Phys. Rev. Lett.* **92**, 130405 (2004).
9. M. Kohl *et al.*,*Phys. Rev. Lett.* **94**, 080403 (2005).
10. M. A. Cazalilla, A. F. Ho, and T. Giamarchi, *Phys. Rev. Lett.* **95**, 226402 (2005).
11. S. Biermann, A. Georges, T. Giamarchi, and A. Lichtenstein, in *Strongly Correlated Fermions and Bosons in Low Dimensional Disordered Systems*, edited by I. V. Lerner *et al.* (Kluwer Academic Publishers, Dordrecht, 2002), p. 81, cond-mat/0201542.
12. E. Arrigoni, E. Fradkin, and S. Kivelson, *Phys. Rev. B* **69**, 214519 (2004).
13. A. P. Mackenzie and Y. Maeno, *Rev. Mod. Phys.* **75**, 657 (2003).
14. T. Ishiguro, in *High Magnetic Fields: Applications in Condensed Matter Physics and Spectroscopy*, Vol. 595 of *Lecture Notes in Physics*, edited by C. Berthier, L. P. Lévy, and G. Martinez (Springer-Verlag, Heidelberg, 2002), pp. 301–313.

EXACT BCS SOLUTION IN THE BCS-BEC CROSSOVER

J. DUKELSKY

Instituto de Estructura de la Materia, CSIC
Serrano 123, 28006 Madrid,Spain

G. ORTIZ

Theoretical Division, Los Alamos National Laboratory
Los Alamos, New Mexico 87545, USA

S. M. A. ROMBOUTS

Universiteit Gent - UGent, Vakgroep Subatomaire en Stralingsfysica, Proeftuinstraat 86, B-9000 Gent, Belgium

The exact Richardson solution of the reduced BCS Hamiltonian is used to study the BCS-to-BEC crossover, as well as the nature of Cooper pairs, in superconducting and Fermi superfluid media. Based on the exact eigenstate we will discuss the Cooper-pair concept proposing a scenario for the BCS-to-BEC crossover in which a mixture of quasifree fermions and pair resonances (BCS) evolves to a system of weakly bound molecules (BEC). In this single unified scenario the Cooper-pair wavefunction has a unique functional form. We propose a new definition of the condensate fraction which, within the limits of the BCS model, gives a qualitative description of recent experiments in ultracold atomic Fermi gases. Finally, we will introduce a new integrable model for asymmetric superfluid systems able to describe different homogeneous and inhomogeneous competing phases such as, breached superconductivity, deformed Fermi superfluidity, and the elusive Larkin-Ovchinnikov-Fulde-Ferrell (LOFF) state.

Keywords: Richardson Model; BCS-BEC Crossover; Cooper pairs.

1. Introduction

Superconductivity or Fermi superfluidity is a phenomenon common to several quantum many body systems like condensed matter systems, atomic nuclei, neutron stars, quantum fluids and ultracold Fermi gases. The microscopic description in terms of a pairing Hamiltonian in the BCS[1] approximation was first found in the context of superconducting metals, and soon after it was applied to finite nuclei.[2] While the pairing Hamiltonian has been extensively used in both fields of physics in the BCS or more sophisticated approximations, the exact solution of the model given by Richardson in the sixties[3] passed almost unnoticed until very recently. The Richardson solution was rediscovered in the process of improving the description of the disappearance of superconducting correlations in ultrasmall aluminum grains,[4] which required a more accurate treatment than could be provided by Number Projected BCS. Since then, much work has been done to generalize the model,[5,6] while preserving its exact solvability, and to apply it to a variety of important physical systems.

The crossover from BCS, characterized by a mixture of free fermions and weakly correlated pairs, to a BEC of molecular dimers can be achieved by either increasing the interaction strength or decreasing the density. For an equal mixture of the two fermion components, the soft BCS-BEC crossover is well described within the BCS approximation solving selfconsistently the gap and number equations.[7] The structure of Cooper pairs in the BCS-BEC crossover has regained attention due to the observation of a large fraction of preformed fermion pairs on the BCS side of the Feshbach resonance in ultracold atomic Fermi gases.[8] We will address this issue from the exact wavefunction of the BCS Hamiltonian,[9] giving a new definition of the Cooper pair in the superconducting medium closer to that of the original Cooper problem and a qualitative diagram for the whole crossover.

An imbalance in the population of two component Fermi gas could lead to a rich and complex phase diagram along the BCS-BEC transition. A density asymmetry, in general, makes pairing less favorable and questions about the nature of the resulting competing phases might arise. To this end, we introduce a new exactly solvable model which displays the competition between different phases like the Larkin-Ovchinnikov-Fulde-Ferrell (LOFF),[10] breached-pair[11] and deformed Fermi-surface superfluid.[12] Its quantum phase diagram as a function of the asymmetry density $\delta\rho = (\rho_a - \rho_b)/(\rho_a + \rho_b)$ and the coupling strength g has been recently studied, at the mean-field level.[13] In our exact solution, albeit in a finite-size lattice, we find different regimes of stability for these various phases.

2. Richardson's Exact Solution of the reduced BCS Hamiltonian

The reduced BCS Hamiltonian, extensively used in condensed matter and nuclear physics to explain the superconducting phenomenon is

$$H = \sum_{\mathbf{k}} \varepsilon_{\mathbf{k}}\, n_{\mathbf{k}} + g \sum_{\mathbf{k},\mathbf{k}'} c^{\dagger}_{\mathbf{k}\uparrow} c^{\dagger}_{-\mathbf{k}\downarrow} c_{-\mathbf{k}'\downarrow} c_{\mathbf{k}'\uparrow} \ , \tag{1}$$

where we have assumed a momentum space basis $\mathbf{k}$ including the spin degree of freedom. H involves all terms with time-reversed pairs $(\mathbf{k}\uparrow, -\mathbf{k}\downarrow)$ from a contact interaction. It is consistent with an effective single-channel description of the BCS-BEC crossover theory[7] in terms of a zero-range potential. The main differences are expected in the BEC region where the reduced BCS Hamiltonian, Eq. (1), cannot account for the reduction in the scattering length of the composite pairs[14] and the corresponding depletion of the molecular condensate.

For simplicity we will consider a symmetric system with $N = N_\uparrow + N_\downarrow$ spin-1/2 (i.e., 2-flavor) fermions in a three-dimensional box of volume V with periodic boundary conditions, interacting through an attractive constant (s-wave-singlet-pairing) potential. The pairing strength g scales with the volume as $g = \frac{G}{V}$. The exact $N = 2M + \nu$ particle eigenstates of H can be written as

$$|\Psi\rangle = \prod_{\ell=1}^{M} \mathsf{S}_\ell^+ |\nu\rangle \ , \text{with } \mathsf{S}_\ell^+ = \sum_{\mathbf{k}} \frac{1}{2\varepsilon_{\mathbf{k}} - E_\ell} c_{\mathbf{k}\uparrow}^\dagger c_{-\mathbf{k}\downarrow}^\dagger , \tag{2}$$

where $|\nu\rangle \equiv |\nu_1, \nu_2 \cdots, \nu_L\rangle$ is a state of ν unpaired fermions ($\nu = \sum_{\mathbf{k}} \nu_{\mathbf{k}}$, with $\nu_{\mathbf{k}} = 1, 0$) defined by $c_{-\mathbf{k}\downarrow} c_{\mathbf{k}\uparrow} |\nu\rangle = 0$, and $n_{\mathbf{k}} |\nu\rangle = \nu_{\mathbf{k}} |\nu\rangle$. L is the total number of single particle states. The GS $|\Psi_0\rangle$ is in the $\nu = 0$ ($N_\uparrow = N_\downarrow$) sector.

Each eigenstate $|\Psi\rangle$ is completely defined by a set of M (in general, complex) spectral parameters (pair energies) E_ℓ which are a solution of the Richardson's equations

$$\frac{1}{g} + \sum_{\mathbf{k}} \frac{1 - \nu_{\mathbf{k}}}{2\varepsilon_{\mathbf{k}} - E_\ell} + 2 \sum_{m(\neq \ell)=1}^{M} \frac{1}{E_\ell - E_m} = 0 \ , \tag{3}$$

and the eigenvalues of H are $\mathcal{E} = \sum_{\mathbf{k}} \varepsilon_{\mathbf{k}} \nu_{\mathbf{k}} + \sum_{\ell=1}^{M} E_\ell$.[3,4] A crucial observation is that if a complex E_ℓ satisfies (3), its complex-conjugate E_ℓ^* is also a solution. Thus, $|\Psi\rangle$ restores time-reversal invariance. The ansatz (2) is a natural generalization of the Cooper-pair problem without an inert Fermi sea, and with all subjected to the pairing interaction. The many-body exact eigenstate (2) is a pair product wavefunction unlike the BCS wavefunction in which all pairs are equal. Moreover, the structure of the pairs is formally equal to the pair wavefunction of the original Cooper problem.

3. The Thermodynamic Limit

The continuum or thermodynamic limit of the Richardson equation was obtained in the limit where the volume and the number of fermion pairs go to infinite (V, $M \to \infty$) while the density $\rho = M/V$ and the scaled pairing strength $g = G/V$ stay finite. Assuming that a fraction of pair energies are complex and organized themselves into an arc Γ in the complex plane which is reflection-symmetric on the real axis, while the remaining pair energies stay real, the Richardson equations transform into the gap and number equations of the BCS theory (for details see Ref. 15).

$$\frac{1}{2} \int_\Omega d\varepsilon \, \frac{g(\varepsilon)}{\sqrt{(\varepsilon - \mu)^2 + \Delta^2}} + \frac{1}{G} = 0 \ , \tag{4}$$

$$\int_\Omega d\varepsilon \ g(\varepsilon) \left[1 - \frac{\varepsilon - \mu}{\sqrt{(\varepsilon - \mu)^2 + \Delta^2}} \right] = \rho \ . \tag{5}$$

This result was not unexpected since it was well known that the BCS approximation gives the exact ground state energy in the thermodynamic limit. In addition, the complete solution of Eqs. (3) in the continuum limit provides an extra equation defining the arc Γ in the complex plane with extreme points $E_F = 2(\mu \pm i\Delta)$, where the chemical potential μ and the gap Δ are obtained selfconsitently from the BCS equations (4,5). Here, we simply present the equation for Γ without derivation

$$0 = -\frac{1}{2}\int_0^\infty \rho(\varepsilon)\left[\frac{2z}{\sqrt{(\varepsilon-\mu)^2+\Delta^2}} + \ln\left[\frac{z-\sqrt{(\varepsilon-\mu)^2+\Delta^2}}{z+\sqrt{(\varepsilon-\mu)^2+\Delta^2}}\right] + \frac{2(\varepsilon-\mu)}{\sqrt{(\varepsilon-\mu)^2+\Delta^2}} \times \right.$$

$$\left. \ln\left[\frac{z+\sqrt{z^2-\Delta^2}}{i\Delta}\right] + \ln\left[\frac{(\varepsilon-\mu)z-\sqrt{(\varepsilon-\mu)^2+\Delta^2}\,\sqrt{z^2-\Delta^2}}{(\varepsilon-\mu)z+\sqrt{(\varepsilon-\mu)^2+\Delta^2}\,\sqrt{z^2-\Delta^2}}\right]\right] d\varepsilon \;, \tag{6}$$

where $z = \sqrt{(E-\mu)^2+\Delta^2}$ and $\rho(\varepsilon)$ is the density of states. Though the equation for Γ seems quite involved, it can be integrated numerically for a given μ and Δ defining the extreme points of the arc.

4. The BCS to BEC Crossover and the Structure of the Cooper pairs

In what follows we will consider a uniform three dimensional Fermi system with single particle energies $\varepsilon_{\mathbf{k}} = \hbar^2 k^2/2m$ and a constant density $\rho = N/V = k_F^3/3\pi^2$. Due to the absence of an upper cutoff, Eq. (4) is singular and the thermodynamic limit becomes a subtle mathematical procedure.

It was shown[7] that this singularity can be cancelled out by considering other physical quantities whose bare counterpart diverges in the same way . For this problem, the physical quantity is the scattering length a_s given by

$$\frac{m}{4\pi\hbar^2 a_s} = \frac{1}{G} + \frac{1}{2}\int_0^\infty d\varepsilon\, \frac{\rho(\varepsilon)}{\varepsilon} \;. \tag{7}$$

The non-singular gap equation after integration is[16]

$$\frac{1}{k_F a_s} = \eta = \sqrt[4]{\mu^2+\Delta^2}\; P_{\frac{1}{2}}\left(-\frac{\mu}{\sqrt{\mu^2+\Delta^2}}\right) \;, \tag{8}$$

where energies are now in units of $\varepsilon_F = \hbar^2 k_F^2/2m$ and lengths in units of $\xi_F = 1/k_F$. $P_\beta(x)$ is the Legendre function of the first kind of degree β. The equations for the conservation of the number of pairs M is

$$-\frac{4}{3\pi} = \eta\mu + (\mu^2+\Delta^2)^{3/4}\; P_{\frac{3}{2}}\left(-\frac{\mu}{\sqrt{\mu^2+\Delta^2}}\right) . \tag{9}$$

It is worthwhile to note here that gap equation (8) and the number equation (9) depend on a single control parameter $\eta = 1/k_F a_s$ that goes from weak coupling BCS ($\eta \to -\infty$) to the extreme BEC limit ($\eta \to \infty$).

Figure 1 shows the BCS-BEC crossover diagram and several arcs corresponding to particular values of η. Since a crossover diagram is not a phase diagram, i.e. there is no symmetry order parameter or non-analyticities sharply differentiating the regions, boundaries are in principle arbitrary.

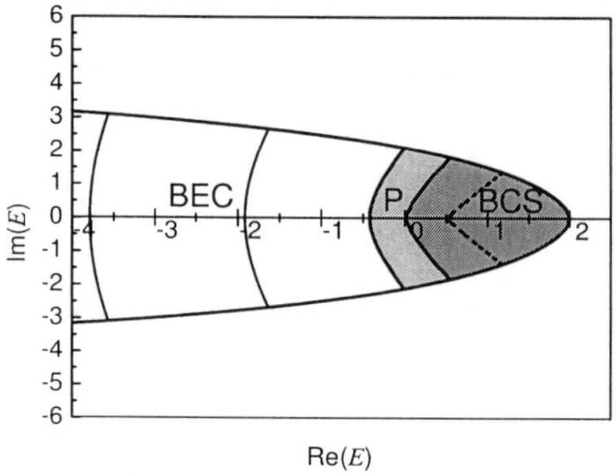

Fig. 1. Ground state crossover diagram displaying the different regimes as a function of η. A few arcs Γ, from Eq. (6), corresponding to the values of $\eta = 0$, 0.37, 0.55, 1 and 2, whose extremes are $E_F = 2(\mu \pm i\Delta)$, are displayed. The dashed line corresponds to the Unitary limit $\eta = 0$.

Here we have adopted the following criteria: The geometry of the arcs Γ serves us to establish a criterium for defining boundary regions in the crossover diagram. In the extreme weak coupling limit, $\eta \to -\infty$, the pair energies are twice the single particle energies ($E_\ell \to 2\varepsilon_{\mathbf{k}}$). As soon as the interaction is switched on, i.e. $k_F a_s$ is an infinitesimal negative number, a fraction f of the pair energies close to 2μ become complex, forming an arc Γ in the complex plane. The fraction $1-$f of fermion pairs below the crossing energy $2\varepsilon_c$ of Γ with the real axis have real pair energies. They are still in a sea of uncorrelated pairs with an effective Fermi energy ε_c. The dark grey region labelled BCS, which extends from $\eta = -\infty$ to $\eta = 0.37$, is characterized by a mixture of complex pair energies with a positive real part and real pair energies. $\eta = 0.37$ is the value at which all pair energies are complex, i.e. f $= 1$, and the effective Fermi sea has disappeared ($\varepsilon_c = 0$). Within the BCS region we plotted the arc with $\eta = 0$ (dashed line) with f $= 0.87$ and $\varepsilon_c = 0.25$. $\eta = 0$ is the universal point at which the solution does not depend in any parameter. The *pseudogap* region P, indicated in light grey, extends from $\eta = 0.37$ to $\eta = 0.55$ where $\mu = 0$. Within this region the real part of the pair energies changes from positive to negative, and P describes a mixture of Cooper resonances and quasibound molecules. The BEC (white) region, $\eta > 0.55$, is characterized by all pair energies having negative real parts, i.e. all pairs are quasibound molecules. Within the BEC region we plotted two arcs with $\eta = 1$ and $\eta = 2$. As η increases further, Γ tends to an almost vertical line with $Re(E) \sim 2\mu$, and $-2\Delta \leq Im(E) \leq 2\Delta$.

Having established a qualitative BCS-BEC crossover diagram, we turn now to the nature of the Cooper pairs. From Eq. (2), the Cooper-pair wavefunction is

$$\varphi_E(\mathbf{r}) = \frac{1}{V}\sum_{\mathbf{k}} \varphi_{\mathbf{k}}^E \, e^{i\mathbf{k}\mathbf{r}} = A\,\frac{e^{-r\sqrt{-E/2}}}{r}\,, \tag{10}$$

with $A^2 = Im(\sqrt{E/2})/2\pi\xi_F^3$, $\varphi_{\mathbf{k}}^E = C/(2\varepsilon_{\mathbf{k}} - E)$, and C being a normalization constant. On the other hand, there are two other definitions of the Cooper pair wavefunction commonly used in the literature. One is associated with the pair correlation function $\varphi_{\mathbf{k}}^{\mathsf{P}} = C_{\mathsf{P}} u_{\mathbf{k}} v_{\mathbf{k}}$, and the other is directly extracted form the number projected BCS wavefunction $\varphi_{\mathbf{k}}^{\mathsf{BCS}} = C_{\mathsf{BCS}} v_{\mathbf{k}}/u_{\mathbf{k}}$. The BCS and Pair wavefunctions have no closed-analytic form in coordinate space and the results presented below correspond to a numerical Fourier integration.

The various definitions of Cooper-pair wavefunctions are depicted in Fig. 2, which compares $\varphi_{\mathsf{BCS}}(r)$, $\varphi_{\mathsf{P}}(r)$, and $\varphi_E(r)$ for interaction strengths which correspond to the BCS, Unitary limit, and BEC regions of Fig. 1. Notice that while φ_{BCS}, and φ_{P} are highly oscillatory in the weak coupling region, this is not the case with φ_E which always has an exponential form. Clearly, a single and unified picture emerges for the crossover when using a many-body state such as (2): φ_E evolves smoothly through the crossover as it should. It is important to mention that the three wavefunctions are *exactly* the same in the strong coupling limit $\eta \to +\infty$.

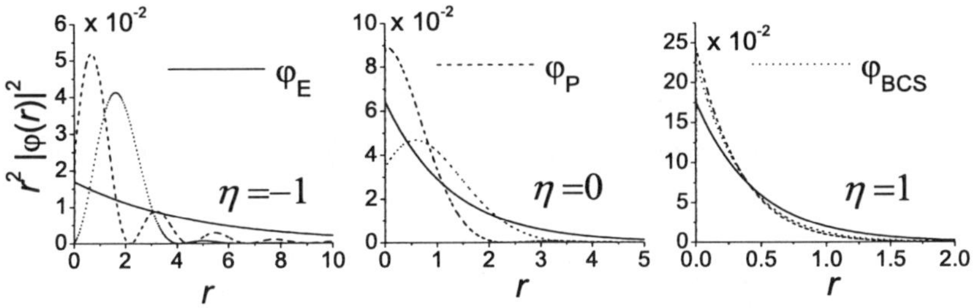

Fig. 2. Cooper-pair wavefunctions for different η. The upper and middle (Unitary limit) panels correspond to the BCS region, while the bottom one is in the BEC region. Except for the Cooper φ_E case, the other two wavefunctions always vanish at $r = 0$. It is only in the limit $\eta \to +\infty$ that the three states *exactly* coincide.

5. Integrable models for asymmetric Fermi superfluids

Consider N_a and N_b fermionic atoms confined to a D-dimensional box of volume V, i.e. $\rho_{a(b)} = N_{a(b)}/V$, with periodic boundary conditions and $g < 0$. (The exact solvability of the problem is not restricted to these latter conditions but for notational convenience we will specialize to this case). The following model Hamiltonian contains the right ingredients to study the competition between the various phases

$$H = \sum_{\mathbf{k}} (\varepsilon_{\mathbf{k}}^a \, n_{\mathbf{k}}^a + \varepsilon_{\mathbf{k}}^b \, n_{\mathbf{k}}^b) + 2g \sum_{\mathbf{k},\mathbf{k}'} a_{\mathbf{k}+\mathbf{Q}}^\dagger b_{-\mathbf{k}}^\dagger b_{-\mathbf{k}'} a_{\mathbf{k}'+\mathbf{Q}}, \tag{11}$$

where $a^\dagger_{\mathbf{k}}(b^\dagger_{\mathbf{k}})$ creates a particle of type $a(b)$ with momentum $\mathbf{k}$ and $n^a_{\mathbf{k}} = a^\dagger_{\mathbf{k}} a_{\mathbf{k}}$, $n^b_{\mathbf{k}} = b^\dagger_{\mathbf{k}} b_{\mathbf{k}}$. The pairing interaction scatters pairs with center-of-mass momentum $\mathbf{Q}$ and "band" energies $\varepsilon^\alpha_{\mathbf{k}} = \epsilon_{\mathbf{k}}/2m_\alpha$ ($\alpha = a, b$), with $\epsilon_{\mathbf{k}}$ representing an arbitrary dispersion (including a non-rotational-invariant one[12]).

The quantum integrability and exact solvability of the Hamiltonian (11) can be derived using an $su(2)$ algebra of pseudospin operators

$$\tau^+_{\mathbf{k},\mathbf{Q}} = a^\dagger_{\mathbf{k}+\mathbf{Q}} b^\dagger_{-\mathbf{k}} = (\tau^-_{\mathbf{k},\mathbf{Q}})^\dagger \ , \ \tau^z_{\mathbf{k},\mathbf{Q}} = \frac{1}{2}(n^a_{\mathbf{k}+\mathbf{Q}} + n^b_{-\mathbf{k}} - 1) \ , \tag{12}$$

and a second, independent, $su(2)$ algebra which also depends on the pair momentum $\mathbf{Q}$

$$S^+_{\mathbf{k},\mathbf{Q}} = a^\dagger_{\mathbf{k}+\mathbf{Q}} b_{-\mathbf{k}} = (S^-_{\mathbf{k},\mathbf{Q}})^\dagger \ , \ S^z_{\mathbf{k},\mathbf{Q}} = \frac{1}{2}(n^a_{\mathbf{k}+\mathbf{Q}} - n^b_{-\mathbf{k}}) \ . \tag{13}$$

These two mutually commuting algebras are often referred to as charge and spin $su(2)$ realizations, respectively. Using the algebraic techniques of the Richardson-Gaudin model,[5,6] one can write down a complete set of integrals of motion $R_{\mathbf{k},\mathbf{Q}}$, with $[R_{\mathbf{k},\mathbf{Q}}, R_{\mathbf{k}',\mathbf{Q}}] = 0$:

$$R_{\mathbf{k},\mathbf{Q}} = \tau^z_{\mathbf{k},\mathbf{Q}} + 2g \sum_{\mathbf{k}'(\neq \mathbf{k})} \frac{1}{\eta_{(\mathbf{k},\mathbf{Q})} - \eta_{(\mathbf{k}',\mathbf{Q})}} \vec{\tau}_{\mathbf{k},\mathbf{Q}} \cdot \vec{\tau}_{\mathbf{k}',\mathbf{Q}} \ , \tag{14}$$

with arbitrary functions η depending upon $\mathbf{k}$ and $\mathbf{Q}$. Their complete set of eigenvectors are of the form

$$|\Psi\rangle = \prod_{\ell=1}^{M} \left(\sum_{\mathbf{k}} \frac{1}{2\eta_{(\mathbf{k},\mathbf{Q})} - E_\ell} \tau^+_{\mathbf{k},\mathbf{Q}} \right) |\nu\rangle \ , \tag{15}$$

where $|\nu\rangle \equiv \prod_{\mathbf{k}} |\nu_{\mathbf{k},\mathbf{Q}}\rangle$ is a quasispin vacuum state defined by $\tau^-_{\mathbf{k},\mathbf{Q}} |\nu\rangle = 0$, and $S^z_{\mathbf{k},\mathbf{Q}} |\nu\rangle = d_{\mathbf{k},\mathbf{Q}} |\nu\rangle$, with $d_{\mathbf{k},\mathbf{Q}} = (2\nu_{\mathbf{k},\mathbf{Q}} - \Omega_{\mathbf{k},\mathbf{Q}})/4$, $\Omega_{\mathbf{k},\mathbf{Q}} = 2$, and the seniority quantum number $\nu_{\mathbf{k},\mathbf{Q}} = 1, 0$, which for the $su(2)$ pair algebra (12) counts the number of unpaired fermions. The complex spectral parameters E_ℓ satisfy the set of non-linear (Richardson) equations

$$\frac{1}{4g} - \sum_{\mathbf{k}} \frac{d_{\mathbf{k},\mathbf{Q}}}{2\eta_{(\mathbf{k},\mathbf{Q})} - E_\ell} + \sum_{m(\neq \ell)} \frac{1}{E_\ell - E_m} = 0 \ . \tag{16}$$

The occupation numbers in momentum space can be obtained from the integrals of motion (14) using the Hellman-Feynman theorem.

The total number of atoms $N = N_a + N_b = 2M + \nu$, where M is the number of atom pairs and ν the number of unpaired ones. Consider now the linear combination $H = 2\sum_{\mathbf{k}} \eta_{(\mathbf{k},\mathbf{Q})} R_{\mathbf{k},\mathbf{Q}} = \sum_{\mathbf{k}} 2\eta_{(\mathbf{k},\mathbf{Q})} \tau^z_{\mathbf{k},\mathbf{Q}} + 2g \sum_{\mathbf{k},\mathbf{k}'} \tau^+_{\mathbf{k},\mathbf{Q}} \tau^-_{\mathbf{k}',\mathbf{Q}} + C$, where $C = 3g\sum_{\mathbf{k}} d_{\mathbf{k},\mathbf{Q}} + g\,(N-L)^2/2 - g\,(N-L)$. Comparing this Hamiltonian with (11) we immediately see that they differ in the kinetic term. Making use of the spin $su(2)$

algebra by adding a term of the form $2\sum_{\mathbf{k}}\xi_{(\mathbf{k},\mathbf{Q})}S^z_{\mathbf{k},\mathbf{Q}}$, the resulting Hamiltonian is (up to an irrelevant constant)

$$H = \sum_{\mathbf{k}}[(\eta_{(\mathbf{k},\mathbf{Q})} + \xi_{(\mathbf{k},\mathbf{Q})})n^a_{\mathbf{k}+\mathbf{Q}} + (\eta_{(\mathbf{k},\mathbf{Q})} - \xi_{(\mathbf{k},\mathbf{Q})})n^b_{-\mathbf{k}}] \\ + 2g\sum_{\mathbf{k},\mathbf{k}'}\tau^+_{\mathbf{k},\mathbf{Q}}\tau^-_{\mathbf{k}',\mathbf{Q}}. \tag{17}$$

Identifying $\eta_{(\mathbf{k},\mathbf{Q})} = \frac{1}{2}[\varepsilon^a_{\mathbf{k}+\mathbf{Q}} + \varepsilon^b_{-\mathbf{k}}]$, and $\xi_{(\mathbf{k},\mathbf{Q})} = \frac{1}{2}[\varepsilon^a_{\mathbf{k}+\mathbf{Q}} - \varepsilon^b_{-\mathbf{k}}]$, we get the Hamiltonian of Eq. (11), after constraining the vectors $\mathbf{k}+\mathbf{Q}$ and $\mathbf{k}$ to be in the same set. The eigenvalue E corresponding to the solutions of Eqs. (16) are given by

$$E = \sum_{\mathbf{k}}\left(\varepsilon^a_{\mathbf{k}+\mathbf{Q}}\nu^a_{\mathbf{k}+\mathbf{Q}} + \varepsilon^b_{-\mathbf{k}}\nu^b_{-\mathbf{k}}\right) + \sum_{\ell}E_\ell, \tag{18}$$

where $\nu^\alpha_{\mathbf{k}}$ denotes the number of unpaired α particles in the state with momentum $\mathbf{k}$. The space dimensionality of the problem enters through the band dispersion $\varepsilon^\alpha_{\mathbf{k}}$, and the effective degeneracies $d_{\mathbf{k},\mathbf{Q}}$ in the exact solution (16). The latter are in turn defined by $\eta_{(\mathbf{k},\mathbf{Q})}$. Assuming space-inversion symmetry ($\varepsilon^\alpha_{\mathbf{k}} = \varepsilon^\alpha_{-\mathbf{k}}$), the degeneracies $\Omega_{\mathbf{k},\mathbf{Q}}$ count the number of states $[\mathbf{k},\mathbf{Q}]$ with the same value of $\eta_{(\mathbf{k},\mathbf{Q})}$.

For an asymmetric system with an excess of the a species ($N_a > N_b$), the atoms fill the lowest states up to $\mathbf{k}^b_F$ with $|\mathbf{k}^a_F| > |\mathbf{k}^b_F|$ at weak coupling. When the interaction is switched on several possible states compete to determine the absolute GS. The position of the unpaired atoms, defining the seniority quantum numbers $\nu_{\mathbf{k},\mathbf{Q}}$, block the available states from scattering pairs of atoms effectively reducing the degeneracies to $d_{\mathbf{k},\mathbf{Q}}$. When $\mathbf{Q} = \mathbf{0}$ the equations reduce to the Richardson model treated in previous sections. In general, configurations are identified by their $g \to 0$ limit, with specific pair and seniority occupations. They can be categorized as follows:

asymmetric BCS (aBCS): $\mathbf{Q} = \mathbf{0}$, a and b particles fill their lowest orbitals up to their corresponding Fermi levels.

breached A: same as aBCS, but the unpaired a particles move up in energy such that pairing correlations can develop around $\mathbf{k}^b_F$.

breached B: same as aBCS, but the unpaired a particles move down in energy such that pairing correlations can develop around $\mathbf{k}^a_F$.

LOFF: finite $\mathbf{Q}$, a and b particles fill their lowest orbitals up to their corresponding Fermi levels.

breached LOFF: finite $\mathbf{Q}$, but now some of the unpaired a particles move to allow more pairing correlations.

We will now explore the competition between the possible phases in a numerical example for $D = 2$. We assume a square lattice with dispersion $\varepsilon^\alpha_{\mathbf{k}} = -2(\cos(k_x) + \cos(k_y))$, with units chosen such that k_x and k_y are multiples of $2\pi/L$, where L is the linear size of the lattice. Being the dispersion equal for both atomic species, we are excluding an asymmetry in the masses or a deformed Fermi surface. However, we don't expect qualitative differences with the results presented below.

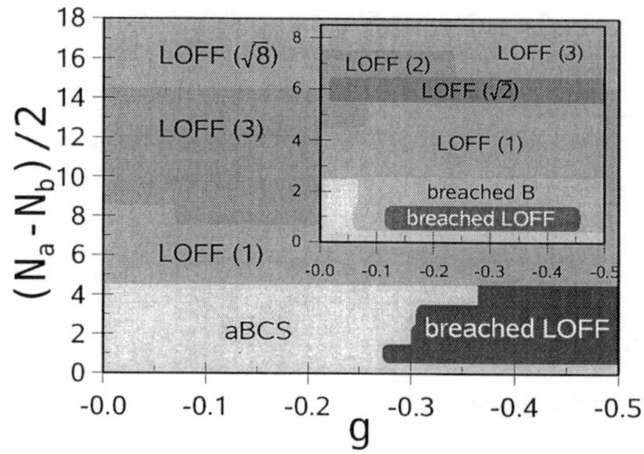

Fig. 3. Quantum phase diagram for a 6×6 lattice at half filling; the inset displays the quarter-filling case (for the LOFF phases, the momentum $\mathbf{Q}$ of the pairs in units of the lattice momentum is indicated in parentheses).

To make these statements more concrete, we have studied numerically model (11) on a 6×6 lattice, and with 18 (quarter-filled) or 36 (half-filled) particles, distributed over a and b states. A constant pairing interaction was used with various values of $\mathbf{Q}$ ($\mathbf{Q} = m\mathbf{q}$, with $m = -2, -1, \ldots, 3$, and $\mathbf{q} = (2\pi/3, 0), (0, 2\pi/3)$, and $(2\pi/3, 2\pi/3)$). In the case of homogeneous breached-pair phases 1, 2, or 3 particles were excited, while 1, or 2 were promoted in the case of breached LOFF. In total 42 different configurations were studied. Starting from the $g = 0$ configurations, the solution of Eq. (16) is obtained by slowly increasing the value of g, and by applying the iteration techniques explained in Ref. 17. In this way exact energies were evaluated up to $g = -0.5$. It is clear from our results that exotic configurations such as LOFF or breached LOFF can have lower energy than the aBCS state. There is a subtle competition between LOFF and the various breached BCS states, and both phenomena can appear simultaneously in an emergent breached LOFF configuration.

6. Summary

In summary, we studied the BCS-BEC crossover problem, as well as the nature of Cooper pairs in a correlated Fermi system, from the exact GS $|\Psi_0\rangle$ of the reduced BCS Hamiltonian. We have analytically determined its exact thermodynamic limit for a uniform $3D$ fermionic gas with quadratic single-particle dispersion. The fundamental difference between the BCS state and the exact many-body wavefunction is the fact that the former, being a mean-field, pairs off all fermions in a unique paired wavefunction while the exact state contains a statistical distribution of quasi-molecular resonant and scattering states (i.e., a mixed state) that depends upon the strength of the interaction between particles. It is only in the extreme BEC regime that the two pictures coincide asymptotically.

We have generalized exactly-solvable Richardson model such that the new integrable Hamiltonian admits several homogeneous and inhomogeneous phases depending upon the relative strength between kinetic and pairing interactions, and the difference in the number of atomic species (given a fixed total number of atoms). The inhomogeneous phases (LOFF) show up as soon as the difference in Fermi momentum between the two species becomes commensurate with the unit lattice momentum. A most significant result is the prediction of a new exotic phase which combines pairs with definite momentum and breached superfluidity/superconductivity that we dubbed breached LOFF. We expect this new phase to be the stable ground state at sufficiently large interaction strengths. These phases can be experimentally differentiated in time-of-flight measurements of the molecular velocity, after sweeping the system through the BCS-to-BEC crossover region.[8] The momentum distribution of unpaired fermions may distinguish the various exotic phases discussed here.

Acknowledgements

JD acknowledges financial support from the Spanish DGI under contract BFM2003-05316-C02-02 and SR from the Fund for Scientific Research-Flanders(Belgium).

References

1. J. Bardeen, L. N. Cooper, and J. R. Schrieffer, *Phys. Rev.* **106**, 162 (1957).
2. A. Bohr, B. R. Mottelson, and D. Pines, *Phys. Rev.* **110**, (1958).
3. R. W. Richardson, *Phys. Lett.* **3**, 277 (1963); *Nucl. Phys.* **52**, 221 (1964).
4. G. Sierra *et al.*, *Phys. Rev.* **B61**, R11 (2001).
5. J. Dukelsky, S. Pittel, and G. Sierra, *Rev. Mod. Phys.* **76**, 643 (2004).
6. G. Ortiz *et al.*, *Nucl. Phys.* **B707**, 401 (2005).
7. A. J. Leggett, in *Modern trends in the theory of condensed matter*, ed. by A. Pekalski and R. Przystawa (Springer Verlag, Berlin, 1980).
8. M. W. Zwierlein *et. al*, *Phys. Rev. Lett.* **92**, 120403 (2004); *Phys. Rev. Lett.* **94**, 180401 (2005).
9. G. Ortiz, and J. Dukelsky, *Phys. Rev.* **A72**, 043611 (2005).
10. A. I. Larkin and Yu. N. Ovchinnikov, *Sov. Phys. JETP* **20**, 762 (1965); P. Fulde and R. A. Ferrell, *Phys. Rev.* **135**, A550 (1964).
11. W. V. Liu and F. Wilczek, *Phys. Rev. Lett.* **90**, 047002 (2003).
12. H. Muther, and A. Sedrakian, *Phys. Rev. Lett.* **88**, 252503 (2002).
13. A. Sedrakian *et al.*, *Phys. Rev.* **A72** 013613(2005); C. -H. Pao *et al.*, cond-mat/0506437; D. T. Son, and M. A. Stephanov, cond-mat/0507586; K. Yang, cond-mat/0508484.
14. D. S. Petrov, C. Salomon and G. V. Shlyapnikov, *Phys. Rev. Lett.* **93**, 090404 (2004).
15. J. M. Roman, G. Sierra, and J. Dukelsky, *Nucl. Phys.* **B634**, 483 (2002).
16. M. Marini *et al.*, *Eur. Phys. J.* **B1**, 151 (1998); T. Papenbrock and G. F. Bertsch, *Phys. Rev.* **C59**, 2052 (1999).
17. S. Rombouts, D. Van Neck, and J. Dukelsky, *Phys. Rev.* **C69**, 061303 (2004).

QUANTUM MONTE CARLO STUDY OF THE GROUND-STATE PROPERTIES OF A FERMI GAS IN THE BCS-BEC CROSSOVER

S. GIORGINI

Dipartimento di Fisica, Università di Trento and BEC-INFM, I-38050 Povo, Italy

G.E. ASTRAKHARCHIK

Dipartimento di Fisica, Università di Trento and BEC-INFM, I-38050 Povo, Italy
and Institute of Spectroscopy, 142190 Troitsk, Moscow region, Russia

J. BORONAT and J. CASULLERAS

Departament de Física i Enginyeria Nuclear, Campus Nord B4-B5
Universitat Politècnica de Catalunya, E-08034 Barcelona, Spain

The ground-state properties of a two-component Fermi gas with attractive short-range interactions are calculated using the fixed-node diffusion Monte Carlo method. The interaction strength is varied over a wide range by tuning the value a of the s-wave scattering length of the two-body potential. We calculate the energy per particle, the one- and two-body density matrix as a function of the interaction strength. Results for the momentum distribution of the atoms, as obtained from the Fourier transform of the one-body density matrix, are reported as a function of the interaction strength. Off-diagonal long-range order in the system is investigated through the asymptotic behavior of the two-body density matrix. The condensate fraction of pairs is calculated in the unitary limit and on both sides of the BCS-BEC crossover.

Keywords: Quantum Monte Carlo; BCS-BEC crossover; condensate fraction.

1. Introduction

Recent experiments on two-component ultracold atomic Fermi gases near a Feshbach resonance have opened the possibility of investigating the crossover from a Bose-Einstein condensate (BEC) to a Bardeen-Cooper-Schrieffer (BCS) superfluid. In these systems the strength of the interaction can be varied over a very wide range by magnetically tuning the two-body scattering amplitude. For positive values of the s-wave scattering length a, atoms with different spins are observed to pair into bound molecules which, at low enough temperature, form a Bose condensate.[1] The molecular BEC state is adiabatically converted into a weakly-interacting ultracold Fermi gas with $a < 0$ and $k_F|a| \ll 1$,[2,3] where standard BCS theory is expected to apply. In the crossover region the value of $|a|$ can be orders of magnitude larger than the inverse Fermi wave vector k_F^{-1} and one enters a new strongly-correlated regime known as unitary limit.[2–4] In dilute systems, for which the effective range of the interaction R_0 is much smaller than the mean interparticle distance, $k_F R_0 \ll 1$, the unitary regime is believed to be universal.[5] In this regime, the only relevant energy scale should be given by the energy of the noninteracting Fermi gas,

$$\epsilon_{FG} = \frac{3}{10}\frac{\hbar^2 k_F^2}{m} . \tag{1}$$

Several properties of an interacting Fermi gas in the BCS-BEC crossover have been already investigated experimentally in harmonically trapped configurations. These include the molecular binding energy,[6] the release energy,[3] the size of the cloud,[2] the frequency of the collective oscillations,[7] the pairing energy,[8] the vortical configurations,[9] the thermodynamic behavior,[10] the momentum distribution[11] and the condensate fraction of pairs of fermionic atoms.[12]

The unitary regime presents a challenge for many-body theoretical approaches because there is not any obvious small parameter to construct a well-posed theory. The first theoretical studies of the BCS-BEC crossover at zero temperature are based on the mean-field BCS equations.[13] More sophisticated approaches take into account the effects of fluctuations,[14] or include explicitly the bosonic molecular field.[15] These theories provide a correct description in the deep BCS regime, but are only qualitatively correct in the unitary limit and in the BEC region. In particular, in the BEC regime the dimer-dimer scattering length has been calculated exactly from the solution of the four-body problem, yielding $a_m = 0.6a$.[16]

Quantum Monte Carlo techniques are the best suited tools for treating strongly-correlated systems. These methods have already been applied to ultracold degenerate Fermi gases in a recent work by Carlson *et al.*[17] In this study the energy per particle of a dilute Fermi gas in the unitary limit is calculated with the fixed-node Green's function Monte Carlo method (FN-GFMC) giving the result $E/N = \xi\epsilon_{FG}$ with $\xi = 0.44(1)$. In a subsequent work,[18] the same authors have extended the FN-GFMC calculations to investigate the equation of state in the BCS and BEC regimes. Their results in the BEC limit are compatible with a repulsive molecular gas, but the equation of state has not been extracted with enough precision.

We report results for the equation of state and other ground-state properties of a Fermi gas in the BCS-BEC crossover region using the fixed-node diffusion Monte Carlo method (FN-DMC). The interaction strength is varied over a very broad range from $-6 \leq -1/k_F a \leq 6$, including the unitary limit and the deep BEC and BCS regimes. Concerning the equation of state, in the unitary and in the BCS limit we find agreement, respectively, with the results of Ref. 17 and with the known perturbation expansion of a weakly attractive Fermi gas.[19] In the BEC regime, we find a gas of molecules whose repulsive interactions are well described by the dimer-dimer scattering length $a_m = 0.6a$. The one- and two-body density matrix are calculated as a function of the interaction strength. From the Fourier transform of the one-body density matrix we obtain the momentum distribution of the gas. From the asymptotic behavior of the two-body density matrix we extract the condensate fraction of pairs of fermionic atoms. The calculated condensate fraction is compared with analytical expansions holding on the BEC and BCS side of the Feshbach resonance. The comparison with mean-field results[13] for the momentum distribution and the condensate fraction is also discussed.

2. Method

We consider a homogeneous two-component unpolarized Fermi gas described by the Hamiltonian

$$H = -\frac{\hbar^2}{2m}\left(\sum_{i=1}^{N/2}\nabla_i^2 + \sum_{i'=1}^{N/2}\nabla_{i'}^2\right) + \sum_{i,i'} V(r_{ii'}) \, , \tag{2}$$

where m is the mass of the particles and $i, j, ...$ $(i', j', ...)$ label spin-up (spin-down) particles. The strength of the interaction is assumed to be determined only by the parameter $1/k_F a$, with $k_F = (3\pi^2 n)^{1/3}$ the Fermi wave-vector fixed by the atomic density $n = N/V$, and a the s-wave scattering length describing the low-energy collisions between the two fermionic species. The interatomic interactions in Eq. (2) are only between atoms with different spin and are modeled by a short-range potential that determines the value of a. In the present study, we use an attractive square-well potential, $V(r) = -V_0$ for $r < R_0$ and $V(r) = 0$ otherwise, with $nR_0^3 = 10^{-6}$. We have verified that in the density range $nR_0^3 = 10^{-7} - 10^{-5}$ the particular form of $V(r)$ is not relevant for the quantities considered in the present study, and the reported results are in this sense universal. The different regimes: BEC ($a > 0$ and $1/k_F a \gg 1$), BCS ($a < 0$ and $1/k_F|a| \gg 1$) and unitary limit ($1/k_F a = 0$), are obtained by varying the potential depth V_0. For the attractive square-well potential the value of the s-wave scattering length is given by $a = R_0[1 - \tan(K_0 R_0)/(K_0 R_0)]$, where $K_0^2 = mV_0/\hbar^2$. We vary K_0 is the range: $0 < K_0 < \pi/R_0$. For $K_0 R_0 < \pi/2$ the potential does not support a two-body bound state and $a < 0$. For $K_0 R_0 > \pi/2$, instead, the scattering length is positive, $a > 0$, and a molecular state appears whose binding energy ϵ_b is determined by the trascendental equation $\sqrt{|\epsilon_b| m/\hbar^2} R_0 \tan(\bar{K} R_0)/(\bar{K} R_0) = -1$, where $\bar{K}^2 = K_0^2 - |\epsilon_b| m/\hbar^2$. Notice that in the limit $R_0 \to 0$ the binding energy is given by $|\epsilon_b| = \hbar^2/ma^2$, corresponding to a zero-range potential. The value $K_0 = \pi/(2R_0)$ corresponds to the unitary limit where $|a| = \infty$ and $\epsilon_b = 0$.

In a FN-DMC simulation the function $f(\mathbf{R}, \tau) = \psi_T(\mathbf{R})\Psi(\mathbf{R}, \tau)$, where $\Psi(\mathbf{R}, \tau)$ denotes the wave function of the system and $\psi_T(\mathbf{R})$ is a trial function used for importance sampling, is evolved in imaginary time $\tau = it/\hbar$ according to the Schrödinger equation

$$\begin{aligned} -\frac{\partial f(\mathbf{R}, \tau)}{\partial \tau} = & - D\nabla_{\mathbf{R}}^2 f(\mathbf{R}, \tau) + D\nabla_{\mathbf{R}}[\mathbf{F}(\mathbf{R}) f(\mathbf{R}, \tau)] \\ & + [E_L(\mathbf{R}) - E_{ref}] f(\mathbf{R}, \tau) \, . \end{aligned} \tag{3}$$

In the above equation $\mathbf{R} = (\mathbf{r}_1, ..., \mathbf{r}_N)$, $E_L(\mathbf{R}) = \psi_T(\mathbf{R})^{-1} H \psi_T(\mathbf{R})$ denotes the local energy, $\mathbf{F}(\mathbf{R}) = 2\psi_T(\mathbf{R})^{-1}\nabla_{\mathbf{R}}\psi_T(\mathbf{R})$ is the quantum drift force, $D = \hbar^2/(2m)$

plays the role of an effective diffusion constant, and E_{ref} is a reference energy introduced to stabilize the numerics. The energy and other observables of the state of the system are calculated from averages over the asymtpotic distribution function $f(\mathbf{R}, \tau \to \infty)$. To ensure positive definiteness of the probability distribution f for fermions, the nodal structure of ψ_T is imposed as a constraint during the calculation. It can be proved that, due to this nodal constraint, the calculated energy is an upper bound to the eigenenergy for a given symmetry.[20] In particular, if the nodal surface of ψ_T were exact, the fixed-node energy would also be exact.

The trial wave function we consider has the general form[17,18,21] $\psi_T(\mathbf{R}) = \Psi_J(\mathbf{R})\Psi_{BCS}(\mathbf{R})$, where Ψ_J contains Jastrow correlations between all the particles,

$$\Psi_J(\mathbf{R}) = \prod_{i<j} f_{\uparrow\uparrow}(r_{ij}) \prod_{i'<j'} f_{\downarrow\downarrow}(r_{i'j'}) \prod_{i,i'} f_{\uparrow\downarrow}(r_{ii'}) , \tag{4}$$

and the BCS-type wave function Ψ_{BCS} is the antisymmetrized product of the pair wave functions $\phi(\mathbf{r}_i - \mathbf{r}_{i'})$

$$\Psi_{BCS}(\mathbf{R}) = \mathcal{A}\left(\phi(\mathbf{r}_1 - \mathbf{r}_{1'})\phi(\mathbf{r}_2 - \mathbf{r}_{2'})...\phi(\mathbf{r}_{N_\uparrow} - \mathbf{r}_{N_\downarrow})\right) . \tag{5}$$

The pair orbital $\phi(\mathbf{r})$ is chosen as

$$\phi(\mathbf{r}) = \beta \sum_{k_\alpha \le k_{max}} e^{i\mathbf{k}_\alpha \cdot \mathbf{r}} + \phi_s(r) , \tag{6}$$

where the sum is performed over the plane wave states $\mathbf{k}_\alpha = 2\pi/L(\ell_{\alpha x}\hat{x} + \ell_{\alpha y}\hat{y} + \ell_{\alpha z}\hat{z})$ up to the largest closed shell $k_{max} = 2\pi/L(\ell^2_{maxx} + \ell^2_{maxy} + \ell^2_{maxz})^{1/2}$ occupied by $N/2$ particles. Here $L = V^{1/3}$ is the size of the cubic simulation box and ℓ are integer numbers. If $\phi_s(r) = 0$ in Eq. (6), it can be shown that Ψ_{BCS} in Eq. (5) coincides with the exact wave function of a free Fermi gas, i.e., the product of Slater determinants of spin-up and spin-down particles.[22] The spherically symmetric function $\phi_s(r)$ in Eq. (6) accounts for s-wave pairing. Our choice for $\phi_s(\mathbf{r})$ is the following. If $1/k_Fa \ge -0.2$, it corresponds to the solution of the two-body problem with the potential $V(r)$. In particular, in the region $a > 0$ we use the bound-state solution $\phi_{bs}(r)$ with energy ϵ_b and in the region $a < 0$ the unbound-state solution corresponding to zero scattering energy: $\phi_{us}(r) = (R_0 - a)\sin(K_0 r)/[r\sin(K_0 R_0)]$ for $r < R_0$ and $\phi_{us}(r) = 1 - a/r$ for $r > R_0$. In the unitary limit, $|a| \to \infty$, $\phi_{bs}(r) = \phi_{us}(r)$. For $1/k_Fa < -0.2$ we use instead $\phi_s(r) = \gamma_1[\exp(-\gamma_2 r) + \exp(-\gamma_2(L-r))]$. Here, γ_1, γ_2, and β in Eq. (6) are variational parameters.

The Jastrow wave function Ψ_J, Eq. (4), is determined as follows. For $1/k_Fa \ge -0.2$, we use $f_{\uparrow\downarrow}(r) = 1$ and $f_{\uparrow\uparrow}(r) = f_{\downarrow\downarrow}(r)$ given by the two-body solution of a fictitious repulsive step potential with range $\tilde{R}$ and scattering length $\tilde{a}$. The boundary conditions $f(r = L/2) = 1$ and $f'(r = L/2) = 0$ determine the wave function in terms of the variational parameters $\tilde{R}$ and $\tilde{a}$. For $1/k_Fa < -0.2$, we use instead $f_{\uparrow\uparrow}(r) = f_{\downarrow\downarrow}(r) = 1$ and for the crossed correlation factor we use again the unbound-state solution, $f_{\uparrow\downarrow}(r) = \phi_{us}(r)$ for $r < \bar{R}$. In order to reduce possible size effects due to the long range tail of $\phi_{us}(r)$, we use $f_{\uparrow\downarrow}(r) = C_1 + C_2(\exp(-\alpha r) + \exp(-\alpha(L-r))$

for $r > \bar{R}$, with $\bar{R} < L/2$ a matching point. The coefficients C_1 and C_2 are fixed by the continuity condition for $f_{\uparrow\downarrow}(r)$ and its first derivative at $r = \bar{R}$, whereas α and $\bar{R}$ are variational parameters. The parameters of the Jastrow function Ψ_J, Eq. (4), are optimized by minimizing the variational expectation value $\langle\psi_T|H|\psi_T\rangle/\langle\psi_T|\psi_T\rangle$. The parameters of the BCS function Ψ_{BCS}, Eq. (5), affect the nodal surface of the trial wave function and are optimized by minimizing the FN-DMC estimate of the energy.

A direct estimate of any operator O in DMC is known as mixed estimate, $\langle O\rangle_m = \langle\psi_T|O|\Psi\rangle/\langle\psi_T|\Psi\rangle$, and is exact only for the Hamiltonian and operators commuting with it. If O is a local operator, one can circumvent this problem by introducing pure estimators. That is not the case for the one- and two-body density matrix. In order to reduce, and even eliminate in practice, a possible bias in the calculation we have used the extrapolated estimator $\langle\Psi|O|\Psi\rangle/\langle\Psi|\Psi\rangle \simeq 2\langle O\rangle_m - \langle O\rangle_v$, with the variational estimator defined as $\langle O\rangle_v = \langle\psi_T|O|\psi_T\rangle/\langle\psi_T|\psi_T\rangle$.

We consider a system with periodic boundary conditions. Residual finite-size effects have been determined carrying out calculations with an increasing number of particles N = 14, 38, and 66. These studies show that the value $N = 66$ is optimal since finite-size corrections are below the reported statistical error in the whole BCS-BEC crossover.

3. Equation of state

The FN-DMC energies are shown in Fig. 1 as a function of the interaction parameter $-1/k_Fa$.[23] We plot the difference $E/N - \epsilon_b/2$, where ϵ_b is the molecular binding energy which is zero for $-1/k_Fa \geq 0$. In the BCS region, $-1/k_Fa > 1$, our results are in agreement with the perturbation expansion of a weakly attractive Fermi gas[19,24]

$$\frac{E}{N\epsilon_{FG}} = 1 + \frac{10}{9\pi}k_Fa + \frac{4(11 - 2\log 2)}{21\pi^2}(k_Fa)^2 + \dots \,. \tag{7}$$

In the unitary limit we find $E/N = \xi\epsilon_{FG}$, with $\xi = 0.42(1)$. This result is compatible with the findings of Refs. 17, 18. The value of the parameter $\beta = \xi - 1$ has been measured in experiments with trapped Fermi gases,[2–4] but the precision is too low to make stringent comparisons with theoretical predictions. In the region of positive scattering length E/N decreases by decreasing k_Fa. At approximately $-1/k_Fa \simeq -0.3$, the energy becomes negative, and by further decreasing k_Fa it rapidly approaches the binding energy per particle $\epsilon_b/2$ indicating the formation of bound molecules.[18] In the BEC region, $-1/k_Fa < -1$, we find that the FN-DMC energies agree with the equation of state of a repulsive gas of molecules

$$\frac{E/N - \epsilon_b/2}{\epsilon_{FG}} = \frac{5}{18\pi}k_Fa_m\left[1 + \frac{128}{15\sqrt{6\pi^3}}(k_Fa_m)^{3/2} + \dots\right] , \tag{8}$$

where the first term corresponds to the mean-field energy of a gas of molecules of mass $2m$ and density $n/2$ interacting with the positive molecule-molecule scattering length a_m, and the second term corresponds to the first beyond mean-field correction [Lee-Huang-Yang[25] (LHY) correction]. If for a_m we use the value calculated by Petrov *et al.*[16] $a_m = 0.6a$, we obtain the curves shown in Fig. 1. If, instead, we use a_m as a fitting parameter to our FN-DMC results in the region $-1/k_F a \leq -1$, we obtain the value $a_m/a = 0.62(1)$. The inset of Fig. 1 shows that in the BEC regime we find evidence of beyond mean-field effects.

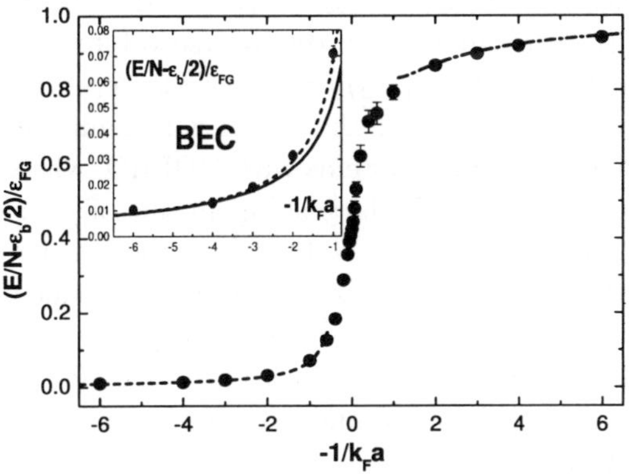

Fig. 1. Energy per particle in the BCS-BEC crossover with the binding energy subtracted from E/N. The dot-dashed line is the perturbation expansion (7) and the dashed line corresponds to the expansion (8) holding in the BEC regime. Inset: enlarged view of the BEC regime $-1/k_F a \leq -1$. The solid line corresponds to the mean-field energy [first term in the expansion (8)], the dashed line includes the beyond mean-field correction [Eq. (8)].

4. Momentum distribution and condensate fraction

The occurrence of off-diagonal long-range order (ODLRO) in interacting systems of bosons and fermions was investigated by C.N. Yang in terms of the asymptotic behavior of the one- and two-body density matrix.[26] In the case of a two-component Fermi gas with $N_\uparrow$ spin-up and $N_\downarrow$ spin-down particles, the one-body density matrix (OBDM) for spin-up particles, defined as

$$\rho_1(\mathbf{r}_1', \mathbf{r}_1) = \langle \psi_\uparrow^\dagger(\mathbf{r}_1')\psi_\uparrow(\mathbf{r}_1)\rangle \,, \tag{9}$$

does not possess any eigenvalue of order $N_\uparrow$. This behavior implies for homogeneous systems the asymptotic condition $\rho_1(\mathbf{r}_1', \mathbf{r}_1) \to 0$ as $|\mathbf{r}_1 - \mathbf{r}_1'| \to \infty$. In the above expression $\psi_\uparrow^\dagger(\mathbf{r})$ ($\psi_\uparrow(\mathbf{r})$) denote the creation (annihilation) operator of spin-up particles. The same result holds for spin-down particles. ODLRO may occur instead in the two-body density matrix (TBDM), that is defined as

$$\rho_2(\mathbf{r}_1', \mathbf{r}_2', \mathbf{r}_1, \mathbf{r}_2) = \langle \psi_\uparrow^\dagger(\mathbf{r}_1')\psi_\downarrow^\dagger(\mathbf{r}_2')\psi_\uparrow(\mathbf{r}_1)\psi_\downarrow(\mathbf{r}_2)\rangle \,. \tag{10}$$

For a homogeneous unpolarized gas with $N_\uparrow = N_\downarrow = N/2$, if ρ_2 has an eigenvalue of the order of the total number of particles N, the TBDM can be written as a spectral decomposition separating the largest eigenvalue yielding for $|\mathbf{r}_1 - \mathbf{r}'_1|$, $|\mathbf{r}_2 - \mathbf{r}'_2| \to \infty$ the asymptotic behavior

$$\rho_2(\mathbf{r}'_1, \mathbf{r}'_2, \mathbf{r}_1, \mathbf{r}_2) \to \alpha N/2 \varphi^*(|\mathbf{r}'_1 - \mathbf{r}'_2|)\varphi(|\mathbf{r}_1 - \mathbf{r}_2|) . \tag{11}$$

The parameter $\alpha \leq 1$ is interpreted as the condensate fraction of pairs, in a similar way as the condensate fraction of single atoms is derived from the OBDM. The complex function $\varphi(r)$ is normalized to the inverse volume $1/V$ and is proportional to the order parameter $\langle\psi_\uparrow(\mathbf{r}_1)\psi_\downarrow(\mathbf{r}_2)\rangle = \sqrt{\alpha N/2}\varphi(|\mathbf{r}_1 - \mathbf{r}_2|)$, whose appearance distinguishes the superfluid state of the Fermi gas. Equation (11) should be contrasted with the behavior of Bose systems with ODLRO, where ρ_1 has an eigenvalue of order N,[27] and consequently the largest eigenvalue of ρ_2 is of the order of N^2.

The momentum distribution $n_\mathbf{k}$, obtained from the Fourier transform of $\rho_1(r)$, is shown in Fig. 2.[23] In the inset of Fig. 2 we compare $n_\mathbf{k}$, calculated using FN-DMC for $1/k_F a = 4$, with the momentum distribution of the atoms in the molecular state $n_\mathbf{k} = 4(k_F a)^3/[3\pi(1 + k^2a^2)^2]$.[28] To reduce finite-size effects in the calculation of the Fourier transform for $1/k_F a = 0$ and -1, we have used the following model for the k-dependence of $n_\mathbf{k}$

$$n_\mathbf{k} = A\left(1 - \frac{(k/k_F)^2 - \mu}{\sqrt{[(k/k_F)^2 - \mu]^2 + \Delta^2}}\right) . \tag{12}$$

The values of μ, Δ and A are free parameters determined by the best fit of the inverse Fourier transform of Eq. (12) to the calculated $\rho_1(r)$, with the constraint $1/V \sum_k n_\mathbf{k} = n/2$. For $A = 1/2$, the above expression reproduces the standard $n_\mathbf{k}$ of BCS theory with μ and Δ, respectively, chemical potential and gap in units of the Fermi energy $\epsilon_F = \hbar^2 k_F^2/2m$. For $1/k_F a = 0$ and -1 the Fourier transform of the BCS model, Eq. (12), reproduces quite well the calculated $\rho_1(r)$ with a χ^2/ν of the order of one. In Fig. 2 we also show the results of $n_\mathbf{k}$ obtained using the BCS mean-field theory,[13] where the values of chemical potential and gap are calculated self-consistently through the gap and number equations.[28] The results of Fig. 2 show that the mean-field theory overestimates the broadening of $n_\mathbf{k}$ in the crossover region $-1 \leq 1/k_F a \leq 1$. Measurements of the momentum distribution of harmonically trapped systems have recently become available in the crossover.[11] The comparison carried out at unitarity, using the local density approximation, shows a good agreement with the observed distribution, but the experimental uncertainty is too large to distinguish between mean-field and FN-DMC results.[11]

The condensate fraction of pairs α has been obtained from the the projected TBDM, defined as[29,30]

$$\rho_2^P(r) = \frac{2}{N}\int d^3\mathbf{r}_1 d^3\mathbf{r}_2 \rho_2(\mathbf{r}_1 + \mathbf{r}, \mathbf{r}_2 + \mathbf{r}, \mathbf{r}_1, \mathbf{r}_2) . \tag{13}$$

By using Eq. (11) one finds that $\lim_{r\to\infty} \rho_2^P(r) = \alpha$. In terms of the order parameter, $F(|\mathbf{r}_1 - \mathbf{r}_2|) = \langle\psi_\uparrow(\mathbf{r}_1)\psi_\downarrow(\mathbf{r}_2)\rangle$, one has instead $\lim_{r\to\infty} \rho_2^P(r) = 2/n \int d^3\mathbf{r}' |F(r')|^2$.

The results for the condensate fraction α are shown in Fig. 3 (details of the calculation are given in Ref. 23). In the BEC regime, the results reproduce the Bogoliubov quantum depletion of a gas of composite bosons $\alpha = 1 - 8\sqrt{n_m a_m^3}/3\sqrt{\pi}$, where $n_m = n/2$ is the density of molecules and $a_m = 0.6a$ is the dimer-dimer scattering length.[16,23] In the opposite BCS regime, the condensate fraction α can be calculated from the result of the BCS order parameter holding for $r \gg a$ (see Ref. 31)

$$F_{BCS}(r) = \frac{\Delta k_F^3}{\epsilon_F} \frac{\sin(k_F r)}{4\pi^2 k_F r} K_0(r/\xi_0) \,, \tag{14}$$

where $\xi_0 = \hbar^2 k_F/m\Delta$ is the coherence length and $K_0(x)$ is the modified Bessel function. If we include the Gor'kov-Melik Barkhudarov correction for the pairing gap[32] $\Delta = (2/e)^{7/3}\epsilon_F e^{-\pi/2k_F|a|}$, we obtain for $\alpha = 2/n\int d^3\mathbf{r} F_{BCS}^2(r)$ the dot-dashed line shown in Fig. 3. On the BEC side of the resonance, we notice that the agreement with the Bogoliubov quantum depletion, extending up to $1/k_Fa \gtrsim 1$, is consistent with the evidence of the LHY beyond mean-field correction in the equation of state.[23] On the BCS side the comparison of the pairing gap Δ with the Gor'kov-Melik Barkhudarov result is discussed in Ref. 18. We have checked that effects due to finite N are smaller than the statistical uncertainty. For example, at $1/k_Fa = 0$, we find $\alpha = 0.57(2)$ for $N = 66$ and $\alpha = 0.61(2)$ for $N = 38$ particles. For $1/k_Fa \leq -1$ the coherence length ξ_0 becomes increasingly larger and these effects start to be relevant. At $1/k_Fa = -1$, the value of α, as obtained from $F_{BCS}(r)$ [Eq. (14)], reduces from 0.10 to 0.08 by cutting off the spatial integral at the simulation box size $r = L/2$. The condensate fraction α can also be estimated using the self-consistent mean-field approach[13] and the result for the order parameter: $F(r) = 1/V\sum_{\mathbf{k}} u_k v_k e^{i\mathbf{k}\cdot\mathbf{r}}$, where u_k and v_k are the usual quasiparticle amplitudes of BCS theory. The result is shown in Fig. 3 with a solid line (see also Ref. 30).

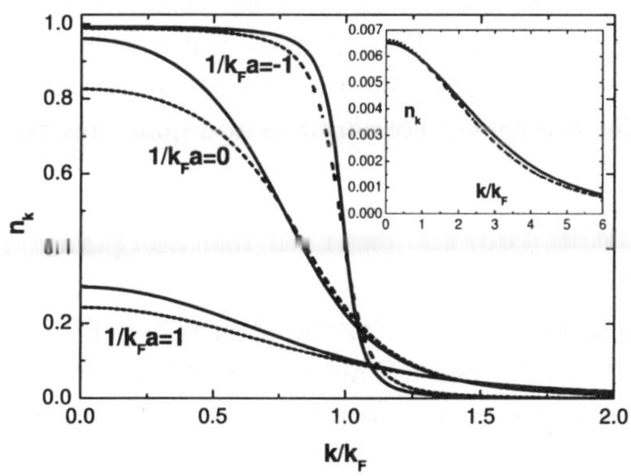

Fig. 2. Momentum distribution $n_{\mathbf{k}}$ for different values of $1/k_Fa$ (solid lines). The dashed lines correspond to $n_{\mathbf{k}}$ calculated using the BCS mean-field approach.[28] Inset: $n_{\mathbf{k}}$ for $1/k_Fa = 4$. The dotted line corresponds to the momentum distribution of the molecular state.

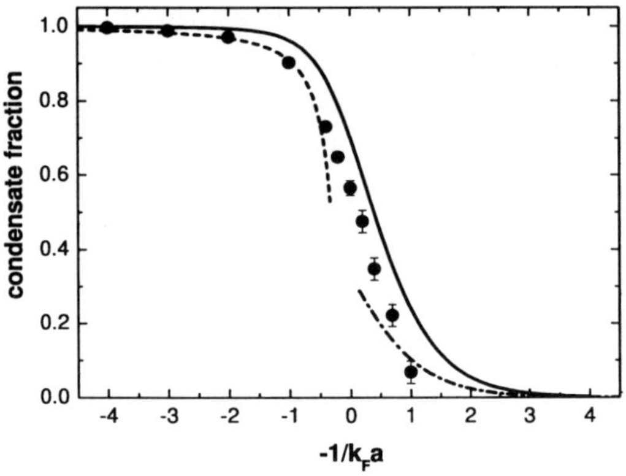

Fig. 3. Condensate fraction of pairs α as a function of the interaction strength: FN-DMC results (symbols), Bogoliubov quantum depletion of a Bose gas with $a_m = 0.6a$ (dashed line), BCS theory using Eq. (14) (dot-dashed line) and self-consistent mean-field theory (solid line).

5. Conclusions

In conclusion, we have presented a study of the equation of state and of correlation functions of a Fermi gas in the BCS-BEC crossover using FN-DMC techniques. In the BCS regime and in the unitary limit our results for the energy per particle are in agreement with known perturbation expansions and with previous FN-GFMC calculations,[17,18] respectively. In the BEC regime, we recover the equation of state of a gas of composite bosons with repulsive effective interactions which are well described by the molecule-molecule scattering length $a_m = 0.6a$ calculated in Ref. 16. We also find evidence of beyond mean-field effects in the equation of state in agreement with the LHY correction. We calculate the momentum distribution and the condensate fraction of fermionic pairs. Significant deviations from the mean-field description point out the role of correlations and quantum fluctuations in the BCS-BEC crossover.

Acknowledgements

GEA and SG acknowledge support by the Ministero dell'Istruzione, dell'Università e della Ricerca (MIUR). JB and JC acknowledge support from DGI (Spain) Grant No. BFM2002-00466 and Generalitat de Catalunya Grant No. 2001SGR-00222.

References

1. S. Jochim *et al.*, *Science* **302**, 2101 (2003); M. Greiner, C.A. Regal and D.S. Jin, *Nature* **426**, 537 (2003); M.W. Zwierlein *et al.*, *Phys. Rev. Lett.* **91**, 250401 (2003).
2. M. Bartenstein *et al.*, *Phys. Rev. Lett.* **92**, 120401 (2004).

3. T. Bourdel *et al.*, *Phys. Rev. Lett.* **93**, 050401 (2004).
4. K.M. O'Hara, S.L. Hemmer, M.E. Gehm, S.R. Granade and J.E. Thomas, *Science* **298**, 2179 (2002).
5. G.M. Bruun, *Phys. Rev.* **A70**, 053602 (2004); S. De Palo, M.L. Chiofalo, M.J. Holland and S.J.J.M.F. Kokkelmans, *Phys. Lett.* **A327**, 490 (2004).
6. C.A. Regal, C. Ticknor, J.L. Bohn and D.S. Jin, *Nature* **424**, 47 (2003).
7. J. Kinast *et al.*, *Phys. Rev. Lett.* **92**, 150402 (2004); M. Bartenstein *et al.*, *ibid.* **92**, 203201 (2004).
8. C. Chin *et al.*, *Science* **305**, 1128 (2004), G.B. Partridge *et al.*, *Phys. Rev. Lett.* **95**, 020404 (2005).
9. M.W. Zwierlein *et al.*, *Nature* **435**, 1047 (2005).
10. J. Kinast *et al.*, *Science* **307**, 1296 (2005).
11. C.A. Regal *et al.*, *Phys. Rev. Lett.* **95**, 250404 (2005).
12. C.A. Regal, M. Greiner and D.S. Jin, *Phys. Rev. Lett.* **92**, 040403 (2004); M.W. Zwierlein *et al.*, *ibid.* **92**, 120403 (2004); M.W. Zwierlein *et al.*, *ibid.***94**, 180401 (2005).
13. A.J. Leggett, in *Modern Trends in the Theory of Condensed Matter*, edited by A. Pekalski and R. Przystawa (Springer-Verlag, Berlin, 1980); P. Nozières and S. Schmitt-Rink, *J. Low Temp. Phys.* **59**, 195 (1985); J.R. Engelbrecht, M. Randeria and C.A.R. Sá de Melo, *Phys. Rev.* **B55**, 15153 (1997).
14. P. Pieri and G.C. Strinati, *Phys. Rev.* **B61**, 15370 (2000); P. Pieri, L. Pisani and G.C. Strinati, *Phys. Rev.* **B70**, 094508 (2004).
15. M. Holland, S.J.J.M.F. Kokkelmans, M.L. Chiofalo and R. Walser, *Phys. Rev. Lett.* **87**, 120406 (2001); Y. Ohashi and A. Griffin, *Phys. Rev.* **A67**, 063612 (2003).
16. D.S. Petrov, C. Salomon and G.V. Shlyapnikov, *Phys. Rev. Lett.* **93**, 090404 (2004).
17. J. Carlson, S.-Y. Chang, V.R. Pandharipande and K.E. Schmidt, *Phys. Rev. Lett.* **91**, 050401 (2003).
18. S.-Y. Chang, V.R. Pandharipande, J. Carlson and K.E. Schmidt, *Phys. Rev.* **A70**, 043602 (2004).
19. K. Huang and C.N. Yang, *Phys. Rev.* **105**, 767 (1957); T.D. Lee and C.N. Yang, *ibid.* **105**, 1119 (1957).
20. P.J. Reynolds, D.M. Ceperley, B.J. Alder and W.A. Lester Jr., *J. Chem. Phys.* **77**, 5593 (1982).
21. S.Y. Chang and V.R. Pandharipande, *Phys. Rev. Lett.* **95**, 080402 (2005).
22. J.P. Bouchaud, A. Georges and C. Lhuillier, *J. Phys. (Paris)* **49**, 553 (1988); J.P. Bouchaud and C. Lhuillier, *Z. Phys.* **B75**, 283 (1989).
23. G.E. Astrakharchik, J. Boronat, J. Casulleras and S. Giorgini, *Phys. Rev. Lett.* **93**, 200404 (2004); *ibid.* **95**, 230405 (2005).
24. Note that for $k_F|a| \ll 1$ the nonanalytic correction to the ground-state energy due to the superfluid gap is exponentially small.
25. T.D. Lee, K. Huang and C.N. Yang, *Phys. Rev.* **106**, 1135 (1957).
26. C.N. Yang, *Rev. Mod. Phys.* **34**, 694 (1962).
27. O. Penrose and L. Onsager, *Phys. Rev.* **104**, 576 (1956).
28. L. Viverit, S. Giorgini, L.P. Pitaevskii and S. Stringari, *Phys. Rev.* **A69**, 013607 (2004).
29. S. De Palo, F. Rapisarda and G. Senatore, *Phys. Rev. Lett.* **88**, 206401 (2002).
30. G. Ortiz and J. Dukelsky, cond-mat/0503664; L. Salasnich, N. Manini and A. Parola, *Phys. Rev.* **A72**, 023621 (2005).
31. J. Bardeen, L.N. Cooper and J.R. Schrieffer, *Phys. Rev.* **108**, 1175 (1957).
32. L.P. Gor'kov and T.K. Melik-Barkhudarov, *Sov. Phys. JETP* **13**, 1018 (1961).

COLLAPSE OF K-Rb FERMI-BOSE MIXTURES IN OPTICAL LATTICES

D. M. JEZEK and H. M. CATALDO

Departamento de Física, Facultad de Ciencias Exactas y Naturales, Universidad de Buenos Aires, RA-1428 Buenos Aires, Argentina.

We study a confined mixture of Rb and K atoms in a one dimensional optical lattice, at low temperature, in the quantal degeneracy regime. This mixture exhibits an attractive boson-fermion interaction, and thus above certain values of the number of particles the mixture collapses. We investigate, in the mean-field approximation, the curve for which this phenomenon occurs, in the space of number of particles of both species. This is done for different types of optical lattices.

Keywords: Atomic gases; optical lattice; mixture.

1. Introduction

Recent experiments on degenerate Fermi-Bose mixtures have opened the possibility of studying in a direct way the effects of quantum statistics in Bose-Einstein condensates (BECs).

From a theoretical point of view a considerable amount of work has been done to study boson-fermion mixtures.[1–3] In particular, an exhaustive and systematic study of the structure of binary mixtures has been performed by Roth in Ref. 1. In this work he has discussed all posible sign combinations of scattering lengths between boson-boson and boson-fermion s-wave interactions.

The aim of this work is to study the stability of the K-Rb mixture being the atoms confined within a harmonic trap plus a one dimensional optical lattice. This mixture exhibits an attractive interaction between fermions and bosons, and thus there exists an upper limit, in the number of particles, for the mixture to be stable. For numbers of particles above some critical ones the system collapses. In a previous work[4] we have analysed some properties of these mixtures. Here we concentrate ourselves on the study of the phenomenon of collapse when this mixture is in the presence of an optical lattice. We observe the effect of varying the height of the barriers and the spacing between consecutive wells.

2. The model

We consider a mixture of condensed bosons (B) and degenerated fermions (F) at zero temperrature confined in an external potential given by the sum of magnetic origin potentials, V_B^H and V_F^H, for bosons and fermions respectively, and of a stationary optical potential V_{opt} modulated along the z axis. The axially symmetric harmonic traps in cylindrical coordinates (r, ϕ, z) read,

$$V_B^H = \frac{1}{2} M_B \left[\omega_{rB}^2 r^2 + \omega_{zB}^2 z^2\right] \tag{1}$$

and

$$V_F^H = \frac{1}{2} M_F \left[\omega_{rF}^2 r^2 + \omega_{zF}^2 z^2\right] \quad , \tag{2}$$

where ω_{rB}, ω_{zB} and ω_{rF} ω_{zF} are the radial (axial) angular trapping frequencies for bosons and fermions, respectively. M_B and M_F are the corresponding masses.

The optical potential has the following expression:

$$V_{opt} = s\ E_r \sin^2(\frac{2\pi z}{d}), \tag{3}$$

where E_r denotes the recoil energy, s an adimensional amplitude parameter and d the spatial period of the lattice.

Thus the total potential for either bosons or fermions is $V_B = V_B^H + V_{opt}$ and $V_F = V_F^H + V_{opt}$, respectively.

Since the number of fermions we are considering is very large, we can use the Thomas-Fermi-Weizsäcker (TFW) approximation to write their kinetic energy density per unit volume as a function of the fermionic local density n_F and its gradients, for more details see Ref. 4. For fully polarized spin 1/2 fermions, it reads:

$$\tau_F(\vec{r}) = \frac{3}{5}(6\pi^2)^{2/3} n_F^{5/3} + \beta \frac{(\nabla n_F)^2}{n_F} \tag{4}$$

in terms of which the femionic kinetic energy is

$$T_F = \frac{\hbar^2}{2M_F} \int d\vec{r}\, \tau_F(\vec{r}) \tag{5}$$

The value of the β coefficient in the Weizsäcker term is fixed to 1/18. This term contributes little to the total fermion kinetic energy, and it is usually ignored.[1]

Neglecting all p-wave interactions, the energy density per unit volume has the form

$$\mathcal{E}(\vec{r}) = \frac{\hbar^2}{2M_B}|\nabla\Psi|^2 + V_B n_B + \frac{1}{2} g_{BB} n_B^2 + g_{BF} n_F n_B + \frac{1}{2M_F}\tau_F + V_F n_F \quad , \tag{6}$$

where $n_B = |\Psi|^2$ denotes the local density of bosons. The boson-boson and boson-fermion coupling constants g_{BB} and g_{BF}, are written in terms of the s-wave scattering lengths a_B and a_{BF} as $g_{BB} = 4\pi\, a_B \hbar^2/M_B$ and $g_{BF} = 4\pi\, a_{BF}\hbar^2/M_{BF}$, respectively. We have defined $M_{BF} \equiv 2M_B M_F/(M_B + M_F)$.

To obtain the ground state for this system, we solve the Gross-Pitaevskii (GP) equation for the bosons coupled to the Thomas-Fermi-Weizsäcker equation for the fermions which arise from the variation of $\mathcal{E}$ with respect to n_B and n_F keeping the particle numbers constant [Euler-Lagrange (EL) equations]:

$$\left(-\frac{\hbar^2\nabla^2}{2M_B} + V_B + g_{BB}\, n_B + g_{BF}\, n_F\right)\Psi = \mu_B \Psi \tag{7}$$

$$\frac{\hbar^2}{2M_F}\left[(6\pi^2)^{2/3} n_F^{5/3} + \beta\frac{(\nabla n_F)^2}{n_F} - 2\beta\Delta n_F\right] + V_F n_F + g_{BF} n_B n_F = \mu_F n_F \quad , \tag{8}$$

where μ_B and μ_F denote the boson and fermion chemical potentials, respectively.

Note that the solution to Eq. (8) is not more complicated than that of the GP equation. This can be readily seen by writing the later in terms of n_B:

$$\frac{\hbar^2}{2M_B}\left[\frac{1}{4}\frac{(\nabla n_B)^2}{n_B} - \frac{1}{2}\Delta n_B\right] + V_B n_B + g_{BB} n_B + g_{BF} n_B n_F = \mu_B n_B \quad , \tag{9}$$

which is formally equivalent to Eq. (8).

We have employed an imaginary time method to find the solution of these expressions written as imaginary time diffusion equations.[5]

3. Numerical results

The system we take into consideration is a ^{87}Rb+ ^{40}K mixture. We have numerically solved Eqs. (9) and (8) using the set of scattering lengths reported by Modugno et al.,[6] namely $a_B = 98.98\, a_0$, and $a_{BF} = -395\, a_0$, being a_0 the Bohr radius. We have used spherically symmetric traps both for bosons and fermions, but due to the one dimensional optical lattice, the mixture only has axial symmetry around the z axis. For bosons we have taken $\omega_B = 2\pi \times 100$ Hz, while for fermions we have taken $\omega_F = \sqrt{M_B/M_F}\, \omega_B$.

In Fig. 1 we display the particle density profiles of bosons (n_B) and fermions (n_F) as functions of z, within a harmonic trap. The typical radius of the Bose condensates we have worked with is $\sim R_B = 5\, \mu$m. The density profiles plotted in this figure correspond to a number of fermions $N_F = 1 \times 10^4$ and the number of bosons is $N_B = 6 \times 10^4$. It may be seen that even for a rather small N_F value, the shape of the boson density differs from the parabolic-type profile yielded by the TF approximation for a fermion-free condensate.

In Fig. 2 (a) and (b) we plot the density profiles for the same number of particles as those used in Fig. 1, but we have added the optical lattice to the trapping potential. The parameters of the optical lattice in Fig. 2 (a) [Fig. 2 (b)] are $s = 4$ and $d = 1.35\ \mu$m ($d = 4\ \mu$m). By comparing these two graphs, it may be seen that the density profile turns out to be much more distorted with respect to that shown in Fig. 1, when the spacing between wells (d) is larger.

Finally, the results related to the collapse are presented in Fig. 3. In this figure we have plotted the stability diagram in the $N_B - N_F$ plane. We show our theoretical prediction for critical numbers (N_F^c, N_B^c), considering the following cases: when the external potential is only given by the harmonic trap (trapping potential), whose

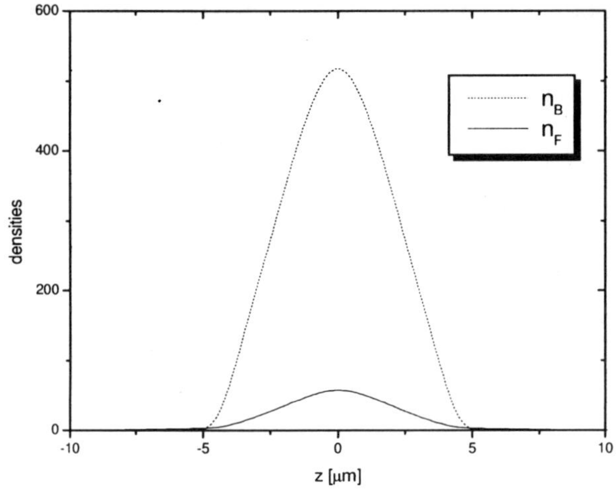

Fig. 1. Boson and fermion density profiles when considering as external potential only the harmonic trap. The number of particles are $N_B = 6 \times 10^4$ and $N_F = 1 \times 10^4$.

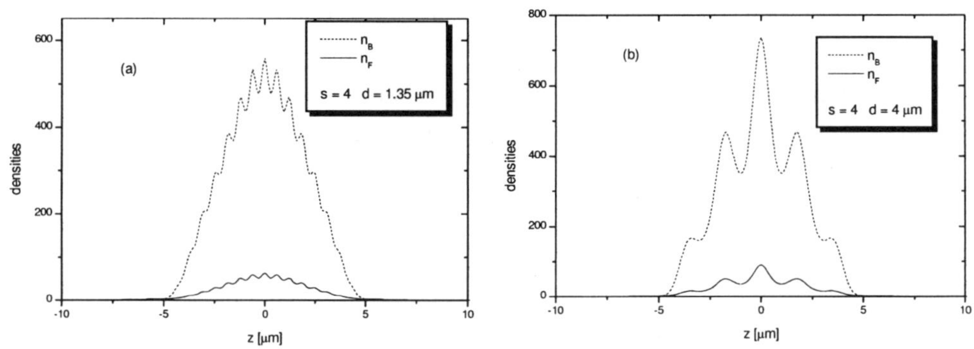

Fig. 2. Boson and fermion density profiles with the external potential including the optical lattice, for the same number of particles as in Fig. 1.

points are indicated by squares in the graph, and when an optical lattice is added with parameters $s = 4$, $d = 1.35$ μm ($s = 4$, $d = 4$ μm), the respective points being indicated by crosses (triangles).

It may be observed that for a spacing between consecutive wells $d = 4$ μm and a height of barriers $s = 4$, the critical number decrease about 20 % with respect to the system without an optical lattice, and thus the system is more unstable. Whereas for the same barrier and a spacing of $d = 1.35$ μm, the curve turns out to be practically superimposed to the previous one. The same observation holds for other values of barrier heights. In summary, the degree of instability turns out to be significantly affected by the spacing of the lattice wells.

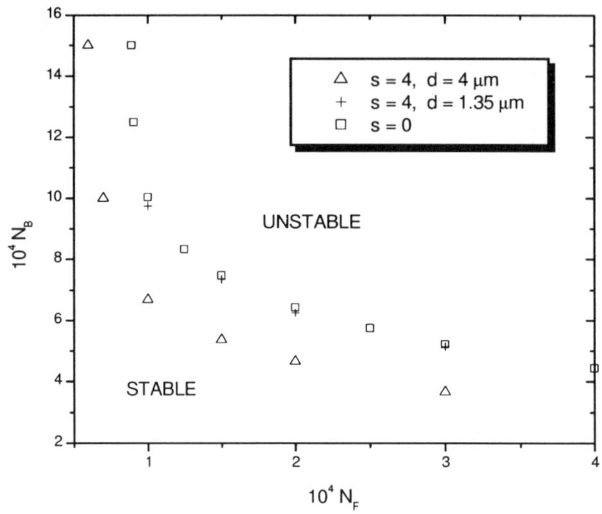

Fig. 3. Stability diagram in the $N_B - N_F$ plane.

In conclusion, we have observed that the inclusion of an optical lattice makes the system to become more unstable, and thus the number of particles should be lowered in a considerable amount, which depends not only on the height of the barrier but also on the spacing between consecutive wells.

Acknowledgements

This work has been performed under Grant PIP 5409 from CONICET, Argentina.

References

1. R. Roth, *Phys. Rev.* **A66**, 013614 (2002).
2. K. Molmer, *Phys. Rev. Lett.* **80**, 1804 (1998).
3. L. Vichi, M. Amoruso, A. Minguzzi, S. Stringari, and M. Tosi, *Eur. Phys. J.* **D11**, 335 (2000).
4. D. M. Jezek, M. Barranco, M. Guilleumas, R. Mayol, and M. Pi, *Phys. Rev.* **A70**, 043630 (2004).
5. M. Barranco, M. Guilleumas, E.S. Hernández, R. Mayol, M. Pi, and L. Szybisz, *Phys. Rev.* **B68**, 024515 (2003).
6. M. Modugno, G. Roati, F. Riboli, G. Roati, G. Modugno, and M. Inguscio, *Phys. Rev.* **A68**, 043626 (2003).

BCS-BEC CROSSOVER IN A SUPERFLUID FERMI GAS

YOJI OHASHI

Institute of Physics, University of Tsukuba, Tsukuba, Ibaraki, 305, Japan
yohashi@sakura.cc.tsukuba.ac.jp

We discuss the superfluid phase transition in a gas of Fermi atoms with a Feshbach resonance. A tunable pairing interaction associated with the Feshbach resonance is shown to naturally lead to the BCS-BEC crossover, where the character of superfluidity continuously changes from the weak-coupling Bardeen-Cooper-Schrieffer (BCS) type to the Bose-Einstein condensation (BEC) of tightly bound molecules, as one decreases the threshold energy 2ν of the Feshbach resonance. We also discuss effects of a trap, as well as the p-wave BCS-BEC crossover adjusted by a p-wave Feshbach resonance.

Keywords: BCS-BEC crossover; superfluid Fermi gas; strong-coupling superfluid theory.

1. Introduction

One of the most exciting topics in the current research of cold atom physics is the BCS-BEC crossover, where the character of superfluidity continuously changes from the weak-coupling BCS type to the BEC of tightly bound molecular bosons as one increases the strength of a pairing interaction.[1–4] In a two-component Fermi gas with a Feshbach resonance, a pairing interaction associated with the Feshbach resonance is tunable by adjusting the threshold energy 2ν of the Feshbach resonance.[5,6] Using this tunable interaction, we can study the BCS-BEC crossover as a function of the threshold energy 2ν.[7] Indeed, this Feshbach mechanism and the resulting BCS-BEC crossover have been recently confirmed experimentally in ^{40}K and ^{6}Li.[8–11] More recently, interesting superfluid properties in the crossover region, such as single-particle excitations,[10] collective modes,[10,11] and vortices,[10] have been also observed.

In this paper, we discuss the BCS-BEC crossover in a gas of Fermi atoms with a Feshbach resonance. Extending the strong-coupling theory developed by Nozières and Schmitt-Rink[2–4] in the context of superconductivity to include the Feshbach resonance, we calculate the superfluid phase transition temperature T_c in the entire crossover region. We also consider effects of a harmonic trap potential, as well as an anisotropic p-wave Feshbach resonance.

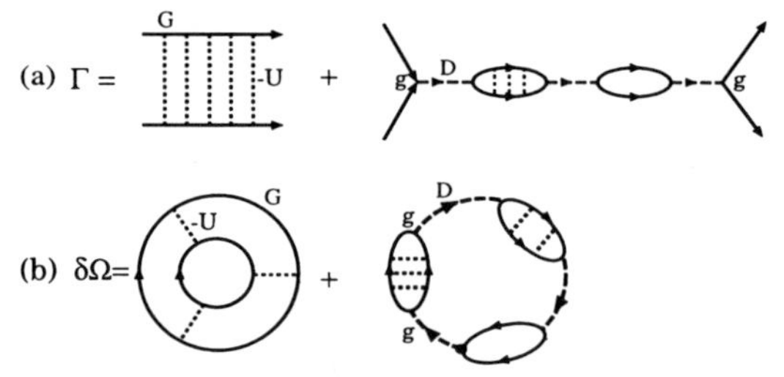

Fig. 1. (a) Particle-particle scattering matrix Γ in the t-matrix approximation. The solid line shows the Fermi single-particle Green's function $G^{-1} = i\omega_n - \xi_{\mathbf{p}}$ (where ω_n is the fermion Matsubara frequency), and the dashed line shows the Bose Green's function D associated with the Feshbach resonance. The dotted line shows the nonresonant interaction $-U$. (b) Fluctuation contribution $\delta\Omega$ to the thermodynamic potential Ω. The thermodynamic potential Ω is given by the sum of the non-interacting part Ω_0 and $\delta\Omega$.

2. BCS-BEC Crossover at T_c

2.1. *Coupled fermion-boson model*

We consider a two-component Fermi gas described by pseudo-spin $\sigma =\uparrow,\downarrow$. The ($s$-wave) Feshbach resonance can be described by the coupled fermion-boson (CFB) model[5–7]

$$H = \sum_{\mathbf{p},\sigma} \varepsilon_{\mathbf{p}} c^\dagger_{\mathbf{p}\sigma} c_{\mathbf{p}\sigma} - U \sum_{\mathbf{p},\mathbf{p}',\mathbf{q}} c^\dagger_{\mathbf{p}+\mathbf{q}/2\uparrow} c^\dagger_{-\mathbf{p}+\mathbf{q}/2\downarrow} c_{-\mathbf{p}'+\mathbf{q}/2\downarrow} c_{\mathbf{p}'+\mathbf{q}/2\uparrow}$$
$$+ \sum_{\mathbf{q}} [\varepsilon^B_{\mathbf{q}} + 2\nu] b^\dagger_{\mathbf{q}} b_{\mathbf{q}} + g \sum_{\mathbf{p},\mathbf{q}} \Big[b^\dagger_{\mathbf{q}} c_{-\mathbf{p}+\mathbf{q}/2\downarrow} c_{\mathbf{p}+\mathbf{q}/2\uparrow} + \text{H.c.} \Big]. \quad (1)$$

Here, $c^\dagger_{\mathbf{p}\sigma}$ is the creation operator of a Fermi atom with the kinetic energy $\varepsilon_{\mathbf{p}} = p^2/2m$ (where m is the mass of a Fermi atom). $b^\dagger_{\mathbf{q}}$ describes molecular bosons associated with the Feshbach resonance. The energy of this molecule is given by $\varepsilon^B_{\mathbf{q}} + 2\nu = q^2/2M + 2\nu$, where 2ν is the threshold energy of the Feshbach resonance. In Eq. (1), g is a coupling constant of the Feshbach resonance describing resonance between atoms and molecules. We also include a nonresonant interaction $-U$, which is taken to be weakly attractive. For simplicity, we first consider a uniform gas. Effects of a harmonic trap will be discussed in Sec. 3.

Since a molecule consists of two Fermi atoms, we take the mass of a molecule $M = 2m$ and impose the conservation for the total number of atoms $N = \sum_{\mathbf{p},\sigma} \langle c^\dagger_{\mathbf{p}\sigma} c_{\mathbf{p}\sigma}\rangle + 2\sum_{\mathbf{q}} \langle b^\dagger_{\mathbf{q}} b_{\mathbf{q}}\rangle$. This constraint can be absorbed into the Hamiltonian by considering the grand canonical form $H - \mu N$, where μ is the Fermi chemical potential. The resulting Hamiltonian has the same form as Eq. (1), except that $\varepsilon_{\mathbf{p}}$ and $\varepsilon^B_{\mathbf{q}}$ are, respectively, replaced with $\xi_{\mathbf{p}} \equiv \varepsilon_{\mathbf{p}} - \mu$ and $\xi^B_{\mathbf{q}} \equiv \varepsilon^B_{\mathbf{q}} - \mu_B$, where $\mu_B \equiv 2\mu$ is the boson chemical potential.

2.2. *Strong-coupling theory including Feshbach resonance*

To calculate T_c adjusted by the Feshbach resonance, we extend the strong-coupling theory developed by Nozières and Schmitt-Rink (NSR)[2] to the CFB model in Eq. (1). Following the NSR theory, the equation for T_c is determined by the Thouless criterion, which states that the superfluid phase transition occurs when the particle-particle vertex function $\Gamma(\mathbf{q}, i\nu_n)$ develops a pole at $\mathbf{q} = \nu_n = 0$ (where ν_n is the boson Matsubara frequency). In the t-matrix approximation in terms of $-U$ and g, $\Gamma(\mathbf{q}, i\nu_n)$ is diagrammatically given by Fig. 1(a). In this figure, the first diagram describes multi-scattering due to the interaction $-U$ and the second one describes the Feshbach resonance. Summing up these diagrams, we obtain[7]

$$\Gamma(\mathbf{q}, i\nu_n) = -\frac{U_{\text{eff}}(\mathbf{q}, i\nu_n)}{1 - U_{\text{eff}}(\mathbf{q}, i\nu_n)\Pi(\mathbf{q}, i\nu_n)}. \tag{2}$$

Here, $U_{\text{eff}}(\mathbf{q}, i\nu_n) \equiv U - g^2 D(\mathbf{q}, i\nu_n)$ involves effects of the Feshbach resonance in the second term, where $D^{-1}(\mathbf{q}, i\nu_n) = i\nu_n - (\xi_{\mathbf{q}}^B + 2\nu)$ is the boson single-particle Green's function. The correlation function $\Pi(\mathbf{q}, i\nu_n)$ describes fluctuations in the Cooper channel,[2–4] given by

$$\Pi(\mathbf{q}, i\nu_n) = \sum_{\mathbf{p}} \frac{1 - f(\xi_{\mathbf{p}+\mathbf{q}/2}) - f(\xi_{\mathbf{p}-\mathbf{q}/2})}{\xi_{\mathbf{p}+\mathbf{q}/2} + \xi_{\mathbf{p}-\mathbf{q}/2} - i\nu_n}, \tag{3}$$

where $f(\varepsilon)$ is the Fermi distribution function. Applying the Thouless criterion to Eq. (2), we obtain the equation for T_c as

$$1 = U_{\text{eff}}(0,0)\Pi(0,0) = \Big(U + \frac{g^2}{2\nu - 2\mu}\Big) \sum_{\mathbf{p}} \frac{\tanh(\varepsilon_{\mathbf{p}} - \mu)/2T_c}{2(\varepsilon_{\mathbf{p}} - \mu)}. \tag{4}$$

Equation (4) has the same form as the ordinary BCS gap equation at T_c. Thus, $U_{\text{eff}}(0,0) = U + g^2/(2\nu - 2\mu)$ may be regarded as an effective pairing interaction. In particular, $U_{\text{F.R.}} \equiv g^2/(2\nu - 2\mu)$ in $U_{\text{eff}}(0,0)$ describes the interaction associated with the Feshbach resonance. Indeed, $U_{\text{F.R.}}$ can be easily obtained from the diagram shown in Fig. 2(a), by assuming that the kinetic energy of each incoming atom equals the Fermi chemical potential μ (which is the characteristic energy of a many fermion gas) and the Feshbach molecule appearing in the intermediate state has the resonance energy 2ν. Since 2ν is experimentally tunable by an external magnetic field,[5] the pairing interaction $U_{\text{eff}}(0,0)$ is also tunable. In particular, it becomes strong when 2ν approaches $2\mu + 0$.

We briefly note that the pairing mechanism shown in Fig. 2(a) is a different type from the conventional 'phonon' mechanism shown in Fig. 2(b). In the intermediate state, while two fermions and one phonon (boson) coexist in Fig. 2(b), only a molecular boson is propagating in Fig. 2(a).

In the weak-coupling BCS regime, one may take $\mu = \varepsilon_F$ in Eq. (4), where ε_F is the Fermi energy. However, the chemical potential μ is know to deviate from ε_F in the strong-coupling regime.[1–4] This effect can be included by considering the equation for the total number N of Fermi atoms,[1–4] given by $N = -\partial\Omega/\partial\mu$ (where

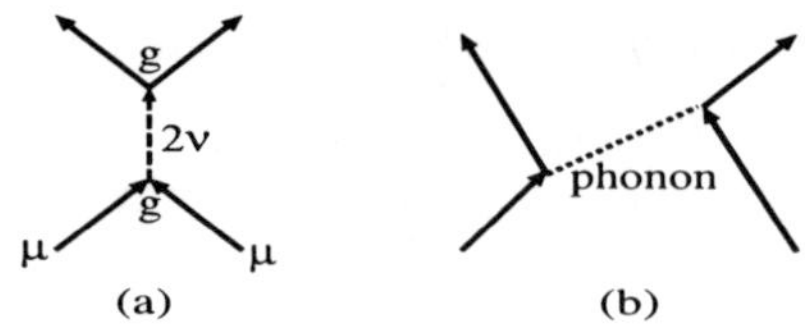

Fig. 2. (a) Pairing mechanism associated with the Feshbach resonance. (b) Conventional 'phonon' mechanism. Solid lines show incoming and outgoing fermions.

Ω is the thermodynamic potential). In calculating Ω, we include fluctuation effects in the Cooper-channel (described by the first term in Fig. 1(b)) and the Feshbach resonance (described by the second term in Fig. 1(b)). Then we obtain

$$N = -\frac{\partial \Omega}{\partial \mu} = N_{\rm F}^0 + 2N_{\rm B}^0 - T\frac{\partial}{\partial \mu}\sum_{\mathbf{q},i\nu_n} e^{i\delta\nu_n}\ln[1 - U_{\rm eff}(\mathbf{q},i\nu_n)\Pi(\mathbf{q},i\nu_n)]. \tag{5}$$

Here, $N_{\rm F}^0 \equiv 2\sum_{\mathbf{p}} f(\xi_{\mathbf{p}})$ and $N_{\rm B}^0 \equiv \sum_{\mathbf{q}} n_{\rm B}(\xi_{\mathbf{q}}^{\rm B} + 2\nu)$ are obtained from the non-interacting part of Ω, where $n_B(\varepsilon)$ is the Bose distribution function. $T_{\rm c}$ and μ are determined self-consistently by solving Eq. (4) together with Eq. (5). As in the ordinary BCS theory, these equations need an energy cutoff ω_c.

We note that one can rewrite Eq. (5) as

$$N = N_{\rm F}^0 + 2\tilde{N}_{\rm B} - T\frac{\partial}{\partial \mu}\sum_{\mathbf{q},i\nu_n} e^{i\delta\nu_n}\ln[1 - U\Pi(\mathbf{q},i\nu_n)]. \tag{6}$$

Here, $\tilde{N}_{\rm B} \equiv -T\frac{\partial}{\partial \mu_B}\sum_{\mathbf{q},i\nu_n} e^{i\delta\nu_n}\ln\tilde{D}^{-1}(\mathbf{q},i\nu_n)$ is the number of *renormalized* Feshbach molecules described by the renormalized Bose Green's function

$$\tilde{D}(\mathbf{q},i\nu_n) \equiv \frac{1}{i\nu_n - (\varepsilon_{\mathbf{q}}^B + 2\nu - 2\mu) + g^2\frac{\Pi(\mathbf{q},i\nu_n)}{1-U\Pi(\mathbf{q},i\nu_n)}}. \tag{7}$$

The last term in the denominator is the self-energy correction describing effects of fluctuations in the Cooper channel on the Feshbach molecules. Noting that Eq. (4) can be rewritten as

$$2\nu - 2\mu = g^2\frac{\Pi(0,0)}{1 - U\Pi(0,0)}, \tag{8}$$

we find that Eq. (7) has a *gapless* pole at $\mathbf{q} = \nu_n = 0$ at $T_{\rm c}$. This means that, although $T_{\rm c}$ is determined by the fermion BCS gap equation (4), this $T_{\rm c}$ is also the BEC phase transition temperature of the (renormalized) Feshbach molecules. Indeed, below $T_{\rm c}$, one can obtain the exact identity,[7] $\langle b_{\mathbf{q}=0}\rangle = -g\Delta/U(2\nu - 2\mu)$, where $\Delta \equiv U\sum_{\mathbf{p}}\langle c_{-\mathbf{p}\downarrow}c_{\mathbf{p}\uparrow}\rangle$ is the BCS order parameter. Namely, both the molecular BEC order parameter $\langle b_0\rangle$ and Δ are always finite in the superfluid phase.

2.3. *BCS limit and BEC limit*

The weak-coupling BCS regime is obtained when $2\nu \gg 2\varepsilon_{\rm F}$. In this regime, since Feshbach molecules are almost absent due to the large threshold energy, one may take $N_{\rm B}^0 \simeq 0$ and $U_{\rm eff}(\mathbf{q}, i\nu_n) \simeq U$. We can also neglect the last term in Eq. (5), because it only contains fluctuation contribution from the *weak* interaction U. As a result, Eq. (5) is reduced to $N \simeq N_{\rm F}^0$, which gives $\mu \simeq \varepsilon_{\rm F}$ as far as $T_{\rm c} \ll \varepsilon_{\rm F}$. Substituting this result into Eq. (4), we obtain the ordinary BCS gap equation at $T_{\rm c}$ with the weak pairing interaction U. The resulting $T_{\rm c}$ in this regime is given by

$$T_{\rm c} = \frac{8\gamma}{\pi e^2} T_{\rm F} e^{\frac{\pi}{2k_{\rm F}\bar{a}_s}} = 0.61 T_{\rm F} e^{\frac{\pi}{2k_{\rm F}\bar{a}_s}} \quad (\gamma = 1.78). \tag{9}$$

Here, $T_{\rm F}$ is the Fermi temperature, and $k_{\rm F}$ is the Fermi momentum. In Eq. (9), we have introduced the two-body scattering length $\bar{a}_s$ in terms of $-U$. The weak-coupling limit corresponds to $\bar{a}_s^{-1} \to -\infty$.

The opposite BEC regime is realized when $2\nu \ll -2\varepsilon_{\rm F}$. In this case, the molecular band is much lower than the fermion band, so that dissociated Fermi atoms are almost absent ($N_{\rm F}^0 \simeq 0$). Because of the absence of Fermi atoms, the correlation function $\Pi(\mathbf{q}, i\nu_n)$ (which describes fluctuations in the Cooper channel) can be also neglected in Eq. (5). Then we obtain

$$\frac{N}{2} = N_{\rm B}^0 = \sum_{\mathbf{q}} \frac{1}{e^{(\varepsilon_{\mathbf{q}}^B + 2\nu - 2\mu)/T} - 1}. \tag{10}$$

This is just the equation for the number of bosons in the case when all the atoms form Feshbach molecules. In the BEC limit, since we can set $\Pi(0,0) = 0$ in Eq. (8), we obtain $2\mu = 2\nu$. Since 2μ ($= \mu_B$) has the meaning of the Bose chemical potential and 2ν is the bottom energy of the molecular band, Eq. (10) with $2\mu = 2\nu$ is found to be the condition for the BEC phase transition of an ideal Bose gas with $N_B = N/2$ Feshbach molecules. Thus, $T_{\rm c}$ in this limit is given by

$$T_{\rm c} = \frac{2\pi}{(\zeta(3/2))^{2/3}} \frac{(N_B/V)^{2/3}}{M} = 0.218 T_{\rm F} \quad (\zeta(3/2) = 2.612). \tag{11}$$

Besides the BEC regime dominated by Feshbach molecules, the BEC regime dominated by Cooper pair bosons also exists when the Feshbach coupling g is very large ($g\sqrt{n} \gg \varepsilon_{\rm F}$, where n is the atomic number density). In such a *broad* Feshbach resonance, the pairing interaction $U_{\rm eff}(0,0)$ can become strong even when the threshold energy 2ν is still much larger than $2\varepsilon_{\rm F}$ ($\gtrsim 2\mu$). In this case, the Feshbach molecules only contribute to the pairing interaction in a *virtual* sense, so that we can neglect $N_{\rm B}^0$ in Eq. (5). We can also neglect the dynamical effect of molecules in $U_{\rm eff}(\mathbf{q}, i\nu_n)$, namely,

$$U_{\rm eff}(\mathbf{q}, i\nu_n) \to U + \frac{g^2}{2\nu} \equiv \tilde{U}. \tag{12}$$

We note that the second term in $\tilde{U}$ has the same form as the interaction associated with the Feshbach resonance in a two-particle system ($\mu = 0$). Eqs. (4) and (5) in this regime have the forms

$$1 = \tilde{U} \sum_{\mathbf{p}} \frac{\tanh(\varepsilon_{\mathbf{p}} - \mu)/2T_{\rm c}}{2(\varepsilon_{\mathbf{p}} - \mu)}, \tag{13}$$

$$N = N_{\rm F}^0 - T \sum_{\mathbf{q}, i\nu_n} e^{i\delta\nu_n} \frac{\partial}{\partial\mu} \ln[1 - \tilde{U}\Pi(\mathbf{q}, i\nu_n)]. \tag{14}$$

In the strong-coupling regime (but still $2\nu \gg 2\varepsilon_{\rm F}$), Eq. (13) gives $\mu = -1/(2m\tilde{a}_s^2)$, where $\tilde{a}_s$ is the scattering length in terms of $-\tilde{U}$, defined by

$$\frac{4\pi\tilde{a}_s}{m} = -\frac{\tilde{U}}{1 - \tilde{U}\sum_{\mathbf{p}}^{\omega_c} \frac{1}{2\varepsilon_{\mathbf{p}}}}. \tag{15}$$

Since $\mu \ll -\varepsilon_{\rm F}$, we can set $N_{\rm F}^0 = 0$ in Eq. (14). Expanding the second term in Eq. (14) around $\mathbf{q} = \nu_n = 0$, we obtain

$$\frac{N}{2} = T \sum_{\mathbf{q}, i\nu_n} e^{i\delta\nu_n} \frac{1}{i\nu_n - (q^2/4m - \mu_C)} = \sum_{\mathbf{q}} n_B(\frac{q^2}{4m} - \mu_C). \tag{16}$$

Here, $\mu_C \equiv 4(\mu + \sqrt{|\mu|/2m\tilde{a}_s^2})$ may be interpreted as the chemical potential of Cooper pair bosons, which vanishes at $T_{\rm c}$ (Note that $\mu = -1/(2m\tilde{a}_s^2)$). Since the Feshbach molecules are absent ($2\nu \gg 2\varepsilon_{\rm F}$), we can understand that Eq. (16) with $\mu_C = 0$ determines $T_{\rm c}$ of an ideal Bose gas of $N/2$ *tightly bound Cooper pairs*. Although the Cooper pair bosons in the present BEC regime are different from the Feshbach molecules in the previous BEC regime ($2\nu \to -\infty$), $T_{\rm c}$ in this regime is also given by Eq. (11).

We obtain a quite different result in a *narrow* Feshbach resonance, where $g\sqrt{n} \lesssim \varepsilon_{\rm F}$. In this case, because of the small coupling g, the pairing interaction $U_{\rm eff}(0,0)$ becomes strong only when $2\nu \sim 2\varepsilon_{\rm F}$ is satisfied. As a result, when one enters the crossover region, Feshbach molecules soon become dominant over Cooper pairs. As result, the BEC regime in a narrow Feshbach resonance is always dominated by Feshbach molecules. The BEC regime dominated by Cooper pairs does not exist in this case.

We briefly note that Eqs. (13) and (14) have the same forms as those in the original NSR theory[2–4] for the *single-channel BCS model*

$$H = \sum_{\mathbf{p},\sigma} (\varepsilon_{\mathbf{p}} - \mu) c_{\mathbf{p}\sigma}^\dagger c_{\mathbf{p}\sigma} - \tilde{U} \sum_{\mathbf{p},\mathbf{p}',\mathbf{q}} c_{\mathbf{p}+\mathbf{q}/2\uparrow}^\dagger c_{-\mathbf{p}+\mathbf{q}/2\downarrow}^\dagger c_{-\mathbf{p}'+\mathbf{q}/2\downarrow} c_{\mathbf{p}'+\mathbf{q}/2\uparrow}. \tag{17}$$

Namely, as expected, one may study a superfluid Fermi gas with a broad Feshbach resonance using the single-channel BCS model in Eq. (17) when $2\nu \gg 2\varepsilon_{\rm F}$. This condition is indeed satisfied in the crossover regime of the recently discovered superfluid ^{40}K and ^{6}Li.[8–11]

2.4. $T_{\rm c}$ *in the BCS-BEC crossover region*

Figure 3 shows the superfluid phase transition temperature $T_{\rm c}$ tuned by the threshold energy 2ν of the Feshbach resonance. In the case of a narrow Feshbach resonance shown in panel (a), $T_{\rm c}$ increases as 2ν decreases, because of the increase of the pairing interaction $U_{\rm eff}(0,0) = U + g^2/(2\nu - 2\mu)$. When $2\nu < 0$, since the system continuously changes into a gas of Feshbach molecules, it approaches $T_{\rm c} = 0.218T_{\rm F}$.

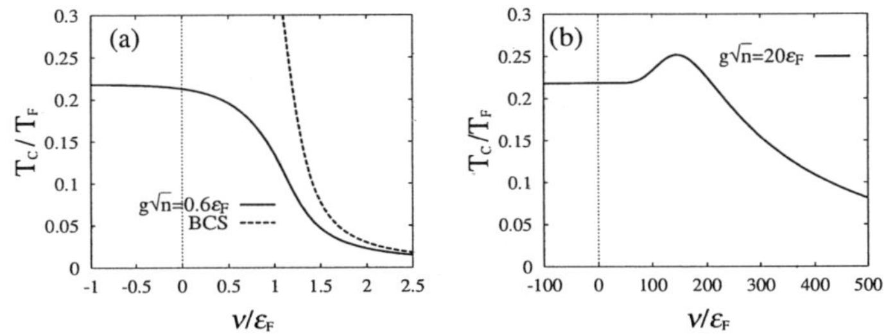

Fig. 3. Superfluid phase transition temperature T_c in the BCS-BEC crossover region as a function of the threshold energy 2ν. (a) Narrow Feshbach resonance. 'BCS' shows the weak-coupling BCS result. (b) Broad Feshbach resonance. We take $Un = 0.3\varepsilon_F$ and $\omega_c = 2\varepsilon_F$.

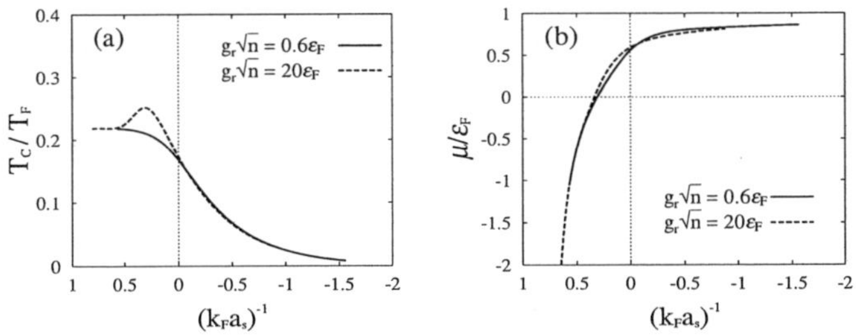

Fig. 4. (a) T_c as a function of the scattering length $(k_F a_s)^{-1}$. The parameters are the same as those in Fig. 3. (b) Fermi chemical potential $\mu(T_c)$ in the BCS-BEC crossover.

A similar crossover behavior can be seen in the case of a broad Feshbach resonance shown in Fig. 3(b). However, because of the large coupling $g\sqrt{n} \gg \varepsilon_F$, the crossover from the weak-coupling BCS regime to the strong-coupling BEC regime occurs when $\nu/\varepsilon_F \sim 100 \gg 1$. Although bosons in the BEC regime continuously change from the Cooper pairs into the Feshbach molecules as one decreases 2ν, this change cannot be seen in the behavior of T_c in this regime.

Figure 4(a) shows T_c as a function of the scattering length a_s defined by Eq. (15) where $\tilde{U}$ is replaced with $U_{\rm eff}(0,0) = U + g^2/(2\nu - 2\mu)$. In contrast to $\tilde{a}_s$, this scattering length a_s depends on the chemical potential μ. When the interaction is measured in terms of a_s, we find that a narrow Feshbach resonance and a broad Feshbach resonance approximately give the same value of T_c for the same value of a_s.[a] We note that, in a broad Feshbach resonance, a_s approximately equals $\tilde{a}_s$ when we consider the interesting BCS-BEC crossover region where $2\nu \gg 2\varepsilon_F \gtrsim 2\mu$.

Figure 4(b) shows the Fermi chemical potential μ in the BCS-BEC crossover at

[a] A similar quasi-universality can be also seen in μ (see Fig. 4(b)). On the other hand, as discussed in Sec. 2.3, dominant bosons in the crossover regime in a broad Feshbach resonance (Cooper pairs) are different from those in a narrow resonance (Feshbach bosons). Thus, physical properties dependent on the detailed character of dominant bosons (such as the density profile in a trapped gas) may depend on the width of the Feshbach resonance, even if one measures the interaction in terms of the scattering length a_s. For more details, see Ref. 16.

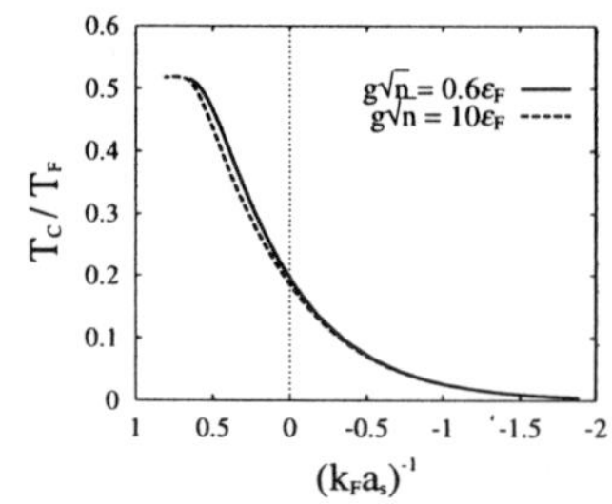

Fig. 5. BCS-BEC crossover behavior of T_c in a trapped Fermi gas. We take $Un = 0.3\varepsilon_F$. Here, $n = N/R_F^3$, where $R_F \equiv \sqrt{2\varepsilon_F/m\omega_0^2}$ is the Thomas-Fermi radius. The scattering length is evaluated at the trap center.

T_c. As expected, μ deviates from the Fermi energy ε_F as one enters the crossover regime, and it becomes negative in the strong-coupling BEC regime. We note that $2\mu < 2\nu$ is satisfied in Fig. 4(b). Thus, $U_{\rm eff}(0,0)$ is always attractive in the entire BCS-BEC crossover.

3. Effects of a trap potential (LDA)

We consider effects of a harmonic trap potential $V(\mathbf{r}) = \sum_{j=x,y,z} m\omega_j^2 r_j^2/2$ in the local density approximation (LDA). This approximation is achieved by replacing the chemical potential μ with $\mu(\mathbf{r}) \equiv \mu - V(\mathbf{r})$ in Eqs. (4) and (5). The resulting equation for T_c depends on $\mathbf{r}$, and T_c is given by the highest $T_c(\mathbf{r})$. Actually, it is obtained at $\mathbf{r} = 0$ (where $\mu(\mathbf{r}) = \mu$), so that the equation for T_c is again given by Eq. (4). The equation for the number of atoms is given by $N = \int d\mathbf{r} N(\mathbf{r})$. Here, $N(\mathbf{r})$ is given by Eq (5), where μ is replaced with $\mu(\mathbf{r})$. When we introduce $\bar{r}_j \equiv (\omega_j/\omega_0) r_j$ ($j = x, y, z$), where $\omega_0 \equiv (\omega_x\omega_y\omega_z)^{1/3}$, the spatial integration in the LDA equation for the number of atoms becomes isotropic in terms of $\bar{\mathbf{r}}$ (Note that $V(\mathbf{r}) = m\omega_0^2\bar{r}^2/2$). Namely, the anisotropy of the trap potential $V(\mathbf{r})$ is not important in considering T_c based on LDA.

Figure 5 shows T_c as a function of a_s in a harmonic trap. As in a uniform gas, T_c in the BCS-BEC crossover is almost independent of the width of the Feshbach resonance. T_c increases with increasing the pairing interaction. In the strong-coupling regime, it approaches T_c of a trapped ideal Bose gas with $N/2$ molecules, given by $T_c = (N/2\zeta(3))^{1/3}\omega_0 = 0.518 T_F$.[7]

4. Unconventional Superfluidity by a *p*-wave Feshbach Resonance

The realization of *p*-wave superfluidity using a *p*-wave Feshbach resonance is a very exciting challenge.[17,18] Recently, *p*-wave Feshbach resonances have been observed in ^{40}K and ^{6}Li.[19–21] When a *p*-wave Feshbach resonance occurs between two different hyperfine states (two-component case; $\sigma = \uparrow, \downarrow$), this system can be described by[17]

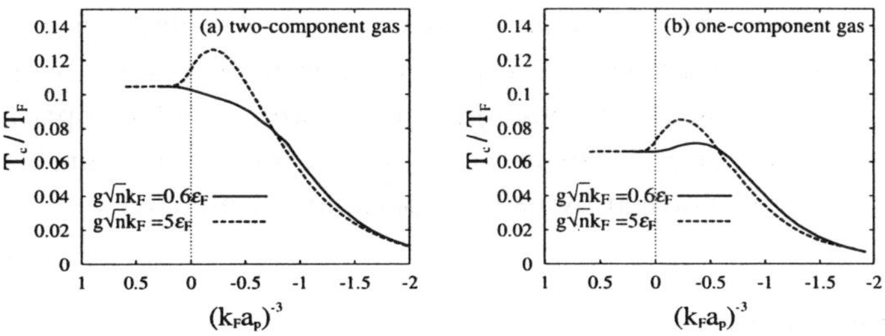

Fig. 6. T_c in the p-wave BCS-BEC crossover as a function of the p-wave scattering length $(k_F a_p)^{-3}$. (a) Two-component case. (b) One-component case.

$$H = \sum_{\mathbf{p},\sigma} \varepsilon_{\mathbf{p}} c^\dagger_{\mathbf{p}\sigma} c_{\mathbf{p}\sigma} - U \sum_{\mathbf{p},\mathbf{p}',\mathbf{q}} \mathbf{p}\cdot\mathbf{p}' c^\dagger_{\mathbf{p}+\mathbf{q}/2\uparrow} c^\dagger_{-\mathbf{p}+\mathbf{q}/2\downarrow} c_{-\mathbf{p}'+\mathbf{q}/2\downarrow} c_{\mathbf{p}'+\mathbf{q}/2\uparrow}$$
$$+ \sum_{\mathbf{q},j=x,y,z} [\varepsilon^B_{\mathbf{q}} + 2\nu] b^\dagger_{\mathbf{q},j} b_{\mathbf{q},j} + g \sum_{\mathbf{p},\mathbf{q},j=x,y,z} p_j \Big[b^\dagger_{\mathbf{q},j} c_{-\mathbf{p}+\mathbf{q}/2\downarrow} c_{\mathbf{p}+\mathbf{q}/2\uparrow} + \mathrm{H.c.} \Big] \quad (18)$$

Here, $b_{\mathbf{q},j}$ $(j = x, y, z)$ describe three kinds of Feshbach bosons, reflecting the three azimuthal angular momenta, $L_z = \pm 1, 0$. In contrast to the s-wave case, a p-wave Feshbach resonance is also possible between atoms in the *same* hyperfine state. The CFB model in this one-component case is given by Eq. (18) where the pseudo-spin index σ is eliminated.

We can extend the strong-coupling theory presented in Sec. 2.2 to the p-wave case.[17] Figure 6 shows the calculated T_c in the p-wave BCS-BEC crossover for a uniform Fermi gas. As in the s-wave case, T_c is found to show quasi-universal behavior, irrespective of the width of the Feshbach resonance, when the interaction is measured in terms of the p-wave scattering length a_p.[17] In the strong-coupling limit, since three kinds of molecules with $L_z = \pm 1, 0$ are dominant, the equation for the number of atoms is given by $N/2 = 3\sum_{\mathbf{q}} n_B(\varepsilon^B_{\mathbf{q}} + 2\nu - 2\mu)$. Because of the factor 3, T_c in the BEC limit is lower than the s-wave case. In the two-component case (Fig. 6(a)), T_c in the BEC limit is given by Eq. (11) where N_B is replaced with $N_B/3$, which gives $T_c = 0.105 T_F$. In the one-component case (Fig. 6(b)), we obtain $T_c = 0.066 T_F$ in the BEC limit. This lower T_c than that in the two-component case is simply due to a larger T_F for a given N, originating from the absence of the spin degeneracy.

5. Summary

To summarize, we have discussed the BCS-BEC crossover in a gas of Fermi atoms with a Feshbach resonance. We showed how the BCS-BEC crossover is realized by varying the threshold energy 2ν of the Feshbach resonance. We pointed out quasi-universality of T_c when plotted in terms of the scattering length, irrespective of the width of the Feshbach resonance. We have also investigated effects of a harmonic trap on T_c, as well as the possibility of unconventional superfluidity associated with a *p*-wave Feshbach resonance. Since physical parameters, such as the interaction, particle density, and lattice parameters (in the presence of an optical lattice), are tunable in cold atom gases, we expect that various fundamental problems of fermion superfluidity may be clarified by using superfluid Fermi gases.

Acknowledgements

I would like thank Allan Griffin for useful discussions. This work was supported by a Grand-in Aid for Scientific Research from the Ministry of Education, Culture, Sports, Science and Technology of Japan.

References

1. A. Leggett, in *Modern Trends in the Theory of Condensed Matter*, edited by A. Pekalski and J. Przystawa (Springer Verlag, Berlin, 1980), p. 14.
2. P. Nozières and S. Schmitt-Rink, *J. Low. Temp. Phys.* **59**, 195 (1985).
3. C. Sá de Melo, M. Randeria and J. R. Engelbrecht, *Phys. Rev. Lett.* **71**, 3202 (1993).
4. M. Randeria, in *Bose-Einstein Condensation*, edited by A. Griffin, D. Snoke and S. Stringari (Cambridge University Press, N.Y., 1995), p.355.
5. E. Timmermans, K. Furuya, P. Milonni and A. Kerman, *Phys. Lett.* **A285**, 228 (2001).
6. M. Holland, S. Kokkelmans, M. Chiofalo and R. Walser, *Phys. Rev. Lett.* **87**, 120406 (2001).
7. Y. Ohashi and A. Griffin, *Phys. Rev. Lett.* **89**, 130402 (2002); *Phys. Rev.* **A67**, 033603 (2003); **A67**, 063612 (2003).
8. C. Regal, M. Greiner, and D. Jin, *Phys. Rev. Lett.* **92**, 040403 (2004).
9. M. Bartenstein, A. Altmeyer, S. Riedl, S. Jochim, C. Chin, J. Denschlag, and R. Grimm, *Phys. Rev. Lett.* **92**, 120401 (2004).
10. M. Zwierlein, C. Stan, C. Schunck, S. Raupach, A. Kerman, and W. Ketterle, *Phys. Rev. Lett.* **92**, 120403 (2004).
11. T. Bourdel, L. Khaykovich, K. Cubizolles, J. Zhang, F. Chevy, M. Teichmann, L. Tarruell, S. J. J. M. F. Kokkelmans, and C. Salomon, *Phys. Rev. Lett.* **93**, 050401 (2004).
12. C. Chin, M. Bartenstein, A. Altmeyer, S. Riedl, S. Jochim, J. Denschlag, and R. Grimm, *Science* **305**, 1128 (2004).
13. J. Kinast, S. Hemmer, M. Gehm, A. Turlapov, and J. Thomas, *Phys. Rev. Lett.* **92**, 150402 (2004).
14. M. Bartenstein, A. Altmeyer, S. Riedl, S. Jochim, C. Chin, J. Denschlag, and R. Grimm, *Phys. Rev. Lett.* **92**, 203201 (2004).
15. M. Zwierlein, J. Abo-Shaeer, A. Schirotzek, C. Schunck, W. Ketterle, *Nature* **435**, 1047 (2005).
16. Y. Ohashi and A. Griffin, *Phys. Rev.* **A72**, 013601 (2005); condmat/0508213.
17. Y. Ohashi, *Phys. Rev. Lett.* **94**, 050403 (2005).
18. T. Ho, and R. Diener, *Phys. Rev. Lett.* **94**, 090402 (2005).
19. C. Regal, C. Ticknor, J. Bohn, and D. Jin, *Phys. Rev. Lett.* **90**, 053201 (2003).
20. J. Zhang, E. van Kempen, T. Bourdel, L. Khaykovich, J. Cubizolles, F. Chevy, M. Teichmann, L. Tarruell, S. Kokkelmans, and C. Salomon, *Phys. Rev.* **A70**, 030702 (2004).

21. C. Schunck, M. Zwierlein, C. Stan, S. Raupach, W. Ketterle, A. Simoni, E. Tiesinga, C. Williams, and P. Julienne, *Phys. Rev.* **A71**, 045601 (2005).

BOSE - EINSTEIN CONDENSATE SUPERFLUID - MOTT INSULATOR TRANSITION IN AN OPTICAL LATTICE

ANA M. REY

Institute for Theoretical Atomic, Molecular and Optical Physics, Harvard-Smithsonian Center of Astrophysics, Cambridge, MA, 02138
arey@cfa.harvard.edu

ESTEBAN A. CALZETTA

Departamento de Fisica, Facultad de Ciencias Exactas y Naturales, Universidad de Buenos Aires- Ciudad Universitaria, 1428 Buenos Aires, Argentina
calzetta@df.uba.ar

BEI-LOK HU

Department of Physics, University of Maryland, College Park, MD 20742
hub@physics.umd.edu

We present in this paper an analytical model for a cold bosonic gas on an optical lattice (with densities of the order of 1 particle per site) targeting the critical regime of the Bose - Einstein Condensate superfluid - Mott insulator transition.

Keywords: Bose-Einstein condensates; phase transitions; Mott transition.

1. Introduction

Since their experimental realization in 1995, Bose - Einstein condensates (BEC) have become one of the most exciting fields in physics.[1] The interest in these systems is also boosted by the possible use of cold atoms in optical lattices in the implementation of quantum information processing (QIP). In most proposals, the physical qubit is a single atom which may be in one of two preferred hyperfine states. This implies a strict control of the number of atoms per site, which in principle may be achieved by driving the system deep into the Mott insulator (MI) regime. However, the gas is usually first condensed in a trap, and then the lattice is imprinted on it. This implies driving the system through the superfluid (SF) - insulator transition. As with other phase transitions, we expect the particle distribution will be determined by events at or just below the critical point; once the hopping parameter is low enough, this distribution will be simply frozen in Ref. 2.

There is no suitable treatment in the literature of the one body density matrix at the transition region for low densities. Not only there is no single approach which is fully reliable throughout, but moreover those which are successful on one asymptotic regime are based on a quite different physical model than the ones which succeed on the other (compare, e.g., Bogoliubov methods against the Tonks - Girardeau gas approach or strong coupling perturbation theory).

We consider a system of N particles distributed over N_s lattice sites, with an integer mean occupation number $n = N/N_s$. In terms of the creation and destruction operators $a_j^\dagger(t)$ and $a_i(t)$, the dynamics is described by the Bose - Hubbard Hamiltonian (BHM) $H = \sum_i \left\{-\sum_j J_{ij}a_i^\dagger a_j + \frac{U}{2}a_i^{\dagger 2}a_i^2\right\}$, where the first term describes hopping between sites, and the second term the in-site repulsion between particles. The matrix J_{ij} is equal to J if the sites i and j are nearest neighbors, and zero otherwise. When the repulsion term dominates, the ground state of the system has definite occupation numbers for each site, and weak correlations among different sites. The system is in the so-called Mott insulator (MI) phase. When the hopping term dominates, atoms condense into a single quantum state extended over the whole lattice; the system is in the superfluid (SF) phase.

2. Calculations and results

In this paper we shall focus on the calculation of the one - body density matrix $\sigma_{1lk} = \left\langle a_l^\dagger(t)\, a_k(t)\right\rangle$ and its Fourier transform N_q. N_q is the expected total number of particles with momentum q (in units of $h/N_s a$, where a is the lattice spacing). N_0 may be identified with the condensate population. To compute the one-body density matrix we shall build a suitable generating functional within the so-called Schwinger-Keldysh or closed time path formulation of field theory.[3]

One possible strategy to deal with this problem is to reduce it to the quantum phase model. This model displays a phase transition, and it has been used to investigate nonequilibrium aspects of the Mott transition.[2] On closer examination, the approximation involved is valid when $Un > J$. Therefore, for $n \sim 1$ it fails at the transition region. In conclusion, while the quantum phase model is the best option on the shelf, it must be generalized to lower densities to be truly reliable in the relevant regime.

Our proposal is to overcome the restriction to large mean occupation numbers by introducing a new set of canonically conjugated variables, so that no square roots appear in the definition of the one - body density matrix or the Hamiltonian. This will make the perturbative expansion starting from a quadratic approximation to the Hamiltonian more straightforward. We have obtained an explicit analytic form for the one-body density matrix involving Elliptic Theta functions;[4] a detailed presentation is given in Ref. 5.

We have compared the analytic results above against an exact calculation of the momentum distribution function for an one dimensional lattice of 9 sites and 9 atoms ($n = 1$). The exact solution was obtained by numerical diagonalization of the Bose Hubbard Hamiltonian. We set $J = 1$ and change U from 0 to 60. We have performed similar calculations for 5 and 7 sites, finding the results to be totally consistent with the $N = 9$ case. We plot the occupation of the homogeneous mode in Fig. 1. We have also plotted the occupation numbers as given by the PNC method (Bogoliubov) calculations, and by first order strong coupling perturbation theory. The solid line is our prediction, the dash-dotted line is the exact numerical solution, the dots correspond to first order strong coupling perturbation theory, and the dashed line to the PNC method.

A preliminary comparison we made against available experimental results[6,7] of the condensate fraction from an array of one-dimensional lattices contained within a three dimensional trap for variable U/J showed fair agreement between the experimental results and the predictions of our model. In these experiments, the central tubes had around $N_s = 60$ populated sites.[7] The mean occupation number was close to $n = 2$ near the center of the trap, and close to $n = 1$ if averaged over all lattices.[8] We have compared the experimental results to the predictions of our model for several values of N_s around 60, and filling fractions $n = 1$ and 2. The results are fairly independent of N_s in this range, and very sensitive to n instead. As a typical representative, we show in Fig. (2) the prediction of our model for the condensate fraction for $N_s = 61$ and $n = 1$. We have superimposed the experimental results as reported in Ref. 6.

Both the comparison against numerical and experimental results show fair agreement at the transition region. Nevertheless, the match is not perfect, indicating that there is still much work to be done.

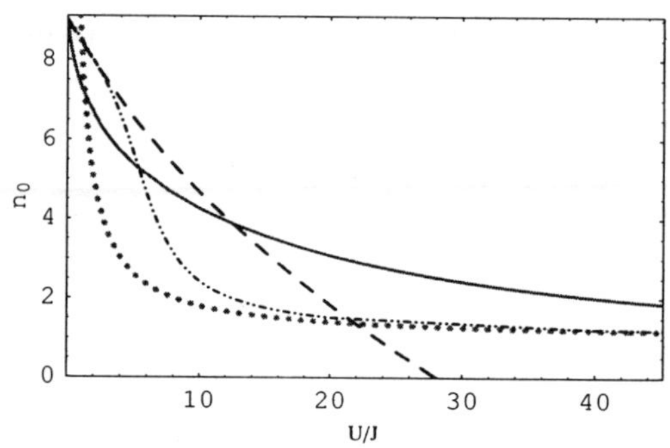

Fig. 1. The occupation number for the homogeneous mode, as a function of U; $n = 1$, $N_s = 9$ and $J = 1$. The solid line is our prediction, the dash-dotted line is the exact numerical solution. The dots correspond to first order strong coupling perturbation theory, and the dashed line to the PNC method.

Acknowledgements

EC acknowledges support from Universidad de Buenos Aires, CONICET and ANPCyT (Argentina); BLH from NSF grant PHY-0426696. A.M. Rey is supported from an the Advanced Research and Development Activity (ARDA) contract and the U.S. National Science Foundation through a grant PHY-0100767 and a grant

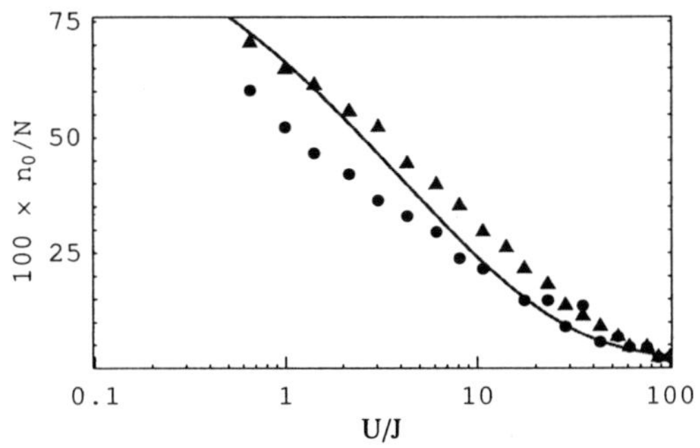

Fig. 2. Condensate fraction (%) plotted against U. Solid line: prediction from our model using the parameters $n = 1$, $N_s = 61$ and $J = 1$, Dots: Experimental points obtained from Fig. 4a in Ref. 6.

from the Institute of Theoretical, Atomic, Molecular and Optical Physics at Harvard University and Smithsonian Astrophysical observatory. We thank M. Köhl for pointing out the relevance of Refs. 6, 7 to this work and for discussing details of the experiment with us.

References

1. K. Southwell (editor), *Nature Insight: Ultracold matter, Nature* **416**, 205 (2002).
2. J. Dziarmaga, A. Smerzi, W. Zurek and A. Bishop, cond-mat/0403607 *Proceedings of NATO ASI Patterns of symmetry breaking*, Krakow, Poland, Sept. 2002.
3. J. S. Schwinger, *J. Math. Phys.* **2**, 407 (1961) L. V. Keldysh, *Zh. Eksp. Teor. Fiz.* **47** , 1515 (1964) [Engl. trans. *Sov. JETP* **20**, 1018 (1965)]; E. Calzetta and B-L. Hu, *Phys. Rev.* **D37**, 2878 (1988).
4. E. T. Whittaker and G. N. Watson, *A Course of Modern Analysis* (Cambridge University Press, Cambridge, 1984).
5. E. Calzetta, B-L. Hu and A. M. Rey, cond-mat/0507256v2 (submitted to *Phys. Rev.* **A**).
6. T. Stoferle et al., *Phys. Rev. Lett.* **92**, 130403 (2004).
7. C. Schori et al., *Phys. Rev. Lett.* **93**, 240402 (2004).
8. M. Köhl, private communication.

PHASE TRANSITIONS AND LOW DIMENSIONAL SYSTEMS

ZERO-TEMPERATURE PHASE DIAGRAM OF DISSIPATIVE RANDOM ISING FERROMAGNETIC CHAINS

L. F. CUGLIANDOLO

Laboratoire de Physique Théorique et Hautes Énergies, Jussieu, 4 Place Jussieu
75252 Paris Cedex 05, France
Laboratoire de Physique Théorique, École Normale Supérieure de Paris, 24 rue Lhomond
75231 Paris Cedex 05, France
leticia@lpt.ens.fr

G. S. LOZANO and H. F. LOZZA

Departamento de Física, FCEN, Universidad de Buenos Aires, Pab. I, Ciudad Universitaria
Buenos Aires, 1428, Argentina
lozano@df.uba.ar, homero@df.uba.ar

We study the zero-temperature critical behavior of dissipative quantum Ising spin chains of finite and infinite length. The spins interact with either constant or random nearest-neighbor ferro-magnetic couplings. They are also subject to a transverse field and coupled to an Ohmic bath of quantum harmonic oscillators. We analyze the coupled system performing Monte Carlo simulations on a classical two-dimensional counterpart model. We find that the coupling to the bath enhances the extent of the ordered phase, as it is known for mean-field spin-glass models. In the case of finite chains we show that a generalization of the Caldeira-Leggett localization transition exists.

Keywords: Spin Chains; Dissipation; Quantum Monte Carlo.

1. Introduction

In recent years, many experiments performed on magnetic systems showed that quantum fluctuations rule the phase transition at very low temperatures.[1] Some features of these quantum phase transitions are captured by simple quantum spin models,[2] where the influence of the environment is dismissed. However, it is known that the coupling to a quantum bath generates highly nontrivial localization phenomena, at least for a single two-level system.[3] Our aim is to study the effect of Ohmic dissipation on the zero-temperature critical behavior of quantum spin chains, both with constant and random nearest-neighbor ferromagnetic interactions.[4]

2. The model

The spin chain coupled to the environment is modeled by

$$\hat{H} = \hat{H}_J + \hat{H}_B + \hat{H}_I + \hat{H}_{CT}\,. \tag{1}$$

The Hamiltonian of the random quantum spin chain is described by

$$H_J = -\sum_{i=1}^{N} J_i \hat{\sigma}_i^z \hat{\sigma}_{i+1}^z - \sum_{i=1}^{N} \Gamma_i \hat{\sigma}_i^x - \sum_{i=1}^{N} h_i \hat{\sigma}_i^z \,, \tag{2}$$

where J_i is the strength of exchange interactions, and may take random values from the interval $[0, 1]$. The intensity of the external magnetic fields are represented by Γ_i and h_i, for the transverse and longitudinal directions, respectively. The Hamiltonian for the quantum bath reads

$$\hat{H}_B = \sum_{l=1}^{\tilde{N}} \frac{1}{2m_l} \hat{p}_l^2 + \sum_{l=1}^{\tilde{N}} \frac{m_l \omega_l^2}{2} \hat{x}_l^2 \,, \tag{3}$$

and its interaction with the spins is

$$\hat{H}_I = -\sum_{i=1}^{N} \hat{\sigma}_i^z \sum_{l=1}^{\tilde{N}} c_{il} \, \hat{x}_l \,. \tag{4}$$

There is also a counter-term that reads

$$\hat{H}_{CT} = \sum_{l=1}^{\tilde{N}} \frac{1}{2m_l \omega_l^2} \left(\sum_{i=1}^{N} c_{il} \, \hat{\sigma}_i^z \right)^2 \,. \tag{5}$$

We derive equilibrium properties from the partition function $Z = \mathrm{Tr}\, e^{-\beta H}$. The integration over the bath variables can be performed explicitly. We use an Ohmic bath with spectral density

$$I_i(\omega) \equiv \frac{\pi}{2} \sum_{l=1}^{\tilde{N}/N} \frac{c_{il}^2}{m_l \omega_l} \delta(\omega - \omega_l) = \begin{cases} 2\pi\alpha\omega & \text{for } \omega < (\Delta\tau)^{-1} \,, \\ 0 & \text{otherwise} \,. \end{cases} \tag{6}$$

This equation defines the parameter α that we use later as a measure of the strength of the bath.

3. Monte Carlo simulations

We analyze the static properties of the system by means of Monte Carlo simulations performed on a classical equivalent partition function

$$Z_J = \sum_{s_i^t = \pm 1} e^{-\mathcal{A}[K_i, B, \alpha; s_i^t]} \,, \tag{7}$$

$$\begin{aligned} \mathcal{A}[K_i, B, \alpha; s_i^t] = & -\sum_{t=0}^{N_\tau - 1} \sum_{i=1}^{N} K_i s_i^t s_{i+1}^t - B \sum_{t=0}^{N_\tau - 1} \sum_{i=1}^{N} s_i^t s_i^{t+1} \\ & - \frac{\alpha}{2} \sum_{t<t'}^{N_\tau - 1} \sum_{i=1}^{N} \left(\frac{\pi}{N_\tau} \right)^2 \frac{s_i^t s_i^{t'}}{\sin^2 (\pi |t - t'| / N_\tau)} \,, \end{aligned} \tag{8}$$

where the effective classical $(1+1)$-dimensional action, $\mathcal{A}$, is obtained after applying the Trotter-Suzuki formula and introducing an imaginary time direction. The classical counterpart model is defined on a rectangular lattice with size $N \times N_\tau$ suitable for Monte Carlo simulations. The thermodynamic limit follows from an extrapolation on the chain length.

4. Results

The description of the system state is done by dimensionless parameters α and $\Delta\tau\Gamma = \beta\Gamma/N_\tau$. The phase transitions in the $(\Delta\tau, \alpha)$ plane can be found by analyzing the Binder ratio

$$g_{av} \equiv \frac{1}{2}\left[3 - \frac{\langle m^4 \rangle}{\langle m^2 \rangle^2}\right], \qquad m \equiv \frac{1}{NN_\tau}\sum_{i=1}^{N}\sum_{t=0}^{N_\tau - 1} s_i^t \,. \tag{9}$$

4.1. *Small N*

The critical values are easily found by plotting g_{av} as a function of α for several values of N_τ. Then, α_c can be located by finding the intersection of the g_{av} curves against α for several N_τ for every choice of $\Delta\tau_c\Gamma_c$.[5]

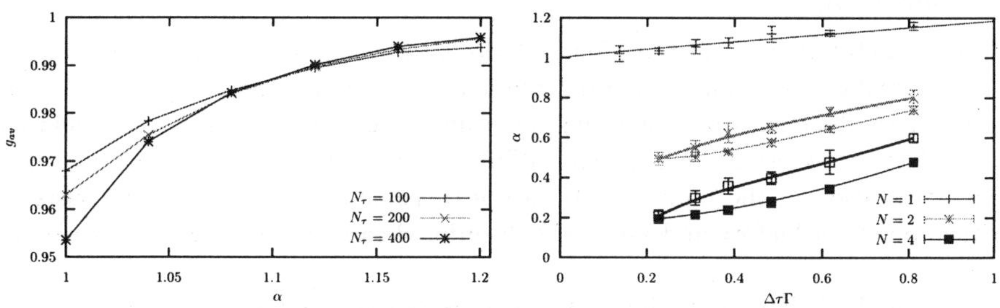

Fig. 1. Left (L): Binder cumulant, Right (R): Critical boundary

Figure 1(L) shows the Binder cumulant as a function of α for $N = 1$ with $\Delta\tau_c\Gamma_c = 0.48$. The localization transition occurs at $\alpha_c = 1.10 \pm 0.02$.

In Fig. 1(R) we present the critical boundary $(\Delta\tau_c\Gamma_c, \alpha_c)$ for small clusters, $N \leq 4$. For $N = 1$ this corresponds to the localization transition in the Caldeira-Leggett model. The data are well fitted by a linear function reaching $\alpha_c = 1$ at $\Delta\tau_c\Gamma_c = 0$.[3] For $N > 1$, we show the transition curve for systems with random exchange interactions (thick lines) and non-random exchange interactions (thin lines).

4.2. *Thermodynamic limit*

For fixed N and generic values of the other parameters the Binder cumulant attains a maximum as a function of N_τ.[6,7] The maximum value g_{av}^{max} is independent of N at criticality (Fig. 2(L)). In addition, g_{av} satisfies an exponetial scaling law.

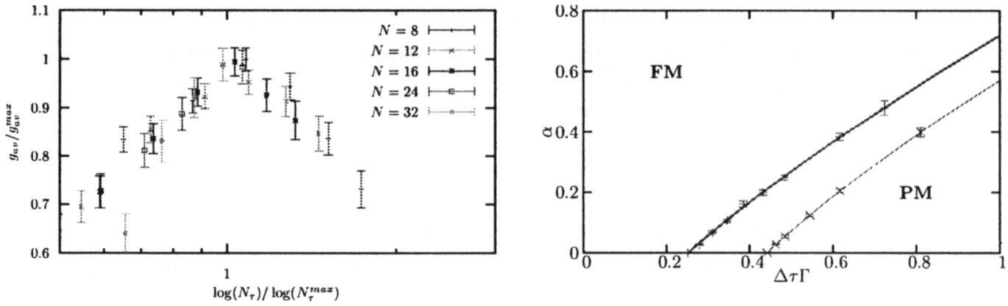

Fig. 2. Left (L): Test of activated scaling, Right (R) Phase diagram for $N = \infty$

Figure 2(R) shows the phase transition line for the disordered infinite chain with a uniform distribution of exchanges on the interval $[0, 1]$ (thick lines) and the non-random infinite chain with $J = 1$ (thin lines). The critical values for $\Delta\tau_c\Gamma_c$ with $\alpha_c = 0$ agree with well established analytical results.[8,9] In general, the stronger is the coupling with the bath, the larger is the ordered phase.

5. Conclusions

With a careful analysis of the Monte Carlo data we succeeded in determining the phase boundary between the paramagnetic and ferromagnetic phases in finite and infinite periodic chains of interacting quantum Ising spins in a transverse field and coupled to an external environment. While for finite number of spins there is no phase transition in the limit $\alpha \to 0$, in the thermodynamic limit one recovers the critical points of McCoy and Wu in the disordered case, and Onsager in the ordered problem.

The method that we used allowed us to determine the Caldeira-Leggett critical value $\alpha_c = 1$ for $N = 1$ with relatively little numerical effort. A similar analysis could be used to study the influence of other types of baths (sub-Ohmic, super-Ohmic) for which less analytical results are available.

The coupling to the bath enhances the extent of the ordered phase, as found in mean-field spin-glasses[10,11] and in the ordered ferromagnetic Ising chain.[12]

Acknowledgments

LFC is a member of the Institut Universitaire de France and acknowledges financial support from an Ecos-Sud travel grant, the ACI project "Optimization algorithms and quantum disordered systems" and the ICTP-Trieste. This research was supported in part by SECYT PICS 03-11609 and PICS 03-05179 and UBACYT/x053.

References

1. T. F. Rosenbaum, *J. Phys.: Condens. Matter* **8**, 9759 (1996), A. Aharony, R. J. Birgeneau, A. Coniglio, M. A. Kastner, and H. E. Stanley, *Phys. Rev. Lett.* **60**, 1330 (1988).

W. Wu, D. Bitko, T. F. Rosenbaum and G. Aeppli, *Phys. Rev. Lett.* **71** 1919 (1993). F. C. Chou, N. R. Belk, M. A. Kastner, R. J. Birgeneau, and A. Aharony *Phys. Rev. Lett.* **75**, 2204 (1995).

2. S. Sachdev, *Quantum phase transitions*, (Cambridge University Press, 1999)
3. A. J. Leggett, S. Chakravarty, A. T. Dorsey, M. P. A. Fisher, A. Garg and W. Zwerger, *Rev. Mod. Phys.* **59** 1, (1987)
4. L. F. Cugliandolo, G. Lozano and H. Lozza, *Phys. Rev.* **B71**, 224421 (2005)
5. *The Monte Carlo Method in Condensed Matter Physics* ed. K. Binder, 2nd ed., (*Topics in Applied Physics*, Vol. 71), Springer-Verlag.
6. H. Rieger and A. P. Young, *Phys. Rev. Lett.* **72**, 4141 (1994); *Phys. Rev.* **B54**, 3328 (1996).
7. M. Guo, R. Bhatt and D. Huse, *Phys. Rev. Lett.* **72**, 4137 (1994); *Phys. Rev.* **B54**, 3336 (1996).
8. B. M. McCoy and T. T. Wu, *Phys. Rev.* **176**, 631 (1968), *ibid* **188**, 982 (1969)
9. L. Onsager, *Phys. Rev.* **65**, 117 (1944)
10. L. F. Cugliandolo, D. R. Grempel, G. Lozano, H. Lozza, C. A. da Silva Santos, *Phys. Rev.* **B66**, 014444 (2002).
11. L. F. Cugliandolo, D. R. Grempel, G. Lozano and H. Lozza, *Phys. Rev.* **B70**, 024422 (2004)
12. P. Werner, K. Voelker, M. Troyer and S. Chakravarty, *Phys. Rev. Lett.* **94**, 047201 (2005)

PHASE TRANSITIONS IN ULTRA-COLD TWO-DIMENSIONAL BOSE GASES

D. A. W. HUTCHINSON and P. B. BLAKIE

Department of Physics, University of Otago,
P. O. Box 56, Dunedin, New Zealand
hutch@physics.otago.ac.nz

We briefly review the theory of Bose-Einstein condensation in the two-dimensional trapped Bose gas and, in particular the relationship to the theory of the homogeneous two-dimensional gas and the Berezinskii-Kosterlitz-Thouless phase. We obtain a phase diagram for the trapped two-dimensional gas, finding a critical temperature above which the free energy of a state with a pair of vortices of opposite circulation is lower than that for a vortex-free Bose-Einstein condensed ground state. We identify three distinct phases which are, in order of increasing temperature, a phase coherent Bose-Einstein condensate, a vortex pair plasma with fluctuating condensate phase and a thermal Bose gas. The thermal activation of vortex-antivortex pair formation is confirmed using finite-temperature classical field simulations.

Keywords: Bose-Einstein condensation; Berezinskii-Kosterlitz-Thouless phase; ultra-cold gases.

1. Introduction

It is now over a decade since Bose-Einstein condensation (BEC) in a dilute atomic gas was first realised.[1] In this time remarkable progress has been made in the control and manipulation of these ultra-cold gases. One possibility offered, through the use of light fields in the form of the optical lattice,[2] by this control is to reduce the dimensionality of the system. By increasing the confining potentials in one or two directions it is possible to freeze out these degrees of freedom creating, effectively, two-[3] or one-dimensional[4] gases. The two-dimensional (2D) gas is of particular interest. In general 2D systems often display unique properties, often remarkably different from those of the comparable bulk system. For example, the 2D electron gas, subjected to a perpendicular magnetic field, displays the rich phenomena of the, essentially single particle, quantum Hall effect and, the even more remarkable, many body fractional quantum Hall effect[5] with the physical manifestation of, the uniquely two-dimensional, fractional statistics[6] in the quasiparticle excitations. More specifically, it is well known that the uniform 2D Bose gas does not undergo BEC[7,8] at finite-temperature. The interacting homogeneous 2D gas does still undergo a normal-superfluid phase transition[9–11] at finite-temperature however, with an order parameter which is only locally phase coherent - the Berezinskii-Kosterlitz-

Thouless (BKT) phase. Experimental evidence for this phase transition has been demonstrated in liquid helium thin film,[12] superconducting Josephson-junction arrays[13] and in spin-polarized atomic hydrogen.[14]

In the trapped 2D Bose gas the density of states is modified sufficiently that, for the ideal gas, condensation into a single state can occur at finite temperature.[15] It has been demonstrated however that long-wavelength phase fluctuations may destroy the global phase coherence of the condensate at a temperature still below the ideal gas critical temperature, T_0 for BEC with only a *quasi-condensate*[16–22] present. The connection between such a quasi-condensate and the correlated vortex-pair plasma of the BKT phase proves unclear however and forms the basis of the material we attempt to address in the remainder of this paper.

2. Free Energy of the Vortex-Antivortex Pair

In order to determine at what temperature thermal activation of vortex-antivortex pairs becomes thermodynamically favourable, we consider the Helmholtz free energy, $F = E - TS$ of N bosonic particles of mass m spatially confined in 2D and via the harmonic potential $V(\mathbf{r}) = m\omega_\perp^2 r^2/2$, where $\omega_\perp$ is the radial trap frequency. We assume the existence of a macroscopic ground state condensate wavefunction $\psi(\mathbf{r})$, which also serves to describe the order parameter of the system in the BEC phase. Isolating the energy contribution due to the condensate, $E_0(T)$, the total internal energy of the system may be written as $E(T) = E_0(T) + \tilde{E}(T)$. The required condensate energy $E_0(T)$ is determined by the functional

$$E_0(T) = \int \left(\frac{\hbar^2}{2m} |\nabla \psi(\mathbf{r})|^2 + V(\mathbf{r})|\psi(\mathbf{r})|^2 + \frac{g}{2}|\psi(\mathbf{r})|^4 \right) \mathrm{d}\mathbf{r} \tag{1}$$

where g is the constant coupling parameter for the particle interactions, and $N_0 = \int |\psi(\mathbf{r})|^2 \mathrm{d}\mathbf{r}$ denotes the number of particles in the condensate.The internal energies for topologically distinct order parameter configurations are then calculated computationally.[23] The entropy contribution to the free-energy difference is due to the multiplicity of order parameter configurations containing a pair of vortices. The statistical weight $W = 2\pi R_{TF}^2/\xi^2$, is obtained by allowing one vortex to reside anywhere within the Thomas-Fermi radius, the partner vortex then having 2π available nearest neighbour sites. Assuming the radius of the area occupied by each vortex to be of the order of the healing length, $\xi = \sqrt{\hbar^2/2m\mu}$, we can obtain an expression for the configurational energy difference $\Delta S(T) = k_B \ln(8\pi\mu^2/\hbar^2\omega_\perp^2)$. The free energy for each configuration can now be evaluated, yielding a critical temperature for the thermal activation of vortex-antivortex pair creation. For the experimental parameters of the Oxford experiment[24] we obtain a critical temperature $T_c \approx 0.5T_0$ although the precise value is dependent upon the total number of atoms and decreases monotonically with N. Of course taking the thermodynamic limit in this case, which one would need to do if one were to try to claim that this is a phase transition, is slightly complicated by the presence of the trap.

This indicates that there are three distinct regions for the harmonically trapped, ultra-cold, dilute Bose gas. At low temperatures there is a phase coherent BEC with a coherence length comparable with the system size as shown by Gies and Hutchinson.[25] At higher temperatures, where the coherence length reduces and off-diagonal order decays algebraically, it becomes thermodynamically favourable to form correlated vortex-antivortex pairs, which we take as a signature of a BKT-like phase. At temperatures approaching the ideal gas critical temperature, T_0 the off-diagonal order decays exponentially. At this point the vortex pairs unbind (although this is not present in our free energy model) and superfluidity is destroyed yielding a thermal gas. This is what is traditionally referred to as the BKT transition.

3. Classical Field Simulations

To confirm these conclusions finite temperature classical field simulations[26,27] were performed. In the classical field approach the system is divided into classical and incoherent regions, determined by the occupation of single particle modes. Highly occupied (classical) modes are described by the projected Gross-Pitaevskii equation

$$i\hbar\frac{\partial\psi(\mathbf{r},t)}{\partial t} = \left[-\frac{\hbar^2}{2m}\nabla^2 + V(\mathbf{r})\right]\psi(\mathbf{r},t) + g\mathcal{P}\left\{|\psi(\mathbf{r},t)|^2\psi(\mathbf{r},t)\right\} \tag{2}$$

where the projector, $\mathcal{P}$, restricts evolution of the classical field to within its subspace. The low occupation, incoherent modes are described using the semiclassical Hartree-Fock approximation, with the classical and incoherent regions being taken to be in thermal equilibrium with one another. At the lowest temperatures no vortices are present in the system. These first emerge, as the temperature is increased, in the low density regions at the edge of the cloud. At higher temperatures it becomes possible for vortex-antivortex pairs to nucleate closer to the trap centre. This is demonstrated in Fig. 1 which is at a temperature of $0.86T_0$. Panel (a) shows a single frame of the full simulation with vortices and antivortices shown by the black crosses and white dashes. These are identified through the singularity in the phase as determined from the simulation. The region containing a slight density dip (due to thermal fluctuations) bounded by the box is then blown up into panel (b). Subsequent time frames are then shown for this expanded region in panels (c), (d) and (e). It can clearly be seen how a vortex-antivortex pair forms in the low density fluctuating region, then separates and stabilises. At later time this pair goes on to annihilate again. The temperature at which vortex-antivortex pairs begin to form in the classical field simulations is entirely consistent with the free energy calculations described above and a strong corroboration of the interpretation of the regime with only algebraic off-diagonal order as a BKT-like phase. Further details of the classical field simulations for the 2D gas and comparison with recent experiments[28] can be found in Ref. 29.

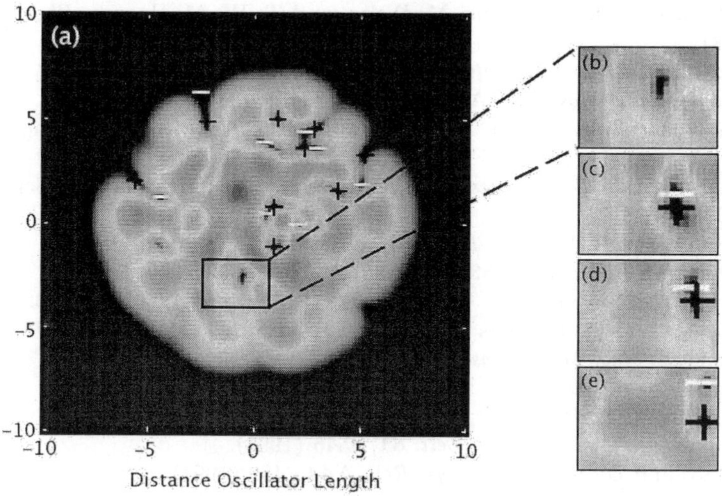

Fig. 1. Panel (a) shows a single frame of the full simulation with vortices and antivortices shown by the black crosses and white dashes. The region bounded by the box is blown up into panel (b). Subsequent time frames are shown for this expanded region in panels (c), (d) and (e) clearly showing the vortex-antivortex pair nucleation.

4. Conclusions

In conclusion we have demonstrated, both using a free energy argument and using classical field simulations that there exist three distinct regions in the phase diagram of the ultra-cold, harmonically confined 2D Bose gas. At very low temperatures compared to the ideal gas condensation temperature T_0 the gas forms a phase coherent BEC. At temperatures of order $0.5T_0$ it becomes thermodynamically favourable for correlated vortex-antivortex pairs to form. The system remains superfluid with algebraic off-diagonal order. This corresponds to a BKT-like superfluid phase. At higher temperatures, approaching T_0, the vortex pairs unbind, the off-diagonal density matrix decays exponentially and the superfluid phase is destroyed yielding a thermal gas. This upper transition is what corresponds to the BKT transition in the uniform system.

Acknowledgements

We would like to thank Tapio Simula for his contributions to this work. We are also indebted to the Marsden Fund and to the University of Otago for financial support.

References

1. For reviews see; E. A. Cornell and C. E Weiman, *Rev. Mod. Phys.* **74**, 875 (2002); W. Ketterle, *Rev. Mod. Phys.* **74**, 1131 (2002).

2. S. Friebel, C. D'Andrea, J. Walz, M. Weitz, and T. W. M. Hansch, *Phys. Rev.* **A57**, R20 (1998).
3. A. Gorlitz *et al.*, *Phys. Rev. Lett.* **87**, 130402 (2001).
4. M. Cristiani *et al.*, *Phys. Rev.* **A65**, 063612 (2002).
5. R. E. Prange and S. M. Girvin, *The Quantum Hall Effect* (Springer-Verlag, New York, 1990).
6. F. Wilczek, *Fractional Statistics and Anyon Superconductivity* (World Scientific, Singapore 1990).
7. N. D. Mermin and H. Wagner, *Phys. Rev. Lett.* **17**, 1133 (1966).
8. P. C. Hohenberg *Phys. Rev.* **158**, 383 (1967).
9. V. L. Berezinskii, *Soviet Phys. JETP* **32**, 493 (1971); *ibid* **34**, 610 (1972).
10. M. Kosterlitz and D. Thouless, *J. Phys. C: Solid St. Phys.* **6**, 1181 (1973).
11. P. Minnhagen, *Rev. Mod. Phys.* **59**, 1001 (1987).
12. D. J. Bishop and J. D. Reppy, *Phys. Rev. Lett.* **40**, 1727 (1978).
13. D. J. Resnick *et al.*, *Phys. Rev. Lett.* **47**, 1542 (1981).
14. A. I. Safanov *et al.*, *Phys. Rev. Lett.* **81**, 4545 (1998).
15. V. Bagnato and D. Kleppner, *Phys. Rev.* **A44** 7439 (1991).
16. V. N. Popov, *Functional Integrals in Quantum Field Theory and Statistical Physics*, (Reidel, Dordrecht, 1983).
17. D. S. Petrov, M. Holzmann, and G. V. Shlyapnikov, *Phys. Rev. Lett.* **84** 2551 (2000).
18. N. Prokof'ev, O. Ruebenacker, and B. Svistunov, *Phys. Rev. Lett.* **87** 270402 (2001).
19. N. Prokof'ev and B. Svistunov, *Phys. Rev.* **A66** 043608 (2002).
20. J. O. Andersen, U. Al Khawaja, and H. T. C. Stoof, *Phys. Rev. Lett.* **88** 070407 (2002).
21. U. Al Khawaja, J. O. Andersen, N. P. Proukakis, and H. T. C. Stoof, *Phys. Rev.* **A66** 013615 (2002).
22. C. Gies, *et al.*, *Phys. Rev.* **A69** 023616 (2004).
23. For full details see; T. P. Simula, M. D. Lee and D. A. W. Hutchinson, *Phil. Mag. Lett.* **85**, 395 (2005).
24. N. L. Smith *et al.*, *J. Phys. B: Atom. Molec. Opt. Phys.* **38**, 223 (2005).
25. C. Gies and D. A. W. Hutchinson, *Phys. Rev.* **A70**, 043606 (2004).
26. P. B. Blakie and M. J. Davis, *Phys. Rev.* **A72**, 063608 (2005).
27. M. J. Davis and P. B. Blakie, *Phys. Rev. Lett.* **96**, 060404 (2006).
28. S. Stock *et al.*, *Phys. Rev. Lett.* **95**, 190403 (2005).
29. T. P. Simula and P. B. Blakie, *Phys. Rev. Lett.* **96**, 020404 (2006).

QUANTUM CRITICAL BEHAVIOUR IN THE INSULATING REGION OF THE 2D METAL INSULATOR TRANSITION*

DAVID NEILSON

Dipartimento di Fisica, Università di Camerino, 62032 Camerino, Italy
School of Physics, University of New South Wales, Sydney 2052, Australia
david.neilson@unicam.it

D.J. WALLACE GELDART

Department of Physics, Dalhousie University, Halifax, NS B3H3J5, Canada
School of Physics, University of New South Wales, Sydney 2052, Australia
Dipartimento di Fisica, Università di Camerino, 62032 Camerino, Italy
wally.geldart@dal.ca

We show the insulating region of the metal-insulator transition phenomena in disordered two-dimensional electron systems contains new information about the quantum critical dynamics at low T because the insulating region and the quantum critical region are two aspects of the localized phase.

Keywords: metal-insulator transition; two-dimensions; quantum critical behaviour.

1. Introduction

Scaling equations based on renormalization group methods have played an important part in theoretical investigations of the metal-insulator transition phenomena observed in 2D electron systems in high purity semiconductor MOSFETs and heterostructures. A scaling picture is relevant if the observed metal-insulator transition anomalies at low temperature are due to a continuous second order transition.[1] The bifurcation of the temperature dependent resistivity observed at a critical carrier density[2–6] is a primary feature of the metal-insulator transition.

The scaling observed in experimental data of the resistivity as a function of temperature T and of electric field in Si and GaAs is suggestive of a quantum critical transition at a critical density n_c in the $T \to 0$ limit. Such a transition is unexpected for non-interacting Fermions, at least in the absence of spin-orbit coupling but it is now accepted that electron-electron interactions can stabilize a metallic phase. It is then possible to have a transition driven by disorder from this metallic phase into an insulating phase as the density is decreased at a fixed low temperature.

*This work was supported by the Natural Sciences and Engineering Research Council of Canada, an Australian Research Council Grant, and a PRIN Grant, Italy.

There is still no consensus on an interpretation of the experimental data as a quantum phase transition. The origins of the experimental behaviour, and the issue of it persisting down to zero temperature limit is still vigorously debated. Altshuler and Maslov[7] have demonstrated that some of the experimental evidence for scaling in silicon can be equally well explained as simply material-dependent effects. In addition, scaling of the resistivity data with respect to parallel magnetic fields has been reported only on the "metallic" branch and not on the "insulating" branch. Even if there is a splitting of the temperature dependent resistivity into two branches, Simmons *et al.*[5] have presented evidence that the down-trending resistivity curves in the "metallic" state near the "critical density", show an upturn when the temperature is lowered further, suggesting that the ground state is always insulating.

Perturbative renormalization group valid in the region of very low resistivity $\mathcal{R} \ll 1$ at low T are well developed.[8,9] $\mathcal{R}$ is the resistance per square in units of h/e^2. It has been shown that the re-entry observed by Pudalov *et al.*[10] and Hamilton *et al.*[11] is accounted for in a weak scaling regime described by the logarithmic approximation.[12] It was found that the $\mathcal{R} \ll 1$ data in the currently attainable temperature range are in a weak scaling regime described by the logarithmic approximation. However a first principles theory is not available in the vicinity of the bifurcation where $\mathcal{R} \sim 1$. It has been demonstrated that in 2D conventional lowest order renormalization group equations truncated at the leading term,

$$d\mathcal{R}/d\log L = \alpha(\gamma_{ee})\mathcal{R}^2 \ , \tag{1}$$

does not lead to bifurcation, so that at the least one more term $\beta(\gamma_{ee})\mathcal{R}^3$ must be included in the expansion.[13] However even then, given that experimentally the bifurcation is only seen at values of $\mathcal{R}$ of the order of unity, one is faced with a poorly behaved perturbation expansion in the region of most interest.

The lack of a detailed theoretical treatment of this strongly interacting problem makes it difficult to experimentally resolve the issue of the existence of a quantum critical point following the established strategy of determining the low temperature limiting behaviour of the $\mathcal{R}(T)$ curves for carrier densities spanning the bifurcation region. The key issue of whether the ground state of the 2D interacting electron system in the presence of weak disorder can be a metal or always remains an insulator remains unresolved.

The issue of the suggested turn up of the metallic curves at very low temperatures is couched in ambiguities not only because there is no theoretical determination of the temperature scale at which overriding insulating behaviour, if it exists, would be expected to switch in, but also because the low temperature up or down movement of the resistivity curves must be measured relative to the separatrix. The temperature dependence of the separatrix is not well established either theoretically or experimentally, so an apparent turn up of a metallic curve does not unambiguously signal a switch to insulating behaviour if the separatrix is rising in the same temperature range.

The observed scaling is also incomplete and up to now determination of z and $z\nu$ has required two independent experiments. These exponents are obtained by fitting to data in the quantum critical sector where $|\delta_n| = |n - n_c|/n_c$ is small. Analysis of temperature and δ_n scaling yields $z\nu$ while analysis of electric field and δ_n scaling yields z and $(z+1)\nu$. Taken together they give values for z and ν, but no consistency checks on the values of z and ν are possible in this procedure.

There is a particular lack of consensus in the intermediate density region. One school argues that, from Landau Fermi liquid theory for the pure system, we know that the quasiparticles are weakly interacting at sufficiently low temperatures. We recall that this result comes from a simple phase space argument. At low temperatures the inverse lifetime of a quasiparticle of energy $E_F + \epsilon$, with $\epsilon << E_F$ varies as $(\epsilon/E_F)^2$, implying stability. If the electron-electron interactions while weak are not completely negligible, then to first order in the electron-electron interactions, there is a logarithmic correction to the Drude conductivity σ_0 at low temperatures $T/T_\tau \ll 1$ given by,

$$\sigma(T) = \sigma_0 + (e^2/\pi h)(1 - F)\log(T/T_\tau) \tag{2}$$

where τ is the elastic scattering time and F is the exchange correction.[14,15] For $1 - F > 0$ the correction to σ_0 is negative for small T, indicating insulator-like behaviour.

However the argument that interactions between electrons are negligible in the low temperature limit is based on Landau Fermi Liquid Theory for the pure system. The corresponding theory in the presence of disorder has not yet been resolved. The disorder introduces additional scattering channels in which quasiparticles can scatter without inducing any accompanying particle-hole excitation. Thus it is by no means clear *a priori* that retaining electron-electron interactions only to first order is adequate in the intermediate range of densities where the effects of interactions are strong.

In this work we investigate hitherto unexplored implications of scaling at a quantum critical point. The condition for criticality is that both δ_n and T must be small. However their ratio, $|\delta_n|^{z\nu}/T$, is arbitrary. The existence of a quantum critical point thus allows a scaling picture not only of the quantum critical sector where δ_n is small relative to T but also of the insulating region where T is small relative to δ_n. Since the critical exponents are properties of the quantum critical point, this has the consequence that data from the insulating regime can provide additional information for the determination of critical exponents in a single experiment. This gives a much better understanding of the scaling properties in the neighbourhood of the quantum critical point and, in addition, provides valuable checks on the theory. In this way the consistency of the assumption of a quantum critical point can be tested.

The physical picture on which this proposition is based is given in the following section. A scaling equation in the insulator critical region is proposed in Sec. 3. Comparison with experimental data is considered in Sec. 4. Our conclusions are given in Sec. 5.

2. Phase Diagram of Quantum Critical Point

We propose the phase diagram in the temperature T and density n (or chemical potential μ) plane has the generic features indicated in Fig. 1. The new feature is the critical line of metallic fixed points at $T = 0$ terminating at a critical end point corresponding to a metallic limit of a Fermi liquid with disorder. This disordered Fermi liquid line terminates at μ_c corresponding to the density at the bifurcation.

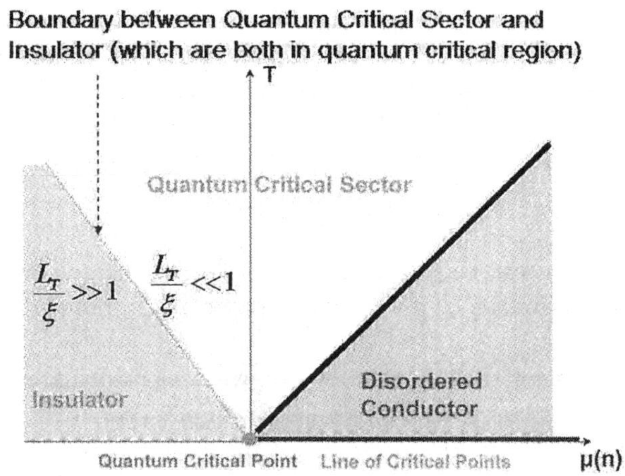

Fig. 1. In our phase diagram a quantum critical point separates insulating and metallic phases at $T = 0$. Thermal properties and critical exponents of the quantum critical sector *and* the insulator critical sector are controlled by this single critical end point, while the critical exponents of the line of critical points of the disordered conductor will in general be different.

Critical behaviour with possible universal scaling may be expected in the region where both $|\delta_n|$ and T are small. We introduce the correlation length $\xi \sim |\delta_n|^{-\nu}$ and the thermal length $L_T \sim 1/T^{1/\nu}$.

There are three regions of particular importance in Fig. 1. The properties of the low T disordered metal at $L_T >> \xi$ with $n > n_c$ are determined by the set of critical exponents characterizing the critical line of metallic fixed points. This is the metallic region of the phase diagram.

The second distinct region is specified by $L_T << \xi$ and will be referred to as the quantum critical sector. In this sector the finite temperature determines the scale of correlations as expected for a quantum critical point.[1] The critical exponents in this sector are determined by the critical end point and they will differ from those of the critical line of metallic fixed points.

The third distinct region of the phase diagram is characterized by $L_T >> \xi$ with $n < n_c$. This is the low T insulating region of the phase diagram. We argue that this low T insulating region is also critical provided both $|\delta_n|$ and T remain small and that T dependence in this region must be controlled by the critical end point.

In other words the quantum critical sector and the low T insulating region differ only in the value of the ratio L_T/ξ but do not otherwise differ in an essential way. This has the result that the critical exponents describing physical properties will be the same in the low T insulator as in the quantum critical sector. In particular the resistivity is given by a single universal function of the L_T/ξ. This ratio is small in the low T insulator and large in the quantum critical sector.

We now indicate the theoretical basis for this extended scaling hypothesis. We can assume that the essential physics of the problem is described by the following effective Hamiltonian or action in the presence of fixed random disorder,

$$S = \int_{\mathbf{k},\omega} \bar{\psi}_{\mathbf{k},\omega} \left[-i\omega + \frac{k^2}{2m} + \mu \right] \psi_{\mathbf{k},\omega} + S_{ee} + S_{dis} \; . \tag{3}$$

The explicitly written first term is the effective action for free particles. The second term describes two-body electron-electron interactions. The final term is the potential scattering of electrons from the distribution of fixed impurities and suitable averages of physical properties are to be taken with respect to this distribution. A detailed theoretical treatment of this strongly interacting problem is not available at high levels of disorder. Our aim is to assume generic conditions for a quantum critical point and to determine the testable implications.

The first two terms in Eq. (3) describe a low T Fermi liquid. The simplest view is that the strong electron-electron interactions maintain this phase until the inevitable insulating phase occurs at sufficiently low density ($n < n_c$).

The particular role of the critical end point in controlling both the quantum critical sector and the insulating regime at low T can be illustrated for the case of non-interacting Fermions without disorder. This example is very simple but illustrates important points. The critical endpoint in this case is at $\mu = 0$ and the critical line $\mu > 0$ corresponds to metallic behaviour. On the critical line the electron dispersion relation in Eq. (3) can be linearized about the Fermi surface so the low energy excitations are linear in $(k - k_F)$. Since the first term in Eq. (3) is also linear in ω it immediately follows that the dynamical critical exponent is $z = 1$ for all finite densities on the critical line. In contrast, precisely at the critical endpoint the dispersion relation cannot be linearized and since the low energy excitations in this case are quadratic it follows that $z = 2$ for scaling invariance at the $T = 0$ quantum critical point. This dynamical critical exponent is a property of the critical end point and explicit calculations show that it controls both the quantum critical sector and the low T insulator.[16]

Features of the scaling near the fixed point in Eq. (3) which were illustrated for the non-interacting case are general and prevail in the presence of disorder and interactions if a quantum critical point exists. Of course both sets of exponents will be altered by electron-electron interaction and disorder renormalizations but the critical exponents associated with the critical end point are different in general from those of the metallic critical line. Our hypothesis is that irrespective of details of the metallic phase, which may be complex, its properties and critical exponents are not relevant to the quantum critical sector and the low T insulator. The critical end point controls both of these.

Some practical implications of this can be illustrated using Fig. 1. Consider starting at low T in the metallic region at density $n > n_c$. Keeping n fixed and increasing T there is a crossover from the metallic phase into the quantum critical sector. The critical exponents and in particular the dynamical critical exponent z of the quantum critical sector are determined by the critical end point and differ from those of the Fermi liquid line.

On the other hand starting at fixed density $n < n_c$ in the quantum critical sector there is no change in the critical exponents as T is lowered to the low T limit of the insulating phase. There is no hard cross over in passing from one to the other along this path in the phase diagram. The dimensionless resistance at points along such a path is given by a single universal function of the L_T/ξ ratio

$$\mathcal{R} = f(L_T/\xi) \ . \tag{4}$$

The conventional treatment of scaling in the quantum critical sector is based on Eq. (4). In order to make additional use of Eq. (4) we require a scaling equation for the resistance in the low T insulating region. This is given in the following section.

3. Scaling Equation for Resistivity in the Insulator

Scaling equations for the resistance can be expressed in the form

$$d\mathcal{R}/dx = f(\gamma_{ee}, \mathcal{R}) \tag{5}$$

where $x = \log(L)$ with L the length of the sample and γ_{ee} symbolically indicates dependence of the electron-electron scattering amplitudes as additional variables. If a scaling picture is valid, f depends on γ_{ee} and $\mathcal{R}$ at the length scale L but not on L explicitly.

It is generally expected that the $\mathcal{R}$ varies exponentially with L in the deep insulating limit

$$\mathcal{R} \sim \exp(L/\xi) \tag{6}$$

We will also need the prefactor of the exponential. For this purpose, some specific principle is needed to form a theoretical $d\mathcal{R}/dx$. We will assume that $d\mathcal{R}/dx$ in the 'deep' insulator can be represented as a positive monotonically increasing function of powers of $\mathcal{R}$ and $\log \mathcal{R}$. The results following indicate that effects due to the variation of γ_{ee} are not significant in this limit so it is omitted.

$$\partial\mathcal{R}/\partial x = f(\mathcal{R}, \log \mathcal{R}) \tag{7}$$

Certainly $\mathcal{R}$ increases very rapidly as L increases. To determine the leading term $(\mathcal{R})^\alpha(\log \mathcal{R})^\beta$ in Eq. (7) at large $\mathcal{R}$, we assume that the rate of increase of $\mathcal{R}$ with

L in the strongly insulating limit is the most rapid possible subject to $\mathcal{R}$ remaining finite for all finite L and diverging only as $L \to \infty$. (Equivalently the resistance becomes infinite only in the strict $T \to 0$ limit.)

This condition can be seen to fix the powers of $(\mathcal{R})^\alpha(\log \mathcal{R})^\beta$ as $\alpha = 1$ and $\beta = 1$. The leading term in the scaling equation at large $\mathcal{R}$ is then

$$\partial\mathcal{R}/\partial x = k\mathcal{R}\log\mathcal{R} \tag{8}$$

where k is a positive constant. The solution of this equation is $\log\mathcal{R} \sim L^k$. The value of k determines the scaling properties of the resistance at large $\mathcal{R}$. If we further assume that $\mathcal{R}$ has a log-normal distribution this fixes $k = 1$.

This determines only the leading variation at large $\mathcal{R}$. For comparison with experimental data it will be necessary to include the first subleading correction proportional to $\mathcal{R}$,

$$\partial\mathcal{R}/\partial x = \mathcal{R}[\log\mathcal{R} - \log\mathcal{R}_0] \tag{9}$$

where the constant of proportionality has been written as $-\log\mathcal{R}_0$ to emphasize how it sets the resistance scale in the solution,

$$\mathcal{R}(L)/\mathcal{R}_0 = \exp(\log[\mathcal{R}(L_s)/\mathcal{R}_0])(L/L_s))\ . \tag{10}$$

$\mathcal{R}(L_s)$ is the resistance at an arbitrary starting point L_s.

To describe the resistance at finite temperature we make the usual replacement $L \to L_{thermal} \propto T^{1/z}$. The resistance in the deep insulator is then

$$\mathcal{R}(T) = \mathcal{R}_0\exp[-(T_1/T)^{1/z}] \tag{11}$$

with

$$T_1 = T_s[\log(\mathcal{R}_s/\mathcal{R}_0]^z \tag{12}$$

In addition to the expected exponential variation we now have the further information that there are no subleading T dependent factors in the low T limit. The prefactor $\mathcal{R}_0$ is independent of temperature (but can depend on density) and z is the critical exponent of the critical end point.

4. Relation to Experiments

4.1. *Insulator critical region data*

We now consider the consequences of Eq. (11) for the interpretation of experimental data in the low temperature insulating regime. Experimental results for the resistivity in the this regime have been reported for a variety of 2D systems. We focus attention particularly on Si.[17,18] Similar results have been reported for GaAs[19–21] and SiGe.[4] The data in the low temperature insulating (large $\mathcal{R}$) range were found to be well described by

$$\mathcal{R}(T) = \mathcal{R}_0(n)\exp[-(T_0(n)/T)^p] \tag{13}$$

where $T_0(n)$ and $\mathcal{R}_0(n)$ are density dependent parameters but are strictly independent of temperature. The exponent p has the value $p \simeq 1/2$ and the density dependent prefactor was found to be $\mathcal{R}_0 \sim h/e^2$ for the lowest density (highest resistance) curves.

These data and especially the fact that the exponent $p \simeq 1/2$ have conventionally been interpreted in terms of phonon-assisted variable range hopping theories. However in contrast to our scaling Eq. (11) for the insulator, the prefactor for variable range hopping theory,[22] and also for Coulomb Gap and correlated multi-electron hopping theories[23] depends on T,

$$\sigma(T) = (\gamma e^2/T)\exp\left[-\left(T_0/T\right)^p\right] \tag{14}$$

which is contrary to experimental observations. Aleiner *et al.*[24] suggested that the use of electron-electron interactions with no phonons[25] might explain the T-independent prefactor of experiments. However an explicit calculation of the prefactor for this case still gives a temperature-dependent prefactor $\sim T^{13/6}$, so this does not resolve the discrepancy between these theories and experiment.

The form of Eq. (13) is consistent with the scaling form Eq. (11) if $T_0(n)$ is specified as T_1 of Eq. (11). At first sight this seems to predict that the fitting parameter $T_0(n)$ depends on the starting point $(\mathcal{R}_s, T_s)$ on each constant density $\mathcal{R}(T,n)$ curve. In fact this is not the case. We have made nonlinear least squares fits to the insulator data with both Eqs. (11) and (13). Excellent fits are obtained and the determined parameters are all consistent. In particular $T_0(n)$ does not depend on the starting point along the density curve. This is a remarkable confirmation that the data in the insulator data region are in a scaling regime and that Eq. (11) is a proper and complete description.

We restricted the experimental data points taken from Ref. 26 for the low-T insulating range at large $\mathcal{R}$ to those lying within the region $L_T/\xi = (T_0(n)/T)^{1/z} > 3$. Our best fits determine the exponent p as $p = 0.49 \pm 0.02$. With $p = 1/z$, the dynamical critical exponent of the critical end point is $z = 2.04 \pm 0.08$.

4.2. *Quantum critical sector data*

Experimental resistivity data have conventionally been analysed by showing that a density dependent temperature scale can be assigned to each curve close to the critical density n_c so that all curves when plotted as functions of T fall onto two universal curves, one for the insulating region and another for the metallic region. The temperature scale was found to vary as a power of $|n - n_c|$ with the same exponent on both sides of the bifurcation. Interpreted in terms of scaling at a quantum critical point,[1] the exponent is $z\nu$. An equivalent statement is

$$\mathcal{R}(T) = f_1\left(\left[T_0(\delta n)/T\right]^{1/z\nu}\right) \tag{15}$$

An alternative procedure is convenient for least squares analysis of the data. An explicit functional form for $\mathcal{R}$ is not known in general. However the hypothesis of a quantum critical point implies scale invariance of the resistivity at a critical value $\mathcal{R}_c$. Hence $d\mathcal{R}/dx = 0$ when $\mathcal{R} = \mathcal{R}_c$. It is generally assumed that $d\mathcal{R}/dx$ vanishes linearly in $\mathcal{R} - \mathcal{R}_c$ with a slope $1/\nu$. Dobrosavljević *et al.*[27] have argued that an improved description is given by applying this generic argument to the vanishing of $\log(\mathcal{R}/\mathcal{R}_c)$ rather than to $\mathcal{R}/\mathcal{R}_c$. The result is

$$\mathcal{R}(T) = \mathcal{R}_c \exp[-\,(T_2/T)^{1/z\nu}] \tag{16}$$

with

$$T_2 = T_s[\log(\mathcal{R}_s/\mathcal{R}_c]^{z\nu} \tag{17}$$

Equation (16) is similar to Eqs. (11) and (13) but they differ in two ways. The T dependence is characterized by the exponent $z\nu$ in the critical sector whereas only z appeared in the insulator. Also the resistivity scale is fixed as $\mathcal{R}_c$ in the critical sector whereas only the density dependent scale $\mathcal{R}_0(n)$ appeared in the insulator. The form quoted in Ref. 27 is obtained if $\log(\rho(T_s)/\rho_c)$ is further approximated by $\delta\rho/\rho_c \approx= -c\delta_n$.

We restricted the experimental data points taken from Ref. 26 for Si in the quantum critical region to those lying within the region $L_T/\xi = [\rho(T_s)/\rho_c]^{\nu}(T_s/T)^{1/z} < 1/3$. Our best fits determine the exponent $z\nu = 2.6 \pm 0.1$. With the value of $z = 2.04 \pm 0.08$ already fixed from the insulator critical region, this permits the determination of ν from data from a single experiment, $\nu = 1.30 \pm 0.02$. A theoretical bound on ν is $\nu \geq 2/d$, where $d = 2$ is the system dimension.

5. Conclusions

It has been conventional to take weak disorder (small $\mathcal{R}$ region) as the starting point for theory, both because this is the range of validity of the perturbative renormalization group approach for which theoretical predictions have been made, and also because pushing the experimental measurements of the metallic curves down to lower T potentially should determine whether or not a metallic ground state limit is approached. Unfortunately it is experimentally extremely difficult to draw unambiguous conclusions from this region of phase space.

Thus weak disorder is not necessarily the optimal choice for the task of determining whether or not a quantum critical point exists for this system. Our proposal instead is to take strong localization as the starting point for theory. The reasons are (i) all of the localization physics is controlled by the same critical endpoint; and (ii) the insulator and quantum critical sectors are both in the quantum critical region.

We have shown that the insulating region in fact contains important new information about the quantum critical dynamics at low T because the insulating region and the quantum critical region are two aspects of the localized phase.

Our proposed scaling equation for the insulator leads to a temperature independent prefactor of the dominant exponential, as is experimentally observed. Our data analysis of $\mathcal{R}(T)$ yields both z and $z\nu$ from a single experiment, so that electric field and density scaling would give an independent consistency check for the determined values of z and ν. A determination that the value of z for the insulator is the same as the value of z for the quantum critical sector would be a strong indicator indeed of the presence of a critical endpoint.

References

1. S.L. Sondhi, S.M. Girvin, J.P. Carini, and D. Shahar, *Rev. Mod. Phys.* **69**, 315 (1997)
2. S.V. Kravchenko, W.E. Mason, G.E. Bowker, J.E. Furneaux, V.M. Pudalov, and M. D'Iorio, *Phys. Rev.* **B51**, 7038 (1995); S.V.Kravchenko, D. Simonian, M.P. Sarachik, W. Mason, and J.E. Furneaux, *Phys. Rev. Lett.* **77**, 4938 (1996)
3. D. Simonian, S.V. Kravchenko, M.P. Sarachik, and V.M. Pudalov, *Phys. Rev. Lett.* **79**, 2304 (1997)
4. P.T. Coleridge, R.L. Williams, Y. Feng and P. Zawadzki, *Phys. Rev.* **B56**, R12764 (1997)
5. M.Y. Simmons, A.R. Hamilton, M. Pepper, E.H. Linfield, P.D. Rose, D.A. Ritchie, A.K. Savchenko, and T.G. Griffiths, *Phys. Rev. Lett.* **80**, 1292 (1998)
6. Y.Y. Proskuryakov, A.K. Savchenko, S.S. Safonov, M. Pepper, M.Y Simmons, D.A. Ritchie, A.G. Pogosov and Z.D. Kvon, *Phys. Stat. Solidi* **B230**, 89 (2002)
7. Boris L. Altshuler and Dmitrii L. Maslov, *Phys. Rev. Lett.* **82**, 145 (1999)
8. A.M. Finkelstein, *Zh. Eksp. Teor. Fiz.* **84c**, 168 (1983) [*Sov. Phys. JETP* **57**, 97 (1983)]
9. C. Castellani, C. Di Castro, P.A. Lee, and M. Ma, *Phys. Rev.* **30**, 527 (1984)
10. V.M. Pudalov, G. Brunthaler, A. Prinz, and G. Bauer, *JETP Lett.* **68**, 534 (1998)
11. A.R. Hamilton, M.Y. Simmons, M. Pepper, E.H. Linfield, P.D. Rose, and D.A. Ritchie, *Phys. Rev. Lett.* **82**, 1542 (1999)
12. D.J.W. Geldart and D. Neilson, *Phys. Rev.* **B70**, 235336 (2004)
13. D.J.W. Geldart and D. Neilson, *Phys. Rev.* **B67**, 205309 (2003)
14. B.L. Altshuler, A.G. Aronov and P.A. Lee, *Phys. Rev. Lett.* **44**, 1288 (1980)
15. H. Fukuyama, *J. Phys. Soc. Jpn.*, **48**, 2169 (1980)
16. S. Sachdev, *Quantum Phase Transitions* (Cambridge University Press, Cambridge, 1999)
17. Whitney Mason, S.V. Kravchenko, G.E. Bowles, and J.E. Furneaux, *Phys. Rev.* **B52**, 7857 (1995)
18. V.M. Pudalov, G. Brunthaler, A. Prinz, and G. Bauer, *JETP Lett.* **68**, 534 (1998)
19. S.I. Khondaker, I.S. Shlimak, J.T. Nicholls, M. Pepper, and D.A. Ritchie *Sol. State Comm.* **109**, 751 (1999)
20. S.I. Khondaker, I.S. Shlimak, J.T. Nicholls, M. Pepper, and D.A. Ritchie *Phys. Rev.* **B59**, 4580 (1999)
21. I. Shlimak, K.J. Friedland, and S.D. Baranovskii, *Sol. State Comm.* **112**, 21 (1999)
22. N.F. Mott, *J. Non-Cryst. Solids* **1** 1 (1968)
23. A.L. Efros and B.I. Shklovskii, *J. Phys.* **C8**, L49 (1975); M. Pollak, *Phys. Stat. Sol.* **b230** 295 (2002)
24. I. L. Aleiner, D.G. Polyakov, and B.I. Shklovskii, in *Proceedings of the 22nd International Conference on the Physics of Semiconductors*, Vancouver, 1994, edited by D.J. Lockwood World Scientific, Singapore, 1994, p. 787
25. L. Fleishman, D.C. Licciardello, and P.W. Anderson, *Phys. Rev. Lett.* **40**, 1340 (1978)
26. V.M. Pudalov, G. Brunthaler, A. Prinz, and G. Bauer, *JETP Lett.* **68** 442 (1998)

27. V. Dobrosavljević, E. Abrahams, E. Miranda, and S. Chakravarty, *Phys. Rev. Lett.* **79**, 455 (1997).

INTERMEDIATE SYMMETRIES IN ELECTRONIC SYSTEMS: DIMENSIONAL REDUCTION, ORDER OUT OF DISORDER, DUALITIES, AND FRACTIONALIZATION

ZOHAR NUSSINOV

Department of Physics, Washington University, St. Louis
MO 63160, USA
zohar@wuphys.wustl.edu

CRISTIAN D. BATISTA

Theoretical Division, Los Alamos National Lab,
Los Alamos, NM 87545, USA
cdb@lanl.gov

EDUARDO FRADKIN

Department of Physics, University of Illinois,
1110 W. Green St., Urbana, IL 61801, USA
efradkin@uiuc.edu

We discuss symmetries intermediate between global and local and formalize the notion of dimensional reduction adduced from such symmetries. We apply this generalization to several systems including liquid crystalline phases of Quantum Hall systems, transition metal orbital systems, frustrated spin systems, (p+ip) superconducting arrays, and sliding Luttinger liquids. By considering space-time reflection symmetries, we illustrate that several of these systems are dual to each other. In some systems exhibiting these symmetries, low temperature local orders emerge by an "order out of disorder" effect while in other systems, the dimensional reduction precludes standard orders yet allows for multiparticle orders (including those of a topological nature).

Keywords: Intermediate symmetries; dimensional reduction; order out of disorder.

1. Introduction and Main Results

Orders are often very loosely classified into two types:

(i) *Global symmetry breaking orders.* In many systems (e.g. ferromagnets), there is an invariance of the basic interactions with respect to global symmetry operations (e.g.rotations in the case of ferromagnets) simultaneously performed on all of the fields in the system. At sufficiently low temperatures (or high coupling), such symmetries may often be "spontaneously" broken.

(ii) *"Topological orders"*.[1] In some cases, even if global symmetry breaking cannot occur, the system still exhibits a robust order of a topological type. This order may only be detected by non-local correlation functions. The most prominent examples of this order are afforded by gauge theories which display local gauge symmetries.

In this article, we investigate *intermediate* (or sliding) symmetries which, generally, lie midway between the global symmetries and local gauge symmetries extremes and provide conditions under which local orders cannot appear. We will review:

(1) a theorem[2] dictating that in many systems displaying intermediate symmetries, local orders are impossible. This theorem also gives upper bounds on multiparticle correlators and suggests for fractionalization in certain instances.

(2) how symmetry allowed orders in such highly degenerate systems may be stabilized by entropic fluctuations. (This is often refered to as the "order out of disorder" mechanism.) In classical (*large S*) renditions of quantum orbital systems, orbital order is stabilized by thermally driven entropic fluctuations.[3,4] These classical tendencies may be fortified by the incorporation of zero point quantum fluctuations.

(3) a duality betwen two prominent systems exhibiting intermediate symmetries. A route by which this duality may be established sheds light on some dualities as direct consequences of geometrical (Z_2) reflections in space-time.[5]

The results reported here appeared, in full detail, elsewhere. Our aim is to give a flavor of these which the reader may then peruse in detail. For pedagogical purposes, we review in Sec. 2, old results conerning lattice gauge theories. In sections thereafter, we discuss new results and constructs concerning systems with non-local intermediate (or sliding) symmetries where some analogies and extensions may be drawn vis a vis the physics of gauge theories which display stronger local symmetries.

2. What are gauge theories and local gauge symmetries?

Throughout this work, we will consider theories defined on a lattice Λ. We start with a review of a very well known topic. In matter coupled gauge theories,[6–9] matter fields ($\{\sigma_i\}$) reside at sites i while gauge fields U_{ij} lie on links between sites i and j. The $\mathbb{Z}_2$ matter coupled to $\mathbb{Z}_2$ gauge field theory is the simplest such theory. On a hypercubic lattice, its action is

$$S = -\beta \sum_{\langle ij \rangle} \sigma_i U_{ij} \sigma_j - K \sum_{\Box} UUUU. \tag{1}$$

The first sum is over all nearest neighbor links $\langle ij \rangle$ while the second is the product of the four gauge fields $U_{ij}U_{jk}U_{kl}U_{li}$ over each minimal plaquette (square) of the lattice. Both matter (σ_i) and gauge (U_{ij}) fields are Ising variables within this theory: $\sigma_i = \pm 1$, $U_{ij} = \pm 1$. The action S is invariant under local $\mathbb{Z}_2$ gauge transformations

$$\sigma_i \to \eta_i \sigma_i, \; U_{ij} \to \eta_i U_{ij} \eta_j \quad (\eta_i = \pm 1) \tag{2}$$

There is a theorem, known as Elitzur's theorem,[10] which disallows quantities not invariant under such these local symmetries (e.g. U, σ) to attain finite expectation values, $\langle \sigma_i \rangle = \langle U_{ij} \rangle = 0$. In a pure gauge theory having no matter coupling ($\beta = 0$ in Eq.(1)), the only symmetry invariant quantities are non-local quantities of a topological nature- the products of gauge fields around closed loops- the "Wilson loops" $W \equiv \langle U_{ij}U_{jk}...U_{pi}\rangle$.[6] In a pure gauge theory, the asymptotic scaling of W with the loop size dictates what phase we are in.[6] In the presence of matter, we also find symmetry invariant open string correlators (e.g., $\langle \sigma_i U_{ij} U_{jk}...U_{pm}\sigma_m\rangle$).[6–9] The simplest realization of Eq.(1) (that in two dimensions) exhibits a non-local (percolation) crossover as a function of β and K.[9]

Electromagnetism is a gauge theory with a local $U(1)$ invariance,

$$\sigma_i \to \eta_i^* \sigma_i, \; U_{ij} \to \eta_i^* U_{ij} \eta_j, \quad (\eta_i = e^{i\theta_i}, \; \theta_i \in \Re). \tag{3}$$

Here, we set $U_{ij} = \exp[iA_{ij}]$, with $A_{ij} = \int_i^j \vec{A} \cdot \vec{dr}$, where $\vec{A}$ is the vector potential. With the complex $U(1)$ fields U_{ij} and σ_{1i}, Eq.(1) is changed by the addition of a complex conjugate. Here, the plaquette term reads $[-\frac{K}{2}(U_i U_{jk} U_{kl} U_{li} + c.c.)] = [-K \cos \Phi_\Box]$ with $\Phi_\Box = A_{ij} + A_{jk} + A_{kl} + A_{li}$ the flux piercing the plaquette. Expanding, in the continuum limit, $(-K \sum_\Box \cos \Phi_\Box) \to (\frac{K}{2} \int dV (\nabla \times \vec{A})^2)$, and familiar continuum electromagnetism appears. Similarly, the first term of Eq.(1) (now $-\frac{\beta}{2} \sum_{\langle ij \rangle} (\sigma_i^* U_{ij} \sigma_j + c.c.)$) becomes, in the continuum, the standard minimal coupling term between charged matter and electromagnetic fields. Here, the Wilson loop becomes the Aharonov-Bohm phase.[11] Higher order groups ($U(1) \times SU(2), SU(3)$) describe the electroweak and strong interactions.

3. What are intermediate symmetries?

An intermediate d-dimensional symmetry of a theory[2] characterized by a Hamiltonian H (or action S) is a group of symmetry transformations such that the minimal non-empty set of fields $\phi_{\mathbf{i}}$ changed by the group operations occupies a d-dimensional subset $\mathcal{C} \subset \Lambda$. The index $\mathbf{i}$ denotes the sites of the lattice Λ. For instance, if a spin theory is invariant under flipping each individual spin then the corresponding gauge symmetry will be zero-dimensional or local. Of course flipping a chain of spins is also a symmetry, but the chain is not the minimal non-trivial subset of spins that can be flipped. In general, these transformations can be expressed as:

$$\mathbf{U}_{lk} = \prod_{\mathbf{i} \in \mathcal{C}_l} \mathbf{g}_{\mathbf{i}k}, \tag{4}$$

where $\mathcal{C}_l$ denotes the subregion l, $\mathcal{C}_l \subset \Lambda$, and $\Lambda = \bigcup_l \mathcal{C}_l$. To make contact with known cases, the local gauge symmetries of Eqs.(2, 3) correspond to $d = 0$ as the region where the local gauge symmetries operate is of dimension $d = 0$. Similarly, e.g. in a nearest neighbor ferromagnet on a $D-$ dimensional lattice, $H = -J \sum_{\langle ij \rangle} \vec{S}_i \cdot \vec{S}_j$, the system is invariant under a global rotation of all spins. As the volume influenced by the symmetry operation occupies a $D-$ dimensional region, we have that $d = D$.

4. Examples of intermediate symmetries

a) *Orbitals-* In transition metal (TM) systems on cubic lattices, each TM atom is surrounded by an octahedral cage of oxygens. Crystal fields lift the degeneracy of the five 3d orbitals of the TM to two higher energy e_g levels ($|d_{3z^2-r^2}\rangle$ and $|d_{x^2-y^2}\rangle$) and to three lower energy t_{2g} levels ($|d_{xy}\rangle$, $|d_{xz}\rangle$, and $|d_{yz}\rangle$). A super-exchange calculation leads to the Kugel-Khomskii Hamiltonian[12,13]

$$H = \sum_{\langle r,r'\rangle} H_{orb}^{r,r'}(\vec{S}_r \cdot \vec{S}_{r'} + \frac{1}{4}). \tag{5}$$

Here, $\vec{s}_r$ denotes the spin of the electron at site r and $H_{orb}^{r,r'}$ are operators acting on the orbital degrees of freedom. For TM-atoms arranged in a cubic lattice,

$$H_{orb}^{r,r'} = J(4\hat{\pi}_r^\alpha \hat{\pi}_{r'}^\alpha - 2\hat{\pi}_r^\alpha - 2\hat{\pi}_{r'}^\alpha + 1), \tag{6}$$

where $\hat{\pi}_r^\alpha$ are orbital pseudospins and $\alpha = x, y, z$ is the direction of the bond $\langle r, r'\rangle$.

(i) In the e_g compounds,

$$\hat{\pi}_r^x = \frac{1}{4}(-\sigma_r^z + \sqrt{3}\sigma_r^x), \qquad \hat{\pi}_r^y = \frac{1}{4}(-\sigma_r^z - \sqrt{3}\sigma_r^x), \quad \hat{\pi}_r^z = \frac{1}{2}\sigma_r^z. \tag{7}$$

This also defines the orbital only "120°-Hamiltonian" given by

$$H_{orb} = J\sum_{r,r'} \sum_{\alpha=x,y,z} \hat{\pi}_r^\alpha \hat{\pi}_{r+\hat{e}_\alpha}^\alpha. \tag{8}$$

Jahn-Teller effects in e_g compounds also lead, on their own, to orbital interactions of the 120°-type.[13] The "120° model" model of Eqs.(7,8) displays discrete ($d = 2$) $[Z_2]^{3L}$ gauge-like symmetries (corresponding to planar Rubick's cube like reflections about internal spin directions- Fig.(1)). The symmetry operators O^α are[2–4]

$$O^\alpha = \prod_{r\in P_\alpha} \hat{\pi}_r^\alpha. \tag{9}$$

Here, $\alpha = x, y, z$ and P_α may denote any plane orthogonal to the cubic $\hat{e}_\alpha$ axis.

(ii) In the t_{2g} compounds (e.g., $LaTiO_3$), we have in H_{orb} of Eq.(8)

$$\hat{\pi}_r^\alpha = \frac{1}{2}\sigma_r^\alpha. \tag{10}$$

This is called the *orbital compass* model. The symmetries of this Hamiltonian are given by Eqs.(9, 10). Rotations of individual lower-dimensional planes about an axis orthogonal to them leave the system invariant. The two dimensional orbital compass model is given by Eqs. (8,10) on the square lattice with $\alpha \in \{x, z\}$ and displays $d = 1$ Z_2 symmetries (wherein the planes P of Eq.(9) become lines).

b) *Spins in transition metal compounds*- Following Ref. 14, we label the three t_{2g} states $|d_{yz}\rangle, |d_{xz}\rangle, |d_{xy}\rangle$ by $|X\rangle, |Y\rangle$, and $|Z\rangle$. In the t_{2g} compounds, hopping is prohibited via intermediate oxygen p orbitals between any two electronic states of orbital flavor α ($\alpha = X, Y$, or Z) along the α axis of the cubic lattice (see Fig.2). As a consequence, as noted in Ref. 14, a uniform rotation of all spins, whose electronic orbital state is $|\alpha\rangle$, in any given plane (P) orthogonal to the α axis $c_{i\alpha\sigma}^\dagger = \sum_\eta U_{\sigma,\eta}^{(P)} d_{i\alpha\eta}^\dagger$ with σ, η the spin directions, leaves Eq.(5) invariant. The net

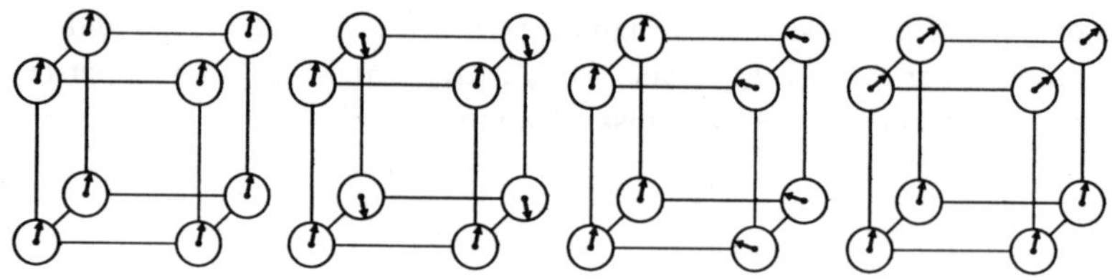

Fig. 1. From Refs. 3 and 4. The symmetries of Eq.(9) applied on the uniform state (at left).

spin of the electrons of orbital flavor $|\alpha\rangle$ in any plane orthogonal to the cubic α axis is conserved. Here, we have $d = 2$ $SU(2)$ symmetries

$$\hat{O}_{P;\alpha} \equiv [\exp(i\vec{S}^{\alpha}_{P} \cdot \vec{\theta}^{\alpha}_{P})/\hbar], \quad [H, \hat{O}_{P;\alpha}] = 0, \tag{11}$$

with $\vec{S}^{\alpha}_{P} = \sum_{i\in P} \vec{S}^{\alpha}_{i}$, the sum of all the spins $\vec{S}^{i,\alpha}$ in the orbital state α in any plane P orthogonal to the direction α (see Fig.2).

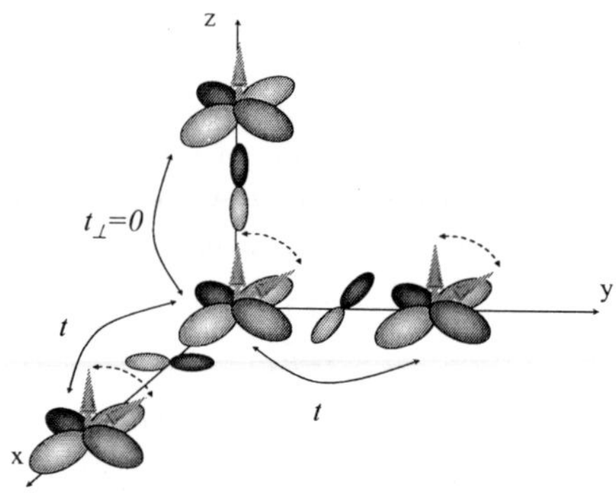

Fig. 2. From Ref. 2. The anisotropic hopping amplitudes leading to the Kugel-Khomskii (KK) Hamiltonian. Similar to Ref. 13 the four lobed states denote the $3d$ orbitals of a transition metal while the intermediate small p orbitals are oxygen orbital through which the super-exchange process occurs. The dark and bright shades denote positive and negative regions of the orbital wave-function. Due to orthogonality with intermediate oxygen p states, in any orbital state $|\alpha\rangle$ (e.g. $|Z\rangle \equiv |d_{xy}\rangle$ above), hopping is forbidden between sites separated along the cubic α (Z above) axis. The ensuing super-exchange (KK) Hamiltonian exhibits a $d = 2$ $SU(2)$ symmetry corresponding to a uniform rotation of all spins whose orbital state is $|\alpha\rangle$ in any plane orthogonal to the cubic direction α.

c) Superconducting arrays: A Hamiltonian for superconducting $(p + ip)$ grains (e.g. of Sr_2RuO_4) on a square grid, was recently proposed,[15]

$$H = -K \sum_{\square} \sigma^z \sigma^z \sigma^z \sigma^z - h \sum_{\mathbf{r}} \sigma^x_{\mathbf{r}}. \tag{12}$$

Here, the four spin product is the product of all spins common to a given plaquette $\square$. The spins reside on the vertices on the plaquette (not on its bonds as gauge fields). These systems have ($d = 1$ Z_2) symmetries similar to those of the two-dimensional orbital compass model. With P any row or column, $\hat{O}_P = \prod_{\vec{r} \in P} \sigma^x_{\vec{r}}$, $[H, \hat{O}_P] = 0$.

d) Other systems: In Refs. 16 and 17 similar symmetries were found in frustrated spin systems. Ring exchange Bose metals, in the absence of nearest neighbor boson hopping, exhibit $d = 1$ symmetries.[18] Continuous *sliding symmetries* of Hamiltonians (actions) invariant under arbitrary deformations along a transverse direction,

$$\phi(x, y) \to \phi(x, y) + f(y), \tag{13}$$

appear in many systems. Amongst others, such systems were discovered in works on Quantum Hall liquid crystalline phases,[19,20] a number of models of lipid bilayers with intercalated DNA strands,[21] and sliding Luttinger liquids.[22]

5. A theorem on dimensional reduction

The absolute mean value of any local quantity (involving only a finite number of fields) which is not invariant under a d-dimensional symmetry group G of the D-dimensional Hamiltonian H is equal or smaller than the absolute mean value of the same quantity computed for a d-dimensional Hamiltonian $\bar{H}$ which is globally invariant under G and preserves the range of the interactions.[2] Non invariant means that the quantity under consideration, $f(\phi_{\mathbf{i}})$, has no invariant component:

$$\sum_k f[\mathbf{g}_{\mathbf{i}k}(\phi_{\mathbf{i}})] = 0. \tag{14}$$

For a continuous group, this is replaced by $\int f[\mathbf{g}_{\mathbf{i}}(\phi_{\mathbf{i}})] d\mathbf{g} = 0$. To determine if spontaneous symmetry breaking occurs, we compute

$$\langle f(\phi_{\mathbf{i}}) \rangle = lim_{h \to 0} lim_{N \to \infty} \langle f(\phi_{\mathbf{i}}) \rangle_{h,N}, \tag{15}$$

where $\langle f(\phi_i) \rangle_{h,N}$ is the mean value of $f(\phi_i)$ computed on finite lattice of N sites and in the presence of a symmetry breaking field h. Since $\Lambda = \bigcup_l C_l$, the site $\mathbf{i}$ belongs at least to one set C_j. It is convenient to rename the fields in the following way: $\phi_{\mathbf{i}} = \psi_{\mathbf{i}}$ if $\mathbf{i} \notin C_j$ and $\phi_{\mathbf{i}} = \eta_{\mathbf{i}}$ if $\mathbf{i} \in C_j$. The mean value $\langle f(\phi_{\mathbf{i}}) \rangle_{h,V}$ is given by:

$$\langle f(\phi_{\mathbf{i}}) \rangle_{h,N} = \frac{\sum_{\{\phi_{\mathbf{i}}\}} f(\phi_{\mathbf{i}}) e^{-\beta H(\phi)} e^{-\beta h \sum_{\mathbf{i}} \phi_{\mathbf{i}}}}{\sum_{\{\phi_{\mathbf{i}}\}} e^{-\beta H(\phi) - \beta h \sum_{\mathbf{i}} \phi_{\mathbf{i}}}} =$$

$$\frac{\sum_{\{\psi_{\mathbf{i}}\}} z_{\{\psi\}} e^{-\beta h \sum_{\mathbf{i} \notin C_j} \psi_{\mathbf{i}}} \left[\frac{\sum_{\{\eta_{\mathbf{i}}\}} f(\eta_{\mathbf{i}}) e^{-\beta H(\phi) - \beta h \sum_{\mathbf{i} \in C_j} \eta_{\mathbf{i}}}}{z_{\{\psi\}}} \right]}{\sum_{\{\psi_{\mathbf{i}}\}} z_{\{\psi\}} e^{-\beta h \sum_{\mathbf{i} \notin C_j} \psi_{\mathbf{i}}}}$$

with,

$$z_{\{\psi\}} = \sum_{\{\eta_{\mathbf{i}}\}} e^{-\beta H(\psi, \eta) - \beta h \sum_{\mathbf{i} \in C_j} \eta_{\mathbf{i}}}. \tag{16}$$

From Eq.(16):

$$|\langle f(\phi_{\rm i})\rangle_{h,N}| \leq | \frac{\sum_{\{\eta_{\rm l}\}} f(\eta_{\rm i}) e^{-\beta H(\bar{\psi},\eta)-\beta h \sum_{{\rm l}\in C_j} \eta_{\rm l}}}{z_{\{\bar{\psi}\}}} |, \tag{17}$$

where $\{\bar{\psi}\}$ is the particular configuration of fields $\psi_{\rm i}$ that maximizes the expression between brackets in Eq.(16). $H(\bar{\psi},\eta)$ is a d-dimensional Hamiltonian for the field variables η which is invariant under the *global* symmetry group G_j of transformations $\mathbf{U}_{jk}$ over the field η. We can define $\bar{H}(\eta) \equiv H(\bar{\psi},\eta)$. The range of the interactions between the η-fields in $\bar{H}(\eta)$ is clearly the same as the range of the interactions between the ϕ-fields in $H(\phi)$. This completes the demonstration of our theorem. Note that the "frozen" variables $\bar{\psi}_{\rm i}$ act like external fields in $\bar{H}(\eta)$ which do not break the global symmetry group of transformations $\mathbf{U}_{jk}$.

Corollary I: Elitzur's theorem.[10] Any local quantity (i.e. involving only a finite number of fields) which is not invariant under a local (or $d=0$) symmetry group has a vanishing mean value at any finite temperature. This is a direct consequence of Eq.(16) and the fact that $\bar{H}(\eta)$ is a zero-dimensional Hamiltonian.[2]

Corollary II.[2] A local quantity which is not gauge invariant under a one-dimensional intermediate symmetry group has a vanishing mean value at any finite temperature for systems with finite range interactions. This is a consequence of Eq.(17) and the absence of spontaneous symmetry breaking in one-dimensional Hamiltonians such as $\bar{H}(\eta) \equiv H(\bar{\psi},\eta)$ with interactions of finite range and strength. Here, $f(\eta_{\rm i})$ is a non-invariant under the global symmetry group G_j [see Eq.(14)].

Corollary III.[2] In finite range systems, local quantities not invariant under continuous two-dimensional symmetries have a vanishing mean value at any finite temperature. This results from [Eq.(17)] with the Mermin-Wagner theorem:[23]

$$lim_{h\to 0} lim_{N\to\infty} \frac{\sum_{\{\eta_{\rm l}\}} f(\eta_{\rm i}) e^{-\beta H(\bar{\psi},\eta)-\beta h \sum_{{\rm l}\in C_j} \eta_{\rm l}}}{z_{\{\bar{\psi}\}}} = 0. \tag{18}$$

We invoked that G_j is a continuous symmetry group of $\bar{H}(\eta) = H(\bar{\psi},\eta)$, $f(\eta_{\rm i})$ is a non-invariant quantity for G_j [see Eq.(14)], and $\bar{H}(\eta)$ is a two-dimensional Hamiltonian that only contains finite range interactions.

The generalization of this theorem to the quantum case is straightforward if we choose a basis of eigenvectors of the local operators linearly coupled to the symmetry breaking field h. Here, the states can be written as a direct product $|\phi\rangle = |\psi\rangle \otimes |\eta\rangle$. Eq.(17) is re-obtained with the sums replaced by traces over the states $|\eta\rangle$:

$$|\langle f(\phi_{\rm i})\rangle_{h,N}| \leq \frac{{\rm Tr}_{\{\eta_{\rm l}\}} f(\eta_{\rm i}) e^{-\beta H(\bar{\psi},\eta)-\beta h \sum_{{\rm l}\in C_j} \eta_{\rm l}}}{{\rm Tr}_{\{\eta_{\rm l}\}} e^{-\beta H(\bar{\psi},\eta)-\beta h \sum_{{\rm l}\in C_j} \eta_{\rm l}}}, \tag{19}$$

In this case, $|\bar{\psi}\rangle$ corresponds to one particular state of the basis $|\psi\rangle$ that maximizes the right side of Eq.(19). Generalizing standard proofs, e.g.,[24] we find a zero temperature quantum extension of Corollary III in the presence of a gap:

Corollary IV.[2] If a gap exists in a system possessing a $d \leq 2$ dimensional continuous symmetry in its low energy sector then the expectation value of any local quantity not invariant under this symmetry, strictly vanishes at zero temperature. Though local order cannot appear, multi-particle (incl. topological) order can exist.

Corollary V. The absolute values of non-symmetry invariant correlators $|G| \equiv |\langle \prod_{i \in \Omega_j} \phi_i \rangle|$ with $\Omega_j \subset C_j$ are bounded from above by absolute values of the same correlators $|G|$ in a d dimensional system defined by C_j in the presence of transverse non-symmetry breaking fields. If no resonant terms appear in the lower dimensional spectral functions (due to fractionalization), this allows for fractionalization of non-symmetry invariant quantities in the higher dimensional system.

6. Consequences of the theorem

(**a**) *Spin nematic order in t_{2g} systems*: If the KK Hamiltonian (Eq.(5)) captures the spin physics of t_{2g} compounds, then no magnetization can exist at finite temperature[2] due to the continuous $d = 2$ symmetries[14] that it displays (Eq.(11)).[2,14] Empirically, low temperature magnetization is detected. Thus, the KK Hamiltonian of Eq.(5) may be augmented by other interactions which lift this symmetry. The simplest quantities invariant under these symmetries are nematic order parameters. In the presence of orbital ordering in the $|\alpha\rangle$ state, superpositions of $\langle \vec{S}_r \cdot \vec{S}_{r+n\hat{e}_\eta} \rangle$, with $\eta = x, y, z$ where $\eta \neq \alpha$ and n an integer, need not vanish. If the KK Hamiltonian embodies the dominant contribution to the spin physics, nematic order might persist to far higher temperatures than the currently measured magnetization.[2]

(**b**) *Orbital order*: The orbital only Hamiltonians discussed earlier exhibit a $d = 2$ *discrete* Z_2 symmetry. The theorem[2] allows such symmetries to be broken. Indeed, as we will review shortly, in these orbital only Hamiltonians, order already appears at the classical level- a tendency which may be enhanced by quantum fluctuations.

(**c**) *Nematic orbital order in two dimensional (p+ip) superconducting arrays and two dimensional orbital systems*: The two dimensional $(p+ip)$ superconducting arrays of Eq.(12) exhibit a $d = 1$ Z_2 symmetry. As these symmetries cannot be broken, no magnetization can arise, $\langle \sigma_\alpha \rangle = 0$. The simplest symmetry allowed order parameter is of the nematic type which is indeed realized classically.[3,4,25]

(**d**) *Fractionalization in spin and orbital systems*: Corollary (V) allows for fractionalization in quantum systems where d= 1, 2. It enables symmetry invariant quasi-particles excitations to *coexist* with non-symmetry invariant fractionalized excitations. Fractionalized excitations may propagate in $d_s = D - d$ dimensional regions (with D the spatial dimensionality of the system). Examples afforded by several frustrated spin models where spinons may drift along lines ($d_s = 1$) on the square lattice[16] and in $d_s = D$ dimensional regions on the pyrochlore lattice.[17]

(**e**) *Absence of charge order*: In systems, such as quantum Hall smectics, in which the system is invariant to the charge density variations of Eq.(13), we have $\langle \phi \rangle = 0$.

7. Order by disorder in symmetry allowed instances

When symmetry breaking is allowed (e.g. the two dimensional Ising symmetry (Eq.(9)) of the 120 ° Hamiltonian), order often transpires by a fluctuation driven mechanism ("order by disorder").[26] Although several states may appear to be equally valid candidate ground state, fluctuations can stabilize those states which have the largest phase space volume for low energy fluctuations about them. These differences are captured in values of the free energies for fluctuations about the contending states. Classicaly, fluctuations are driven by thermal effects. Quantum tunneling processes may fortify such tendencies.

If the Pauli matrices σ in Eq.(8) are replaced by the spin S generators and the limit $S \to \infty$ is taken then we will obtain the classical 120° model. Here, the *free energy* has strict minima for six uniform orientations[34] $\vec{S}_i = \pm S\hat{a}, \vec{S}_i = \pm S\hat{b}, \vec{S}_i = \pm S\hat{c}$. Out of the exponentially large number of ground states (supplanted by an additional global $U(1)$ rotational symmetry which emerges in the ground state sector), only six are chosen. Interfaces between uniform states (such as that borne by the application of d=2 Z_2 reflections on a uniform state, see fig.(1)) leads to a surface tension additive in the number of symmetry operations. Being of an entropic origin, the surface tension between various uniform domains is temperature independent and does not diverge at low temperatures.[3,4] Orbital order already appears within the classical (formally, $S \to \infty$) limit[3,4] and is not exclusively reliant on subtle quantum zero point fluctuations (captured by $1/S$ calculations) for its stabilization. Indeed, orbital order is detected up to relatively high temperatures ($\mathcal{O}(100K)$).[27]

8. Dualities as space-time reflections

Explicit operator representations show that the two dimensional variant of the orbital compass model (Eqs.(8, 10) is dual to the Xu-Moore model of $(p+ip)$ superconducting arrays (Eq.(12).[5] We now examine this duality in the discrete Euclidean path integral formulation. This examination illustrates how geometrical reflections may lead to dualities.[5] In a basis quantized along σ^z, the zero temperature Euclidean action of the two dimensional orbital compass model is

$$S = -K_x \sum_{\Box \in (x\tau) \text{ plane}} \sigma^z_{r,\tau}\sigma^z_{r,\tau+\Delta\tau}\sigma^z_{r+\hat{e}_x,\tau}\sigma^z_{r+\hat{e}_x,\tau+\Delta\tau} - (\Delta\tau)J_z \sum_r \sigma^z_r \sigma^z_{r+e_z}. \quad (20)$$

A schematic of this action in Euclidean space-time is shown in Fig.(3). If we relabel the axes and replace the spatial index x with the temporal index τ, we will immediately find the classical action corresponding to the the Hamiltonian of Eq.(12) depicting $p + ip$ superconducting grains in a square grid. This suggests that the planar orbital compass system and the $(p + ip)$ Hamiltonian (Eq.(12)) are dual to each other as indeed occurs at all temperatures. Strong-weak coupling dualities that these Hamiltonians (and others) display can be similarly established.[5]

This work was supported, in part, by the DOE at LANL (CDB), and by the National Science Foundation through grants NSF DMR 0442537 at UIUC (EF).

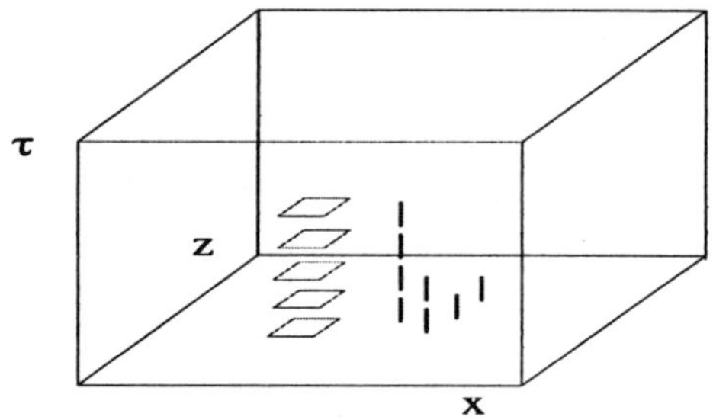

Fig. 3. From Ref. 5. The classical Euclidean action corresponding to the Hamiltonian of Eq.(12) at zero temperature in a basis quantized along the σ^z direction. The transverse field leads to bonds parallel to the imaginary time axis while the plaquette interactions become replicated along the imaginary time axis. Taking an equal time slice of this system, we find the four spin term of Eq.(12) and the on-site magnetic field term. If we interchange τ with z, we find the planar orbital compass model in the basis quantized along the σ^x direction.

References

1. X-G. Wen, *Quantum Field Theory of Many-Body Systems* (Oxford University Press, Oxford, 2004), and references therein.
2. C. D. Batista and Z. Nussinov, *Phys. Rev.* **B72**, 045137 (2005)
3. Z. Nussinov, M. Biskup, L. Chayes, J. v. d. Brink, *Europhys. Lett.* **67**, 990 (2004)
4. M. Biskup, L. Chayes, Z. Nussinov, *Comm. Math. Phys.* **255**, 253 (2005)
5. Z. Nussinov and E. Fradkin, *Phys. Rev.* **B71**, 195120 (2005)
6. J. B. Kogut, *Rev. Mod. Phys.* **51**, 659 (1979)
7. E. Fradkin and S. H. Shenker, *Phys. Rev.* **D19**, 3682 (1979)
8. F. Wegner, *J. Math. Phys.* **12**, 2259 (1971); K. G. Wilson, *Phys. Rev.* **D7**, 2911 (1974); J. Kogut and L. Susskind, *Phys. Rev.* **D11**, 395 (1975); R. Balian, J. M. Drouffe, and C. Itzykson, *Phys. Rev.* **D11**, 2098 (1975)
9. Z. Nussinov, *Phys. Rev.* **D72**, 054509 (2005)
10. S. Elitzur, *Phys. Rev.* **D12**, 3978 (1975)
11. Y. Aharonov and D. Bohm, *Phys. Rev.* **11**, 485 (1959)
12. K. I. Kugel and D. I. Khomskii, *Sov. Phys. JETP* **37**, 725 (1973)
13. J. v. d. Brink, G. Khaliullin and D. Khomskii, *Orbital effects in manganites*, T. Chatterij (ed.), *Colossal Magnetoresistive Manganites*, Kluwer Academic Publishers (2002)
14. A. B. Harris et al., *Phys. Rev. Lett.* **91**, 087206 (2003)
15. C. Xu and J. E. Moore, *Phys. Rev. Lett.* **93**, 047003 (2004).
16. C. D. Batista and S. A. Trugman, *Phys. Rev. Lett.* **93**, 217202 (2004).
17. Z. Nussinov, C. D. Batista, B. Normand, and S. A. Trugman, cond-mat/0602528
18. A. Paramekanti, L. Balents, and M. P. A. Fisher, *Phys. Rev.* [illegible]
19. M. J. Lawler and E. Fradkin, *Phys. Rev.* **B70**, 165310 (2004)
20. L. Radzihovsky and A. T. Dorsey, *Phys. Rev. Lett.*, **88**, 216802 (2002)
21. C. S. O'Hern and T. C. Lubensky, *Phys. Rev.* **E58**, 5948 (1998)
22. V. J. Emery et al., *Phys. Rev. Lett.*, **85**, 2160 (2000)
23. N. D. Mermin and H. Wagner, *Phys. Rev. Lett.* **17**, 1133 (1966).
24. Assa Auerbach, *Interacting Electrons and Quantum Magnetism*, Springer-Verlang 1994 (chapter 6, in particular)
25. A. Mishra et al., *Phys. Rev. Lett.* **93**, 207201 (2004)
26. E. F. Shender, *Sov. Phys. JETP* **56**, 178 (1982); C. L. Henley, *Phys. Rev. Lett.* **62**,

2056 (1989).

27. Y. Murakami *et al*, *Phys. Rev. Lett.* **80**, 1932 (1998); Y. Endoh *et al*, *Phys. Rev. Lett.* **82**, 4328 (1999); S. Ishihara and S. Maekawa, *Phys. Rev. Lett.* **80**, 3799 (1998); I.S. Elfimov, V.I. Anisimov and G.A. Sawatzky, *Phys. Rev. Lett.* **82**, 4264 (1999). Y. Tokura and N. Nagaosa, *Science* **288**, 462 (2000), and references therein.

INFORMATION GEOMETRY AND PHASE TRANSITIONS

MARIELA PORTESI *

Instituto de Física La Plata (IFLP, CONICET–UNLP) and
Departamento de Física, Facultad de Ciencias Exactas, Universidad Nacional de La Plata,
CC 67, (1900) La Plata, Argentina

ANGEL L. PLASTINO †

Instituto de Física La Plata (IFLP, CONICET–UNLP) and
Departamento de Física, Facultad de Ciencias Exactas, Universidad Nacional de La Plata,
CC 67, (1900) La Plata, Argentina

FLAVIA PENNINI‡

Departamento de Física, Universidad Católica del Norte,
Av. Angamos 0610, Antofagasta, Chile

We present, from an information theoretic viewpoint, an analysis of phase transitions and critical phenomena in quantum systems. Our study is based on geometrical considerations within the Riemannian space of thermodynamic parameters that characterize the system. A metric for the space can be derived from an appropriate definition of distance between quantum states. For this purpose, we consider generalized α-divergences that include the standard Kullback–Leibler relative entropy. The use of other measures of information distance is taken into account, and the thermodynamic stability of the system is discussed from this geometric perspective.

Keywords: Information geometry; Quantum generalized divergence; Metric tensor.

1. Introduction

We consider the geometrical structure of the space of thermodynamic parameters that characterize a given physical system. For this purpose it is essential to provide an appropriate quantity that gives an idea of 'how far' two states are, depending on the values of a set of parameters. That quantity should play the role of a distance between states and the whole machinery of geometry could be applied. This point of view has been applied to various models in classical as well as quantum systems, and it has been seen to detect critical regions and phase transitions.

Some related previous works in this field are the ones by Janyszek and Mrugała,[1]

* portesi@fisica.unlp.edu.ar

† plastino@fisica.unlp.edu.ar

‡ fpennini@ucn.cl, Permanent address: IFLP and DF–FCE–UNLP, CC. 67, (1900) La Plata, Argentina

Brody and Rivier,[2] Ruppeiner,[3] among others, where *standard* information measures were taken into account. Also Amari and Nagaoka[4] consider information-geometry applications in the area of mathematical probabilities, employing the so-called α-divergences to introduce the notion of distance between two different probability distributions of *exponential* shape. The ("counter-")problem of treating the geometry of escort distributions employing the standard divergence has been addressed by Abe.[5] Also, Trasarti-Battistoni[6] studied some general issues concerning geometrical approaches to non-extensive statistical mechanics.

The present work is organized as follows: in Sec. 2 we summarize known results on standard statistical mechanics and Riemannian geometry. Generalized information measures are introduced in Sec. 3, together with some geometrical aspects in generalized thermostatistics. Some conclusions are drawn at the end.

2. Standard Statistical Mechanics and Riemannian Geometry

Given a physical system, the equilibrium density operator ρ (the one that maximizes the von Neumann–Shannon entropy $S = -\mathrm{Tr}\rho \ln \rho$, subject to a set of constraints provided by given mean values) can be represented in an r-dimensional space whose axis correspond to the parameters $\beta^1, \ldots, \beta^r$ (e.g. temperature, pressure, magnetic field) associated to the operators $F_1, \ldots, F_r$. The density operator has the form $\rho = Z^{-1} \exp(-\sum_{i=1}^{r} \beta^i F_i)$, where Z is the partition function.

The informational definition of distance between two states is given in terms of the relative or cross entropy, also called Kullback–Leibler (KL) divergence (or information gain), $K(\rho||\sigma) = \mathrm{Tr}\rho(\ln \rho - \ln \sigma)$. Taking the symmetric KL divergence $K(\rho||\sigma) + K(\sigma||\rho)$ as measure of distance between two neighbor states in the space of parameters, described by $\rho = \rho(\{\beta\})$ and $\sigma = \rho(\{\beta + d\beta\})$, the metric tensor and other geometric magnitudes are deduced (by expansion around $\{\beta\}$), arriving at an expression of the form $dD^2 = g_{ij} d\beta^i d\beta^j$ where the coefficients of the metric tensor of the space of parameters are given by $g_{ij}(\{\beta\}) = -\langle \partial_i \partial_j \ln \rho \rangle$, with $\langle X \rangle \equiv \mathrm{Tr}(\rho X)$ and $\partial_i X \equiv \partial X / \partial \beta^i = X_{,i}$, for $i, j = 1, \ldots, r$.

Various alternative forms for the metric coefficients can be given. For instance, they can be seen as elements of the matrix of partial second derivatives of $f \equiv \ln Z$: $g_{ij} = \partial_i \partial_j \ln Z = f_{,ij}$. In terms of the mean values $m_i = \langle F_i \rangle = -\partial_i \ln Z$, they read $g_{ij} = -\partial_i m_j = -\partial_j m_i$. The case of dealing with a set of commuting operators (or with classical systems) is derived directly without the need of symmetrizing the KL divergence, with the result: $g_{ij} = \langle \rho^{-1} \partial_i \rho \, \partial_j \rho \rangle = \langle \partial_i \ln \rho \, \partial_j \ln \rho \rangle$. The metric tensor constitutes the Fisher information matrix $G = (g_{ij})$. Finally, in terms of thermodynamic fluctuations (covariances or second moments), the connection is $g_{ij} = \langle (F_i - m_i)(F_j - m_j) \rangle$.

The Riemann–Christoffel (RC) tensor and the scalar curvature of the Riemannian space of parameters are derived from the fundamental tensor G. There are

$r^2(r^2-1)/12$ independent components of the RC tensor, given by the contractions:

$$R_{ijkl} = \frac{1}{4} g^{st} (f_{,sil} f_{,tjk} - f_{,sik} f_{,tjl}), \qquad f \equiv \ln Z(\{\beta\}) \tag{1}$$

where g^{st} is the fundamental contravariant tensor. The particular case $r = 2$ gives simply $G = H(f)$ (H indicates the Hessian matrix), and then the scalar curvature defined as $R = g^{jk} g^{il} R_{ijkl}$ gives $R = (2/g) R_{1212}$ with $g = \det(G)$ and

$$R_{1212} = -\frac{1}{4g} \begin{vmatrix} f_{,11} & f_{,12} & f_{,22} \\ f_{,111} & f_{,112} & f_{,122} \\ f_{,112} & f_{,122} & f_{,222} \end{vmatrix}$$

The above considerations have been applied to discuss the ideal gas, van der Waals gas (in both cases, the parameters are T and P), Ising model and other magnetic systems (mean field, Potts 1D, etc.) (with T and H as parameters). It has been seen that: (i) a non interacting model has a flat geometry ($R = 0$), (ii) R diverges at the critical point for interacting systems, (iii) R contains information on third moments. Thus, the curvature R represents a measure of the thermodynamic stability of a system, useful for the characterization of phase transitions.

3. Generalized Information Measures and Geometry

In the nonextensive formalism,[7] the q-entropy $S_q(\rho) \equiv (1 - \mathrm{Tr}\,\rho^q)/(q-1)$, $q \in \mathbb{R}$, is extremized by the density operator $\rho = e_q\big(-\sum_{i=1}^{r} \beta^i (F_i - m_i)\big)/\bar{Z}_q$, where $\{\beta^1, \dots, \beta^r\}$ is the set of *optimized* Lagrange multipliers (OLM),[8] and $m_1, \dots, m_r$ are the corresponding *generalized* expectation values given by $m_i = \langle F_i \rangle_q \equiv \mathrm{Tr}\,(\rho^q F_i)/\mathrm{Tr}\,\rho^q$. The quantity $\bar{Z}_q \equiv \mathrm{Tr}\, e_q\big(-\sum_{i=1}^{r} \beta^i (F_i - m_i)\big)$ is the pseudo-partition function and, of course, $\mathrm{Tr}\,\rho = 1$. Here $e_q(x)$ stands for the q-exponential function.[7] The limiting case $q \to 1$ reproduces the standard situation given above.

The quantum q-divergence[5] is defined as $K_q(\rho\|\sigma) = \mathrm{Tr}\rho^q(\rho^{1-q} - \sigma^{1-q})/(1-q)$. Taking the symmetric generalized KL divergence $K_q(\rho\|\sigma) + K_q(\sigma\|\rho)$ as measure between $\rho = \rho(\beta)$ and $\sigma = \rho(\beta + d\beta)$ (obviously here $\rho(\beta)$ is the OLM density matrix) and expanding up to second-order in β-variations, we obtain the *generalized fundamental tensor* $g_{ij}^{(q)} = q\mathrm{Tr}(\rho\, \partial_j \ln \rho\, \partial_i \ln \rho)$, which we can cast in a compact form as $g_{ij}^{(q)} = -\partial_j m_i - (q-1)\partial_j \ln \bar{Z}_q\, \partial_i \ln \bar{Z}_q$. Defining $\bar{f} = \ln \bar{Z}_q$, the RC tensor reads[9,10]

$$\begin{aligned} R_{ijkl}^{(q)} = &-(q-1)(\bar{f}_{,ik}\bar{f}_{,jl} - \bar{f}_{,il}\bar{f}_{,jk}) - g^{(q)\,st} \Big\{ \frac{1}{4}(m_{s,jl}m_{t,ik} - m_{s,jk}m_{t,il}) + \\ &+ \frac{1}{2}(q-1)\left[\bar{f}_{,s}(m_{t,ik}\bar{f}_{,jl} - m_{t,il}\bar{f}_{,jk}) + (\bar{f}_{,ik}m_{s,jl} - \bar{f}_{,il}m_{s,jk})\bar{f}_{,t}\right] + \\ &+ (q-1)^2(\bar{f}_{,ik}\bar{f}_{,jl} - \bar{f}_{,il}\bar{f}_{,jk})\bar{f}_{,s}\bar{f}_{,t} \Big\} \end{aligned} \tag{2}$$

The generalized scalar curvature can be calculated in the same way as indicated in the previous section. In the case $r = 2$ we obtain[9,10]

$$R^{(q)} = \frac{2}{(g^{(q)})^2}\left\{\frac{1}{4}\begin{vmatrix} m_{1,1} & m_{1,2} & m_{2,2} \\ m_{1,11} & m_{1,12} & m_{1,22} \\ m_{1,12} & m_{1,22} & m_{2,22}\end{vmatrix} + (q-1)\left[\frac{1}{4}\begin{vmatrix} \bar{f}_{,1}^2 & \bar{f}_{,1}\bar{f}_{,2} & \bar{f}_{,2}^2 \\ m_{1,11} & m_{1,12} & m_{1,22} \\ m_{1,12} & m_{1,22} & m_{2,22}\end{vmatrix} - \right.\right.$$
$$-J(m_1,m_2)\,|H(\bar{f})| + \frac{1}{2}\left[J(\bar{f},m_2)\left(J(\bar{f}_{,1},m_{1,2}) - J(\bar{f}_{,2},m_{1,1})\right) - \right.$$
$$\left.\left.\left. - J(\bar{f},m_1)\left(J(\bar{f}_{,1},m_{2,2}) - J(\bar{f}_{,2},m_{2,1})\right)\right]\right]\right\} \tag{3}$$

with $g^{(q)} = J(m_1,m_2) + (q-1)\left[\bar{f}_{,1}\,J(\bar{f},m_2) - \bar{f}_{,2}\,J(\bar{f},m_1)\right]$, where $J(x,y)$ denotes Jacobian determinant. As $q \to 1$, $R^{(q)}$ reduces to the standard curvature R.

Summing up, we have considered a (one-index) family of generalized information measures to introduce a metric in the space of parameters in quantum many-body models, and we calculated, within an OLM framework, the generalized (q-dependent) expressions for the metric and Riemann–Christoffel tensors, as well as the scalar curvature of the space. The formalism has been applied in some detail to an ideal (non-interacting) system.[9,10] We cite, as possible extensions -currently under study-: application to other physical systems (van der Waals gas and magnetic systems), the use of different sets of parameters, also other generalized measures of informational distance, and discussion in other nonextensive frameworks.

Acknowledgements

Financial support from CONICET and UNLP, Argentina, is acknowledged. MP also acknowledges ANPCyT (PICT No. 03-11903/2002), Argentina.

References

1. H. Janyszek and R. Mrugała, *Phys. Rev.* **A39**, 6515 (1989).
2. D. Brody and N. Rivier, *Phys. Rev.* **E51**, 1006 (1995).
3. G. Ruppeiner, *Phys. Rev.* **A20**, 1608 (1979); *Rev. Mod. Phys.* **67**, 605 (1995); *Ibid.* **68**, 313 (1996).
4. S. Amari and H. Nagaoka, *Methods of Information Geometry* (AMS and Oxford University Press, 2000).
5. S. Abe, *Phys. Rev.* **E68**, 031101 (2003).
6. R. Trasarti-Battistoni, cond-mat/0203536.
7. C. Tsallis, *J. Stat. Phys.* **52**, 479 (1988); see http://tsallis.cat.cbpf.br/biblio.htm
8. S. Martínez, F. Nicolás, F. Pennini and A. Plastino, *Physica* **A286**, 489 (2000).
9. M. Portesi, F. Pennini and A. Plastino, cond-mat/0511490.
10. M. Portesi, A. Plastino and F. Pennini, Physica **A**, in press (2006); cond-mat/0510434.

MAPPING REACTION PATHS IN PHASE-SPACE

JULIEN TAILLEUR

Physique et Mécanique des Milieux Hétérogñes CNRS UMR7636,
École Supérieure de Physique et de Chimie Industrielles de la Ville de Paris,
10 rue Vauquelin, 75231 Paris, FRANCE *

SORIN TĂNASE-NICOLA

FOM Institute for Atomic and Molecular Physics,
Kruislaan 407, 1098 SJ Amsterdam, THE NETHERLANDS †

JORGE KURCHAN

Physique et Mécanique des Milieux Hétérogñes CNRS UMR7636,
École Supérieure de Physique et de Chimie Industrielles de la Ville de Paris,
10 rue Vauquelin, 75231 Paris, FRANCE ‡

Given a dynamics in configuration or phase-space, it is often important to map the barriers, the separatrices emanating from them, and the current distributions of the reaction paths. We describe a strategy to do this efficiently.

Keywords: Reaction; separatrix; barrier.

1. From stable to unstable structures

Simulations of complex systems are in most cases concentrated on the exploration of stable structures. Thus, for example, Monte Carlo methods give us the equilibrium configuration (or at least long-lived metastable states), and optimisation programmes search for minima in a cost function.

There are however instances in which the relevant information is contained in unstable regions. For example, in nucleation problems we wish to understand the form of the critical droplet which separates the basins of attraction of the metastable and stable states. Similarly, in Physical Chemistry an important problem is to map the reaction path taking from one quasiequilibrium configuration to another – in a classical low-temperature case this goes through a saddle point in phase-space.[1] Another example is that of Classical (e.g. Celestial) Mechanics, where the knowledge of separatrices or homoclinic trajectories is relevant to understand the stability

*tailleur@pmmh.espci.fr
†S.Tanase@amolf.nl
‡jorge@pmmh.espci.fr

properties of a system. In all these cases, it is precisely the instability of the object of interest that makes its simulation problematic.

2. A Strategy

We shall describe a strategy[2,3] to study unstable structures, originally inspired in the use of Supersymmetric Quantum Mechanics as an instrument to derive Morse theory.[4,5]

To find minima of a function V, a natural method is *simulated annealing*:

$$\dot{x}_i = -f_i + \sqrt{2T} \underbrace{\eta_i}_{\text{white noise}} \tag{1}$$

where

$$f_i = \frac{\partial V}{\partial x_i} \tag{2}$$

By slowly decreasing the temperature T, the system may avoid high-lying metastable states, and eventually find the lowest minimum. The probability distribution associated to the Langevin process (1) evolves according to the Fokker-Planck equation:

$$\begin{aligned} \dot{P}(x) &= -HP(x) \\ H &= \sum_i \frac{\partial}{\partial x_i}\left(T\frac{\partial}{\partial x_i} + f_i\right) \end{aligned} \tag{3}$$

The question is now: how can we modify this dynamics to make it search the for barriers instead of stable states? Let us start by extending the Fokker-Planck operator, for the moment with no justification, by adding a set of Fermion operators a_i and enlarging the Hilbert space accordingly:

$$H_s = \sum_i \frac{\partial}{\partial x_i}\left(T\frac{\partial}{\partial x_i} + f_i\right) + \sum_{ij} a_i^\dagger \frac{\partial f_i}{\partial x_j} a_j \tag{4}$$

Clearly, H_s preserves the total Fermion number. We can generalise(3) by considering the evolution of a state with k Fermions:

$$\dot{\Phi}^{(k)} = -H_s \Phi^{(k)}$$

It turns out that for small temparatures T, the dynamics (3) for different fermion sectors converge to different objects:[2]

- 0 FERMION: we recover the Langevin / Fokker-Planck equation: $\Phi^{(0)} = P$ converges to *local minima*
- 1 FERMION: $\Phi^{(1)} = \sum_i J_i(x) a_i^\dagger |-\rangle$ converges to *reaction paths* joining two minima passing through saddles with one unstable direction, as in Fig. 1.

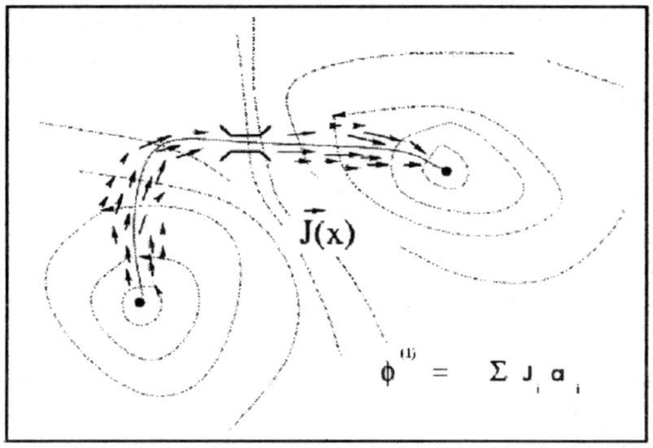

Fig. 1. The one-fermion field $J(\vec{x})$ converges to a reaction current distribution

- 2 FERMIONS:$\Phi^{(2)} = \sum_i J_{ij}(x) a_i^\dagger a_j^\dagger |-\rangle$ converges to the two-dimensional surface spanned by paths that go downhill from index-two saddles (the index is the number of unstable directions).
- k FERMIONS converges to the k-dimensional surface spanned by paths that go downhill from index k saddles.

In addition, *if f_i derives from a potential* H_s posesses a (super)symmetry:

$$[H_s, Q] = [H_s, \bar{Q}] = 0 \tag{6}$$

$$H_s = \frac{1}{T}(\bar{Q} + Q)^2 \tag{7}$$

where the charges are defined as:

$$\bar{Q} = -\mathrm{i} \sum_i \left(T \frac{\partial}{\partial x_i} + \frac{\partial V}{\partial x_i} \right) a_i^\dagger \quad ; \quad Q = -\mathrm{i} T \sum_i \frac{\partial}{\partial x_i} a_i, \tag{8}$$

and satisfy:

$$\bar{Q}^2 = Q^2 = 0. \tag{9}$$

The operators Q and $\bar{Q}$ produce mappings between eigenvectors of H_s (much as angular momentum operators generate the multiplets in atomic physics), and this in turn establishes relations between all the index k saddles, their unstable manifolds and topology of the space. These relations are *Morse theory*.[6]

Here we are interested for example in reaction paths passing through saddles of index one. The discussion above implies that, if we manage to simulate

$$\dot{\Phi}^{(1)} = -H_s \Phi^{(1)}$$

the system will converge *directly* to the reaction paths, and this will be the generalisation of simulated annealing we are looking for. Indeed there are several methods to do this, here we shall describe the ('Diffusion Monte Carlo') method originally devised for simulations of electrons in Atomic and Solid State physics: One considers a number of non-interacting 'clones' of the particles, each one carrying a vector $\vec{v}$, which evolves with

$$\dot{v}_i = -\sum_j \frac{\partial f_i}{\partial x_j} v_j \tag{11}$$

The particles perform ordinary diffusion following Eq. (1), but they clone with a rate proportional to $\propto d|v|/dt$ (see Fig. 2).

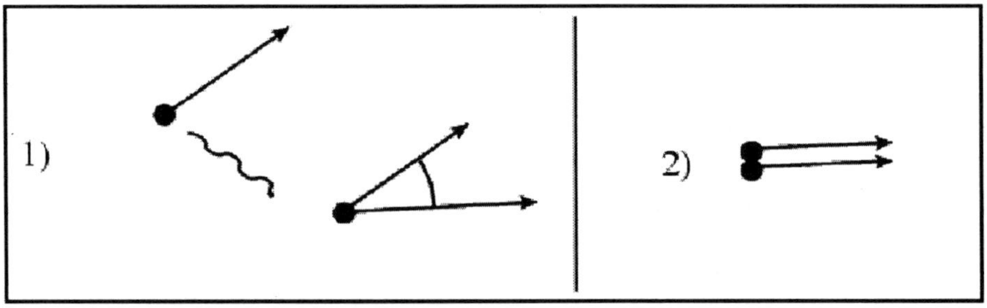

Fig. 2. The walkers' dynamics: diffusion with cloning.

The long-time joint population distribution $\mathcal{F}(\vec{x},\vec{v})$ of walkers with a vector $\vec{v}$ yields the current components through averaging over $\vec{v}$:

$$J_a(\vec{x}) = \int dv_a \; v_a \; \mathcal{F}(\vec{x},\vec{v}) \tag{12}$$

The $\mathcal{F}(\vec{x},\vec{v})$ obtained by the walkers may be such that the integral (12) vanishes: in that case the distribution lies outside the fermionic space. Remarkably enough, even in those cases $\mathcal{F}(\vec{x},\vec{v})$ has an interesting meaning.

2.1. *Introducing inertia*

Before showing some results, let us consider a sligtly more general and less standard situation that arises when the system evolves in *phase-space* following Hamilton's equations with friction and noise, thus incorporating the effect of inertia:

$$\begin{aligned} \dot{q}_i &= \frac{\partial \mathcal{H}}{\partial p_i} = p_i \\ \dot{p}_i &= -\frac{\partial \mathcal{H}}{\partial q_i} - \gamma p_i + \sqrt{2\gamma T}\eta_i \end{aligned} \tag{13}$$

The probability evolution is then given by Kramers' operator:

$$\frac{\partial P(\mathbf{q},\mathbf{p};t)}{\partial t} = -H_K P(\mathbf{q},\mathbf{p};t) = \sum_{i=1}^{N} \left(\frac{\partial}{\partial p_i} J_{p_i} + \frac{\partial}{\partial q_i} J_{q_i} \right)$$

$$H_k = \sum_{i=1}^{N} \frac{\partial}{\partial p_i}\left(\gamma T\frac{\partial}{\partial p_i} + \gamma p_i + \frac{\partial V}{\partial q_i}\right) - \frac{\partial}{\partial q_i}\frac{\partial \mathcal{H}}{\partial p_i} \tag{14}$$

which also defines the phase-space currents

$$J_{q_i} = \frac{\partial \mathcal{H}}{\partial p_i}P(\mathbf{q},\mathbf{p},t) \qquad J_{p_i} = -\left(\gamma T\frac{\partial}{\partial p_i} + \gamma\frac{\partial \mathcal{H}}{\partial p_i} + \frac{\partial \mathcal{H}}{\partial q_i}\right)P(\mathbf{q},\mathbf{p},t). \tag{15}$$

Again, in this case we can obtain an extended Hamiltonian by introducing to sets of fermions a_i, b_i, with $i = 1, ..., N$.

$$H_s = H_K + \sum_{i,j=1}^{N} \frac{\partial^2 V}{\partial q_i \partial q_j} b_i^\dagger a_j + \gamma \sum_i b_i^\dagger b_i - \sum_i a_i^\dagger b_i \tag{16}$$

which posesses a (much less studied) supersymmetry generated by two nilpotent operators Q' and $\bar{Q}'$ (see Ref. 3 for details). Just as in the pure dissipative case, one can reconstruct Morse theory on the basis of the trajectories associated with Hamilton's equations with damping.[3]

From what we know of the Langevin/Fokker-Planck case, we might have expected that the information on reaction paths is encoded in the one-fermion wavevectors

$$\Phi^{(1)} = \sum_i \left(J_{q_i}^{\text{red}} a_i^\dagger|-\rangle + J_{p_i}^{\text{red}} b_i^\dagger|-\rangle\right) \tag{17}$$

Surprisingly, these eigenvectors are not given in terms of the phase-space current, but rather of *reduced current* defined as:

$$J_{q_i}^{\text{red}} \equiv J_{q_i} + T\frac{\partial P(\mathbf{q},\mathbf{p})}{\partial p_i} \qquad J_{p_i}^{\text{red}} \equiv J_{p_i} - T\frac{\partial P(\mathbf{q},\mathbf{p})}{\partial q_i}. \tag{18}$$

The reduced current has the following good properties:

- It differs from the current by a term without divergence, hence the fluxes over closed surfaces coincide.
- It is zero in equilibrium. In a case with metastable states, it is small everywhere, while the true current is large within the states.

One can show that, also in this case, a dynamics like the one in Fig. 2 converges to the reduced current, with the particles carrying a $2N$ dimensional vector (v_i^q, v_i^p) evolving through (11) with f_i given by the phase-space drift terms in (13).

As an example, consider the double-well Hamiltonian of Fig. 3:

$$\mathcal{H} = \frac{p^2}{2} - 2q^2 + q^4 \tag{19}$$

In Fig. 4 we show a typical passage activated by the noise, and the corresponding reduced current for the problem. The spiral form of the trajectories reflects the presence of damping. The frame to the left of Fig. 4 was obtained by direct simulation (waiting for the activated passage to happen), while the right frame was obtained

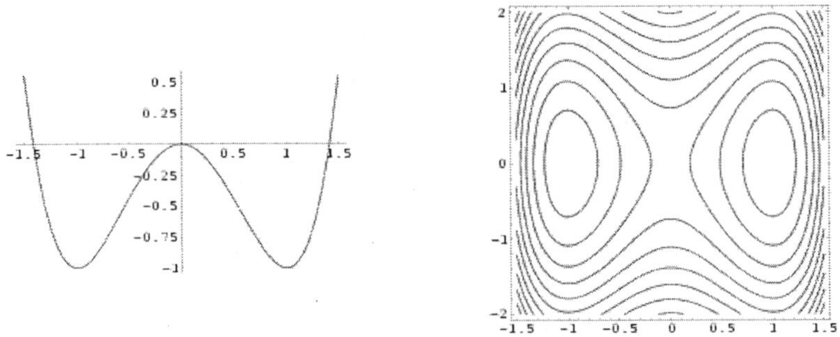

Fig. 3. The double-well potential in configuration space (left), and in phase-space (right, contour lines)

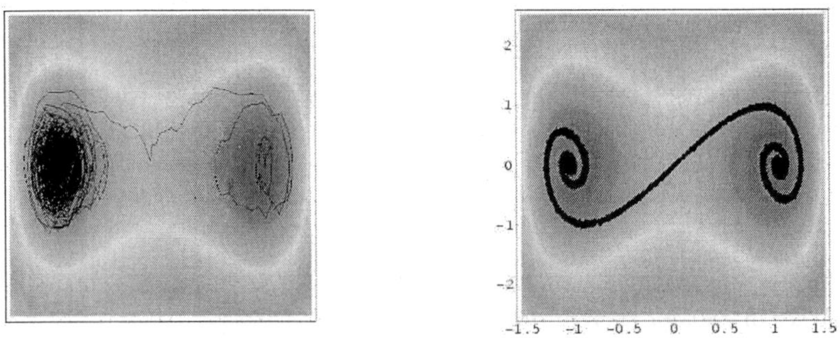

Fig. 4. An activated trajectory (left). The reduced current (instanton) associated with the passage (right).

(much faster) by applying the clone dynamics. In Fig. 5 we see the 'one fermion' dynamics in action: an initial distribution of clones evolves into the distribution of reduced current, which contains all the information on the passage.

2.2. *Pure Hamiltonian (frictionless) dynamics*

The case of pure Hamiltonian dynamics is particularly interesting. It is obtained from the previous one, taking the limit:

$$\gamma \longrightarrow 0 \qquad T \longrightarrow \infty \qquad \gamma T = \epsilon \quad \text{finite but small} \tag{20}$$

so that the equations of motion read:

$$\dot{q}_i = \frac{\partial \mathcal{H}}{\partial p_i} \tag{21}$$

$$\dot{p}_i = -\frac{\partial \mathcal{H}}{\partial q_i} + \sqrt{2\epsilon}\eta_i \tag{22}$$

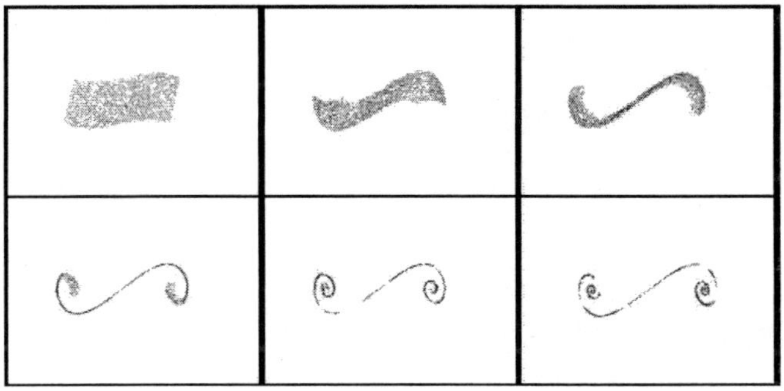

Fig. 5. The clones converging to the reduced current distribution.

This dynamics leads to a flat exploration of phase-space, since it corresponds to infinite temperature, even for small ϵ. However, the associated clone dynamics (Fig. 2) is interesting, as we shall see. The operator H_s in the limit (20) commutes with the charges:

$$Q_H = -i\sum_{i=1}^{N}\left(\frac{\partial}{\partial q_i}a_i + \frac{\partial}{\partial p_i}b_i\right) \qquad \bar{Q}_H = [K^\dagger, Q]_- = -i\sum_{i=1}^{N}\left(\frac{\partial}{\partial q_i}b_i^\dagger - \frac{\partial}{\partial p_i}a_i^\dagger\right) \quad (23)$$

and can be written in the form

$$H = \{Q_H, A\} = \{\bar{Q_H}, \bar{A}\} \quad (24)$$

where $A, \bar{A}$ can easily be obtained from (8).

The dynamics with cloning in this case converges to structures such that the integral (12) vanishes, and hence do not map immediately into one-fermion functions – and hence we must regard the supersymmetric construction as just a source of inspiration. However, they still have a geometric meaning: they converge with resolution $\sim \sqrt{\epsilon}$ to separatrices in integrable systems and homoclinic trajectories when chaos sets in.

As an example, let us consider billiard of Fig. 6, given by $r = 1+\delta\cos(2\theta)$, whose chaoticity increases with δ (it is *not* an ellipse). As usual with billiard systems, we present the results with a Poincaré section given by the Birkhoff variables $(s, \cos\varphi)$ shown in Fig. 7. Figure 8 shows three Poincaré sections for increasing levels of chaoticity. The dark region is the attractors for the clones: it is concentrated on the separatrix (this is true even for an integrable case), and on the principal homoclinic orbit associated to the unstable point, once chaos sets in. If, on the other hand, one projects away the main structures, for example by absorbing the walkers that are close to the main separatrix, one can zoom into more detailed structures (Fig. 9). It is remarkable that the clones converge *very rapidly* to very unstable structures. Figure 10 shows an enlarged image of the homoclinic trajectory, together with some nearby ones.

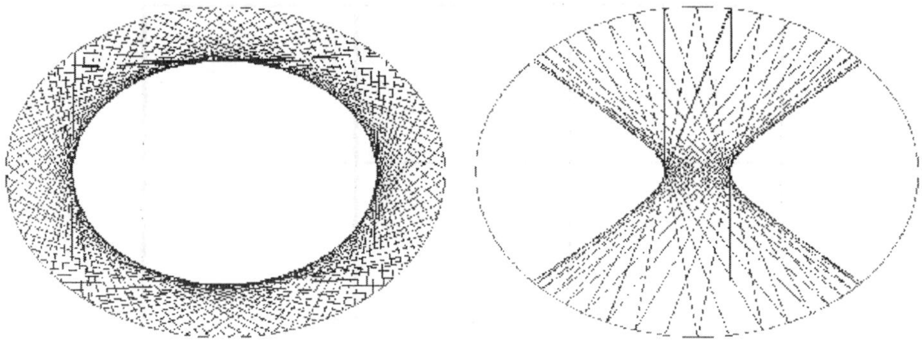

Fig. 6. The billiard. The two types of trajectories are delimited by a separatrix.

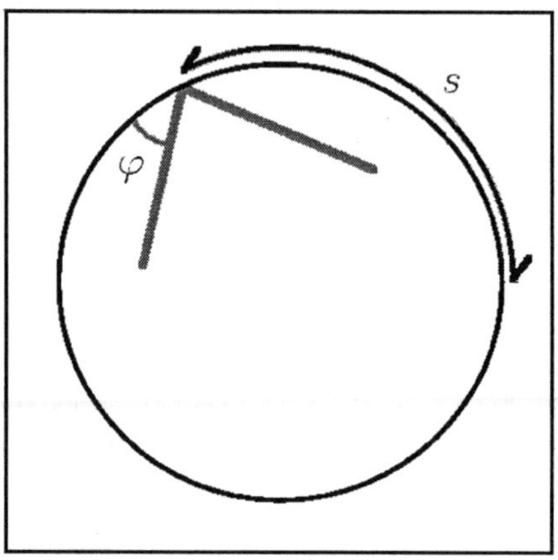

Fig. 7. The canonical variables (s, φ) for the billiard

3. Perspectives

The dynamics considered here is a natural generalisation of simmulated annealing that converges to unstable structures. We believe that it may prove useful for the determination of reaction paths and barriers, as well as for the study of chaotic transport in near integrable (such as planetary) systems. Another intriguing application is the search of localised 'breather' solutions in nonlinear systems.

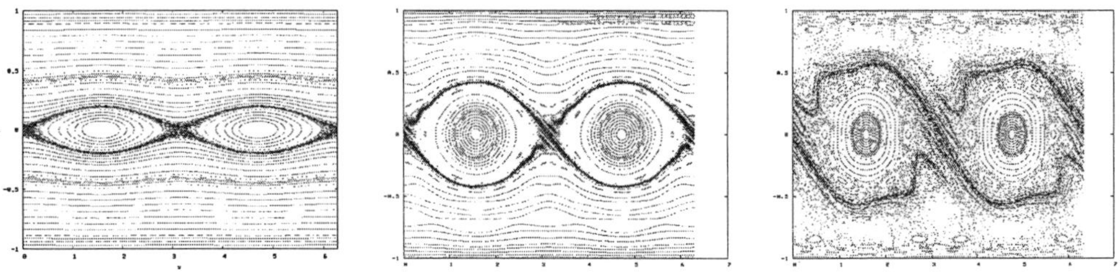

Fig. 8. Poincaré Section of the trajectories for, from left to right, $\delta = 0.01$ $\delta = 0.05$ and $\delta = 0.1$. The dark region is the one populated by the walkers, and correspond to the main separatrix and homoclinic trajectories.

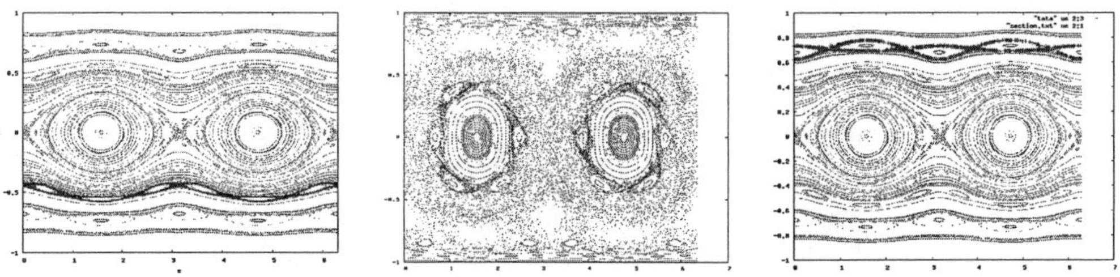

Fig. 9. Same as in Fig. 8, but with the walkers made to zoom into secondary structures.

References

1. D. J.Wales, *Energy Landscapes* (Cambridge University Press, 2003).
2. S. Tanase-Nicola , J. Kurchan; *Phys. Rev. Lett.* **91**, 188302 (2003); *J. Stat. Phys.* **116**, 1201 (2004).
3. J. Tailleur, S. Tanase-Nicola and J. Kurchan; *Kramers equation and supersymmetry* cond-mat/0503545, J. Stat. Phys., to appear.
4. E. Witten, *J. Diff. Geom.* **17**, 661 (1982).
5. J. Milnor, *Morse Theory* (Princeton University Press, 1963).

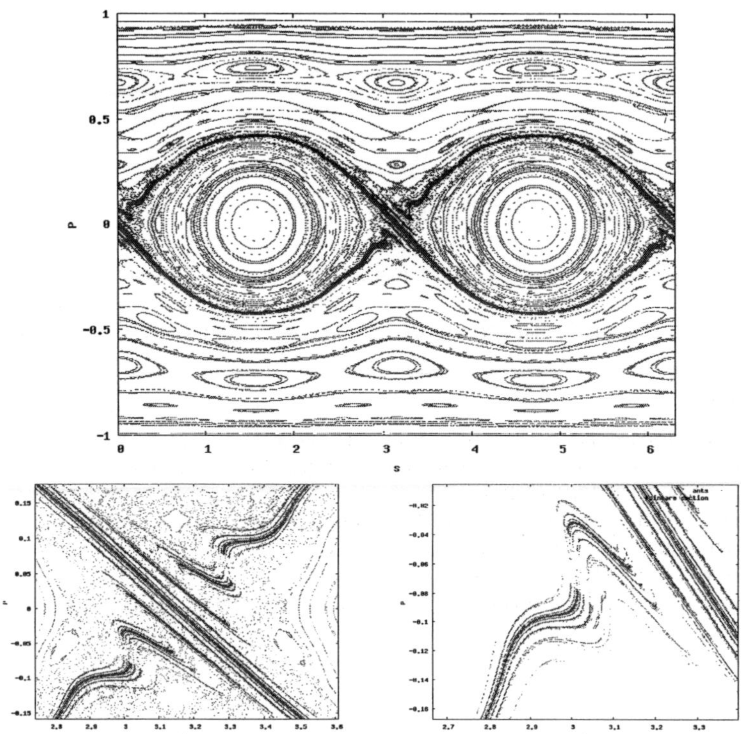

Fig. 10. A detail of the homoclinic (and nearby) trajectories. Different trajectories bypass small tori.

QUANTUM FLUIDS IN NANOPORES

NATHAN M. URBAN
Department of Physics, Pennsylvania State University
University Park, Pennsylvania 16802-6300, USA
nurban@phys.psu.edu

MILTON W. COLE
Department of Physics, Pennsylvania State University
University Park, Pennsylvania 16802-6300, USA
mwc@psu.edu

We describe calculations of the properties of quantum fluids inside nanotubes of various sizes. Very small radius (R) pores confine the gases to a line, so that a one-dimensional (1D) approximation is applicable; the low temperature behavior of 1D ^{4}He is discussed. Somewhat larger pores permit the particles to move off axis, resulting eventually in a transition to a cylindrical shell phase—a thin film near the tube wall; we explored this behavior for H_2. At even larger $R \sim 1$ nm, both the shell phase and an axial phase are present. Results showing strong binding of cylindrical liquids ^{4}He and ^{3}He are discussed.

Keywords: Quantum fluids; nanotubes; phases.

1. Introduction

The discovery of carbon nanotubes has provided a playground for theoretical physics analogous to that (~1970) based on the discovery of adsorption on flat, well-characterized surfaces. In the former case, excitement arises from the tantalizing possibility that one-dimensional (1D) physics can be tested by studying adsorbed gases near nanotubes, just as studies of monolayer films provided tests of 2D physics.

Many groups have explored the properties of quantum fluids on the external surface of nanotube bundles and the interstitial regions within the bundles, stimulated by both the intriguing geometry and several experimental results.[1–9] Our group has predicted several novel phase transitions for interstitial quantum fluids, including a high temperature (liquid-vapor) condensation and a BEC that exhibits 4D (!) thermodynamic properties.[2,3] This paper summarizes instead diverse results concerning quantum fluids *inside* single nanotubes, obtained with a variety of methods. These studies are far from complete, with significant theoretical questions yet to be answered. The following section discusses the case of ^{4}He in 1D, with applications to small radius (R) pores. Section 3 explores the behavior of absorbed H_2 as R increases, so that the 1D approximation breaks down. Section 4 discusses the nature of films in large pores ($R \sim 1$ nm), where one encounters both a "cylindrical shell" phase of the film on the surface and a so-called "axial" phase, which is very much like the 1D system in the small R case.

Throughout this paper we omit the details of both the adsorption potential and the techniques used in the calculations. Those can be found in existing publications, as well as a thesis and longer article currently being drafted.[4] Our emphases are new results, qualitative behavior, and significant open questions.

2. Behavior in the 1D Limit

The 1D ^{4}He system is interesting for several reasons. One is that the liquid is barely bound (by about 1.7 mK) and very rarefied (mean spacing about 2.7 nm!); in fact, the venerable Lennard-Jones (LJ) pair potential is too weakly attractive to produce this bound state.[5,6] A related fact is that the threshold interaction strength for 1D binding of the liquid state coincides with that of the 1D dimer.[6] Recently, L.W. Bruch and C. Carraro (private communications) have shown that the cohesive energy of the 1D many-body system has the 1D dimer binding energy as a lower limit, the two energies possibly coinciding. There is a closely related, intriguing aspect to the dimer problem. Consider the three-dimensional (3D) dimer problem, focusing on the s-wave channel. The radial Schrödinger equation for that problem coincides with the Schrödinger equation for the 1D dimer. A key difference between $D = 1$ and $D = 3$ is the requirement that the wave function $\psi(r)$ vanish at 3D separation $r = 0$. However, this difference is inconsequential for a 1D system involving hard-core interactions. Hence, the 3D wavefunctions and spectra coincide, at least for the s-channel, with those of the 1D problem. Putting all of this information together, it might be "expected" theoretically that all of the three energies agree, with the common value 1.7 mK.

Recently, we have studied the thermal properties of 1D liquid ^{4}He, using the path integral method.[9] If one were to anticipate the behavior theoretically, one might treat the system with the Landau model, based on elementary excitations above the ground state. In the limit that the low-lying excitations are phonons, with $T = 0$ speed $s(\rho)$ at density ρ, this model predicts that the energy per particle ΔE (relative to the ground state energy) satisfies

$$\Delta E(T) = F(k_B T)^2 \,, \tag{1}$$

$$F = \frac{\pi/6}{\hbar s \rho} \,. \tag{2}$$

Preliminary results of the path integral calculations, in Fig. 1, are consistent with this prediction at low T and high ρ; e.g., values of the coefficient F in Eq. 1 fit to the data at $\rho = 2.5/\text{nm}$ (and higher) agree with the value predicted by Eq. 2. At $\rho = 2/\text{nm}$ and below, instead, the values of F begin to disagree and the departure from the quadratic dependence occurs at lower T. The latter is not surprising because the T^2 dependence of ΔE is expected only far below the 1D Debye temperature, $\Theta = \hbar s\rho/(k_B\pi)$. Since $\Theta \sim 8$ K at $\rho = 2/\text{nm}$ and $\Theta \sim 30$ K at $\rho = 2.5/\text{nm}$, different behavior of ΔE is expected for the two densities at 5 K.

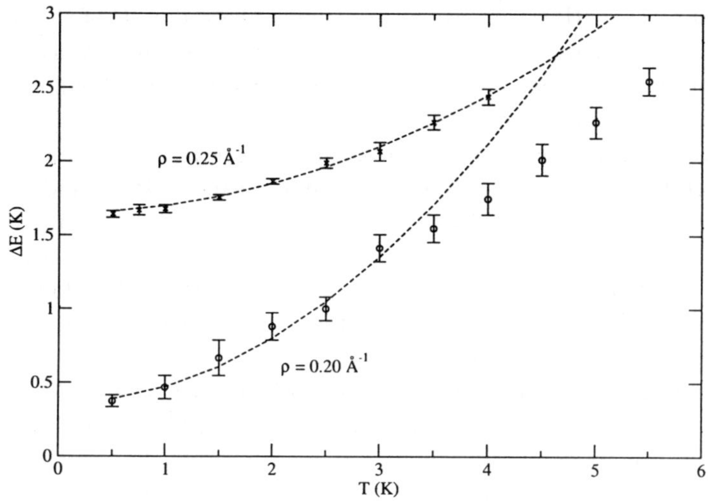

Fig. 1. Energy per molecule of 1D H_2 at $\rho = 2.0$ and 2.5/nm. Quadratic curves are based on the Landau-like model. (From Ref. 4.)

Departure from simple model behavior at *very* low ρ is also expected on general grounds, since the speed of sound becomes imaginary below the spinodal density, $\rho \sim 0.245$/nm (compared to an equilibrium density 0.36/nm). The phonon theory has no meaning at such low ρ. The observed departure from Landau-like predictions occurs at much higher density than that, however. This empirical behavior remains to be understood. At very low density, one might be inclined to use a virial expansion. Šiber has evaluated the first virial correction to the 1D ^{4}He system. His results show significant departure from the classical specific heat (1/2 Boltzmann per atom) at relatively high T, even for very low density. This has nothing to do with exchange, which does not contribute because of the hard core repulsion.[10] One expects higher order virial terms to contribute in addition, making this system particularly interesting to explore experimentally.

3. Spreading Away From the Axis

The $T = 0$ properties of 1D H_2 were first calculated by Boronat, Gordillo, and Casulleras.[7] Here, we report preliminary results of finite T behavior as the radius is increased from a very small value to $R = 0.7$ nm.

Assuming a Lennard-Jones interaction between the gas adsorbate molecules and the inner pore surface, as adsorbate density is increased the gas will begin to adsorb onto the pore as a cylindrical shell film, at a distance from the surface on the order of the LJ parameter σ. However, we expect that if the radius R of the pore is less than this characteristic distance, the gas-surface repulsion will heavily restrict the transverse motion of the gas, and the two-dimensional (2D) shell will collapse into a 1D line of molecules on the pore axis.

Geometrically, this axial compression at small pore radii exhibits itself in the nanopore potential as a transition from a minimum near $r \sim R-\sigma$ and a maximum at $r = 0$ (i.e., an off-axis potential well) when $R \gtrsim \sigma$, to a simple minimum at $r = 0$ when $R \lesssim \sigma$ (an on-axis potential well). For the case of an infinitesimally thin tube, this transition is analytically known to occur at $R \simeq 1.212\sigma$.[11] We do not know an analogous analytic result for the case studied here, a pore within bulk material, but numerically the transition point is similar.

For concreteness, we chose to study pores in MgO glass. The H_2-MgO LJ interaction parameters are given by $\sigma = 2.014$ Å, $\epsilon = 45.91$ K.[12] The H_2-H_2 interaction potential was taken to be of the Silvera-Goldman form.[13] The system was studied numerically via the path integral Monte Carlo algorithm.[14,15] Temperatures between $T = 0.5$ K and 3 K, radii between $R = 2$ and 7 Å, and densities of $\rho = (2.55 \times 10^{-3}/\text{Å})/R^2$ and $\rho = (7.13 \times 10^{-2}/\text{Å})/R^2$. (The densities correspond to choosing a fixed number $N = 1$ or $N = 28$ of particles in a simulation cell of radius R and fixed length 125 Å, with periodic longitudinal boundary conditions.)

The results obtained were very similar for both densities and all ranges of temperature and radii studied. A representative plot of the H_2 radial density distribution and the pore potential is given in Fig. 2 for radii between $R = 2.5$ and 3.25 Å.

The lowest pore size depicted, $R = 2.5$ Å, is just beyond the transition point (2.45 Å for the tubular case) to an off-axis minimum, where the H_2 should switch from a 1D axial phase to a 2D cylindrical shell. This is reflected in the potential energy curve (Fig. 2), which is very shallow for $R = 2.5$ Å. Correspondingly, we see that at this radius the H_2 molecules are still concentrated on the axis. However, with a remarkably small increase in pore radius, by 0.25–0.5 Å, the distribution rapidly shifts to peak off axis, signifying the onset of the shell phase. It is interesting that the transition from axial to shell phases coincides in R with the shift of the classical potential minimum from on- to off-axis, even in the presence of quantum effects and interparticle interactions. One expects larger effects of interactions at higher density.

It should be noted that the radii depicted in Fig. 2 are unphysically small for MgO pores of realistic size. We chose to study them despite this problem in order to theoretically study an axial to cylindrical shell transition in a pore geometry, which can only occur for small pores which are highly confining. In addition, there are some systems of larger radii in which the first layer to be adsorbed becomes a rigidly bound film, confining a fluid phase to a very localized vicinity of the pore axis.[16] Such possibilities help to justify the study of quasi-1D fluids.

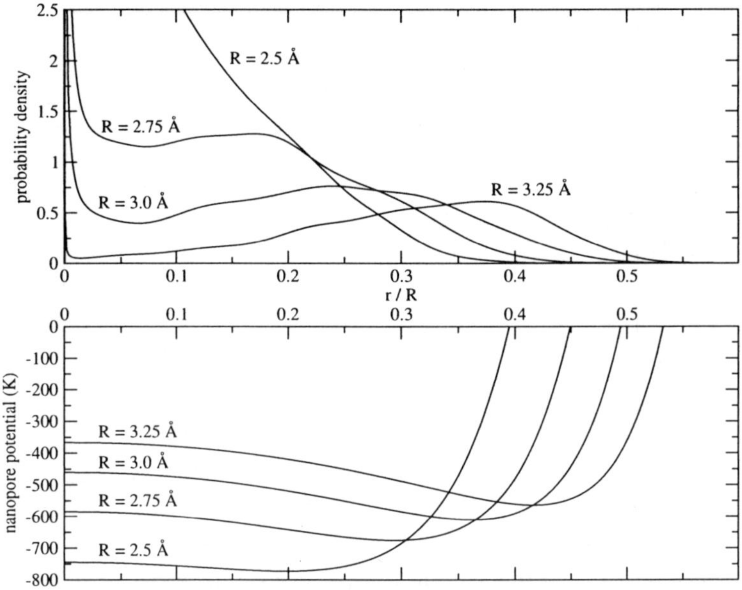

Fig. 2. H_2 radial probability density distribution (at $T = 0.5$ K, $\rho = (2.55 \times 10^{-3}/\text{Å})/R^2$) and MgO pore potential, for pore radii $R = 2.5$, 2.75, 3.0, and 3.25 Å, as functions of dimensionless radius r/R. (Radial densities near $r = 0$ are exaggerated due to finite size effects after normalizing the 3D radial distribution by $1/(2\pi r)$ to obtain the 1D radial density.)

4. Large Pore Phenomena

Relatively few simulation studies have been carried out for quantum fluids in "typical" size nanotubes, $R \sim 0.6$ to 1 nm. Path integral calculations of Gatica *et al.*[17] reported the behavior of H_2 over a limited range of R at $T = 10$ K. One of the more interesting phenomena is the pore-filling transition, shown for H_2 in Fig. 3.[17] At low μ, all of the molecules are localized within a thin layer, at $r = 0.3$ nm, located near the distance of closest approach to the nanotube. Above a threshold value of μ, the axial phase appears and grows rapidly with increasing μ. This axial phase can be thought of as an independent 1D phase. A close analogy is the behavior of the second layer film of He or H_2 on the surface of graphite, often treated by assuming that the only role of the first layer is to provide a holding potential.[18]

Analytical and numerical problems associated with matter in cylindrical geometry are often computationally demanding, motivating the use of simplifying models that (we hope) capture the essential physics. Recently, we have explored such a description of the shell phase; the model assumes that all particles are constrained to lie on a cylindrical surface, $r = R$. One might expect that by varying R between $R = 0$ and $R = \infty$, one interpolates smoothly between 1D and 2D behaviors. This is naïve; instead, an intriguing "anomaly" arises: a significant enhancement of the binding occurs when the diameter of the cylinder, $d = 2R$, is comparable to the equi-

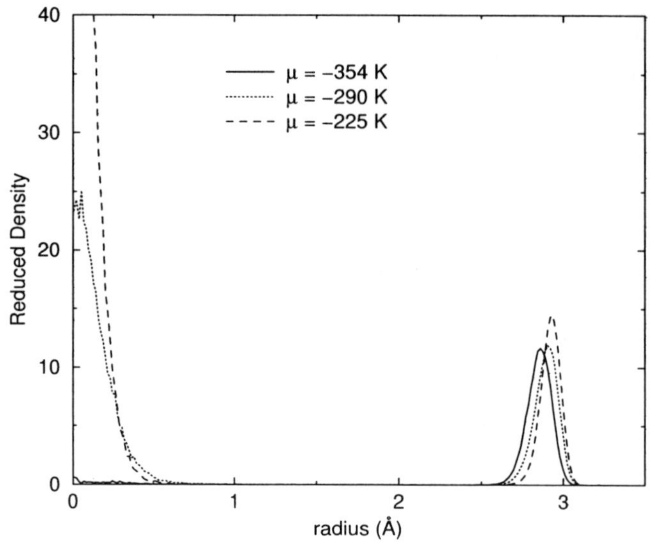

Fig. 3. Pore-filling transition of H_2 in a tube of radius 0.6 nm, from Gatica *et al.*[17] Results are densities as a function of r at 10 K and indicated chemical potentials.

librium separation $r_{\min}$ in the pair potential. The condition $d \sim r_{\min}$ corresponds (for LJ interactions) to $\sigma/R \sim 1.7$. Indeed, this argument does explain the R value corresponding to the maximum cohesive energy (seen in Fig. 4) of the "cylindrical liquids" ^{4}He and ^{3}He. The ^{3}He case is perhaps the most dramatic, because its liquid state does not exist in either 1D or 2D, while the cylindrical liquid ^{3}He is found to have cohesive energy as high as 1.26 K for $R = 0.18$ nm.

These are variational results, obtained with Jastrow and Slater-Jastrow wave functions for ^{4}He and ^{3}He, respectively. Qualitatively similar, enhanced binding behavior was found for related problems involving similar binding problems on a cylindrical surface: He or H_2 dimers, a crystalline lattice confined to a cylinder, and the virial coefficient of a classical fluid.[19] The origin of this general behavior is that two interacting particles can maximally exploit a divergent "specific area" when the interatomic separation is favorable. This occurs when the particles are on opposite sides of the cylinder, with separation $|\mathbf{r}_2 - \mathbf{r}_1| = d = r_{\min}$. The specific area is defined as the cylindrical area residing within a separation interval $[r, r+dr]$, divided by dr. For completeness, we note that analogous results for the dimer binding have been found by Aichinger *et al.*, using both the simple model of confinement on a cylinder and more realistic study of dimers inside a nanotube.[20] The optimal binding value found for R is very different in the two cases.

Acknowledgments

We are grateful to our collaborators (Massimo Boninsegni, Louis Bruch, Mercedes Calbi, Carlo Carraro, Silvina Gatica, and Milen Kostov) for many contributions to

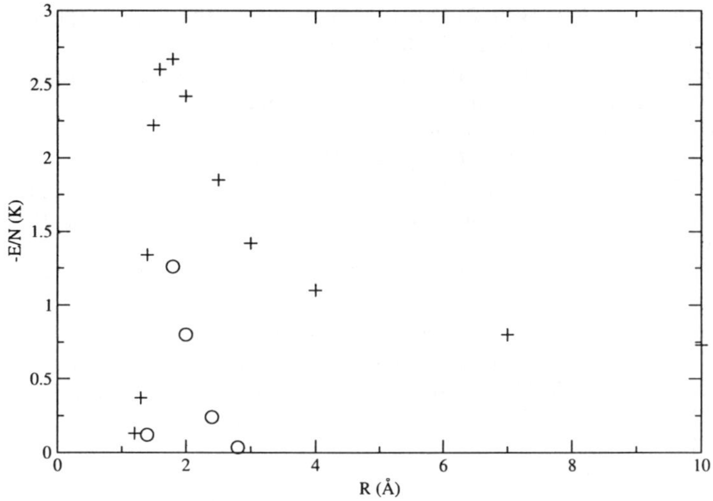

Fig. 4. Cohesive energy per atom of cylindrical liquids, as a function of R. Pluses are ^{4}He data (from Kostov *et al.*[19]) and circles are ^{3}He data (from C. Carraro, unpublished).

this project and to NSF for its support of this research.

References

1. S. Ramachandran, T.A. Wilson, D. Vandervelde, D.K. Holmes, and O.E. Vilches, *J. Low Temp. Phys.* **134**, 115 (2004); K.A. Williams, P.C. Eklund, M.K. Kostov, and M.W. Cole, *Phys. Rev. Lett.* **88**, 165502 (2002); J.C. Lasjaunias, K. Biljakovic, J.L. Sauvajol, and P. Monceau, *Phys. Rev. Lett.* **91**, 025901 (2003); J.V. Pearce, M.A. Adams, O.E. Vilches, M.R. Johnson, and H.R. Glyde, *Phys. Rev. Lett.* **95**, 185302 (2005).
2. M.M. Calbi, F. Toigo, and M.W. Cole, *Phys. Rev. Lett.* **86**, 5062 (2001); C. Carraro, *Phys. Rev. Lett.* **89**, 115702 (2002).
3. F. Ancilotto, M.M. Calbi, S.M. Gatica, and M.W. Cole, *Phys. Rev.* **B70**, 165422 (2004).
4. N.M. Urban, M. Boninsegni, and M.W. Cole, unpublished.
5. M. Boninsegni and S. Moroni, *J. Low Temp. Phys.* **118**, 1 (2000); J. Boronat, M.C. Gordillo and J. Casulleras, *ibid.* **126**, 199 (2002).
6. E. Krotscheck, M.D. Miller, and J. Wojdylo, *Phys. Rev.* **B60**, 13028 (1999); E. Krotscheck and M.D. Miller, *Phys. Rev.* **B60**, 13038 (1999).
7. M.C. Gordillo, J. Boronat, and J. Casulleras, *Phys. Rev.* **B61**, R878 (2000).
8. M. Boninsegni, S. Lee and V.H. Crespi, *Phys. Rev. Lett.* **86**, 3360 (2002).
9. M. Boninsegni, N. Prokof'ev, B. Svistunov, *Worm Algorithm for Continuous-space Path Integral Monte Carlo Simulations*, to appear in *Phys. Rev. Lett.* (cond-mat/0510214).

10. L.W. Bruch, *Phys. Rev.* **B70**, 016501 (2004); A. Šiber, *Phys. Rev.* **B70**, 016502 (2004).
11. G. Stan and M.W. Cole, *Surf. Sci.* **395**, 280 (1998).
12. G. Vidali, G. Ihm, H.Y. Kim, and M.W. Cole, *Surf. Sci. Rep.* **12**, 133 (1991).
13. I.F. Silvera and V.V. Goldman, *J. Chem. Phys.* **69**, 4209 (1978).
14. D.M. Ceperley, *Rev. Mod. Phys.* **67**, 279 (1995).
15. M. Boninsegni, *J. Low Temp. Phys.* **141**, 27 (2005).
16. J. Taniguchi, A. Yamaguchi, H. Ishimoto, H. kegami, T. Matsushita, N. Wada, S.M. Gatica, M.W. Cole, F. Ancilotto, S. Inagaki, and Y. Fukushima, *Phys. Rev. Lett.* **94**, 065301 (2005).
17. S.M. Gatica, G. Stan, M.M. Calbi, J.K. Johnson, and M.W. Cole, *J. Low Temp. Phys.* **120**, 337 (2000).
18. M. Pierce and E. Manousakis, *Phys. Rev. Lett.* **81**, 156 (1998).
19. M.K. Kostov, M.W. Cole, G.D. Mahan, C. Carraro and M.L. Glasser, *Phys. Rev.* **B67**, 075403 (2003); M.M. Kostov, M.W. Cole, and G.D. Mahan, *Phys. Rev.* **B66**, 075407 (2002); M.M. Calbi, S.M. Gatica, M.J. Bojan, and M.W. Cole, *Phys. Rev.* **E66**, 061107 (2002).
20. M. Aichinger, S. Kilic, E. Krotscheck, and L. Vranje, *Phys. Rev.* **B70**, 155412 (2004); S. Kilic, E. Krotscheck, and R. Zillich, *J. Low Temp. Phys.* **116**, 245 (1999).

AN EXTENDED CONCEPT OF UNIVERSALITY IN A STATISTICAL MECHANICS MODEL

CARLOS WEXLER

Department of Physics and Astronomy, University of Missouri, Columbia, Missouri, 65211, USA
wexlerc@missouri.edu

CINTIA M. LAPILLI

Department of Physics and Astronomy, University of Missouri, Columbia, Missouri, 65211

PETER PFEIFER

Department of Physics and Astronomy, University of Missouri, Columbia, Missouri, 65211

Nature uses phase transitions as powerful regulators of processes ranging from climate to the alteration of phase behavior of cell membranes, building on the fact that thermodynamic properties of a solid, liquid, or gas are sensitive fingerprints of intermolecular interactions. The only known exceptions from this sensitivity are critical points, where two phases become indistinguishable and thermodynamic properties exhibit universal behavior: systems with widely different intermolecular interactions behave identically. Here we report a new, stronger form of universality, in which different members of a family of two-dimensional systems—the discrete p-state clock model—behave identically both near and away from critical points, if the temperature exceeds a value T_{eu} ('extended universality). We show that all thermal averages are identical to those of the continuous planar rotor model ($p = \infty$) above T_{eu}, that phase transitions above T_{eu} are identical to the Berezinskii-Kosterlitz-Thouless (BKT) transition, and that transitions below T_{eu} are distinctly non-BKT. The results generate a comprehensive map of the three phases of the model and, by virtue of the discrete rotors behaving like continuous rotors, an emergent symmetry, not present in the Hamiltonian. This symmetry, or many-to-one map of intermolecular interactions onto thermodynamic states, demonstrates previously unknown limits for macroscopic distinguishability of different microscopic interactions.

Keywords: phase transitions, universality, magnetism

1. Introduction

Universality near a critical point states that the critical exponents, describing how thermodynamic observables go to zero or infinity at the transition, depend only on the range of interaction, symmetries of the Hamiltonian, and dimensionality of the system. It arises as the system develops fluctuations, e.g., droplets and bubbles, of all sizes near the critical point, which wash out the details of the interaction and render the system scale-invariant.[1–5] By contrast, extended universality occurs over a whole range of temperatures, yields not just identical critical exponents but identical values of *all observables ('collapse of observables')* for different systems, and is not contingent on large fluctuations. To the best of our knowledge, no such collapse of thermodynamic observables has been observed before.

The p-state clock model, also known as p-state vector Potts model or Z_p model,[6] in two dimensions, describes how dipoles or spins, $\mathbf{s}_i$, align/misalign at low/high energy. The Hamiltonian reads

$$H_p = -J_0 \sum_{\langle i,j \rangle} \mathbf{s}_i \cdot \mathbf{s}_j = -J_0 \sum_{\langle i,j \rangle} \cos(\theta_i - \theta_j)\,, \tag{1}$$

where each spin can make p *discrete* angles $\theta_i = 2\pi n_i/p$, $n_i \in \{1, \ldots, p\}$, the sum is over nearest neighbors, and the coupling is ferromagnetic, $J_0 > 0$. The p orientations are imposed by the underlying substrate or molecular shapes. The model interpolates between spin up/down of the Ising model[7] ($p = 2$) and the continuum of directions of the planar rotor, or XY, model[8–12] ($p = \infty$). It has been extensively studied for how the strong phase transition in the Ising model, with broken symmetry in the low-temperature ferromagnetic phase and no broken symmetry in the high-temperature paramagnetic phase, gives way to the weak, vortex-unbinding BKT transition,[9–12] without broken symmetry, in the rotor model. Elitzur *et al.*[13] showed that the model has a rich phase diagram: for $p = 2, 4$ it belongs to the Ising universality class; for $p > 4$ three phases exist—a low-temperature ordered and a high-temperature disordered phase, like in the Ising model, plus a quasi-liquid intermediate phase. Duality transformations[13,14] and renormalization group (RG)[15,16] gave insight into the phases in terms of a related model,

$$H_{\{h_p\}} = -J_0 \sum_{\langle i,j \rangle} \cos(\theta_i - \theta_j) + \sum_i \sum_p h_p \cos(p\,\theta_i)\,, \tag{2}$$

where the θ_i's are continuous and the h_p's are symmetry-breaking fields, mimicking the constraint to p spin directions in the clock model. The clock model is obtained by letting $h_p \to \infty$ for a selected p. José *et al.*[16] showed, in a self-dual approximation of (2), that the fields were relevant for $p < 4$ and irrelevant for $p > 4$. But (1) is not self-dual for $p > 4$, and RG approximations examining the effect of the discreteness of the angular variables are delicate near $p = 6$. As a result, the transition points of (1) in the three-phase region are not precisely known.

2. The Phase Diagram

We establish the phase diagram, the collapse of observables, and the associated temperature T_{eu} by Monte Carlo (MC) simulations.[17] The simulations were performed on a square lattice of size $N = L \times L$ with $L =$ 8–72 and periodic boundary conditions. Averages involved sampling of 10^5–10^6 configurations, with equilibration runs of $p\times$(1,000–5,000) MC steps, a step being one attempt to change every spin.

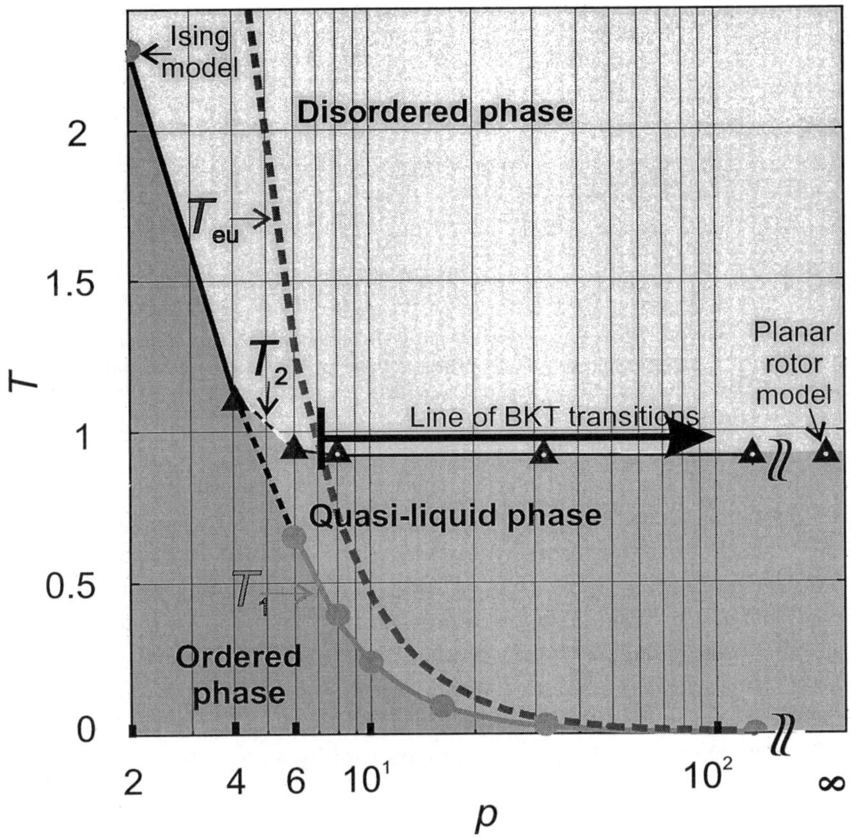

Fig. 1. Phase diagram of the p-state clock model. The Ising model, $p = 2$, exhibits a single second-order phase transition, as does the $p = 4$ case, which is also in the Ising universality class. For $p > 4$, a quasi-liquid phase appears, and the transitions at T_1 and T_2 are both second-order. The line $T_{eu}(p)$ separates the phase diagram into a region where the thermodynamic observables do depend on p, below T_{eu}, and a region where their values are p-independent, above T_{eu} (collapse of observables, extended universality). For $p \geq 8$, we observe $T_{eu} < T_2 = T_{BKT} \simeq 0.89$. Throughout, temperatures are in units of J_0/k_B, where k_B is Boltzmann's constant, and energies are in units of J_0 The separation into two regions—with and without collapse of observables, resolves long-standing questions of similarities and differences between the clock model and planar rotor model (see text).

Figure 1 summarizes our results. The Ising model shows the expected phase transition at $T_c^{\text{Ising}} = 2/\ln[1+\sqrt{2}] \simeq 2.27$. The case $p = 4$ also shows a single transition, at $T_c = T_c^{\text{Ising}}/2 \simeq 1.13$. Most interesting is the case $p > 4$. It hosts the two transitions predicted by Elitzur *et al.*,,[13] and the collapse of observables at $T > T_{eu}$ At T_{eu} the system switches from a p-dependent state ($T < T_{eu}$: discrete symmetry) to a state indistinguishable from $p = \infty$ ($T > T_{eu}$: continuous symmetry). For $p \leq 4$ there is no collapse.

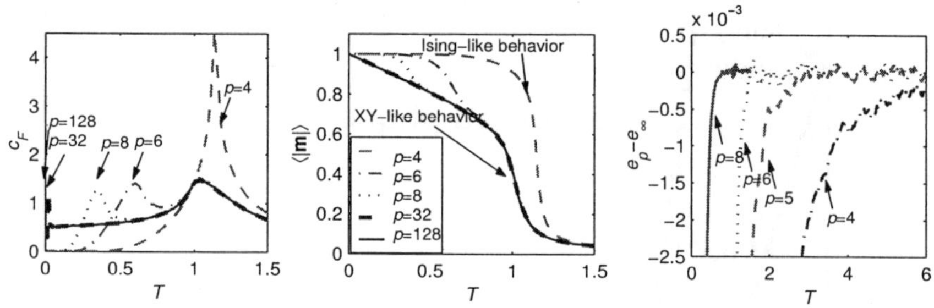

Fig. 2. Thermal properties. Heat capacity (left panel), magnetization (center panel), and difference of internal energy per spin relative to the planar rotor, or XY, model (right panel). The data correspond to a system size of $L = 72$ ($N = 5{,}184$ spins). All curves coalesce above T_{eu} for $p \geq 5$ (collapse of observables, extended universality).

We characterized the transitions as follows.[18] We use Binder's fourth-order cumulants[17] in magnetization, $U_L \equiv 1 - \frac{1}{3}\langle m^4 \rangle / \langle m^2 \rangle^2$, and energy, $V_L \equiv 1 - \frac{1}{3}\langle e^4 \rangle / \langle e^2 \rangle^2$. The high-temperature transition, T_2 is obtained from the fixed point of U_L. The latent heat, proportional to $\lim_{L\to\infty}[\frac{2}{3} - \min_T V_L]$, vanishes, signaling a second-order transition. The low-temperature transition, T_1, is obtained from the temperature derivative of the magnetization, $\partial\langle|\mathbf{m}|\rangle/\partial T$, and $\partial U_L/\partial T$, which diverge as $L \to \infty$ (thermodynamic limit). Finite-size scaling (FSS) of the minima of these derivatives yields $T_1 = \lim_{L\to\infty} T_{1,L}$. We find $T_1 = 4\pi^2/(\tilde{T}_2 p^2)$, with $\tilde{T}_2 = 1.67 \pm 0.02$: the ordered phase disappears rapidly as $p \to \infty$.

3. Collapse of Thermodynamic Observables

Figure 2 shows selected thermodynamic observables: the heat capacity, c_F, and magnetization, $\langle \mathbf{m} \rangle$, per spin. The Ising-like behavior for $p = 4$ and the three phases for $p > 4$ are evident. Figure 3 proves the collapse of thermodynamic observables: c_F and $\langle \mathbf{m} \rangle$ are manifestly p-independent for $p > 4$ and $T > T_{eu}$ with

$$T_{eu} = \left(\frac{2\pi}{p}\right)^2 \frac{1}{T_{BKT}}, \tag{3}$$

where $T_{BKT} \simeq 0.89$ and the internal-energy differences abruptly vanish at $T = T_{eu}$.

The specific form of $T_{eu}(p)$ can be understood as follows. (i) The large-p, small-$(\theta_i - \theta_j)$ expansion of (1) yields a characteristic temperature, $\sim (2\pi/p)^2$ such that all averages become p-independent whenever $T/(2\pi/p^2) \gg 1$ implying an asymptotic collapse of observables. (ii) Elitzur *et al.*[13] noted that discreteness of the angles θ_i becomes irrelevant for the critical properties of (1), for sufficiently large p, implying the collapse of observables at *critical points*, T_2. (iii) A similar irrelevance of the discreteness of angles, imposed by $h_p \to \infty$ was observed[16] for (2), subject to $T > (2\pi/p)^2/T_k$ where $T_k = 1.35$ is the BKT point of the self-dual approximation of (2). (iv) For the *full* collapse of observables in the clock model, these partial results suggest that a necessary condition for collapse is $T > (2\pi/p)^2/T_{BKT}$. The fit of our data for $T_{eu}(p)$[18] yielding (3), validates this expectation and shows that the condition is necessary and sufficient.

The collapse/non-collapse above/below the curve $T_{eu}(p)$ makes far-reaching predictions for the transitions T_1 and T_2, which we now test. We begin with T_2. We observe that $T_2 > T_{eu}$ for $p \geq 8$, which implies that the transition T_2 must be BKT for in that case Previous work advanced only the ***plausibility*** of such universality. To test our assertion, beyond the equality $T_2 = T_{BKT}$ we equate BKT behavior to the following planar-rotor properties:[11,12] (i) discontinuous jump to zero of the helicity modulus, $\Upsilon(T^-_{BKT}) = 2T_{BKT}/\pi$; (ii) exponentially diverging correlation length, $\xi \sim \exp[c/\sqrt{T - T_{BKT}}]$; (iii) temperature-dependent power-law decay of two-point correlation functions, with exponent $\eta(T_{BKT}) = 1/4$; (iv) decay of the magnetization, with exponent $\tilde{\beta} = 3\pi^2/128$.[19]

Our simulations fully confirm these properties at T_2 and $p \geq 8$.[18] We illustrate this for the discontinuity of the helicity modulus. Following the Minnhagen-Kim stability argument,[20] we evaluate the change in free energy when a twist Δ is applied to the spins:

$$f = \frac{\Upsilon}{2}\Delta^2 + \frac{\Upsilon_4}{4!}\Delta^4 + \cdots . \tag{4}$$

Figure 3 shows Υ and Υ_4 (fourth-order helicity) as a function of T and system size L for $p = 8$. At T_2, $\lim_{L\to\infty} \Upsilon_4 < 0$, so if $\lim_{L\to\infty} \Upsilon$ went to zero continuously as $T \to T_2^-$, the free energy would turn negative and the system would become unstable as $T \to T_2$ This contradiction implies that Υ goes to zero discontinuously. The same result is obtained for all $p \geq 8$, as predicted by the collapse of observables. Conversely, the non-collapse of observables at T_2 for $p < 8$ suggests that the transition at $p = 6$ differs from BKT. This is indeed the case: Υ does not vanish, and Υ_4 converges to zero as $L \to \infty$. The nonzero helicity modulus and its continuity at T_2 make the transition manifestly non-BKT, according to our criterion. Other critical properties computed at T_2 also differ from the BKT values.[18] Furthermore, visibly $T_2 > T_{BKT}$ (Fig. 1). Thus, contrary to prior conjectures that the transition at T_2 and $p = 6$ is BKT-like,[13,16,21,22] we find that it differs significantly from BKT. Specifically, our analysis shows that a twist at T_2^+ costs much more energy, $f = \frac{1}{2}\Upsilon\Delta^2 + \mathcal{O}(\Delta^6)$, than in the BKT case, $f = \mathcal{O}(\Delta^6)$.

We turn to the low-temperature transition, T_1, which also has been argued to be BKT-like for $p \geq 6$.[22,23] The non-collapse of observables at T_1, for all finite p, suggests, and our simulations substantiate,[18] that this transition, too, differs significantly from BKT. E.g., our FSS analysis of the temperature derivatives of the magnetization and its fourth-order cumulant gives a power-law dependence $T_{1,L} = T_1 + \mathcal{O}(1/L)$,[18] as expected when the low-temperature phase exhibits long-range order. In contrast, BKT would give $T_{1,L} = T_1 + \mathcal{O}[(\ln L)^{-2}]$.[19,22,23]

Thus the collapse of thermodynamic observables, and the associated universality away from critical points, has remarkable consequences on the phase diagram of the clock model. When present, $T > T_{eu}$, the collapse causes the spins to lose

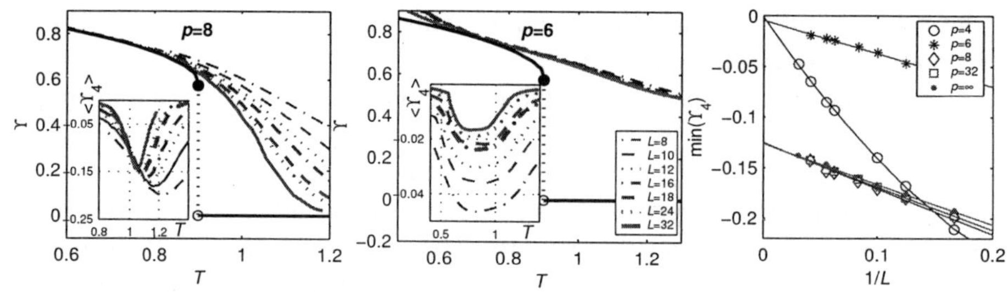

Fig. 3. Helicity modulus, Υ, and fourth-order helicity, Υ_4, for $p = 8$ (left panel) and $p = 6$ (center panel) across the phase transition T_2. The bottom curve in the two panels, for reference, is the modulus for the planar rotor, which jumps from $2T_{BKT}/\pi$ (full circle) to zero at $T = T_{BKT}$. For all $p \geq 8$, Υ vanishes in the thermodynamic limit ($L \to \infty$) for $T > T_{BKT}$ (not shown). Right panel: The extrapolation of Υ_4 to the thermodynamic limit yields two classes of results: Υ_4 converges to the universal value -0.126 ± 0.005 for $p \geq 8$, and to zero for $p = 4, 6$. This implies that Υ undergoes a jump at T_2 if and only $p \geq 8$ (see text), in agreement with the fact that $T_2 > T_{eu}$ for $p \geq 8$.

their identity as discrete-symmetry variables and become indistinguishable from the continuous-symmetry variables of the planar rotor. At critical points, T_2 for $p \geq 8$, it guarantees that all critical properties are *identical* to those of the BKT transition. Away from critical points, it guarantees that the quasi-liquid phase and disordered phase are *identical* to those of the planar rotor. When the collapse is absent, $T < T_{eu}$, the spins retain their discrete symmetry, and all critical points, T_2 for $p < 8$ and T_1 for $p < \infty$, are distinctly non-BKT. In this case, the critical properties vary with the distance from the onset of extended universality, $T_{eu} - T_2$, or $T_{eu} - T_1$: if this distance is small, the transition is close to BKT, even if it is not from quasi-long-range order to disordered; if the distance is large, the transition is vastly different.

4. Extended Universality as an Emergent Property

Just as universality at a critical point is accompanied by invariance of the system under the scale transformation $\mathbf{r}_i \to \lambda \mathbf{r}_i$, even though the Hamiltonian has no such symmetry, extended universality is accompanied by invariance under the transformation $\theta_i \to \theta_i + \alpha$, α arbitrary, even though (1) is invariant only under discrete rotations. Thus both universalities are generated by an emergent symmetry, not present in the Hamiltonian. What makes extended universality different is that the symmetry is present over a *whole range of temperatures, not just at the critical point.* This suggests that thermal averages at $T > T_{eu}$ should be expressible in terms of a coarse-grained Hamiltonian invariant under rotation by α, and that this representation is exact. The construction of such a representation, and accordingly the origin of extended universality, is an open problem. Related questions are: If observables show p-dependent ferromagnetic ordering below T_1, but all p-dependence is lost above T_{eu}, what is the nature of the region $T_1 < T < T_{eu}$? Is the transition from uncollapsed to collapsed at T_{eu}, at fixed p, a phase transition in itself? If so, what is the nature of the nonanalyticity at T_{eu}?

5. Experimental Effects of Extended Universality

A range of experimental systems, from thin magnetic films to monolayers of adsorbed molecules,[24] have been modeled by dipoles restricted to p orientations. Our results imply that p-state characteristics can be observed only at low temperatures, $T < T_{eu}(p)$. On the high side, $T > T_{eu}(p)$, we expect the results to be relevant for vortex dynamics in membranes[25] and supercritically adsorbed gases in fuel storage.[26,27] The collapse of observables may be studied directly in a monolayer of rotaxane, a molecular wheel threaded by a molecular axle, on a single-crystal surface.[28] If the wheels have p-fold symmetry, modulo a polar group mediating the interaction between neighboring wheels, their dynamics should be governed by (1). E.g., for $p = 8$, the hallmark of the collapse will be that the heat capacity peaks at $T = 0.37$ and coalesces with the $p = \infty$ curve at $T = T_{eu} = 0.69$. The collapse may also induce a change in the NMR signal of the polar group, as the group switches from a discrete rotor at $T < T_{eu}$ to a continuous rotor at $T > T_{eu}$.

Acknowledgements

We thank H. Fertig, G. Vignale, H. Taub, and K. Knorr for useful discussions. Acknowledgment is made to the University of Missouri Research Board and Council, to the Petroleum Research Fund, and to the National Science Foundation, for support of this research.

References

1. L.D. Landau, and E.M. Lifshitz, *Statistical Physics* (MIT Press, London, 1966).
2. L.P. Kadanoff *et al.*, *Rev. Mod. Phys.* **39**, 395 (1967).
3. A.A. Migdal, *Sov. Phys.* JETP **42**, 743 (1976).
4. L.P. Kadanoff, in *Proceedings of 1970 Varenna Summer School on Critical Phenomena*, ed. M.S. Green (Academic Press, New York, 1970).
5. See e.g. E. Wigner, *Physics Today* March 1964, p. 34.
6. R. Potts, *Proc. Camb. Phil. Soc.* **48**, 106 (1952).
7. L. Onsager, *Phys. Rev.* **B65**, 117 (1944).
8. N.D. Mermin, and H. Wagner, *Phys. Rev. Lett.* **17**, 1133 (1966).
9. V.L. Berezinsky, *Sov. Phys.* JETP **32**, 493 (1970).
10. J.M. Kosterlitz, and D.J. Thouless, *J. Phys.* **C6**, 1181 (1973).
11. J.M. Kosterlitz, *J. Phys.* **C7**, 1046 (1974).
12. D.R. Nelson, and J.M. Kosterlitz, *Phys. Rev. Lett.* **39**, 1201 (1977).
13. S. Elitzur, R.B. Pearson, and J. Shigemitsu, *Phys. Rev.* **D19**, 3698 (1979).
14. R. Savit, *Rev. Mod. Phys.* **52**, 453 (1980).
15. J. Villain, *J. Physique* **36**, 581 (1975).
16. J.V. José, L.P. Kadanoff, S. Kirkpatrick, and D.R. Nelson, *Phys. Rev.* **B16**, 1217 (1977).

17. K. Binder, and D.W. Heermann, *Monte Carlo Simulation in Statistical Physics* (Springer, Berlin, 4$^{\rm th}$ ed., 2002).
18. Details will be published elsewhere.
19. S.T. Bramwell, and P.C.W. Holdsworth, *J. Phys.: Condens. Matter* **5**, L53 (1993).
20. P. Minnhagen, and B.J. Kim, *Phys. Rev.* **B67**, 172509 (2003).
21. M.S.S. Challa, and D.P. Landau, *Phys. Rev.* **B33**, 437 (1986).
22. Y. Tomita, and Y. Okabe, *Phys. Rev. Lett.* **86**, 572 (2001).
23. E. Rastelli, S. Regina, and A. Tassi, *Phys. Rev.* **B69**, 174407 (2004).
24. E.g., S. Faßbender *et al.*, *Phys. Rev.* **B65**, 165411 (2002).
25. M.M. Tomczak *et al.*, *Biophys. J.* **82**, 874(2002).
26. P. Pfeifer *et al.*, *Phys. Rev. Lett.* **88**, 115502 (2002).
27. S. Patchkovskii *et al.*, *Proc. Nat. Acad. Sci.* USA **102**, 104439 (2005).
28. E.g., Y.H. Jang, S.S. Jang, and W.A. Goddard III, *J. Am. Chem. Soc.* **127**, 4959 (2005).

ELECTRONS AND FERMION SYSTEMS

THEORETICAL INVESTIGATION OF 3*d* NANOSTRUCTURES ON Cu SURFACES: THE INFLUENCE OF LOCAL ENVIRONMENT

SONIA FROTA-PESSÔA

Instituto de Física, Universidade de São Paulo CP 66318, 05315-970,
São Paulo, SP, Brazil
sfpessoa@macbeth.if.usp.br

ANGELA B. KLAUTAU

Departamento de Física, Universidade Federal do Pará, Belém, PA, Brazil
aklautau@ufpa.br

Here, we use a first-principles approach, implemented directly in real space, which allows us to investigate the behavior of 3*d* nanostructures deposited onto metallic surfaces. To illustrate the flexibility of the approach, results of the exchange coupling J for 3d dimers on Cu(001) are presented. Calculations indicate that Cr and Mn dimers have antiferromagnetic alignment, while Fe, Co and Ni are stable in the ferromagnetic configuration. We also use the method to investigate the electronic structure of Fe, Co and Ni nanoclusters on Cu(001) surfaces. Our main purpose is to understand, from first principles, how electronic charge around the 3d sites is affected by the changes on the local environment associated with the lower coordination number of surfaces sites. Charge transfers, as well as charge character (*s*, *p* or *d*) around ferromagnetic Fe, Co and Ni clusters are investigated. If we consider a region defined by the volume per atom in the corresponding metal, we find that large charge transfers are present at cluster sites. These transfers are mainly due to a drastic decrease in the number of *s* and *p* electrons around the site, while the number of 3*d* electrons around the site remains practically unchanged.

Keywords: nanostructures; electronic structure; magnetism.

1. Introduction

In the last few years, considerable attention has been devoted to the study of nanostructured magnetic materials, due to their unusual properties and potential technological applications. From granular materials to nanoclusters deposited onto metallic surfaces, the systems are characterized by small regions with different characteristics from the matrix which supports them, breaking the translation symmetry of the original matrix. As the nanostructure gets larger and more complex, the powerful k-space methods used to treat periodic systems become inapplicable or inappropriate and other approaches must be used. A very suitable alternative is the use of similar first-principles methods which are implemented in real-space and thus do not require periodicity. The real-space linear muffin-tin orbit (RS-LMTO-ASA) approach,[1–3] based on the well known LMTO-ASA formalism,[4,5] allows us to perform first-principles self-consistent density functional calculations directly in real space and can be very useful when investigating the properties of nanostructures deposited on metallic surfaces and other complex non-periodic metallic systems.

The RS-LMTO-ASA procedure is very similar to the k-space LMTO-ASA formalism, but the solution of the eigenvalue problem is done directly in real space, with the help of the recursion method.[6] When applied to crystalline (periodic) metallic systems, it gives results which are similar to those obtained by the usual k-space methods. For defects embedded in metallic systems it yields results which are in good agreement with those from the KKR-GF formalism.[1] The RS-LMTO-ASA has two very interesting features which make its application very convenient. It is an order (N) method, whereby the computational effort grows linearly with N, the number of inequivalent atoms considered in the calculation. In most k-space calculations this effort grows as N^3, severely restricting their application to complex systems. The other interesting feature of the method is the fact that both the Hamiltonian and the wave function expansion, written in terms of a basis set of *s*, *p* and *d* electrons, are very similar in form to those used in regular TB parameterized approaches. Therefore, the results obtained using the first-principles RS-LMTO-ASA approach have a very simple physical interpretation in terms of *s*, *p* and *d* electrons at each site. The method can be used in metallic systems to study various properties, such as exchange coupling,[7,8] isomer shift,[9,10] electric field gradients and hyperfine fields,[11] spin and orbital moments,[12,13] spectral densities around surfaces,[14] interface mixing on multilayers,[15] non-collinear magnetism,[16] etc....

Here we use the RS-LMTO-ASA approach to study the electronic structure of 3d-transition metal nanostructures deposited onto Cu(001) surfaces. In Sec. 2 we introduce the method and in Sec. 3 we discuss the free surface. In Sec. 4.1 we study the exchange coupling for 3d dimers on Cu(001), while in Sec. 4.2 we obtain results for Fe, Co and Ni clusters on this surface. Conclusions are presented in Sec. 5.

2. The real space method

The RS-LMTO-ASA is a first-principles, self-consistent scheme, based on the LMTO-ASA formalism[4,5] and the recursion method.[6] It is a linear method, and the solutions are accurate around a given energy E_ν, usually taken at the center of gravity of the occupied part of the *s*, *p* and *d* bands being considered. In the RS-LMTO-ASA scheme we work in the orthogonal representation of the LMTO-ASA formalism,[4,5] and expand the orthogonal Hamiltonian in terms of tight-binding (TB) parameters, neglecting terms of order of $(E - E_\nu)^3$ and higher. Within this approximation the Hamiltonian H is of the TB form, connecting each site to those within a few layers around it. In this context the eigenvalue problem has a simple form,[5,14] which can be solved in real space, applying the recursion method:[6]

$$(H - E)u = 0, \tag{1}$$

$$\Psi_E = \sum_{RL} [\varphi_{l\nu}(r_R) + (E - E_\nu)\dot{\varphi}_{l\nu}(r_R)] Y_L(\hat{r}_R) u_{RL}(E) \tag{2}$$

The functions $\varphi_{l\nu}(r)$ and $\dot{\varphi}_{l\nu}(r)$ are solutions of the Schrödinger equation with quantum numbers $L = (l, m)$ and its first derivative with respect to energy, calculated at energy E_ν. They are defined inside the Wigner-Seitz (WS) sphere associated with site R and taken as zero outside this region. The orthogonal Hamiltonian H can be written in terms of parameters of the tight binding representation as:

$$H = E_\nu + \bar{h} - \bar{h}\bar{o}\bar{h}, \tag{3}$$

where

$$\bar{h} = \bar{C} - E_\nu + \bar{\Delta}^{1/2}\, \bar{S}\, \bar{\Delta}^{1/2}. \tag{4}$$

Here $\bar{h}$ is a Hermitian matrix, the quantities $\bar{C}$, $\bar{\Delta}$ and $\bar{o}$ are potential parameters and $\bar{S}$ is the structure constant in the tight-binding LMTO-ASA representation. To solve the eigenvalue problem in real space we consider a large cluster to simulate the system, and use the recursion method[6] with the Beer-Pettifor terminator[17] to complete the recursion chain. The present calculations were performed within the local spin density approximation (LSDA), using the exchange and correlation potential of von Barth and Hedin.[18]

The RS-LMTO-ASA scheme has been successfully applied to study crystalline systems,[1] impurities and defects in simple[10,12,19] and complex metallic hosts[20] and other systems such as metallic surfaces,[3,14] defects in surfaces[3,13,21] and multilayers.[15] The basic procedure is the same in all cases, but the electrostatic potential VES and the Fermi level must be determined according to the system being studied. A detailed description of how these quantities should be determined in each situation can be found in the corresponding references cited above.

3. The free metallic surface

Here we will discuss the behavior of the charge distribution in the vicinity of a metallic surface. In the present work we will be often mentioning charge transfers and charge character and it is important to define these quantities before we proceed. In the RS-LMTO-ASA approach the wave function is expanded in terms of s, p and d basis functions around each site. The radial basis functions of Eq. (2) are defined within a WS sphere around the corresponding site R. In a monatomic system the volume of the WS sphere is equal to the volume per atom in the system. In a close-packed system, the ASA is a good approximation and one can consider that the system is composed of a superposition of such spheres. We can now define the electronic charge transfer ΔQ for a given site as the excess of electrons (relative to charge neutrality) inside the WS sphere around it. Note that with this definition the charge transfer is negative if the sphere, initially taken as neutral, loses electrons and positive if electrons are transferred into it. The charge within the sphere at site R, consists of electrons with s, p or d character and the amount of each character ($L = s, p$ or d) depends on the coefficients u_{RL} in the expansion of Eq. (2).

In a simple monatomic system (e.g. fcc Cu), each of the sites is neutral and there is no charge transfer. But if we consider a WS sphere around the corresponding isolated atom, some of the charge, especially s and p charge related to more extended atomic wave functions, will be found outside the WS sphere. These are extreme cases; the fully coordinated bulk and the atom with no neighbors would be interesting to understand what happens around the free surface, where the sites have a reduced coordination. In Fig. 1a we show a schematic representation of the surface and the WS spheres around it: atoms are placed in the dark ones, while the light ones are the so called empty spheres, which are present to provide a basis set at these sites, allowing the charge to flow into this region. We expect that, as

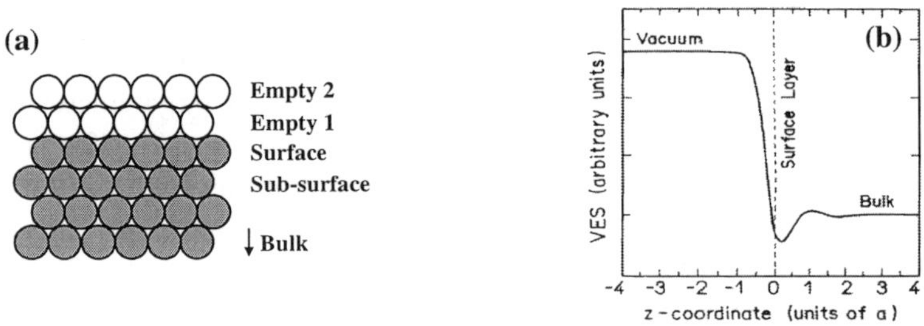

Fig. 1. (a) Schematic representation of a metallic surface and (b)Variation of VES with depth. The surface layer is at $z = 0$ and subsequent layers are placed at every positive integer and half integer value of z.

in the atomic case, some electrons (tails of atomic functions) will be found in the empty spheres around the surface layer, which will be negatively charged (note that since electrons flow into the empty spheres, by our definition the electronic charge transfer ΔQ is positive). Since the total charge must be conserved, the surface layer is often positively charged and acts as a parallel plate capacitor, creating a potential barrier (work function) which prevents the electrons from escaping from the system. As electrons flow from the surface into empty space, the potential barrier increases, making it more difficult for them to leave the surface. The amount of transferred charge is calculated self-consistently by the interplay of these factors. A typical curve for the electrostatic potential VES seen by the electrons in the surface region is shown in Fig. 1b. The value of VES will be constant deep in the bulk and out in the vacuum, while a step-like structure is seen at the surface. It is interesting to compare the charge inside the WS sphere around a surface site with that of the bulk, and this is done in Table 1, where we also show the charges inside the two rows of empty spheres shown in Fig. 1. We see from the table that there are approximately 0.198 electrons, mostly s and p, inside the first empty sphere, while further out, in the second sphere, the number is negligible. The Cu sites at the surface lose mainly s and p electrons (a reduction of 20% relative to bulk sites) to both the empty sphere and the subsurface layer, while the number of d electrons,

Table 1. Charge distribution for the Cu(001) surface.

	Bulk	Sub-surface	Surface	Empty 1	Empty 2
ΔQ	0.00	0.07	- 0.27	0.198	0.002
$s+p$	1.48	1.53	1.17	0.164	0.002
d	9.52	9.54	9.56	0.034	0.000

which are associated with narrower orbitals, remains similar to that of the bulk.

4. Results and discussion

We use the RS-LMTO-ASA approach to investigate Cr, Mn, Fe, Co and Ni dimers on Cu(001) surfaces. For ferromagnetic Fe, Co and Ni we have also obtained results for small (5 and 9-atom) clusters on Cu(001). The procedure used to embed the $3d$ atoms in the Cu(001) surface is described in detail elsewhere.[21] A cluster of around 4000 atoms was used to represent the system and no lattice relaxation was included. A cutoff parameter $LL = 20$ was taken in the recursion chain. During the embedding, the $3d$ sites and their first neighbors were included self-consistently, while for all the other sites, free surface potential parameters were used.

4.1. *3d dimers on Cu(001) surfaces*

Here, to illustrate the flexibility of the real space approach, we obtain the exchange coupling between the sites of a $3d$ dimer in Cu(001). To simulate the dimer, we substitute two adjacent empty-sphere sites in the layer close to the surface (Empty 1 in Fig. 1a) by $3d$ sites. It can be shown that, under appropriate conditions, some features of itinerant magnetism can be treated in terms of an effective Heisenberg model, where the effective exchange coupling J_{ij}^{eff} between magnetic sites i and j can be obtained from first-principles calculations.[7,8] Within the RS-LMTO-ASA formalism, for the ferromagnetic (FM) state, the effective coupling J_{ij}^{eff} is equal to the quantity J_{ij}, given in terms of the Greens functions G_{ij} as:[7]

$$J_{ij} = \frac{\Im m\, Tr}{4\pi} \int_{-\infty}^{E_F} dE \left[\delta_i(E) G_{ij}^{\uparrow\uparrow}(E) \delta_j(E) G_{ji}^{\downarrow\downarrow}(E)\right], \tag{5}$$

where the trace is over orbital indices, and $G_{ij}^{\sigma\sigma}$ is the propagator for electrons with spin σ between sites i and j. The quantity $\delta_i(E)$ has units of energy and may be associated with an energy-dependent local exchange splitting at site i. In general, when biquadratic and higher orders exchange are relevant, it can be shown that J_{ij} does not give the exchange constant, but actually defines the stability of the FM state (D_{FM}).[22] We note that Eq. (5) can also be used to investigate the stability of antiferromagnetic (AFM) states[8] and is, except for a sign convention, equivalent to the stability condition D_{AFM} in the literature.[22] In all cases, when $J > 0$, the given spin configuration is stable against spin rotations, while for $J < 0$ it is unstable. In a simple Heisenberg model, the effective exchange constant J_{ij}^{eff} can be obtained from first principles in terms of the stability condition of Eq. (5) by taking $J_{ij}^{eff} = J_{ij}$ for the FM configuration and $J_{ij}^{eff} = -J_{ij}$ for the AFM one.

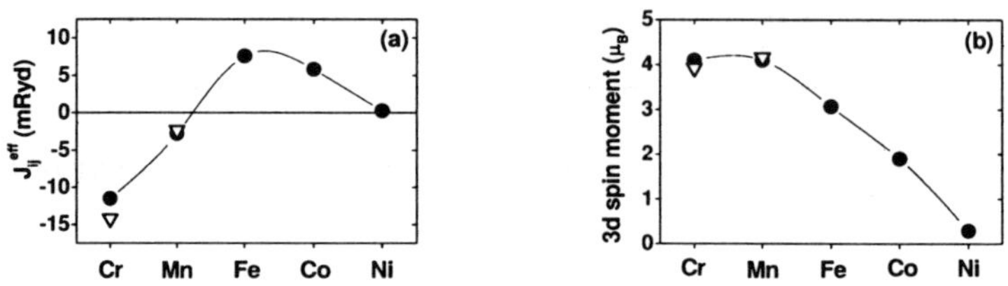

Fig. 2. J_{ij}^{eff} for $3d$ sites of the dimer for stable (full dots) and unstable (open triangles) configurations are shown in Fig. 2a. In Fig. 2b the corresponding spin magnetic moments are given.

Here we have calculated the stability condition J_{ij} for ferromagnetic Cr, Mn, Fe, Co and Ni dimers on Cu(001). We found that the FM state is stable for Fe, Co and Ni ($J_{ij} > 0$), but unstable in the case of Cr and Mn($J_{ij} < 0$). Therefore, for Cr and Mn dimers, we calculated J_{ij} for the AFM configuration and found that it is stable. In Fig. 2a we give J_{ij}^{eff} (full dots) for the stable FM (Fe,Co and Ni) or AFM (Cr and Mn) configurations. Values of J_{ij}^{eff} for the unstable FM configuration (empty triangles) are shown for Cr and Mn dimers. Corresponding magnetic moments for the dimer $3d$ sites are shown in Fig. 2b. Previous calculations indicate that the Ni adatom in Cu(001) does not develop a magnetic moment,[13] while the Ni monolayer over Cu(001) is known to be magnetic. The present results suggest that the Ni sites in the dimer already exhibit a moment.

The Heisenberg equation for the $3d$ sites in the dimer is given by $H^{dimer} = -2J_{ij}^{eff}\,\hat{e}_i.\hat{e}_j$, where $\hat{e}_i$ is a unit vector in the direction of the moment at site i in the dimer.[8] In this context the exchange coupling J_{ij}^{eff} is a constant and does not change as one rotates the system from an unstable to a stable state. Therefore the difference between the values of J_{ij}^{eff} calculated for the stable and the unstable configurations gives us an idea of how closely the Heisenberg model can be expected to be obeyed.

4.2. *Fe, Co and Ni clusters on Cu(001) surfaces*

Here we consider flat nanostructures consisting of 5 and 9-atom clusters of Fe, Co and Ni on Cu(001). We focus on the influence of local environment, here defined by the number of neighbors around a given site, upon the magnetic moment, charge transfers and charge distribution at that site. In the previous section we have shown that Fe, Co and Ni dimers on Cu(001) should be stable in the ferromagnetic configuration. Since there is no frustration, we expect the larger 5 and 9-atom clusters also to be ferromagnetic. We do not include Cr and Mn clusters in our study, since the antiferromagnetic configuration obtained for the dimers suggests that a more complex non-collinear magnetic-moment arrangement may be present. Actually, calculations for small Mn clusters (up to 10 atoms) supported on a fcc Cu(111), performed using the recently developed non-collinear RS-LMTO-ASA approach, indicate that the stable configuration in several of these systems is non-collinear.[16]

The flat $3d$ nanoclusters were simulated by replacing either 5 or 9 of the empty spheres adjacent to the surface by Fe, Co or Ni atom sites. For the 5-atom cluster we have atoms at a central site and at the four sites of the first shell around it on the Cu(001) surface. To obtain the 9-atom cluster, $3d$ atoms are also placed at the four sites of the second shell of neighbors. With respect to charge transfer and charge character, the 5-atom nanocluster has two inequivalent sites, the central site ($S5_0$) and the sites in the first shell ($S5_1$). For the 9-atom nanostructure we have 3 inequivalent sites, the central site ($S9_0$), the ones in the first shell ($S9_1$), and those in the second shell ($S9_2$). In our analysis, we include results of previous RS-LMTO-ASA calculations for adatoms,[13] as well as the present results for Fe, Co and Ni dimers and flat clusters on Cu(001). In all cases the sites have four Cu atoms of the surface layers as first neighbors. Therefore here the differences in local environment will be characterized by N, the number of Fe, Co or Ni neighbors around each site. The number of neighbors N for each type of site considered here is given in Table 2.

Table 2. $3d$ sites and corresponding number of neighbors in the cluster, N.

adatom (N=0)	dimer (N=1)	$S5_1$ (N=1)	$S9_2$ (N=2)	$S9_1$ (N=3)	$S5_0$ (N=4)	$S9_0$ (N=4)

Before we proceed it is interesting to consider the charge transfer and charge character around Fe, Co and Ni sites in the bulk. There, all sites are equivalent and charge transfers ΔQ are expected to be zero. The number of s and p electrons is very similar in Fe, Co and Ni, close to a total of 1.45 s and p electrons per site. The number of d electrons is given by $V - (s + p)$, where V, the number of electrons in the incomplete s, p, d shell, is 8, 9 and 10 for Fe, Co and Ni, respectively. This gives approximately 6.55 d electrons for Fe, 7.55 for Co and 8.55 for Ni. Experimentalists using the XMCD technique are often interested in the number of d holes, determined by subtracting the number of d electrons from 10. In Fig. 3a we show the charge transfer ΔQ and the number of s and p electrons, as a function of the number of neighbors N of Table 2, for Fe (empty squares), Co (empty triangles) and Ni (empty circles) sites in the systems considered here. As expected from the analysis of Sec. 3, charge transfer is negative at all sites, since they tend to lose electrons (mainly s and p) to the empty sphere region. The adatom, with no neighbors of its own kind ($N = 0$), has the largest charge transfer, almost one electron in the case of Fe. As N increases, the charge transfer gets smaller in magnitude, and for $N = 4$ it is of the same order as those calculated for a clean Cu(001) surface (see Table 1) or a Co monolayer on Cu(001). It is clear that as N grows, a smaller number of s and p electrons escape into empty space and therefore more charge of s and p character should be found around the site. This is seen in Fig. 3a, where the number of

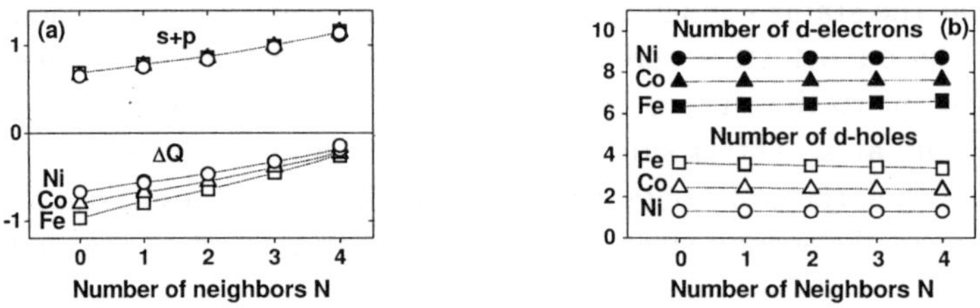

Fig. 3. (a) The charge transfer ΔQ and the number of $s+p$ electrons, as a function of N, for Fe (empty squares), Co (empty triangles) and Ni (empty circles) sites. (b) The number of d electrons (full symbols) and d holes (empty symbols) around the $3d$ sites as a function of N.

electrons with $s+p$ character grows from around 0.65 for the adatom to a little less than 1.2 for central sites with $N = 4$. The number of $s+p$ electrons at these central sites with $N = 4$ is again very similar to that calculated for the clean surfaces, but still smaller than 1.45, the number expected for sites in the bulk. We note that the behavior of the $s+p$ curve as a function of N in Fig. 3a is almost the same for Fe, Co and Ni. As mentioned before, the d states, associated with narrower orbitals, are less likely to escape into free space. In Fig. 3b, we show the number of d electrons (full symbols) and d holes (empty symbols) around the $3d$ sites as a function of N. The changes in the number of d electrons is small, of order of 2% − 3% relative to the bulk. The number of d holes thus also stays roughly constant.

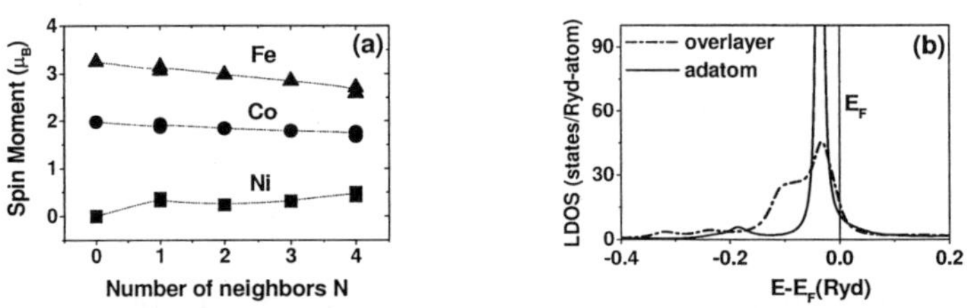

Fig. 4. (a) Spin magnetic moments at Fe, Co and Ni sites, as a function of N. (b) The LDOS of the Ni adatom (dotted line) and of a Ni overlayer (full line) on Cu(001).

Finally, in Fig. 4a we show values for the spin magnetic moments at Fe, Co and Ni sites as a function of N. The moments of Fe and Co are close to saturation and decrease slightly as the number N of $3d$ neighbors increases. For Ni, the adatom is non magnetic, the dimer ($N = 1$) shows a moment which decreases slightly for $N = 2$ and then increases again, reaching at $N = 4$ a value similar to that of the Ni monolayer on Cu(001). To understand the behavior of Ni, one can think in terms of the Stoner criterion for magnetism, which says that a Ni site should develop a moment when the LDOS at the Fermi Level in the nonmagnetic state (N_{E_F}) reaches 13.8 states/Ryd-spin.[5] The d LDOS for transition-metal adatoms on simple-metal surfaces shows a very narrow Lorenzian-like peak, characteristic of virtual bound states. Cu, Ag and Au, with their d bands practically filled, can be seen as simple metals in this context. As N increases, the LDOS gets broader and more structured due to hybridization with the $3d$ bands of the transition metal neighbors. This is illustrated in Fig. 4b, where we compare the LDOS of the Ni adatom ($N = 0$) with that of Ni overlayer ($N = 4$) on Cu(001). The LDOS of the adatom reaches very large values (around 150 states/Ryd-spin) in the region of the peak, while the LDOS of the Ni overlayer is broader and more structured. In general, the value of N_{E_F} for the adatom depends on the d occupation at the site. For Ni, the band is more than 4/5 full and N_{E_F} is smaller for $N = 0$ (adatom) than for $N = 4$ (the overlayer; cf. Fig. 4b). The value for the adatom is actually below the Stoner limit and it does not develop a moment. Now consider Co and Fe in a situation similar to that of Fig. 4b. Since Co and Fe have considerably smaller d occupations, the Fermi level would be placed closer to the center of the peak, yielding extremely high values of N_{E_F} for the adatom ($N = 0$) and strongly favoring magnetism. It is clear that for Fe and Co, as the band gets broader and more structured, N_{E_F} decreases and the tendency towards magnetism in turn decreases. These simple arguments help to explain the results of Fig. 4a, where the moment of Fe and Co sites tends to decrease with N, while for Ni it tends to increase.

5. Conclusions

In this paper we have briefly described the RS-LMTO-ASA approach, which allows us to perform first-principles, self-consistent calculations within the LSD formalism directly in real space. We have applied the approach to investigate spin moments and magnetic ordering of $3d$ (Cr, Mn, Fe, Co and Ni) dimers on Cu(001). We found that Fe, Co and Ni dimers exhibit FM ordering while the AFM arrangement is stable for Cr and Mn dimers. We have also performed calculations for ferromagnetic 5 and 9-atom Fe, Co and Ni clusters on Cu(001) and investigated the influence of local environment (here defined in terms of the number N of $3d$ neighbors) on local magnetic moments, charge transfers and charge character (s, p or d) at each $3d$ (Fe, Co or Ni) site. We find that Fe and Co moments tend to decrease with increasing N, while for Ni they show an overall increase. The Ni adatom is non magnetic, but the dimer already exhibits magnetism. These tendencies are explained within a simple model, in terms of the Stoner criterium for magnetism.

We show that large charge transfers ΔQ can be present at adatom-cluster sites, due to their reduced coordination. These transfers are mainly due to a drastic decrease in the number of $s + p$ electrons around the sites, while the number of $3d$ electrons remains practically unchanged. The linear increase in the number of $s + p$ electrons around the Fe, Co and Ni sites as N increases is accompanied by a decrease in the magnitude of the charge transfer at the same sites.

Acknowledgements

We are grateful for the financial support of CNPq, Brazil and thank R. B. Muniz for useful discussions. Computational facilities of the LCCA, University of S. Paulo and of the CENAPAD at the University of Campinas, SP, Brazil, were used.

References

1. P. R. Peduto, S. Frota-Pessôa and M. S. Methfessel, *Phys. Rev.* **B44**, 13283 (1991).
2. S. Frota-Pessôa, *Phys. Rev.* **B46**, 14570 (1992).
3. P. R. Peduto and S. Frota-Pessôa, *Brazilian Journal of Physics* **27**, 574 (1997); A. B. Klautau, P. R. Peduto and S. Frota-Pessôa, *J. Magn. Magn. Mat.* **186**, 223 (1998).
4. O. K. Andersen, *Phys. Rev.* **B12**, 3060 (1975).
5. O. K. Andersen, O. Jepsen, and D. Glötzel, in *Highlights of Condensed Matter Theory*, edited by F. Bassani, F. Funi and M. P. Tosi (North-Holland, Amsterdam, 1985).
6. R. Haydock, in *Solid State Physics*, edited by H. Ehrenreich, F. Seitz, and D. Turnbull (Acad. Press, New York, 1980), Vol. 35, p. 215.
7. S. Frota-Pessôa, R. B. Muniz and J. Kudrnovský, *Phys. Rev.* **B62**, 5293 (2000).
8. S. Frota-Pessôa, *J. Magn. Magn. Mat.* **226-230**, 1021 (2001).O. Beutler, R. Kirsch, M. J. Prandolini, W. D. Brewer, J. Kapoor, P. J. Jensen, S. Frota-Pessôa and D. Riegel, *Europhys. Lett.* **70**, 520 (2005).
9. L. A. Terrazos and S. Frota-Pessôa, *Phys. Rev.* **B56**, 13035 (1997).
10. S. Frota-Pessôa, L. M. de Mello, H. M. Petrilli, and A. B. Klautau, *Phys. Rev. Lett.* **71**, 4206 (1993).
11. S. Ferreira and S. Frota-Pessôa, *Phys. Rev.* **B51**, 2045 (1995); R. Kirsch, M. J. Prandolini, O. Beutler, S. Frota-Pessôa, *Europhys. Lett.* **59**, 430 (2002).
12. S. Frota-Pessôa, *Phys. Rev.* **B69**, 104401 (2004); W.D. Brewer, A. Scherz, C. Song, H. Wende, K. Baberschke, P. Beneok, and S. Frota-Pessôa, *Phys. Rev. Lett.* **93** 077205-1 (2004).
13. A. B. Klautau and S. Frota-Pessôa, *Surf. Sci.* **579** (2005) 27; A. B. Klautau and S. Frota-Pessôa, *Phys. Rev.* **B70**, 193407 (2004).
14. S. B. Legoas *et al.*, A. A. Araujo, B. Laks, A. B. Klautau, and S. Frota-Pesôa, *Phys. Rev.* **B61** (2000) 10417.
15. S. Frota-Pessôa, A. B. Klautau and S. B. Legoas, *Phys. Rev.* **B66**, 132416 (2002); A. B. Klautau, S. B. Legoas, R. B. Muniz and S. Frota-Pessôa, *Phys. Rev.* **B60**, 3421 (1999).
16. A. Bergman et al., unpublished.
17. N. Beer and D. G. Pettifor, in *The Electronic Structure of Complex Systems*, edited by W. Temmermann and P. Phariseau (Plenum Press, New York, 1984).
18. V. von Barth and L. Hedin, *J. Phys.* **C5** (1972) 1629.
19. R. N. Nogueira and H. M. Petrilli, *Phys. Rev.* **B63** 2405 (2001); A. Bergman, E. Holmström, A. M. N. Niklasson, L. Nordström, S. Frota-Pessôa and O. Eriksson, *Phys. Rev.* **B70**, 174446 (2004).
20. S. B. Legoas and S. Frota-Pesôa, *Phys. Rev.* **B61**, 12566 (2000).
21. A. B. Klautau and S. Frota-Pessôa, *Surf. Sci.* **497**, 385 (2002).
22. A. T. Costa, R. B. Muniz and D. Mills, *Phys. Rev. Lett.* **94**, 137203-1 (2005).

INFRARED-ABSORPTION SPECTRUM OF ELECTRON BUBBLES IN LIQUID HELIUM

M. PI, M. BARRANCO, V. GRAU, and R. MAYOL

Universitat de Barcelona, Facultat de Física, Departament E.C.M.
Diagonal 647, 08028 Barcelona, Spain.

Within finite temperature Density Functional Theory, we have calculated the energy of the transitions from the ground state to the first two excited states in the electron bubbles in liquid helium at pressures from zero to about the solidification pressure. For ^{4}He at low temperatures, our results are in very good agreement with infrared absorption experiments. We have found that the $1s-2p$ transition energies are sensitive not only to the size of the electron bubble, but also to its surface thickness. We also present results for the infrared transitions in the case of liquid ^{3}He.

Keywords: electron bubbles, cavitation, surface tension, liquid helium

1. Introduction

Excess electrons in liquid helium are known to form electron bubbles because of the strong repulsion between the helium atoms, which are very weakly polarizable, and the intruder electrons. Experimental and theoretical spectroscopic studies on electron bubbles have been carried out for many years since the pioneering works of Northby and Sanders,[1] and of Dexter and coworkers.[2,3] Some twenty years later, Grimes and Adams carried out a detailed analysis of the infrared-absorption spectrum of the electron bubble in liquid ^{4}He. In a first work,[4] they have used a photoconductive mechanism to detect the transitions which operated only over a limited region of the pressure-temperature $(P-T)$ plane because it appeared to be associated with trapping of bubbles on vortices in the superfluid. Later on, they have improved their experimental apparatus and have observed the electronic transitions in direct infrared absorption.[5] Recent experiments on cavitation in liquid helium have renewed the interest in the study of single electron bubbles.[6]

The simplest model to address electron bubbles in liquid helium supposes that they are confined in a spherical well potential of radius R. The total energy of the electron-helium (e-He) system is then written as a function of R

$$U(R) = E_e + 4\pi R^2 \sigma + \frac{4}{3}\pi R^3 P \;, \tag{1}$$

where E_e is the ground state electronic energy, P is the pressure applied to the system, and σ is the surface tension of the liquid. For an infinite spherical well potential, $E_e = \pi^2\hbar^2/(2m_e R^2)$. This model is able to qualitatively reproduce the experimental infrared-absorption energies, and is simple enough to allow to calculate shape fluctuations of electron bubbles and line shapes.[7] It can be refined[2–5] by taking into account the finite depth of the well V_0, which is about 1 eV, Ref. 8. Once $U(R)$ has been minimized with respect to R and the radius of the equilibrium bubble R_{eq} has been determined, it is easy to obtain the energies of the (n, l) excited states and to compute the transition energies to the ground state if the Franck-Condon principle holds, i.e., if the absorption of a photon by the electron bubble is a process much faster than the time needed for the helium bubble to adapt itself to the electron wave function in the excited state.

Another key ingredient entering Eq. (1) is the surface tension σ, which for ^{4}He at zero T and P is about 0.274 K Å^{-2}, Ref. 9. This yields $R_{eq} = 18.9$ Å. The surface tension is only known along the liquid-vapor coexistence line. Consequently, to use Eq. (1) one has to rely on model calculations of $\sigma(P,T)$. The situation has been discussed in detail by Grimes and Adams, who concluded that in order to perfectly fit their experimental results with Eq. (1), one has to take the depth of the well V_0 P-dependent and a surface tension σ independent of P, which seems difficult to justify. For instance, early calculations[3,10] indicated that the surface tension σ nearly doubles from $P = 0$ to 25 atm. It is worthwhile to stress that when both V_0 and σ were made P-dependent following the expected pressure dependences, the agreement between theory[3] and experiment was only qualitative, see Fig. 2 of Ref. 4.

In this work we present a theoretical description of infrared-absorption $1s - 1p$ and $1s - 2p$ excitation energies of electron bubbles in liquid helium. It is based on the use of a finite-temperature density functional (DF) approach in conjunction with a realistic electron-helium effective potential. DF methods have become increasingly popular in recent years as useful computational tools to study the properties of classical and quantum inhomogeneous fluids.[11,12] In the frame of DF theory, the properties of an electron bubble approaching the surface of liquid ^{4}He have been studied by Ancilotto and Toigo[13] using the so-called Orsay-Paris (OP) zero temperature finite-range DF[14] and the pseudopotential proposed in Ref. 15 as e-He interaction.

Density functional theory has also proven to be the most successful approach in addressing cavitation in liquid helium so far.[16,17] It incorporates in a self-consistent way the equation of state of bulk liquid and surface tension of the liquid-gas interface as a function of temperature. It allows for a flexible description of the electron bubble, incorporating surface thickness effects. Within DF theory one avoids the use of macroscopic concepts such as surface tension and pressure at a nanoscopic scale; however, it is a continuous, not an atomic description of the system. Used in conjunction with a Hartree-type e-He potential, we have recently shown that this approach quantitatively reproduces the existing experimental data on cavitation of electron bubbles in liquid helium below saturation pressure.[18] Consequently, it is a tested framework to address other properties of electron bubbles.

2. Density-Functional Approach and Electron-Helium Interaction

Our starting point is the finite temperature zero-range DF of Ref. 19 that reproduces thermal properties of liquid ^{4}He such as the experimental isotherms and the ^{4}He liquid-gas coexistence line up to $T = 4.5$ K, and the T dependence of the surface tension of the liquid free surface. We have taken the Hartree-type e-He effective potential derived by Cheng et al.[20] (see also Ref. 21) as e-He interaction. This allows us to write the free energy of the system as a functional of the ^{4}He particle density ρ, the excess electron wave function Ψ, and T:

$$F[\rho,\Psi,T] = \int d\vec{r}\, f(\rho,T) + \frac{\hbar^2}{2m_e}\int d\vec{r}\,|\nabla\Psi(\vec{r})|^2 + \int d\vec{r}\,|\Psi(\vec{r})|^2 V(\rho) \,, \tag{2}$$

where $f(\rho,T)$ is the ^{4}He free energy density per unit volume written as

$$f(\rho,T) = f_{vol}(\rho,T) + \beta\frac{(\nabla\rho)^2}{\rho} + \xi(\nabla\rho)^2 \,. \tag{3}$$

In this expression, $f_{vol}(\rho,T)$ consists of the free energy density of a Bose gas, plus a phenomenological density and temperature dependent part f_{int} that takes into account the effective interaction of helium atoms in the bulk liquid. The parameters of f_{int} and those of the density gradient terms in Eq. (3) have been adjusted so as to reproduce physical quantities like the equation of state of the bulk liquid and the surface tension of the liquid free surface.

The e-He interaction $V(\rho)$ is written as a function of the local helium density[20]

$$V(\rho) = \frac{\hbar^2 k_0^2}{2m_e} + \frac{2\pi\hbar^2}{m_e}\rho\, a_\alpha - 2\pi\alpha e^2\left(\frac{4\pi}{3}\right)^{1/3}\rho^{4/3} \,, \tag{4}$$

where $\alpha = 0.208$ Å^3 is the static polarizability of a ^{4}He atom, and k_0 is determined from the helium local Wigner-Seitz radius $r_s = (3/4\pi\rho)^{1/3}$ by solving the trascendental equation

$$\tan[k_0(r_s - a_c)] = k_0\, r_s \,, \tag{5}$$

with a_c and a_α being the scattering lengths arising from a hard-core and from a polarization potential. We have taken[20] $a_\alpha = -0.06$ Å, $a_c = 0.68$ Å.

For given P and T values we have solved Euler-Lagrange equations which result from the variation of the constrained grand potential density $\tilde{\omega}(\rho,\Psi,T) = \omega(\rho,\Psi,T) - \varepsilon|\Psi|^2$, where

$$\omega(\rho,\Psi,T) = f(\rho,T) + \frac{\hbar^2}{2m_e}|\nabla\Psi|^2 + |\Psi|^2 V(\rho) - \mu\rho \,. \tag{6}$$

It yields

$$\frac{\delta f}{\delta\rho} + |\Psi|^2\frac{\partial V}{\partial\rho} = \mu \tag{7}$$

$$-\frac{\hbar^2}{2m_e}\Delta\Psi + V(\rho)\Psi = \varepsilon\Psi \,, \tag{8}$$

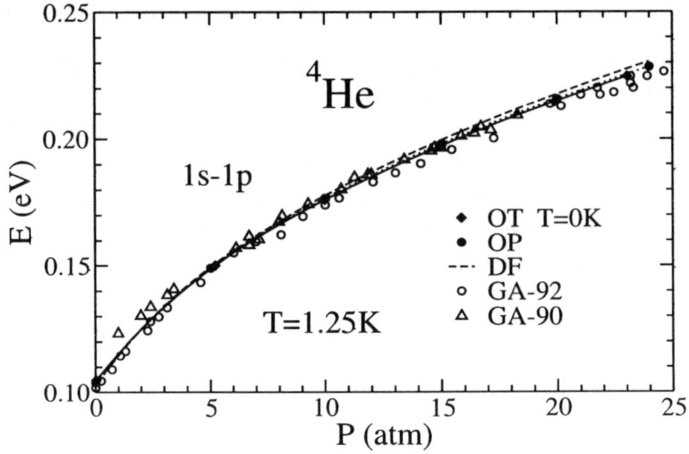

Fig. 1. ^{4}He $1s - 1p$ transition energies (eV) as a function of P (atm) for $T = 1.25$ K. The open triangles are the observed points from the electron bubble photocurrent,[4] and the open circles correspond to direct infrared absorption measurements.[5] The dashed line represents the results obtained using the zero-range DF discussed in Sec. 2. Filled circles connected with a dotted line are the results obtained using the OP finite-range DF at $T = 1.25$ K, and filled diamonds connected with a solid line are results obtained using the zero temperature OT finite-range DF.

where ε is the lowest eigenvalue of the Schrödinger equation obeyed by the electron. These equations are solved assuming spherical symmetry, imposing for ρ that $\rho'(0) = 0$ and $\rho(r \to \infty) = \rho_b$, where ρ_b is the density of the bulk liquid, and that the electron is in the $1s$ state. Fixing ρ_b and T amounts to fix P and T, since the pressure can be obtained from the bulk equation of state $P = -f_{vol}(\rho_b, T) + \mu\rho_b$, and the ^{4}He chemical $\mu = \partial f_{vol}(\rho, T)/\partial\rho|_T$ is known in advance. We have used a multidimensional Newton-Raphson method[22] for solving Eqs. (7) and (8), after having discretized them using 13-point formulas for the r derivatives. A fine mesh of step $\Delta r = 0.1$ Å has been employed, and the equations have been integrated up to $R_\infty = 150$ Å to make sure that the asymptotic bulk liquid has been reached. After obtaining the equilibrium configuration ($1s$ state), the spectrum of the electron bubble ε_{nl} is calculated from Eq. (8) keeping frozen the helium density (Franck-Condon principle).

3. Results

3.1. *Liquid ^{4}He*

Figure 1 shows the $1s - 1p$ transition energies (eV) as a function of P (atm) for $T = 1.25$ K. It can be see that the agreement between theory and experiment is good from $P = 0$ to the solidification pressure, and consequently, the physical process seems well undestood. However, a minor discrepancy appears at high pressures, and especially in the description of the $1s - 2p$ transition energies, as it can be seen in Fig. 2. This is not surprising, since one would expect that the $2p$ state is

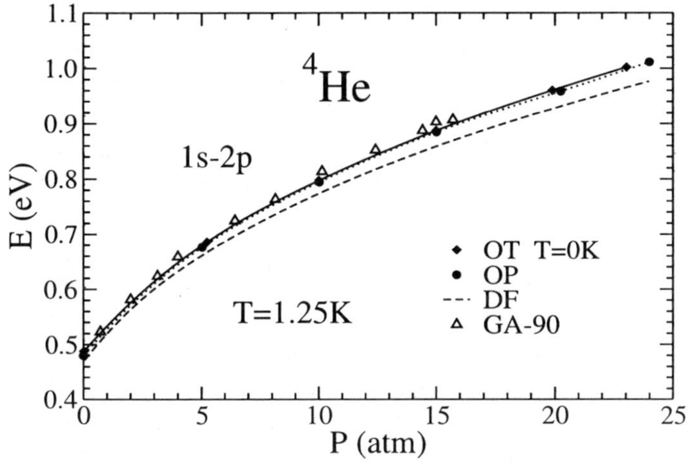

Fig. 2. ^{4}He $1s-2p$ transition energies (eV) as a function of P (atm) for $T = 1.25$ K. The open triangles are the observed points from the electron bubble photocurrent.[4] The dashed line represents the results obtained using the zero-range DF discussed in Sec. 2. Filled circles connected with a dotted line are the results obtained using the OP finite-range DF at $T = 1.25$ K, and filled diamonds connected with a solid line are results obtained using the zero temperature OT finite-range DF.

more sensitive to fine details of the bubble structure, in particular to its thickness, because the $2p$ wave function penetrates deeper into the liquid.

We have improved the method of Sec. 2 to achieve a better agreement with experiment. The improvement is based on the observation that zero-range DF's as the one described in Sec. 2 are fitted to reproduce the experimental surface tension of liquid helium at $T = 0$,[19,23] and then, the T-dependence of σ[19] and the thickness t of the free surface -defined as the difference between the distances at which the density equals $0.1\rho_b$ and $0.9\rho_b$, where ρ_b is the bulk density at the given (P,T)- come out as predictions of the formalism. It turns out that at $T = 0$, zero-range DF's overestimate t by about 1 Å. Indeed, recent measurement of the ^{4}He surface thickness yield values around 6 Å, Refs. 24, 25, whereas the value predicted in Refs. 19, 23 is about 7 Å. In the case of ^{3}He, the experimental value is about 7.5 Å,[26] whereas the prediction using zero-range DF's is about 8.5-9 Å, Refs. [illegible]. We recall that the surface thickness of liquid helium is a quantity rather difficult to determine experimental and theoretically,[28] and the dispersion of the values assigned to it is large.[23–30] Only recently, the mentioned values ($\sim$ 6-6.5 Å for ^{4}He and $\sim$ 7.5 Å for ^{3}He) have emerged as likely accurate determinations of the surface thickness of liquid helium free surface.

In the nineties, a new class of DF's has appeared that retains some of the simplicity of the original zero-range DF's, incorporating finite-range effects that are absolutely necessary to address a wide class of physical phenomena, like elementary excitations in bulk liquid and large inhomogeneities caused by the presence of impurities in bulk liquid and droplets, and also by substrates. For ^{4}He, two such DF's are the OP functional and a generalization of it called Orsay-Trento (OT) functional.[31] It is remarkable that they reproduce the experimental surface tension at $T = 0$ without having imposed it in their construction, and also the surface thickness. In particular, OP yields $\sigma = 0.277$ K Å^{-2}, $t = 5.8$ Å, and OT yields $\sigma = 0.272$ K Å^{-2}, $t = 6$ Å.[14,31] A recent diffusion Monte Carlo calculation yields $\sigma = 0.281(3)$ K Å^{-2}, t=6.3(4) Å.[30]

It is then quite natural to ask oneself if the remaining disagreement between theory and experiment shown in Figs. (1) and (2) might arise from a worse description of the electron bubble surface in the case of zero-range DF's. To answer this question, we have repeated the calculations using the zero temperature OT functional -thermal effects on the excitation energies are expected to be small for $T = 1.25$ K, and this is what we have found, see below-, and also using a straightforward generalization at finite T of the OP functional, which is possible because of the similarities between OP and the zero-range DF of Sec 2. It is worth to know that a generalization of the OT density functional at finite temperatures is also available.[32]

Density profiles of electron bubble at different pressures and $T = 1.25$ K, obtained with the OP and the zero-range functional, are shown in Fig. 3. Similar results have been obtained using the OT functional, which yields density profiles with more pronounced oscillations. The excess electron squared wave functions $|\Psi|^2$ corresponding to the $1s, 1p$, and $2p$ states are also represented. For the zero-range DF we also give the value of the surface thickness t, which goes from 6.1 Å at $P = 0$ atm to 4.3 Å at $P = 20$ atm. In the case of OP, and especially of OT, it is difficult to define the thickness because of the density oscillations in the surface region (see also Ref. 33). For reference, we indicate that the surface thickness of the electron bubble at $T = 1.25$ K and $P = 0$ atm obtained with OP is 5 Å, one Å smaller that the value yielded by the zero-range DF.

Figures (1) and (2) also show the results obtained with OP and OT. It can be seen that in this case, the agreement between theory and experiment is excellent. It is worthwhile to mention that Eloranta and Apkarian[33] have also obtained the absorption energies up to $P \sim 12$ atm using the OT functional and the pseudopotential of Jortner et al.[34] as e-He interaction. Their results for the $1s - 2p$ transition energies are not in as good agreement with experiment as ours, likely because of the different e-He interaction we have used. Besides, the relevance of a correct description of the surface thickness to reproduce the experimental data has been overlooked in their work, which is mainly about the dynamics of electron bubble expansion in liquid ^{4}He.

3.2. *Liquid ^{3}He*

We have also studied the infrared-absorption spectrum for electron bubbles in liquid ^{3}He. The application of DF theory to describe electron bubble explosions in ^{3}He

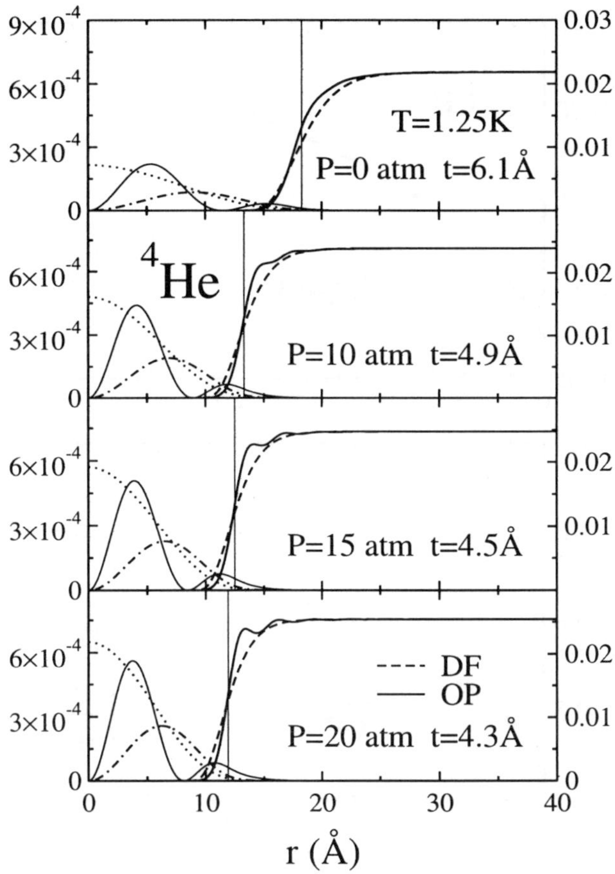

Fig. 3. ^{4}He electron bubble density profiles in Å^{-3} (right scale) and excess electron squared wave functions $|\Psi|^2$ in Å^{-3} (left scale), as a function of the radial distance r (Å) for $T = 1.25$ K and different pressures. The dashed lines correspond to the zero-range DF, and the solid lines to the OP functional. The vertical thin lines indicate the equilibrium radius R_{eq} yielded by the simple electron bubble model Eq. (1). The value of the surface thickness t is also given for the zero-range DF. The electron squared wave functions $|\Psi|^2$ correspond to the OP calculation; dotted line, $1s$ state; dot-dashed line, $1p$ state; thin solid line, $2p$ state.

proceeds as shown in Sec. 2. We have used the zero-range DF proposed in Ref. 37 to describe the inhomogeneous liquid, and also a finite-range DF, called from now on FR, obtained from the zero-range one following the procedure indicated in Ref. 14 for the ^{4}He case. The e-He interaction is given again by Eq. (4) with the parameters corresponding to ^{3}He, namely $\alpha = 0.206$ Å^3, and same values for a_α and a_c.

We represent in Fig. 4 the infrared-absorption energies for the $1s-1p$ and $1s-2p$ transitions. It can be seen that the results are qualitatively similar to those found for ^{4}He, i.e., a fair insensitivity of the $1s - 1p$ transition energy to the detailed structure of the bubble surface, and an underestimation of the $1s - 2p$ transition energy by the zero-range DF. The half density radius of the electron bubble $R_{1/2}$

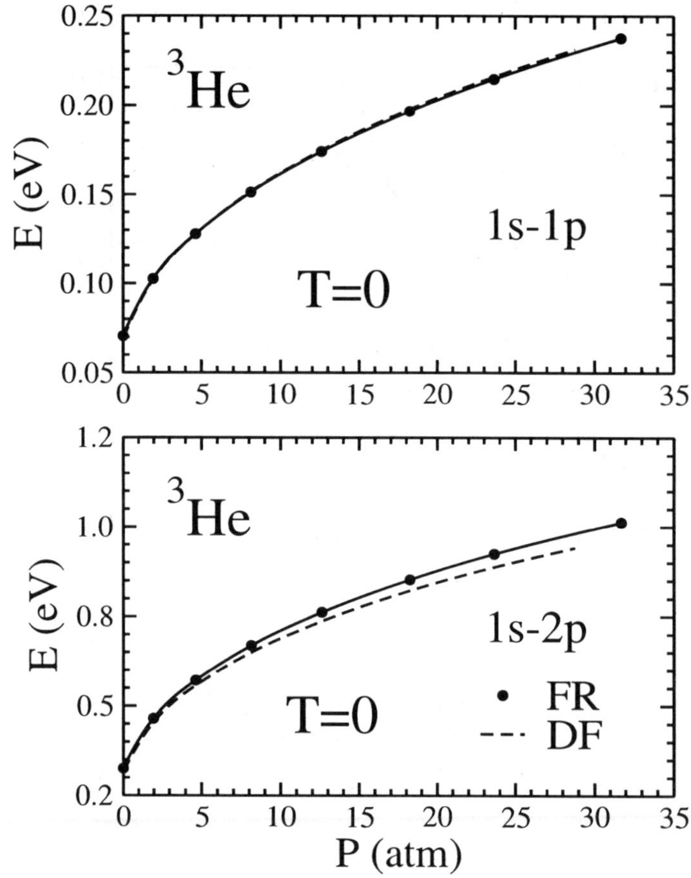

Fig. 4. ^{3}He $1s - 1p$ and $1s - 2p$ transition energies (eV) as a function of P (atm) for $T = 0$ K. The dashed line is the result obtained using the zero-range DF of Ref. 27. Filled dots connected with a solid line are the results obtained using the zero temperature finite-range DF.

goes from 22.5 Å at $P = 0$ to 11.8 Å at $P = 22.3$ atm.

The smaller excitation energies in the case of ^{3}He are due to the smaller surface tension for this isotope, $\sigma = 0.113$ K Å^{-2},[35] which causes that electron bubble radii are larger for ^{3}He than for ^{4}He (e.g., $R_{eq} = 23.5$ Å at $P = 0$ instead of 18.9 Å), yielding smaller excitation energies, as shown for instance by the simple model of Eq. (1). In particular, at $P = 0$ the model gives for the ratio of the $1s - 1p$ excitation energies the value $(\sigma_3/\sigma_4)^{1/2}$, in good agreement with the results obtained within the DF approach.

4. Summary

We have demonstrated the suitability of the density functional approach to quantitatively address electron bubbles in liquid He. On the one hand, we have shown that the DF approach, in conjunction with a realistic electron-helium interaction, is able to reproduce without any further assumption, the low temperature infrared spectrum of electron bubbles experimentally determined by Grimes and Adams.[4,5] On the other hand, the method yields results that agree with the experimental findings of Maris and co-workers on cavitation of electron bubbles below saturation pressure,[6,36] in a wide range of temperatures.[18]

We have shown that the analysis of infrared-absorption transitions of electron bubbles constitute a stringent test for the theoretical models aiming a detailed description of the free surface of liquid helium, in particular of its surface thickness. This is at variance with electron bubble cavitation, which seems to be sensitive only to global properties of the surface, like its tension. Indeed, we have checked that the cavitation pressure for electron bubbles in liquid ^{4}He yielded by the zero-range DF is -2.07 bar, whereas OP yields -2.13 bar, and OT yields -2.08 bar. These differences are far smaller than the experimental error bars, and are partly due to the small differences in the surface tensions predicted by these functionals.

There are some related problems that can be studied as a natural extension of the work carried out until present within the DF frame. In particular, the effect on the critical cavitation pressure of quantized vortices pinned to excess electrons.[6] The infrared absorption above $T \sim 2$ K is another open problem whose quantitative description requires to relax the Franck-Condon principle. Both problems, presently unaffordable by any microscopic approach, can be addressed within DFT and time-dependent DFT.[33,37] As these are not trivial issues at all, they may be the next test grounds to assess the capabilities and limitations of the DF theory applied to liquid helium.

Acknowledgments

We would like to thank F. Ancilotto, E. S. Hernández, and H. J. Maris for useful discussions. This work has been performed under grants FIS2005-01414 from DGI (Spain) and 2005SGR00343 from Generalitat de Catalunya.

References

1. J. A. Northby and T. M. Sanders, *Phys. Rev. Lett.* **18**, 1184 (1967).
2. W. B. Fowler and D. L. Dexter, *Phys. Rev.* **176**, 337 (1968).
3. T. Miyakawa and D. L. Dexter, *Phys. Rev.* **A1**, 513 (1970).
4. C. C. Grimes and G. Adams, *Phys. Rev.* **B41**, 6366 (1990).
5. C. C. Grimes and G. Adams, *Phys. Rev.* **B45**, 2305 (1992).
6. J. Classen, C.-K. Su, M. Mohazzab, and H. J. Maris, *Phys. Rev.* **B57**, 3000 (1998).
7. H. J. Maris and W. Guo, *J. Low Temp. Phys.* **137**, 491 (2004).
8. W. T. Sommer, *Phys. Rev. Lett.* **12**, 271 (1964); M. A. Woolf and G. W. Rayfield, *ibid.* **15**, 235 (1965).
9. P. Roche, G. Deville, N. J. Appleyard, and F. I. B. Williams, *J. Low Temp. Phys.* **106**, 565 (1997).
10. D. Amit and E. P. Gross, *Phys. Rev.* **145**, 130 (1966).

11. R. Evans, in *Liquids at interfaces*, J. Charvolin, J. F. Joanny, and J. Zinn-Justin, eds. (Elsevier, 1989), p. 1
12. M. Barranco, R. Guardiola, E.S. Hernández, R. Mayol, J. Navarro, and M. Pi, *J. Low Temp. Phys.*, in print (2006).
13. F. Ancilotto and F. Toigo, *Phys. Rev.* **B50**, 12 820 (1994).
14. J. Dupont-Roc, M. Himbert, N. Pavloff, and J. Treiner, *J. Low Temp. Phys.* **81**, 31 (1990).
15. N. R. Kestner, J. Jortner, M. H. Cohen, and S. A. Rice, *Phys. Rev.* **140**, A56 (1965).
16. Q. Xiong and H. J. Maris, *J. Low Temp. Phys.* **77**, 347 (1989).
17. M. Barranco, M. Guilleumas, M. Pi, and D. M. Jezek, in *Microscopic Approaches to Quantum Liquids in Confined Geometries*, E. Krotscheck and J. Navarro, eds. (World Scientific, Singapore, 2002), p. 319.
18. M. Pi, M. Barranco, R. Mayol, and V. Grau, *J. Low Temp. Phys.* **139**, 397 (2005).
19. A. Guirao, M. Centelles, M. Barranco, M. Pi, A. Polls, and X. Viñas, *J. Phys. Condens. Matter* **4**, 667 (1992).
20. E. Cheng, M. W. Cole, and M. H. Cohen, *Phys. Rev.* **B50**, 1136 (1994); Erratum *ibid.* **B50**, 16 134 (1994).
21. M. Rosenblit and J. Jortner, *Phys. Rev.* **B52**, 17 461 (1995).
22. W. H. Press, S. A. Teulosky, W. T. Vetterling, and B. P. Flannery, *Numerical Recipes in Fortran 77: The Art of Scientific Computing* (Cambridge University Press, Cambridge, 1999).
23. S. Stringari and J. Treiner, *J. Chem. Phys.* **87**, 5021 (1987); *Phys. Rev.* **B36**, 8369 (1987).
24. J. Harms, J. P. Toennies, and F. Dalfovo, *Phys. Rev.* **B58**, 3341 (1998).
25. K. Penanen, M. Fukuto, R. Heilmann, I. F. Silvera, and P. S. Pershan, *Phys. Rev.* **B62**, 9621 (2000).
26. J. Harms, J. P. Toennies, M. Barranco, and M. Pi, *Phys. Rev.* **B63**, 184513 (2001).
27. M. Barranco, M. Pi, A. Polls, and X. Viñas, *J. Low Temp. Phys.* **80**, 77 (1990).
28. D. O. Edwards and W. F. Saam, *Progress in Low Temp. Phys.*, Vol. VII A, p. 285 (1978).
29. D. V. Osborne, *J. Phys.: Condens. Matter* **1**, 289 (1989).
30. J. M. Marín, J. Boronat, and J. Casulleras, *Phys. Rev.* **B71**, 144518 (2005).
31. F. Dalfovo, A. Lastri, L. Pricaupenko, S. Stringari, and J. Treiner, *Phys. Rev.* **B52**, 1193 (1995).
32. F. Ancilotto, F. Faccin, and F. Toigo, *Phys. Rev.* **B62**, 17035 (2000).
33. J. Eloranta and V. A. Apkarian, *J. of Chem. Phys.* **117**, 10139 (2002).
34. J. Jortner, N. R. Kestner, S. A. Rice, and M. H. Cohen, *J. Chem. Phys.* **43**, 2614 (1965).
35. M. Iino, M. Suzuki, A. J. Ikushima, and Y. Okuda, *J. Low Temp. Phys.* **59**, 291 (1985).
36. C.-K. Su, C. E. Cramer, and H. J. Maris, *J. Low Temp. Phys.* **113**, 479 (1998).
37. L. Giacomazzi, F. Toigo, and F. Ancilotto, *Phys. Rev.* **B67**, 104501 (2003).

FLUCTUATIONS AND PAIRING IN FERMI SYSTEMS: A CROSSING-SYMMETRIC APPROACH

KHANDKER F. QUADER
Physics Department, Kent State University
Kent, OH 44242, U.S.A.
quader@scorpio.kent.edu

Treating competing fluctuations, e.g., density, spin, current, need a tractable, self-consistent approach. One method that treats particle-particle and particle-hole correlations self-consistently is the diagrammatic "crossing-symmetric equations" method. In a general calculation for pairing, non-local interaction plays an important role in enhancing certain quantum fluctuations and thereby determining the pairing symmetry.

Keywords: Crossing-symmetry; non-local; pairing

1. Introduction

In recent years there has been a tremendous interest in systems that exhibit novel pairing states with uncoventional pair symmetries, such as p, d or extended-$s$wave. Examples of systems with exotic magnetic and unconventional superconducting phases are weak ferromagnets (FM) (e.g. UGe_2,[1] $ZrZn_2$,[2] $URhGe$, [3] heavy fermion compounds (e.g. UPt_3, $CePd_2Si_2$,[4] and high temperature superconductors (e.g. $YBaCuO$, $LaSrCuO$). The nature of these phases are not well understood. The pairing mechanism or the pairing boson are believed to be fermionic in origin, rather than due to lattice phonons. Possible non-phonon pairing boson can arise from ferromagnetic or antiferromagnetic spin fluctuations, that become dominant near a ferromagnetic or antiferromagnetic instability, or from charged excitations such as excitons or plasmons.

We study the general nature of fermion pairing in strongly interacting 3D systems with underlying *finite-range local*, *non-local* interactions. For this we make use of a microscopic theory that allows for the consideration of the full momentum dependence of the renormalized many-body interactions: the diagrammatic *crossing symmetric technique*.[5] Our goal here is to find in a self-consistent manner *quantum fluctuations* and competing *pairing bosons* that are fermionic in origin. Pairing in higher angular momentum ($l \neq 0$) channels can effectively lower the strong mutual fermionic repulsion. Hence unconventional pair symmetries are believed possible in these systems.

Since it proves to be extremely difficult to calculate excited state properties of a Fermi system using the full scheme, we use a *tractable* version,[6–10] and the ideas

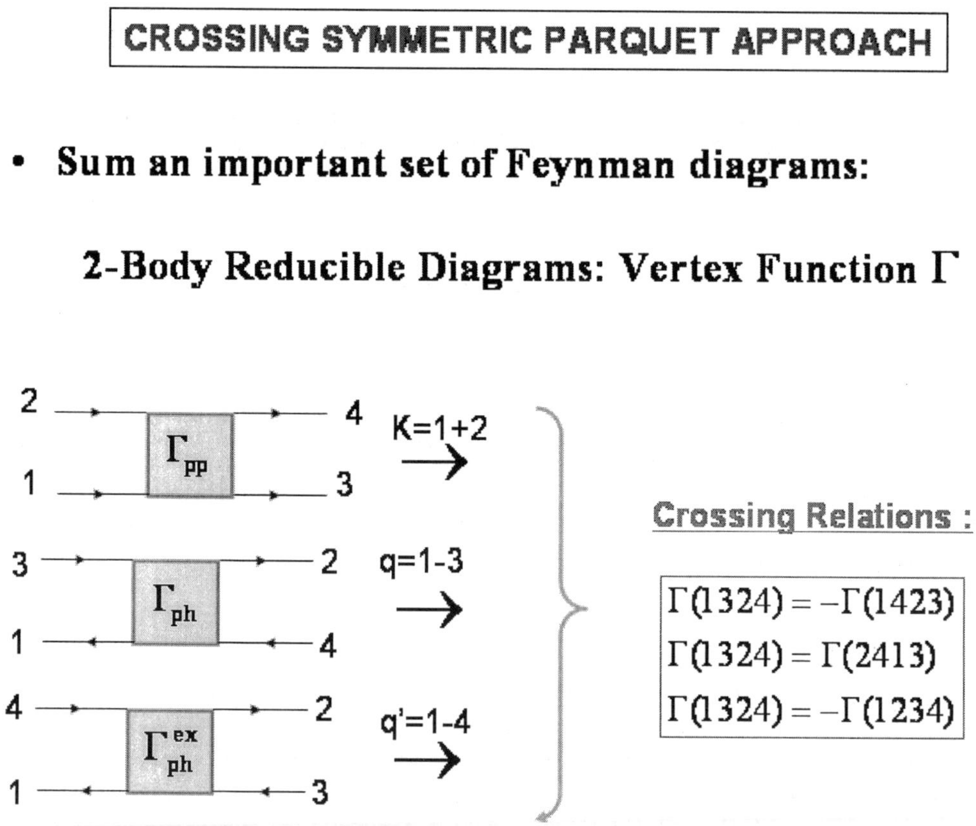

Fig. 1. 2-body reducible vertex functions Γ in the particle-particle (s), particle-hole (t) and exchange particle-hole (u) channels. These give rise to coupled non-linear integral equations that are crossing-symmetric; a few crossing relations are shown. K, q, q' are the momentum-energy transfers in the 3 channels.

of a generalized Fermi Liquid Theory (defined for finite momentum). This allows us to obtain renormalized quasiparticle interaction functions, $F_l(q)$ and *pairing amplitudes* with various angular momentum content (s, p, d, etc). Among the key finding in the paramagnetic state, is the remarkable role of the non-local interaction in switching from p-wave pairing (for small, *local*, finite-range interaction) to s-wave pairing for repulsive non-local interaction; and to a narrow region of d-wave pairing for an attractive one. The s-wave pairing, accompanied by large effective mass enhancements, is driven by transverse current fluctuations competing with ferromagnetic spin fluctuations; the d-wave pairing by "antiferromagnetic-like" spin fluctuations.

2. Crossing-symmetric Equations

At the core of our tractable scheme is the "parquet" method[5] that sums a large class of important Feynman diagrams, namely, the 2-body planar diagrams; Fig. 1. Each of the vertices represent a large set of Feynman diagrams, and the lines with arrows the relevant fermion particle and hole Green's functions. On considering the s (particle-particle), t (particle-hole), and u (exchange particle-hole) channels in a conserving, self-consistent fashion, and using crossing symmetry relations (as shown in Fig. 1), one is led to a set of crossing symmetric non-linear coupled integral equation(s) for the fully reducible 4-point vertex function Γ. It has however proved difficult to solve the full "parquet" equations and/or the resulting crossing-symmetric equation for the full vertex self-consistently to obtain excited state properties.

3. Tractable Crossing-symmetric Equations

Progress in obtaining exited state properties can be made on recognizing the well-defined Babu-Brown limit on the Fermi surface. A judicious exploitation of the p-h processes near the fermi surface, partial resummation of certain diagrams, quasiparticle renormalization, and the imposition of crossing symmetry leads to a *tractable* version of the coupled non-linear integral equations.[6–10] The renormalized quasiparticle interactions in the Fermi liquid retain the full crossing symmetry of the underlying Hamiltonian while reproducing the low energy physics of a Fermi liquid. The approach insures that the Ward identities for charge, spin, current, etc. are satisfied. Thus, this reformulation in terms of Landau interaction functions, scattering amplitudes, and a fully 1p-1h irreducible part, is exact on the Fermi surface.

We extrapolate away from the the Fermi surface by summing the p-h diagrams to all orders in both the t and u channels with renormalized, fully *momentum-dependent* interaction functions.[7] This necessarily requires non-diagonal phase space functions, $\chi(q), \chi(q')$, in addition to diagonal Lindhard-type functions. The formulation is particularly well-suited for calculations of different quantum fluctuations because the renormalized interaction can be separated into a part that is fully 1p-1h irreducible, termed the "direct" interaction, $D(q)$, and a part that includes the exchange 1p-1h reducible processes, $U(q')$ termed the "induced" interaction. The resulting equations for the interaction functions, $F(q)$, scattering amplitudes, $A(q)$, and $D(q)$ are shown schematically in Fig. 2.

In physical terms, the "induced" part, $U(q')$ represent exchange of competing quantum fluctuations, such as density, spin-density, current, spin-current, and higher order fluctuations. The "direct" part contains p-p diagrams to all order (e.g. t-matrix) as well as processes that reflect possible non-locality of the underlying Hamiltonian. We have extended it to include the full momentum and spin polarization dependence of the interaction functions.[7,11]

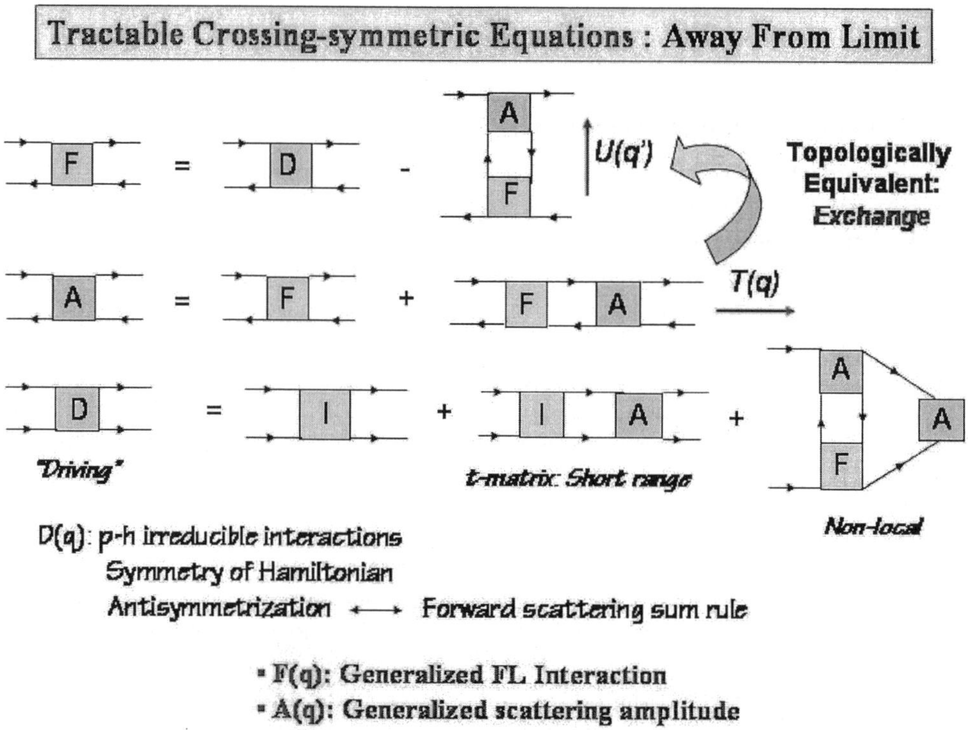

Fig. 2. Tractable crossing symmetric equations for $F(q)$, $A(q)$, and $D(q)$, and their diagram forms away from the Fermi surface. q and q' are the momentum transfers in the p-h and exchange p-h channels respectively.

Within this formulation of the theory, given a "direct" interaction $D(q)$, one can obtain the renormalized interactions, $F(q)$, the scattering amplitudes $A(q)$ and the effective mass m^*/m. Thus the chief approximation is in the choice of $D(q)$. It is important, however, to ensure that $D(q)$ is properly antisymmetrized so as to preserve crossing symmetry. For a properly constructed "direct" term projected in orthogonal channels (e.g. angular momentum, ℓ, and spin-symmetric(s)/antisymmetric(a) channels), one obtains a set of coupled non-linear integral equations for $F_l^{s,a}(q)$'s and $A_l^{s,a}(q)$'s in different angular momentum channels that have to be solved self-consistently; see Fig. 3. The dimensionality of the problem to be solved depends on the number of moments retained for $F(q)$ and $A(q)$.

Coupled Non-linear Equations: ph & ex-ph Channels

$$1.\ F_l^s(q) = D_l^s(q) + P_{l,q}\Big[\sum_{l',m',n'} \Big\{ \frac{1}{2} F_{n'}^s(q')\chi_{l'}^{n',m'}(q')A_{m'}^s(q') + \frac{3}{2} F_{n'}^a(q')\chi_{l'}^{n',m'}(q')A_{m'}^a(q')\Big\}\Big]$$

projection operator

$$2.\ F_l^a(q) = D_l^a(q) + P_{l,q}\Big[\sum_{l',m',n'} \Big\{ \frac{1}{2} F_{n'}^s(q')\chi_{l'}^{n',m'}(q')A_{m'}^s(q') - \frac{1}{2} F_{n'}^a(q')\chi_{l'}^{n',m'}(q')A_{m'}^a(q')\Big\}\Big]$$

$$3.\ A_l^s(q) = F_l^s(q) - \sum_{m,n} F_n^s(q)\chi_l^{n,m}(q)A_m^s$$

$$4.\ A_l^a(q) = F_l^a(q) - \sum_{m,n} F_n^a(q)\chi_l^{n,m}(q)A_m^a$$

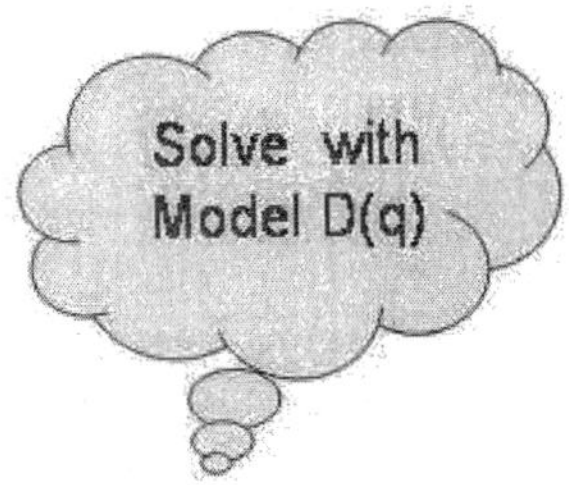

Fig. 3. Double Legendre expansions of $F(q)$, $A(q)$ and $D(q)$ (Fig 5 above) in q, q' variables give rise to these coupled non-linear equations of the ph and exchnage ph channesl

4. Model Calculations

In our present calculation, for the "direct" term, $D(q)$, we use the following model: Fourier transform of an effective quasiparticle interaction. Since we allow for the intermediate states to be *off* the Fermi surface, the scattering processes are necessarily a function of two indepenedent variables, viz., (q, q'), or two angular variables that depend on the interaction channel, s, t, u. This requires a *double Legendre expansion* of $D(q), F(q), A(q)$; see Fig Accordingly the phase space functions can be *non-diagonal* Lindhard-type functions.

The antisymmetrized form of $D(q)$ is parametrized by a zero-range interaction term, U_o between up ($\uparrow$) and down ($\downarrow$) -spin quasiparticles; a finite-range interaction term U_f between two up ($\uparrow$) -spin quasiparticles; and a non-local term, U_{nl} between

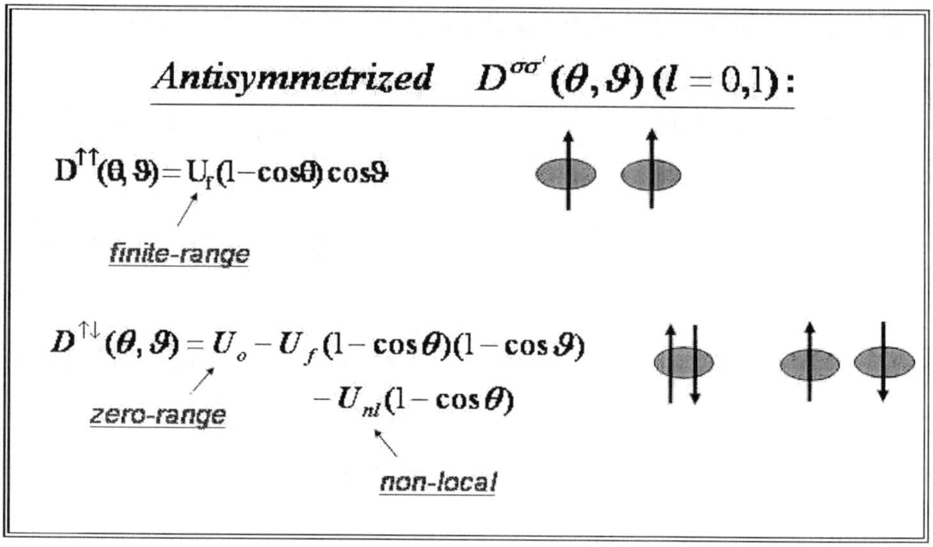

Fig. 4. Antisymmetrized "Direct" interaction showing zero-range (U_o), finite-range (U_f), and the non-local (U_{nl} components used in our model calculations

up (↑) and down (↓) -spin quasiparticles. The $\ell = 0, 1$ terms in $D^{\sigma,\sigma'}(\theta,\phi)$ are shown in Fig. 4; θ, ϕ are the scattering angles in the s-channel. . In the Landau limit: $q \to 0$, $q' \to 0$, $\cos\theta \to \cos\Theta_L = \cos\Theta'_L$, and $\cos\phi \to 1, -1$, where Θ_L and Θ'_L are the generalized Landau angles in the t and u -channel respectively. The range of our 3 parameters are such that $U_f, U_{nl} \sim O(0.1)U_o$.

4.1. *Pairing Interactions*

Pair-couplings, λ_ℓ, in different angular momentum channels, can be classified as *singlets* or *triplets*, depending whether ℓ is *even* or *odd*. To obtain the pair-couplings, we construct the *triplet* and *singlet* pairing channel scattering amplitudes, $T^{t,s}(\theta,\phi)$ from the scattering amplitudes $A^{s,a}(\theta,\phi)$, obtained by solving our coupled non-linear equations. The construction of pairing interactions is shown in Fig. 5.

Pairing Channel Scattering Amplitudes

$$T^{t,s}(\theta,\vartheta) = \sum_{l,m} T^{s,t}_{l,m}(\theta = \pi, \vartheta)\,(-1)^l P_m \cos(\vartheta)$$

$$T^{t}(\theta,\vartheta) = A^{s}(\theta,\vartheta) + A^{a}(\theta,\vartheta) \quad : \text{Triplet}$$

$$T^{s}(\theta,\vartheta) = A^{s}(\theta,\vartheta) - 3A^{a}(\theta,\vartheta) \quad : \text{Singlet}$$

Pairing Interaction in Angular Momentum Channels

$$\lambda_l = \frac{1}{8}\int_{-1}^{+1} d(\cos\vartheta)\, T^{t,s}(\theta = \pi, \vartheta)\, P_l(\cos\vartheta)$$

Fig. 5. Construction of the pairing interactions, λ_ℓ from the scattering amplitudes, $A^{s,a}(\theta,\phi)$ obtained from our calculations

5. Solutions and Results

For our calculations, we found that retaining between 10 and 15 moments in the double moment Legendre expansions of $F(q), A(q)$ were sufficient for stable results; higher moments were found to be not significant.

For most of the calculations, we kept the finite-range interaction fixed at $U_f \sim O(0.1)U_o$, and varied the strength of the zero-range interaction, U_o and the non-local interaction, U_{nl}. Fig. 6 shows 3D plots of the s, p, d pairing interaction strengths, $\lambda_o, \lambda_1, \lambda_2$ as functions of U_o and U_{nl}.

The results are summarized in a schematic phase diagram in Fig. 7. Our calculation of the spin fluctuations $F^a_o(q)$ show that it is peaked at low-q for repulsive non-local interactions, reminiscent of ferromagnetic excitations, and at high-q for attractive non-local interactions, reminiscent of antiferromagnetic spin fluctuations. We believe that the *p-wave* pairing is driven mainly by ferromagnetic spin fluctua-

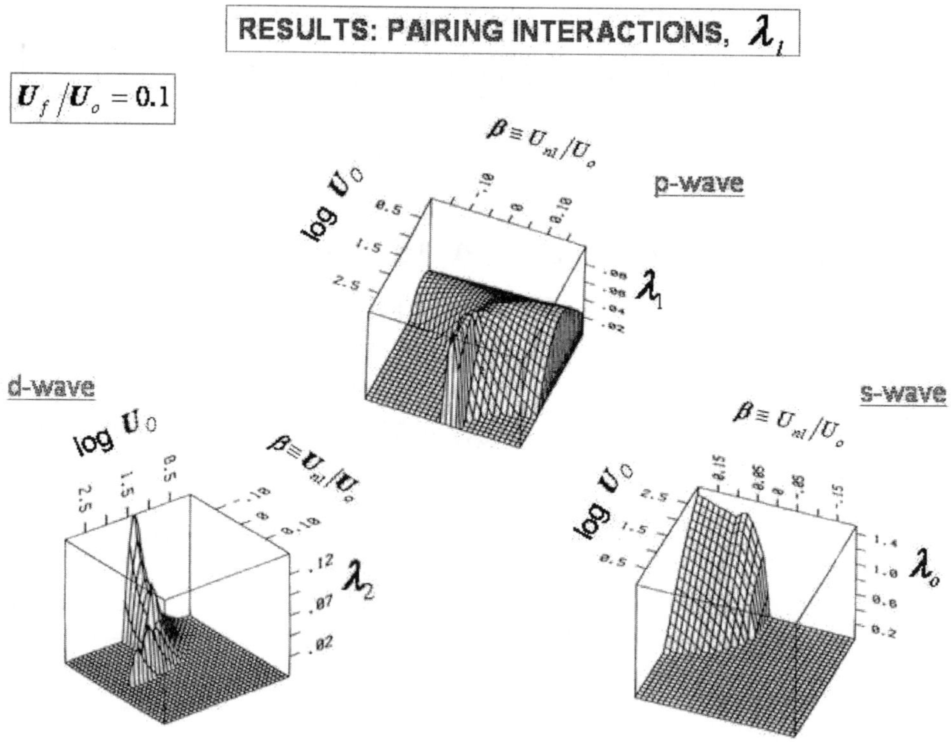

Fig. 6. 3D plots of our results for the s-wave (λ_0), p-wave (λ_1), and d-wave (λ_2) pairing interactions shown as functions of zero-range (U_o) and nono-local (U_{nl}) interactions, and a fixed finite-range (U_f) interaction

tions ("paramagnons"), F_0^a, for both repulsive and attractive values of the non-local interaction. We find that the strength of transverse current fluctuations (and consequently the effective mass) *increase* as the U_o is increased for repulsive U_{nl}. This competes with diminishing strength of spin fluctautions to favor *s-wave* pairing. For attractive U_{nl} and a narrow region of U_o, we find *d-wave* pairing. We believe that this is due to the development of strong anti-ferromagnetic type spin fluctuations accompanied by very small current fluctuations. Since transverse current fluctuations increases with U_o for repulsive non-local interactions, effective mass increases, and for attractive non-local interactions, it decreases.

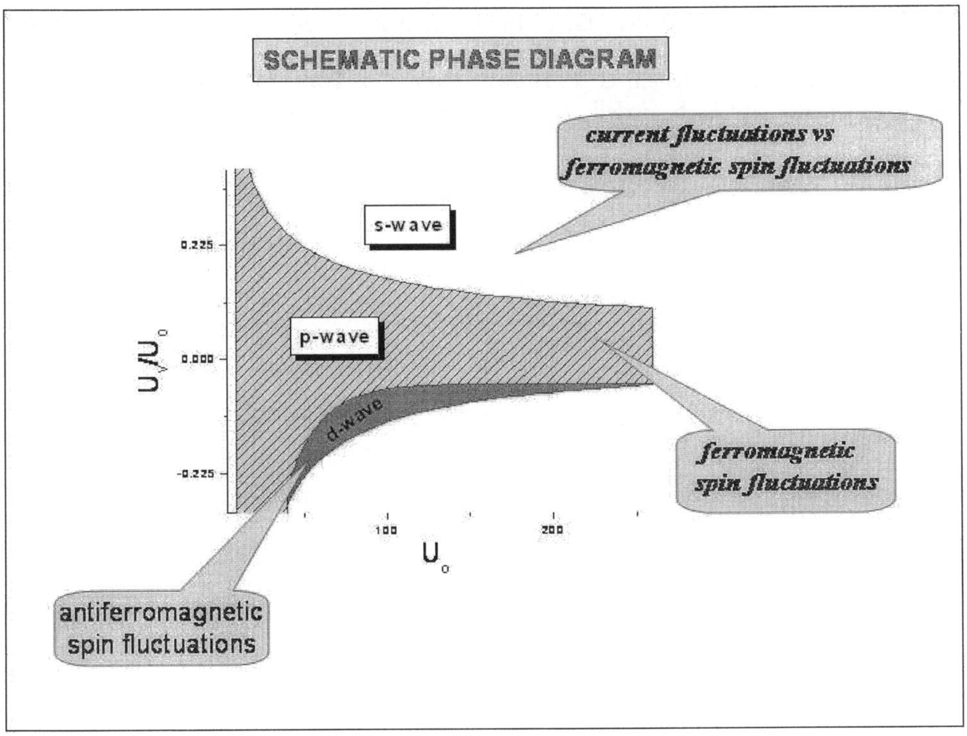

Fig. 7. Schematic Phase Diagram of our pairing results shown for different values of repulsive zero-range interaction, U_o, and repulsive and attractive values of the non local interaction, U_V. The finite-range interaction, U_f is fixed at $\approx O(0.1)U_o$

References

1. N S. Saxena, et al., *Nature* (London) **406**, 587 (2000).
2. D.N. Aoki, et al., *Nature* (London) **413**, 613 (2001).
3. C. Pfleiderer et al.,*Nature* **412**, 59 (2001).
4. N. D. Mathur et al., *Nature* **394**, 39 (1998).

5. A. D. Jackson, A. Lande, and R. A. Smith, *Phys. Reports* **86**, 55 (1982); N. E. Bickers and D. J. Scalapino, *Ann. Phys.* **193**, 206 (1989).
6. K. F. Quader, "The Induced Interaction Way to Quasiparticle Interactions: Minimal Fermion Parquet", in *Windsurfing the Fermi Sea*, eds. T. T. S. Kuo, and J. Speth (Elsevier, 1987), Vol. 2, p. 390.
7. T. L. Ainsworth and K. S. Bedell, *Phys. Rev.* **B35**, 8425 (1987).
8. S. Babu and G. E. Brown, *Ann. Phys.* 78, 1 (1973).
9. T. L. Ainsworth, K. S. Bedell, G. E. Brown, and K. F. Quader, *J. Low Temp. Phys.* **50**, 315 (1983).
10. K. F. Quader, K. S. Bedell, G. E. Brown, *Phys. Rev.* **B36**, 156 (1987).
11. K. F. Quader and K. S. Bedell, *J. Low Temp. Phys.*, **58**, 89 (1985).

THEORY OF ELECTRON SPECTROSCOPIES IN STRONGLY CORRELATED SEMICONDUCTOR QUANTUM DOTS

MASSIMO RONTANI

CNR-INFM National Research Center on nanoStructures and bioSystems at Surfaces (S3)
Via Campi 213/A, Modena, 41100, Italy
rontani@unimore.it

Quantum dots may display fascinating features of strong correlation such as finite-size Wigner crystallization. We here review a few electron spectroscopies and predict that both inelastic light scattering and tunneling imaging experiments are able to capture clear signatures of crystallization.

Keywords: Quantum dots; configuration interaction; electron solid.

1. Quantum Dots as Tunable Correlated Systems

Semiconductor quantum dots[1,2] (QDs) are nanostructures where electrons (or holes) are confined by electrostatic fields. The confinement field may be provided by compositional design (e.g., the QD is formed by a small gap material embedded in a larger-gap matrix) or by gating an underlying two-dimensional (2D) electron gas. Different techniques lead to nanometer size QDs with different shapes and strengths of the confinement. As the typical de Broglie wavelength in semiconductors is of the order of 10 nm, nanometer confinement leads to a discrete energy spectrum, with energy splittings ranging from fractions to several tens of meV.

The similarity between semiconductor QDs and natural atoms, ensuing from the discreteness of the energy spectrum, is often pointed out.[3–7] Shell structure,[5,6,8] correlation effects,[9] Kondo physics,[10,11] are among the most striking experimental demonstrations. An intriguing feature of these *artificial* atoms is the possibility of a fine control of a variety of parameters in the laboratory. The nature of ground and excited few-electron states has been shown to vary with artificially tunable quantities such as confinement potential, density, magnetic field, inter-dot coupling.[1,2,7,12–14] Such flexibility allows for envisioning a vast range of applications in optoelectronics (single-electron transistors,[15] lasers,[16] micro-heaters and micro-refrigerators based on thermoelectric effects[17]) life sciences,[18] as well as in several quantum information processing schemes in the solid-state environment (e.g. Ref. 19).

Almost all QD-based applications rely on (or are influenced by) electronic correlation effects, which are prominent in these systems. The dominance of interaction in artificial atoms is evident from the multitude of strongly correlated few-electron states measured or predicted under different regimes: Fermi liquid, Wigner molecule (the precursor of Wigner crystal in 2D bulk), charge and spin density wave, incompressible state reminescent of fractional quantum Hall effect in 2D (for reviews see Refs. 12, 20). The origin of strong correlation effects in QDs is the following: While the kinetic energy term of the Hamiltonian scales as r_s^{-2}, r_s being the parameter measuring the average distance between electrons, the Coulomb energy term scales as r_s^{-1}. Contrary to natural atoms, in QDs the ratio of Coulomb to kinetic energy can be rather large (even larger than one order of magnitude), the smaller the carrier density the larger the ratio. This causes Coulomb correlation to severely mix many different Slater determinants.

The way of tuning the strength of correlation in QDs we focus on in this paper is to dilute electron density. At low enough densities, electrons evolve from a "liquid" phase, where low-energy motion is equally controlled by kinetic and Coulomb energy, to a "crystallized" phase, reminescent of the Wigner crystal in the bulk, where electrons are localized in space and arrange themselves, in absence of disorder, in a geometrically ordered configuration (Wigner molecule[12]), so that electrostatic repulsion is minimized.[7,12] So far, there are no direct experimental confirmations of the existence of these fascinating states.

In this paper we focus on theoretical properties of electron states in Wigner molecules and on their possible signatures in experimentally available spectroscopies. We predict that both inelastic light scattering and wave function imaging techniques, like scanning tunneling spectroscopy (STS), may provide direct access to the peculiar behavior of crystallized electrons. Specifically, the former is able to probe the "normal modes" of the molecules, while the latter is sensitive to the spatial order of the electron phase.

The structure of the paper is as follows. After an exposition of our theoretical and computational approach (Sec. 2), we briefly review inelastic light scattering and imaging spectroscopies (Sec. 3), showing a few results for the four-electron Wigner molecule (Secs. 4 and 5).

2. Theoretical Approach to the Interacting Problem: Configuration Interaction

The theoretical understanding of QD electronic states in a vaste class of devices is based on the envelope function and effective mass model.[21] Here, changes in the Bloch states, the eigenfunctions of the bulk semiconductor, brought about by "external" potentials other than the perfect crystal potential, are taken into account by a slowly varying (envelope) function which multiplies the fast oscillating periodic part of the Bloch states. This decoupling of fast and slow Fourier components of the wave function is valid provided the modulation of the external potential is slow on the scale of the lattice constant. Then, the theory allows for calculating such envelope functions from an effective Schrödinger equation where only the external potentials appear, while the unperturbed crystalline potential enters as a renormalized electron mass, i.e. the effective value m^* replaces the free electron mass. Therefore, single-particle (SP) states can be calculated in a straightforward way once compositional and geometrical parameters are known. This approach was proved to be remarkably accurate by spectroscopy experiments for weakly confined QDs;[12,13] for strongly confined systems, such as certain classes of self-assembled QDs, atomistic methods might be necessary.[22]

The QD "external" confinement potential originates either from band mismatch or from the self-consistent field due to doping charges. In both cases, the total field modulation which confines a few electrons is smooth and can often be approximated by a parabolic potential in two dimensions. Then, the fully interacting effective-mass Hamiltonian of the QD system reads as:[12]

$$H = \sum_i^N H_0(i) + \frac{1}{2}\sum_{i\neq j} \frac{e^2}{\kappa|\boldsymbol{r}_i - \boldsymbol{r}_j|}, \tag{1}$$

with

$$H_0(i) = \frac{\boldsymbol{p}_i^2}{2m^*} + \frac{1}{2}m^*\omega_0^2 r_i^2. \tag{2}$$

Here, N is the number of free conduction band electrons localized in the QD, e and κ are respectively the electron charge and static relative dielectric constant of the host semiconductor, $\boldsymbol{r}$ is the position of the electron, $\boldsymbol{p}$ is its canonically conjugated momentum, ω_0 is the natural frequency of a 2D harmonic trap.

The eigenstates of the SP Hamiltonian (2) are known as Fock-Darwin (FD) orbitals.[1] The typical QD lateral extension is given by the characteristic dot radius $\ell_{\mathrm{QD}} = (\hbar/m^*\omega_0)^{1/2}$, ℓ_{QD} being the mean square root of r on the FD lowest-energy level. As we keep N fixed and increase ℓ_{QD} (decrease the density), the Coulomb-to-kinetic energy dimensionless ratio[23] $\lambda = \ell_{\mathrm{QD}}/a_{\mathrm{B}}^*$ [$a_{\mathrm{B}}^* = \hbar^2\kappa/(m^*e^2)$ is the effective Bohr radius of the dot] increases as well, driving the system into the Wigner regime.

We solve numerically the few-body problem of Eq. (1), for the ground (or excited) state at different numbers of electrons, by means of the configuration interaction (CI) method:[24] We expand the many-body wave function in a series of Slater determinants built by filling in the FD orbitals with N electrons, and consistently with global symmetry constraints. Specifically, the global quantum numbers of the few-body system are the total angular momentum in the direction perpendicular to the plane, M, the total spin, S, and its projection along the z-axis, S_z. In the Slater-determinant basis the few-body Hamiltonian (1) is a large, block diagonal sparse matrix that we diagonalize by means of a newly developed parallel code.[25] Note that, before diagonalization, we are able to build separate sectors of the Fock space corresponding to different values of (M, S, S_z). We use, as a single-particle basis, up to 36 FD orbitals, and we are able to diagonalize matrices of linear dimensions up to $\approx 10^6$ (see Ref. 24). The output of calculations consists of both energies and wave functions of the selected correlated states. We then post-process wave functions to obtain those response functions connected to the spectroscopy of interest, which we will discuss in Sec. 3.

3. Quantum Dot Spectroscopies

There are two main classes of electron spectroscopies in QDs (cf. Fig. 1). In single-electron tunneling spectroscopies one is able to inject just one electron into the interacting system, in virtue of the Coulomb blockade effect.[15] In this way one accesses quantities related mainly to the ground states of both N and $N+1$ electrons, like the dot chemical potential $E_0(N+1) - E_0(N)$, where $E_0(N)$ is the N-electron ground state energy. While such transport spectroscopies were very successful in QD studies, they suffer severe limitations when probing excited states. On the other hand, optical techniques such as far-infrared and inelastic light scattering spectroscopies, which leave the probed system uncharged, allow for an easier access to excitation energies $E^*(N)$. Below we focus on two specific examples of these families of experiments (for reviews see Refs. 1, 2, 12, 20).

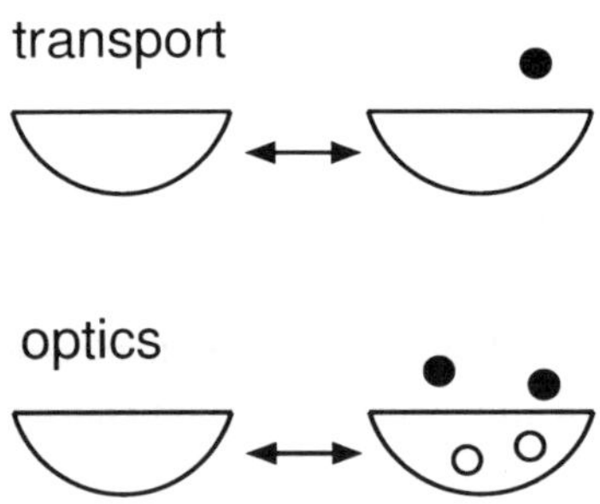

Fig. 1. Transport vs. optical electron spectroscopies in quantum dots.

3.1. *Inelastic light scattering*

Inelastic light scattering experiments in QDs probe low-lying neutral excitations. Depending on the relative orientation of the polarizations of the incoming and scattered photons, one is able to access either charge or spin density fluctuations.[26] Different excited states can be probed by varying the frequency of the laser with respect to the optical energy band gap of the host semiconductor, or by changing the momentum tranferred from photons to electrons.[9,27–29]

Only recently QDs with very few electrons were studied.[9,29] Here we focus on charge density excitations probed when the laser frequency is far from the resonance with the optical gap.[30,31] In this limit, the scattering cross section $d\sigma/d\omega$ at a given energy ω is proportional to the momentum-resolved dynamical response function:

$$\frac{\mathrm{d}\sigma}{\mathrm{d}\omega} \approx \sum_n |M_{n0}|^2 \, \delta\left(\omega - \omega_n + \omega_0\right), \text{ with } M_{n0} = \textstyle\sum_{ab\sigma} \left\langle a|e^{i\mathbf{q}\cdot\mathbf{r}}|b\right\rangle \left\langle n|c^{\dagger}_{a\sigma}c_{b\sigma}|0\right\rangle . \quad (3)$$

Here $|0\rangle$ and $|n\rangle$ are the ground and excited interacting few-body states, as obtained by the CI computation, with energies ω_0 and ω_n, respectively, $c^{\dagger}_{a\sigma}$ is a fermionic operator creating an electron in the a-th FD orbital with spin σ, $\mathbf{q}$ is the wave vector transferred in the inelastic photon scattering event. Formula (3) describes how ground and excited states are selectively coupled by charge density fluctuations of momentum $\mathbf{q}$.

3.2. *Wave function imaging via tunneling spectroscopy*

The imaging experiments, in their essence, measure quantities directly proportional to the probability for transfer of an electron through a barrier, from an emitter, where electrons fill in a Fermi sea, to a dot, with completely discrete energy spectrum. In multi-terminal setups one can neglect the role of electrodes other than the emitter, to a first approximation. The measured quantity can be the current,[32,33] the differential conductance,[6,34,35] or the QD capacitance,[5,36,37] while the emitter can be the STS tip,[32,34,35] or a n-doped GaAs contact,[5,6,33,36,37] and the barrier can be the vacuum[32,34,35] as well as a AlGaAs spacer.[5,6,33,36,37]

The measured quantities are generally understood as proportional to the density of carrier states at the resonant tunneling (Fermi) energy, resolved in either real[32,34,35] or reciprocal[33,36] space. However, Coulomb blockade phenomena and strong inter-carrier correlation —the fingerprints of QD physics— complicate the above simple picture. Below we review the theoretical framework we have recently developed[38,39] to clarify the quantity actually probed by imaging spectroscopies.

According to the seminal paper by Bardeen,[40] the transition probability (at zero temperature) is given by the expression $(2\pi/\hbar)\,|\mathcal{M}|^2\,n(\epsilon_f)$, where $\mathcal{M}$ is the matrix element and $n(\epsilon_f)$ is the energy density of the final QD states. The standard theory would predict the probability to be proportional to the total density of QD states at the resonant tunneling energy, ϵ_f, possibly space-resolved since $\mathcal{M}$ would depend on the resonant QD orbital.[41] Let us now assume that: (i) Electrons in the emitter do not interact and their energy levels form a continuum. (ii) Electrons from the emitter access through the barrier a single QD at a sharp resonant energy, corresponding to a well defined interacting QD state. (iii) The QD is quasi-2D, the electron motion being separable in the xy plane and z axis, which is parallel to the tunneling direction. (iv) Electrons in the QD all occupy the same confined orbital along z, $\chi_{\mathrm{QD}}(z)$. Then one can show[38] that the matrix element $\mathcal{M}$ may be factorized as

$$\mathcal{M} \propto TM, \quad (4)$$

where T is a purely single-particle matrix element while the integral M contains the whole correlation physics.

The former term is proportional to the current density evaluated at any point z_{bar} in the barrier:

$$T = \frac{\hbar^2}{2m^*}\left[\chi_{\text{E}}^*(z)\frac{\partial\chi_{\text{QD}}(z)}{\partial z} - \chi_{\text{QD}}(z)\frac{\partial\chi_{\text{E}}^*(z)}{\partial z}\right]_{z=z_{\text{bar}}}, \tag{5}$$

where $\chi_{\text{E}}(z)$ is the resonating emitter state along z evanescent in the barrier. The term (5) contains the information regarding the overlap between emitter and QD orbital tails in the barrier, $\chi_{\text{E}}(z)$ and $\chi_{\text{QD}}(z)$, respectively. Since T is substantially independent from both N and xy location, its value is irrelevant in the present context.

On the other hand, the in-plane matrix element M conveys the information related to correlation effects. If we now specialize to STS and assume an ideal, point-like tip, then $M \propto \varphi_{\text{QD}}(\boldsymbol{r})$ where $\varphi_{\text{QD}}(\boldsymbol{r})$ is the quasi-particle (QP) wave function of the interacting QD system:

$$\varphi_{\text{QD}}(\boldsymbol{r}) = \langle N-1|\hat{\Psi}(\boldsymbol{r})|N\rangle. \tag{6}$$

Here $\hat{\Psi}(\boldsymbol{r})$ is the fermionic field operator destroying an electron at position $\boldsymbol{r} \equiv (x, y)$, $|N-1\rangle$ and $|N\rangle$ are the QD interacting ground states with $N-1$ and N electrons, respectively, calculated via CI (Sec. 2). We omit spin indices for the sake of simplicity.

Therefore, the STS differential conductance is proportional to $|\varphi_{\text{QD}}(\boldsymbol{r})|^2$, which is the usual result of the standard non-interacting (or mean-field) theory[35,41] for the highest-energy occupied (or lowest-energy unoccupied) orbital, provided the SP (mean-field) orbital is replaced by the QP wave function unambiguously defined by Eq. (6).

4. Inelastic Light Scattering Spectroscopy in the Wigner Limit

We focus on the four-electron system at density low enough ($\lambda = 10$) to induce the crystallization of a Wigner molecule. Figure 2(a) displays our CI results for the low-energy region of the excitation spectrum for different values of the total orbital angular momentum and spin multiplicities (the so called *yrast* line[12]). The ground state is the spin triplet state with $M = 0$, which is found to be the lowest-energy state in the range $0 \leq \lambda \leq 20$. The absence of any level crossing as the density is progressively diluted (i.e. λ increases) implies that the crystallization process evolves in a continuos manner, consistently with the finite-size character of the system. Nevertheless, several features of Fig. 2(a) demonstrate the formation of a Wigner molecule in the dot.

First, the two possible spin multiplets other than $S = 1$, namely $S = 0, 2$, lie very close in energy to the ground state. This is understood by invoking electron crystallization. In fact, in the Wigner limit, the Hamiltonian of the system turns into a classical quantity, since the kinetic energy term in it may be neglected with respect to the Coulomb term. Therefore, only commuting operators (the electron

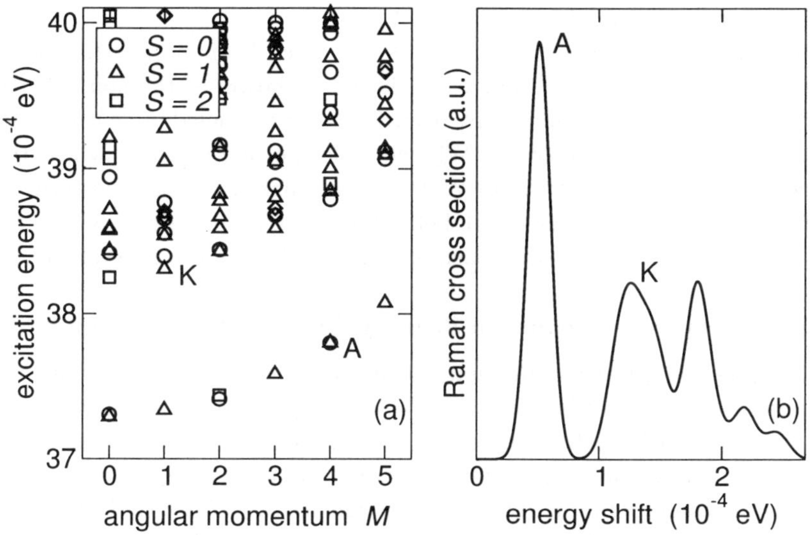

Fig. 2. (a) Excitation energies of the four-electron Wigner molecule vs. total orbital angular momentum M, for different values of the total spin S, with $\lambda = 10$. (b) Corresponding Raman scattering cross section for charge excitations in the off-resonance limit. Here we set $|\mathbf{q}| = 2 \times 10^5$ cm^{-1} and introduce a fictitious gaussian broadening of Raman peaks ($\sigma = 0.9 \times 10^{-5}$ eV). GaAs parameters were used throughout the paper.

positions) appear in the Hamiltonian. In this regime the spin, which has no classical counterpart, becomes irrelevant:[24,42] spin-dependent energies show a tendency to degeneracy. This can also be understood in the following way: if electrons sit at some lattice sites with unsubstatial overlap of their localized wave functions, then the total energy must not depend on the relative orientation of neighboring spins.

Second, an exact replica of the sequence of the two lowest-energy singlet and triplet states occurs for both $M = 0$ and for $M = 4$ [(cf. the triplet state labeled A in Fig. 2(a)]. Such a period of four units on the M axis identifies a magic number,[3,43] whose origin is brought about by the internal spatial symmetry of the interacting wave function.[44,45] In fact, when electrons form a stable Wigner molecule, they arrange themselves into a *four-fold* symmetry configuration, where charges are localized at the corners of a square.[[illegible],46,47]

A third distinctive signature of crystallization is the appearance of a *rotational band*,[43] which in Fig. 2(a) is composed of those lowest-energy levels, separated by energy gaps of about 0.1 meV, that increase monotonically as M increases. This band is called "rotational" since it can be identified with the quantized levels $E_{\text{rot}}(M)$ of a rigid two-dimensional top, given by the formula

$$E_{\text{rot}}(M) = \frac{\hbar^2}{2I} M^2,$$

where I is the moment of inertia of the top.[43] These excitations may be thought of as the "normal modes" of the Wigner molecule rotating as a whole in the xy plane around the vertical symmetry axis parallel to z.

Figure 2(b) displays the calculated off-resonance Raman spectrum for charge density fluctuations ($\Delta S = 0$, where ΔS is the variation of the spin with respect to the ground state). The dominant peak is the one labeled A, corresponding to the $\Delta M = 4$ normal mode of the Wigner molecule of Fig. 2(a), which therefore represents a clear, visible feature of Wigner crystallization. Notice also the appearance of the so called Kohn mode[1] (labeled K) at higher energy, which is a dipolar ($\Delta M = 1$) collective motion of the center of mass of the electron system.

5. Wigner Molecule Formation Seen by Imaging Spectroscopy

Figure 3 shows the QP wave function, corresponding to the injection of the fourth electron into the QD, as a function of x (at $y = 0$), for two different values of λ. As λ increases (from 0.5 to 10), the density decreases going from the non-interacting limit ($\lambda = 0.5$) deep into the Wigner regime ($\lambda = 10$). At high density ($\lambda = 0.5$, approximately corresponding to the electron density $n_e = 3.3 \times 10^{12}$ cm^{-2}) the wave function substantially coincides with the non-interacting FD p-like orbital. By in-

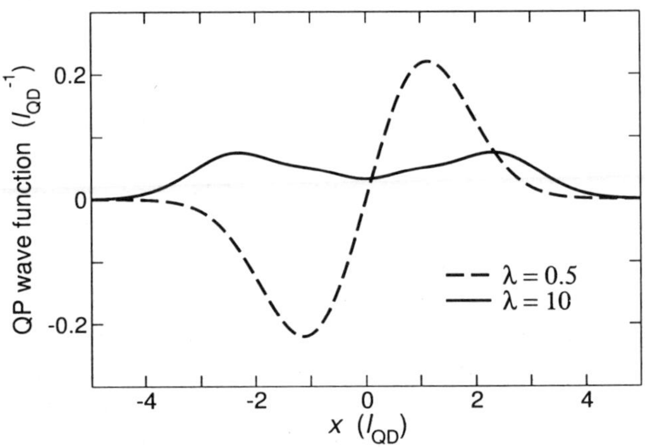

Fig. 3. Quasi-particle wave function vs. x ($y = 0$) for different values of the dimensionless Coulomb-to-kinetic energy ratio λ. The STS differential conductance, proportional to the wave function square modulus, corresponds to the tunneling process $N = 3 \to N = 4$. The length unit is the characteristic dot radius ℓ_{QD}.

creasing the QD radius (and λ), the QP wave function weight moves towards larger values of r (the s-like symmetry of the $\lambda = 10$ curve comes from a transition of the three-electron ground state[24] around $\lambda \approx 6$). By measuring lengths in units of ℓ_{QD}, as it is done in Fig. 3, this trivial effect should be totally compensated. However, we see for $\lambda = 10$ ($n_e \approx 1.1 \times 10^9$ cm^{-2}) that the now much stronger correlation is responsible for an unexpected weight reorganization, which is related to the formation of a "ring" of crystallized electrons in the Wigner molecule (cf. Sec. 4). The shape of the $\lambda = 10$ curve of Fig. 3 is consistent with the onset of a solid phase with four electrons sitting at the apices of a square, as discussed in Sec. 4. Note also the dramatic weight loss of QPs as λ is increased: the stronger the correlation, the more effective the "orthogonality" between interacting states.

6. Conclusions

The Wigner molecule is an intriguing electron phase, peculiar of QDs, which still lacks for experimental confirmation. We predict that both inelastic light scattering and imaging spectroscopies are able to probe distinctive features of crystallization.

Acknowledgements

G. Goldoni and E. Molinari in Modena have been deeply involved in the elaboration of the results presented here. We thank for fruitful discussions V. Pellegrini, C. P. García, A. Pinczuk, S. Heun, A. Lorke, G. Maruccio, S. Tarucha, B. Wunsch. This paper is supported by Supercomputing Project 2006 (Iniziativa Trasversale INFM per il Calcolo Parallelo), MIUR-FIRB Italia-Israel RBIN04EY74, Italian Ministry of Foreign Affairs (Ministero degli Affari Esteri, DGPCC).

References

1. L. Jacak, P. Hawrylak, and A. Wójs, *Quantum Dots* (Springer, Berlin, 1998).
2. T. Chakraborty, *Quantum Dots - A Survey of the Properties of Artificial Atoms* (North-Holland, Amsterdam, 1999).
3. P. A. Maksym and T. Chakraborty, *Phys. Rev. Lett.* **64**, 108 (1990).
4. M. A. Kastner, *Phys. Today* **46** (March), 24 (1993).
5. R. Ashoori, *Nature* **379**, 413 (1996).
6. S. Tarucha, D. G. Austing, T. Honda, R. J. van der Hage, and L. P. Kouwenhoven, *Phys. Rev. Lett.* **77**, 3613 (1996).
7. M. Rontani, G. Goldoni, and E. Molinari, in *New directions in mesoscopic physics (towards nanoscience)*, ed. R. Fazio *et al.*, *NATO Science Series II: Physics and Chemistry* **125**, 361 (2003).
8. M. Rontani, F. Rossi, F. Manghi, and E. Molinari, *Appl. Phys. Lett.* **72**, 957 (1998).
9. C. P. García, V. Pellegrini, A. Pinczuk, M. Rontani, G. Goldoni, E. Molinari, B. S. Dennis, L. N. Pfeiffer, and K. W. West, *Phys. Rev. Lett.* **95**, 266806 (2005).
10. D. G. Goldhaber-Gordon, H. Shtrikman, D. Mahalu, D. Abusch-Magder, U. Meirav, and M. A. Kastner, *Nature* **391**, 156 (1998).
11. S. M. Cronenwett, T. H. Oosterkamp, and L. P. Kouwenhoven, *Science* **281**, 540 (1998).
12. S. M. Reimann and M. Manninen, *Rev. Mod. Phys.* **74**, 1283 (2002).
13. M. Rontani, S. Amaha, K. Muraki, F. Manghi, E. Molinari, S. Tarucha, and D. G. Austing, *Phys. Rev. B* **69**, 85327 (2004).
14. T. Ota, M. Rontani, S. Tarucha, Y. Nakata, H. Z. Song, T. Miyazawa, T. Usuki, M. Takatsu, and N. Yokoyama, *Phys. Rev. Lett.* **95**, 236801 (2005).
15. H. Grabert and M. H. Devoret (eds.), *Single charge tunneling, NATO ASI series B: Physics* **294** (Plenum, New York, 1992).
16. N. Kirstaedter *et al.*, *Electron. Lett.* **30**, 1416 (1994).

17. H. L. Edwards, Q. Niu, G. A. Georgakis, and A. L. de Lozanne, *Phys. Rev.* **B52**, 5714 (1995).
18. X. Michalet *et al.*, *Science* **307**, 538 (2005).
19. D. Loss and D. P. DiVincenzo, *Phys. Rev.* **A57**, 120 (1998).
20. M. Rontani, F. Troiani, U. Hohenester, and E. Molinari, *Solid State Commun.* **119**, 309 (2001).
21. G. Bastard, *Wave mechanics applied to semiconductor heterostructures* (Les Editions de Physique, Les Ulis, 1998).
22. L.-W. Wang and A. Zunger, *Phys. Rev.* **B59**, 15806 (1999).
23. R. Egger, W. Häusler, C. H. Mak, and H. Grabert, *Phys. Rev. Lett.* **82**, 3320 (1999).
24. Full details on our algorithm and its perfomances in M. Rontani, C. Cavazzoni, D. Bellucci, and G. Goldoni, *J. Chem. Phys.* (2006), available also as cond-mat/0508111.
25. See the code website `http.//www.s3.infm.it/donrodrigo`.
26. G. Abstreiter, M. Cardona, and A. Pinczuk, in *Light scattering in solids IV*, eds. M. Cardona and G. Güntherodt (Springer-Verlag, Berlin, 1984), p. 5.
27. D. J. Lockwood, P. Hawrylak, P. D. Wang, C. M. Sotomayor Torres, A. Pinczuk, and B. S. Dennis, *Phys. Rev. Lett.* **77**, 354 (1996).
28. C. Schüller, K. Keller, G. Biese, E. Ulrichs, L. Rolf, C. Steinebach, D. Heitmann, and K. Eberl, *Phys. Rev. Lett.* **80**, 2673 (1998).
29. T. Brocke, M.-T. Bootsmann, M. Tews, B. Wunsch, D. Pfannkuche, Ch. Heyn, W. Hansen, D. Heitmann, and C. Schüller, *Phys. Rev. Lett.* **91**, 257401 (2003).
30. P. Hawrylak, *Solid State Commun.* **93**, 915 (1995).
31. M. Rontani, G. Goldoni, F. Manghi, and E. Molinari, *Europhys. Lett.* **58**, 555 (2002).
32. B. Grandidier, Y. M. Niquet, B. Legrand, J. P. Nys, C. Priester, D. Stiévenard, J. M. Gérard, and V. Thierry-Mieg, *Phys. Rev. Lett.* **85**, 1068 (2000).
33. E. E. Vdovin, A. Levin, A. Patanè, L. Eaves, P. C. Main, Yu. N. Khanin, Yu. V. Dubrovskii, M. Henini, and G. Hill, *Science* **290**, 122 (2000).
34. O. Millo, D. Katz, Y. Cao, and U. Banin, *Phys. Rev. Lett.* **86**, 5751 (2001).
35. T. Maltezopoulos, A. Bolz, C. Meyer, C. Heyn, W. Hansen, M. Morgenstern, and R. Wiesendanger: *Phys. Rev. Lett.* **91**, 196804 (2003).
36. O. S. Wibbelhoff, A. Lorke, D. Reuter, and A. D. Wieck, *Appl. Phys. Lett.* **86**, 092104 (2005).
37. D. Reuter, P. Kailuweit, A. D. Wieck, U. Zeitler, O. Wibbelhoff, C. Meier, A. Lorke, and J. C. Maan, *Phys. Rev. Lett.* **94**, 026808 (2005).
38. M. Rontani and E. Molinari, *Phys. Rev.* **B71**, 233106 (2005).
39. M. Rontani and E. Molinari, *Jpn. J. Appl. Phys.* (2006) [cond-mat/0507688].
40. J. Bardeen, *Phys. Rev. Lett.* **6**, 57 (1961).
41. J. Tersoff and D. R. Hamann, *Phys. Rev.* **B31**, 805 (1985).
42. M. Rontani, C. Cavazzoni, and G. Goldoni, *Computer Physics Commun.* **169**, 430 (2005).
43. M. Koskinen, M. Manninen, B. Mottelson, and S. M. Reimann, *Phys. Rev.* **B63**, 205323 (2001).
44. W. Y. Ruan, Y. Y. Liu, C. G. Bao, and Z. Q. Zhang, *Phys. Rev.* **B51**, 7942 (1995).
45. P. A. Maksym, H. Imamura, and G. P. Mallon, *J. Phys.: Condens. Matter* **12**, R299 (2000).
46. F. Bolton and U. Rössler, *Superlatt. Microstruct.*, **13**, 139 (1993).
47. V. M. Bedanov and F. M. Peeters, *Phys. Rev.* **B49**, 2667 (1994).

NUCLEATION OF VORTICES IN THIN SUPERCONDUCTING DISKS

M. B. SOBNACK, F. V. KUSMARTSEV, D. R. GULEVICH

Department of Physics, Loughborough University
Loughborough LE11 3TU, United Kingdom
M.B.Sobnack@lboro.ac.uk

J. C. H. FUNG

Department of Mathematics, The Hong Kong University of Science and Technology
Clear water Bay, Kowloon, Hong Kong, China
majfung@ust.hk

We study the nucleation of vortices in a thin mesoscopic superconducting disk in an applied magnetic field perpendicular to the disc. We write down an expression for the free energy of the system with an arbitrary number of vortices and anti-vortices. For a given applied field, we minimize the free energy to find the optimal position of the vortices and anti-vortices. We also calculate the magnetization of the disk as a function of the applied field and hence the determine the different configurations possible in which a fixed number of fluxoids can penetrate the disk.

Keywords: Vortices, anti-vortices, magnetization

1. Introduction

Over the years, there have been numerous studies (see, for example, Refs. 1–10) of vortex nucleation in both thin superconducting films of infinite extent and in thin bounded films with traverse dimensions R. Fetter[5] and Buzdin[7] both studied the problem of flux penetration in a thin superconducting disk. More recently, Chibotaru *et al.*[10] looked at flux penetration in thin mesoscopic superconductors of various shapes and concluded that whether or not anti-vortices penetrate the samples depends on the geometry of the sample.

In this paper, we study the effect of an applied magnetic field $\mathbf{H}$ on the nucleation of vortices (and anti-vortices) in a small superconducting disk of radius R and thickness d. We restrict the study to the case $R < \Lambda = \lambda^2/d$ (λ is the usual London penetration depth), and $d \ll \Lambda$ and to the case of low field regime, with $H = |\mathbf{H}|$ near the lower critical field H_{c1}. Under such conditions, the interaction between vortices is logarithmic,[11] and we use the method of images, in analogy with electrostatic interactions between 2D charges, to write down the Gibbs free energy G of an arbitrary configuration of vortices and anti-vortices. This is briefly described in Sec. 2 (a full derivation will be reported elsewhere). We minimize the free energy for different configurations and applied field H. We also calculate the magnetization of the disk as a function of H. We conclude in Sec. 3.

2. Free Energy

The method we use is essentially that of Buzdin.[7] Consider a disk-shaped superconducting film of radius $R < \Lambda$ and thickness $d \ll \Lambda$ lying in the $x-y$ plane. We use cylindrical coordinates (r, θ, z) centered on the axis of symmetry of the disk. The restriction on R is to suppress the screening effects of the superconducting currents $\mathbf{j_s}$ so that the magnetic field will be almost the same inside and outside the disk. The Gibbs free energy in a constant applied field $\mathbf{H} = \nabla \wedge \mathbf{A}_{\text{app}} = H\mathbf{k}$ is

$$G = \frac{1}{8\pi} \int \left[(\mathbf{B} - \mathbf{H})^2 + \lambda^2 (\nabla \wedge \mathbf{B})^2 \right] \mathrm{d}^3 r, \tag{1}$$

where $\mathbf{B} = \nabla \wedge \mathbf{A}$ is the local magnetic field. Consider vortices $(q_i > 0)$/anti-vortices $(q_i < 0)$ of flux $\phi_i = q_i \phi_0, q_i \in \mathbb{Z}$, where $\phi_0 = hc/2e$ is the flux quantum, located at points $\mathbf{r_i}$ in the plane of the disk. Then London's equation for the superconducting currents is

$$\mathbf{j_s} = \frac{c}{4\pi\lambda^2}(\mathbf{A}_\phi - \mathbf{A}), \text{ where } \mathbf{A}_\phi = \sum_i \hat{\boldsymbol{\theta}} \Phi_i(\mathbf{r} - \mathbf{r_i}), \text{ with } \Phi_i(\mathbf{r}) = \frac{\phi_i}{2\pi r}. \tag{2}$$

For a small disk, the screening is weak and one can write $\mathbf{A} \simeq \mathbf{A}_{\text{app}}$. This, together with Eq. (2), leads to

$$G = \frac{d}{8\pi\lambda^2} \int (\mathbf{A}_\phi - \mathbf{A}_{\text{app}})^2 \mathrm{d}^2 r \tag{3}$$

for the free energy of the disk. The correct boundary condition on the edge of the disk is that the current $\mathbf{j_s}$ should have only a tangential component. This is achieved, in analogy with electrostatics, by adding for each vortex of flux ϕ_i at $\mathbf{r_i}$ in the disk an image anti-vortex of flux $-\phi_i$ at $\mathbf{r}'_\mathbf{i} = (R/r_i)^2 \, \mathbf{r_i}$ beyond the disk and leads to

$$\mathbf{A}_\phi = \sum_i \left[\Phi_i(\mathbf{r} - \mathbf{r_i}) - \Phi_i(\mathbf{r} - \mathbf{r}'_\mathbf{i}) \right] \hat{\boldsymbol{\theta}}. \tag{4}$$

It is then not difficult, though lengthy, to show that the free energy of an arbitrary configuration of vortices is

$$\frac{8\pi\lambda^2}{d} G = \sum_i \phi_i f(\mathbf{r_i}, \mathbf{r}'_\mathbf{i}) - \sum_i \phi_i \sum_j \frac{\phi_j}{2\pi} \ln \frac{R}{r_j} - \frac{1}{2} H \sum_i (R^2 - r_i^2)\phi_i + \frac{\pi}{8} H^2 R^4, \tag{5}$$

where

$$f(\mathbf{r_i}, \mathbf{r}'_\mathbf{i}) = \sum_j \frac{\phi_j}{2\pi} \left[\ln \frac{\lambda}{|\mathbf{r_i} - \mathbf{r_j}|} - \ln \frac{\lambda}{|\mathbf{r_i} - \mathbf{r}'_\mathbf{j}|} \right].$$

The summation is over the vortices in the disk $(r_i < R)$ and the contribution of the image vortices is taken care of in the definition of the field f. f diverges whenever $\mathbf{r_i} = \mathbf{r_j}$ and, as is usual, one uses the coherence length ξ as a cutoff.

It is easy to show that the energy of a configuration of N vortices, each of flux $\phi_i = q\phi_0 (q_i = q$ for all i), at $r_i = r$ and a vortex (anti-vortex) of flux $\phi = L\phi_0$ ($L > 0$ for a vortex and $L < 0$ for an anti-vortex) is

$$g(N,L) = g(N,0) + L^2 \ln\frac{R}{\xi} - 2LNq\ln z - Lh, \tag{6}$$

where

$$g(N,0) = \frac{1}{4}h^2 + Nq^2\ln\frac{R}{\xi} - N(N-1)q^2\ln z + Nq^2\ln(1-z^2) - Nqh(1-z^2)$$
$$+ \frac{1}{2}Nq^2\sum_{n=1}^{N-1}\ln\frac{1-2z^2\cos(2\pi n/N)+z^4}{4\sin^2(\pi n/N)}. \tag{7}$$

In the above equations, $z = r/R$, $g = (16\pi^2\lambda^2/d\phi_0^2)G$ is the dimensionless free energy and $h = \pi R^2 H/\phi_0$ is the dimensionless applied magnetic field. The magnetization M of the disk follows from $\partial(G + M\cdot H)/\partial H = 0$ and this gives the reduced magnetization m as $m(N,L) = -\partial g(N,L)/\partial h$. Note that Eqs. (6) and (7), with $q = 1$ and $L = 1$, reproduce Buzdin's[7] results.

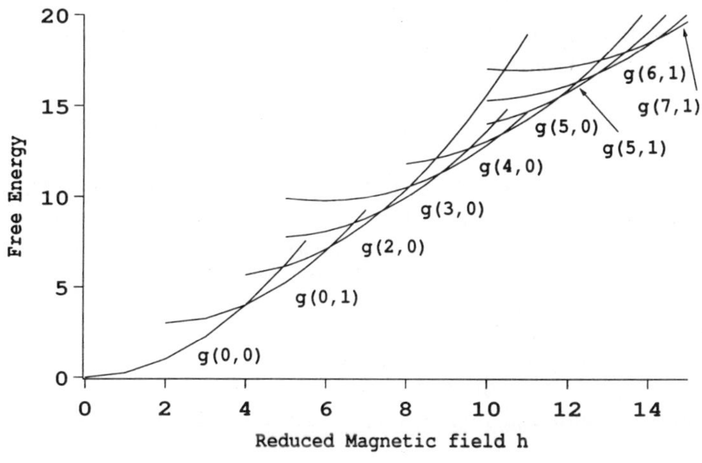

Fig. 1. Free energy $g(N,L)$ as a function of the applied field h for different configurations.

We minimize $g(N,L)$ for z for various h and N and L and calculate $m(N,L)$ as a function of h. Fig. 1 gives the free energy $g(N,L)$ as a function of h for up to $N = 7$ off-center vortices and for up to 1 ($L = 1$) vortex at the center of the disk. We take $\ln(R/\xi) = 4$ and $\phi = \phi_0$ (*i.e.*, $q = 1$). As expected, the results show that for a given configuration of vortices, the energy increases with h until h reaches a critical value (the points of intersection of the lowest $g(N,L)$) at which an extra vortex can penetrate the disk. The entrance of each extra vortex is associated with a jump in the magnetization of the disk at the critical fields, as shown in Fig. 2.

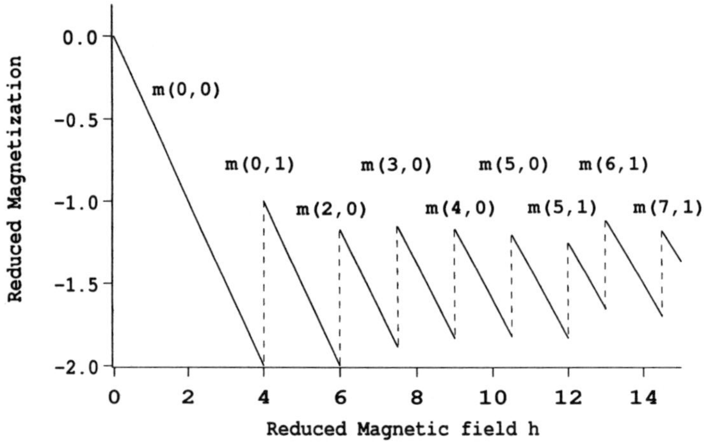

Fig. 2. Magnetization $m(N, L)$ of the disk as a function of h for up to $N = 7, L = 1$.

3. Conclusions

We have presented some preliminary study of the nucleation of vortices in a disk as a function of the applied magnetic field. Work to extend this study to finite temperatures where anti-vortices may nucleate in the disk is currently under way.

References

1. M. R. Beasley, J. E. Mooij and T. P. Orlando, *Phys. Rev. Lett.* **42**, 1165 (1979).
2. S. Doniach and B. A. Huberman, *Phys. Rev. Lett.* **42**, 1169 (1979); B. A. Huberman and S. Doniach, *Phys. Rev. Lett.* **43**, 950 (1979).
3. B. I. Halperin and D. R. Nelson, *J. Low Temp. Phys.* **36**, 599 (1979).
4. L. A. Turkevich, *J. Phys.* **C12**, L385 (1979).
5. A. L. Fetter, *Phys. Rev.* **B22**, 1200 (1980).
6. A. I. Buzdin, *Phys. Rev.* **B47**, 11416 (1993).
7. A. I. Buzdin, *Phys. Lett.* **A196**, 267 (1994).
8. A. K. Geim, S. V. Dubonos, J. G. S. Lok, M. Henini and J. C. Mann, *Nature* **396**, 144 (1998).
9. A. K. Geim, S. V. Dubonos, J. J. Palacios, I. V. Grigorieva, M. Henini and J. J. Schermer, *Phys. Rev. Lett* **85**, 1528 (2000).
10. L. V. Chibotaru, A. Ceulemans, V. Bruyndoncx and V. Moshchalkov, *Nature* **408**, 833 (2000).
11. J. Pearl, *Appl. Phys. Lett.* **5**, 65 (1964).

NUCLEAR SYSTEMS

A NEW REALISTIC MANY-BODY APPROACH FOR THE DESCRIPTION OF HIGH-ENERGY SCATTERING PROCESSES OFF COMPLEX NUCLEI

M. ALVIOLI, C.CIOFI DEGLI ATTI

Department of Physics, University of Perugia and Istituto Nazionale di Fisica Nucleare, Sezione di Perugia, Via A. Pascoli, I-06123, Perugia, Italy

H. MORITA

Sapporo Gakuin University, Bunkyo-dai 11, Ebetsu 069-8555, Hokkaido, Japan

A new linked cluster expansion for the calculation of ground state observables of many-body nuclei with realistic interactions has been developed, in order to single out the major contributions to the relevant quantities when Nucleon-Nucleon correlations are taken into account in the wave function. Using the $V8'$ potential the ground state energy, density and momentum distribution of complex nuclei have been calculated and found to be in good agreement with the results obtained within the Fermi Hyper Netted Chain, and Variational Monte Carlo approaches. Using the same cluster expansion, with wave function and correlations parameters fixed from the calculation of the ground-state observables, we have calculated various high energy scattering processes off complex nuclei. We made use of the Glauber multiple scattering theory, which can be readily included into the cluster expansion we have developed, to take into account final state interaction effects in the semi-inclusive reaction $A(e, e'p)X$, and calculated the distorted momentum distribution, which is a necessary ingredient to estimate the cross section.

Keywords: Cluster expansion; realistic N-N interactions; hadron-nucleus scattering.

1. Introduction

The knowledge of the nuclear wave function, in particular its most interesting and poorly known part, *viz* the correlated one, is not only a prerequisite for understanding the details of bound hadronic systems, but is becoming at present a necessary ingredient for a correct description of medium and high energy scattering processes off nuclear targets; these in fact represent a current way of investigating short range effects in nuclei as well as those QCD effects (e.g. color transparency, hadronization, dense hadronic matter, etc) which manifest themselves in the nuclear medium. The problem of using a realistic many-body description is not trivial, for one has first to solve the many body problem and then has to find a way to apply it to scattering processes. The difficulty mainly arises because even if a reliable and manageable many-body description of the ground state is developed, the problem remains of the calculation of the final state. In the case of *complex nuclei*, much remains to be done in order to achieve a consistent treatment of Initial State Correlations (ISC) and final state interaction (FSI), which is feasible in the case of ***few-body systems*** while to date is still lacking for nuclei with mass number larger than $A = 4$; the structure of the best trial wave function resulting from very sophisticated calculations (e.g. the variational Monte Carlo ones) is so complicated, that its use in the calculation of various processes at intermediate and high energies appears to be not easy task. We have developed an economic method for the calculation of the ground-state properties (energies, densities and momentum distributions) of complex nuclei within a framework which can be easily applied to the treatment of various scattering processes, keeping the basic features of ISC as predicted by the structure of realistic Nucleon-Nucleon (NN) interactions.

2. The cluster expansion

We write the nuclear Hamiltonian in the usual form, i.e.:

$$\hat{H} = \hat{T} + \hat{V} = -\frac{\hbar^2}{2\,M_N}\sum_{i=1}^{A}\nabla_i^2 + \sum_{i<j}\hat{v}_2(\boldsymbol{x}_i,\boldsymbol{x}_j)\,, \tag{1}$$

where the vector $\boldsymbol{x}$ denotes the set of nucleonic degrees of freedom, and the two-body potential $\sum_{i<j}\hat{v}_2(\boldsymbol{x}_i,\boldsymbol{x}_j) = \sum_{n=1}^{N} v^{(n)}(r_{ij})\,\hat{O}_{ij}^{(n)}$, with $r_{ij} = |\boldsymbol{r}_i - \boldsymbol{r}_j|$ the relative distance of nucleons i and j, contains the well known spin- and isospin-dependent operator $\hat{O}_{ij}^{(n)}$. The evaluation of the expectation value of the nuclear Hamiltonian (1) is object of intensive activity which, in the last few years, has produced considerable results; our goal is to present an economical, but effective method for the calculation of the expectation value of any quantum mechanical operator $\hat{A}$

$$\langle\,\hat{A}\,\rangle = \frac{\langle\psi_o\,|\,\hat{A}\,|\,\psi_o\rangle}{\langle\psi_o\,|\psi_o\rangle} = \frac{\langle\phi_o\,|\,\hat{F}^\dagger\,\hat{A}\,\hat{F}\,|\,\phi_o\rangle}{\langle\phi_o\,|\,\hat{F}^\dagger\,\hat{F}\,\phi_o\rangle}\,; \tag{2}$$

where ϕ_o is a Shell-Model (SM), mean-field wave function, and $\hat{F}$ is a symmetrized *correlation operator*, which generates correlations into the mean field wave function. According to the two-body interaction in Eq. (1), the correlated ground state wave function ϕ_o is cast in the following form:

$$\hat{F}(\boldsymbol{x}_1,\boldsymbol{x}_2\ldots\boldsymbol{x}_A)\,\phi_o(\boldsymbol{x}_1,\boldsymbol{x}_2\ldots\boldsymbol{x}_A) = \hat{S}\prod_{i<j}^{A}\hat{f}(r_{ij})\,\phi_o(\boldsymbol{x}_1,\boldsymbol{x}_2\ldots\boldsymbol{x}_A) \tag{3}$$

Table 1. Benchmark calculation for the ground state potential, kinetic, total energy and energy per particle for ^{16}O, at the first order of η-expansion (Eq. (5)), compared with the FHNC results for the corresponding quantities calculated with the same wave function.

	V	**T**	**E$_o$**	**E$_o$/A**
η−exp[1]	-390.37	323.50	-66.87	-4.18
FHNC[5]	-390.30	325.18	-65.12	-4.07

with

$$\hat{f}(r_{ij}) = \sum_{n=1}^{N} \hat{f}^{(n)}(r_{ij}) \qquad\qquad \hat{f}^{(n)}(r_{ij}) = f^{(n)}(r_{ij})\,\hat{O}_{ij}^{(n)}\,. \tag{4}$$

In the following, we are going to describe the cluster expansion (η – *expansion*) technique we used to evaluate Eq. (2); the solution can be found by applying the variational principle, with the variational parameters contained both in the correlation functions and in the mean field single particle wave functions. The expectation value $\langle \hat{A} \rangle$ defined in (2) can be cluster-expanded and, at first order of the η–expansion,[1-3] it reads as follows

$$\langle \hat{A} \rangle_1 = \langle \phi_o \,|\sum_{i<j} \left(\hat{f}_{ij} \hat{A} \hat{f}_{ij} - \hat{A} \right) |\, \phi_o \rangle - \langle \hat{A} \rangle_o \langle \phi_o \,|\sum_{i<j} \left(\hat{f}_{ij} \hat{f}_{ij} - 1 \right) |\, \phi_o \rangle\,, \tag{5}$$

where $\langle \hat{A} \rangle_o$ is given by $\langle \phi_o \,|\hat{A}|\, \phi_o \rangle$. The $2nd$ order term can straightforwardly be obtained by the same technique used to derive Eq. (5). Given the two-body interaction as in Eq. (1), the expectation value of the Hamiltonian can be written as:

$$E_o \;=\; \int d\boldsymbol{k}\, k^2\, n(\boldsymbol{k}) \;+\; \sum_n \int d\boldsymbol{r}_1 d\boldsymbol{r}_2\; v^{(n)}(r_{12}) \rho^{(2)}_{(n)}(\boldsymbol{r}_1, \boldsymbol{r}_2)\,, \tag{6}$$

where $\rho^{(2)}_{(n)}(\boldsymbol{r}_1, \boldsymbol{r}_2)$ is the (spin and isospin dependent-) Two Body Density (TBD) matrix,[1,2] and $n(\boldsymbol{k})$ is the nucleon momentum distribution, defined in terms of the non-diagonal One Body Density Matrix (OBDM), $\rho^{(1)}(\boldsymbol{r}_1, \boldsymbol{r}'_1)$, as

$$n(\boldsymbol{k}) \;=\; \frac{1}{(2\pi)^3} \int d\boldsymbol{r}_1 d\boldsymbol{r}'_1\, e^{i\,\boldsymbol{k}\cdot(\boldsymbol{r}_1 - \boldsymbol{r}'_1)}\; \rho(\boldsymbol{r}_1, \boldsymbol{r}'_1)\,. \tag{7}$$

The knowledge of the OBDM allows one to calculate other relevant quantities like e.g. the density distribution $\rho(\boldsymbol{r}) \;=\; \rho^{(1)}(\boldsymbol{r}_1 \;=\; \boldsymbol{r}'_1 \equiv \boldsymbol{r})$. The results of calculation of the ground state energy, the charge density and the two-body density and momentum distribution using the realistic $V8'$ interaction[4] is discussed in detail in Refs. 1–3; in Table 1 some results for a benchmark calculation of the quantity in Eq. (6) are shown and it can be seen that the agreement with the reference FHNC calculation is very good. Results for the momentum distribution and *radial* TBD, defined as $\rho^{(2)}_{(n)}(r) = \int d\boldsymbol{R}\, \rho^{(2)}_{(n)}(\boldsymbol{r}_1, \boldsymbol{r}_2)$, where $\boldsymbol{r}$ and $\boldsymbol{R}$ are the relative and center of mass vectors of particles 1 and 2,[1,3] are shown in Fig. 1; we obtained the best wave functions for ^{16}O and ^{40}Ca, used in 1 and to be used in various applications, from the results achieved variationally within the FHNC[5] approach, i.e., we used the correlation functions provided by the FHNC method and then optimized the single particle wave functions in order to reproduce the experimentally observed charge densities and radii for these two nuclei; we found that ^{16}O can be described both with with harmonic oscillator or Saxon-Woods mean field functions while, in the case of ^{40}Ca, Saxon-Woods functions have to be used to produce satisfactory densities and momentum distributions, thought the agreement with the FHNC results for the ground state energy is not as satisfactory as in the ^{16}O case.[1,2] The method we

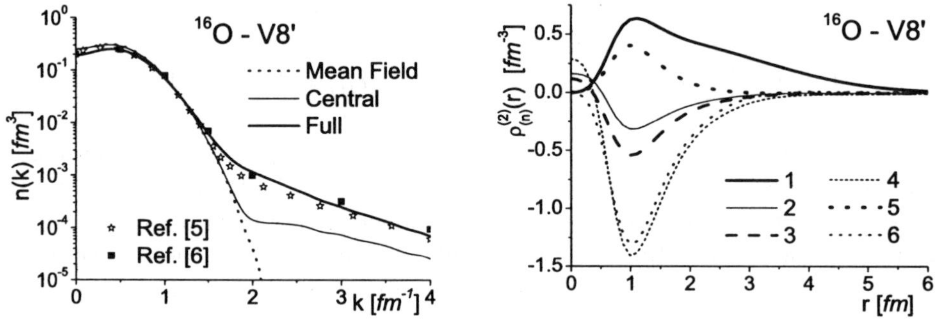

Fig. 1. The momentum distribution (Eq. (7); *left*) and the *radial* two-body density (Eq. (1; *right*) of ^{16}O calculated within the cluster expansion. The η-expansion (*thick solid*) result for the momentum ditribution is compared with the VMC[6] (*squares*) and the FHNC[5] (*stars*) ones; the mean field (*dots*) and the Jastrow (*thin solid*) results is also show, in the left figure. The six components of $\rho^{(2)}_{(n)}$,[1,2] which couple with the first six components of the $V8'$ potential are shown in the right figure.

have developed appears to be a very effective, transparent and parameter-free one. An extensive discussion of the applications the method of the cluster expansion to scattering problems is out of the scope of the present report and will be presented elsewhere. In this report, we only would like to show what the effect of final

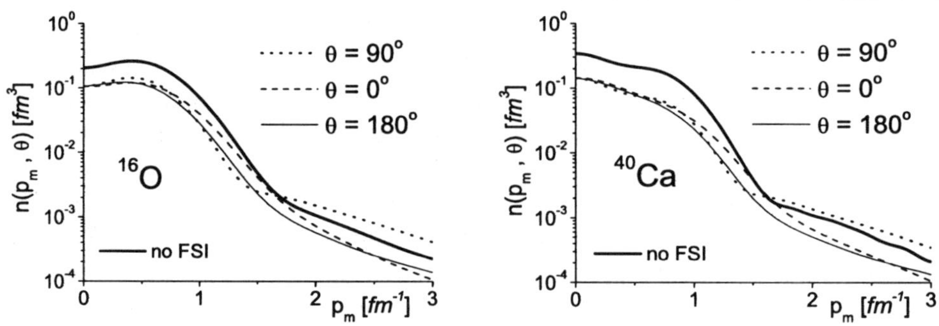

Fig. 2. The *distorted* momentum distribution of Eq. (8) for ^{16}O (*left*) and ^{40}Ca (*right*). *Thick solid* line: no final state interaction taken into account; *dotted*, *dashed* and *thin solid* lines: FSI in the Glauber framework for *perpendicular*, *parallel* and *antiparallel* kinematics, respectively; θ represents the angle between the missing momentum $\boldsymbol{p}_m$ and the three-momentum transfer, $\boldsymbol{q}$.

state interaction à la Glauber[7,8] is on the nucleon momentum distribution in the semi-inclusive $A(e,e'p)X$ reaction; defining the *distorted* momentum distribution as

$$n_D(\mathbf{p}_m) = \frac{1}{(2\pi)^3}\int e^{i\mathbf{p}_m\cdot(\mathbf{r}_1-\mathbf{r}_1')}\rho_D(\mathbf{r}_1,\mathbf{r}_1')\,d\mathbf{r}_1 d\mathbf{r}_1'\,, \tag{8}$$

with $\boldsymbol{p}_m = \boldsymbol{q}-\boldsymbol{p}$ is the missing momentum defined as the difference between the between the three-momentum transfer and the detected proton momentum and

$$\rho_D(\mathbf{r}_1,\mathbf{r}_1') = \frac{\langle\psi_o\,|\,S^\dagger\,\hat\rho(\mathbf{r}_1,\mathbf{r}_1')\,S'\,|\,\psi_o\rangle'}{\langle\psi_o\,|\,\psi_o\rangle}\,, \tag{9}$$

where $\hat\rho_1(\tilde{\boldsymbol{r}}_1,\tilde{\boldsymbol{r}}_1')$ is the OBDM operator, S is the Glauber S-matrix, i.e.

$$S \to S_G(\mathbf{r}_1\ldots\mathbf{r}_A) = \prod_{j=2}^{A} G(\mathbf{r}_1,\mathbf{r}_j) \equiv \prod_{j=2}^{A}\left[1-\theta(z_j-z_1)\Gamma(\mathbf{b}_1-\mathbf{b}_j)\right] \tag{10}$$

with asymptotic values of the parameters appearing in the Γ functions.[7] The results for the distribution of Eq. (8) are compared to the non-distorted ones in Fig. 2; it can be seen how the FSI strongly affects the results at all kinematics.

3. Summary

To sum up, we have shown that, using realistic models of the NN interaction, a proper approach based on cluster expansion techniques can produce reliable approximations for those diagonal and non diagonal density matrices which appear in various medium and high energy scattering processes off nuclei, so that the role of nuclear effects in these processes can be reliably estimated without using free parameters to be fitted to the data. Our approach has been applied to semi-inclusive $A(e,e'p)X$ reaction;[2,3] we also calculated the total neutron-nucleus cross section[9,10] within the same Glauber approach, obtaining, once inelastic shadowing effects are also taken into account, a very good agreement with data over a wide range of mass number A, which confirms that NN correlations cannot be neglected in the nucleus wave function.

References

1. M. Alvioli, C. Ciofi degli Atti, H. Morita, *Phys. Rev.* **C72** 054310 (2005)
2. M. Alvioli, PhD Thesis, University of Perugia (2003).
3. M. Alvioli, C. Ciofi degli Atti, H. Morita, *Fizica* **B13** 585 (2004).
4. B. S. Pudliner, V. R. Pandharipande, J. Carlson, S. C. Pieper and R. B. Wiringa, *Phys. Rev.* **C56**, 1720 (1999).
5. A. Fabrocini, F. Arias de Saavedra and G. Co', *Phys. Rev.* **C61**, 044302 (2000) and *Private Communication.*
6. S.C. Pieper, R.B. Wiringa and V.R. Pandharipande, *Phys. Rev.* **C46**, 1741 (1992).
7. C. Ciofi degli Atti, and D. Treleani, *Phys. Rev* **C60**, 024602 (1999).
8. H. Morita, M. A. Braun, C. Ciofi degli Atti and D. Treleani, *Nucl. Phys.* **A699**, 328c (2002).
9. M. Alvioli, C. Ciofi degli Atti, I. Marchino and H. Morita, *in preparation*
10. M. Alvioli, C. Ciofi degli Atti, I. Marchino and H. Morita; invited talk at the *XXIV International Workshop on Nuclear Theory, Rila, Bulgaria* (2005); nucl-th/0510079.

VARIATIONAL DESCRIPTION OF FEW–NUCLEON SYSTEMS: BOUND AND SCATTERING STATES

ALEJANDRO KIEVSKY, MICHELE VIVIANI, LAURA E. MARCUCCI and SERGIO ROSATI

Physics Department, University of Pisa and Istituto Nazionale di Fisica Nucleare, Via Buonarroti 2, 56100 Pisa, Italy

kievsky@pi.infn.it

Three– and four–nucleon systems are described using the hyperspherical harmonic (HH) method. Bound and scattering states are expanded in the HH basis and the corresponding binding energies and S-matrices are obtained using a variational principle. Modern nucleon-nucleon potentials plus three-nucleon interactions are considered. The calculated quantities as binding energies, cross section and polarization observables are accurate at the level of 1% or better.

Keywords: Few–nucleon systems; binding energies; elastic scattering.

1. Introduction

The study of the three– and four–nucleon systems is interesting from different points of view. On one side many reactions involving three and four nucleons, like ^{2}H(p,γ)^{3}He, ^{2}H(d,p)^{3}H, ^{2}H(d,n)^{3}He, are of extreme astrophysical interest. Moreover the $A = 3, 4$ systems have become increasingly important as testing grounds for models of the nuclear force. In particular, the $A = 4$ system is the simplest system that presents the complexity, thresholds and resonances, that characterize nuclear systems. The theoretical description of $A = 3, 4$ systems still constitutes a challeging problem from the standpoint of nuclear few-body theory. In this context the Hyperspherical Harmonic (HH) method with and without the inclusion of correlation factors, has been used to describe both, bound and scattering states in $A = 3, 4$ systems considering modern nucleon–nucleon (NN) potentials plus three-nucleon forces.[1–3] The inclusion of correlation factors accelerates the convergence of the expansion since improves the description of the short range part of the wave function. However, the HH basis has also been used without the inclusion of correlation factors, showing the necessity of a very large basis in order to obtain converged results.[4,5] In this paper we briefly review the results obtained with the HH expansion for bound and scattering states in the $A = 4$ system.

2. Hyperspherical Harmonic basis for three– and four–nucleon systems

The $A = 3,4$ bound state wave functions can be put in the form[4,5]

$$\Psi_A = \rho^{-(3A-4)/2} \left[\sum_{\alpha=1}^{N_c} \sum_{[K]} u^{\alpha}_{[K]}(\rho) B^{\alpha}_{[K]}(\Omega_A) \right] , \tag{1}$$

where $B^{\alpha}_{[K]}(\Omega_A)$ is a complete antisymmetric spin-isospin HH function for $A = 3,4$ corresponding to channel α and $[K]$ denotes a set of quantum numbers that define a HH basis element with K the grand angular quantum number. In general the hyperradial functions can be expanded in a complete basis, as for example, Laguerre polynomials multiplied by an exponential factor

$$u^{\alpha}_{[K]} = \sum_m A^{\alpha}_{m[K]} \mathcal{L}^{(n)}_m(\beta\rho)\, \mathrm{e}^{-\beta\rho} \tag{2}$$

with β a nonlinear parameter and $n = 5,8$ for $A = 3,4$, respectively. Therefore it is possible to define an antisymmetric basis element $|\alpha\, m\, [K] >$ and the description of the $A = 3,4$ bound state can be obtained solving the following generalized problem

$$\sum_{\alpha,m,[K]} A^{\alpha}_{m[K]} < \alpha'\, m'\, [K'] | H - E | \alpha\, m\, [K] > = 0 \ . \tag{3}$$

For scattering states in which a nucleon impacts on the $A-1$ system the $A = 3,4$ scattering wave function can be put in the form[6]

$$\Psi_{N,A-1} = \rho^{-(3A-4)/2} \left[\sum_{\alpha=1}^{N_c} \sum_{[K]} u^{\alpha}_{[K]}(\rho) B^{\alpha}_{[K]}(\Omega_A) \right] + \Psi_{asym} \tag{4}$$

where Ψ_{asym} describes the asymptotic configuration in which the nuclear interaction between the incoming nucleon and the $A-1$ target is negligible. Ψ_{asym} is a linear combination of the ingoing and outgoing solutions of the relative Schroedinger equation times the $A-1$ bound state. The relative weight between the ingoing and outgoing solutions for an incoming nucleon in a L, S, J^{π} state is the S–matrix element $\mathcal{S}^J_{LS,LS'}$. It can be determined using the Kohn variational principle

$$[\mathcal{S}^J_{LS,LS'}] = \mathcal{S}^J_{LS,LS'} + \mathrm{i} < \Psi_{N,A-1} | H - e | \Psi_{N,A-1} > \ . \tag{5}$$

3. Results

In Table 1 binding energies as well as kinetic energy and occupation probabilities calculated using the HH expansion are given for the alpha particle using different potential models and are compared to the Faddeev-Yakuyvosky[7,8] (FY), Green Function Montecarlo[9] (GFMC) and no-core shell model[10] (NCSM) techniques. The potential of Argonne AV18, the CD-Bonn potential and the chiral potential N3LO has been considered as well as the AV18 plus the Urbana three-body force (UR). The last column of the table shows the maximum grand angular quantum number

K used in the expansion which gives an indication of the size of the basis. For example for the AV18 potential approximately 8000 HH basis states have been used whereas for the N3LO potential only 1000 HH are necessary. With this very soft potential the extension of the HH technique to systems with $A > 4$ appears feasible.

Table 1. The HH results for binding energies, mean values of the kinetic energy and occupation probabilities obtained with different potential models are compared to different techniques.

potential	Method	B(MeV)	T(MeV)	$P_P(\%)$	$P_D(\%)$	[K]
AV18	HH	24.22	97.84	0.35	13.74	72
	FY[7]	24.25	97.80	0.35	13.78	
	FY[8]	24.22	97.77			
AV18+UR	HH	28.46	113.30	0.73	16.03	72
	FY[7]	28.50	113.21	0.75	16.03	
	GFMC[9]	28.34(4)	110.7(7)			
CD-Bonn	HH	26.12	77.62	0.22	10.72	52
	FY[7]	26.16	77.59	0.23	10.77	
N3LO	HH	25.09	68.79	0.172	8.951	24
	FY[7]	25.41				
	NCSM[10]	25.36				

In Figs. 1,2 differential cross section and the vector analyzing power A_y are shown at five different energies. The HH results using the AV18+UR (solid line) and AV18 (dashed line) potentials are compared to the available data. As can be seen from the figures, the calculations reproduce the cross section although some disagreements are observed at backward angles. Conversely, there is a noticeable underestimation of A_y.

References

1. A. Kievsky, M. Viviani, and S. Rosati, *Nucl. Phys.* **A551**, 241 (1993)
2. A. Kievsky, M. Viviani, and S. Rosati, *Nucl. Phys.* **A577** , 511 (1994)
3. M. Viviani, A. Kievsky, and S. Rosati, *Few-Body Syst.* **18**, 25 (1995)
4. M. Viviani, A. Kievsky, and S. Rosati, *Phys. Rev.* **C71**, 024006 (2005)
5. A. Kievsky, L. E. Marcucci, M. Viviani, and S. Rosati, *Few-Body Syst.* **22**, 1 (1997)
6. A. Kievsky, *Nucl. Phys.* **A624**, 125 (1997)
7. A. Nogga, H. Kamada, W. Glöckle, and B. R. Barrett, *Phys. Rev.* **C65**, 054003 (2002)
8. R. Lazauskas and J. Carbonell, *Phys. Rev.* **C70**, 044002 (2004)
9. R. B. Wiringa, S. C. Pieper, J. Carlson, and V. R. Pandharipande, *Phys. Rev.* **C62**, 014001 (2000)
10. P. Navrátil and E. Caurier, *Phys. Rev.* **C69**, 014311 (2004)
11. B. M. Fisher *et al.*, in *Proc. XVII Int. IUPAP Conf. on Few-Body Problems in Physics*, ed. W. Glöckle and W. Tornow (Elsvier, Amsterdam, 2004), pp. S76-S78
12. K. F. Famularo *et al.*, *Phys. Rev.* **93**, 928 (1954)
13. M. Viviani, A. Kievsky, S. Rosati, E. A. George, and L. D. Knutson, *Phys. Rev Lett.* **86**, 3739 (2001)
14. M. T. Alley and L. D. Knutson, *Phys. Rev.* **C48**, 1890 (1993)

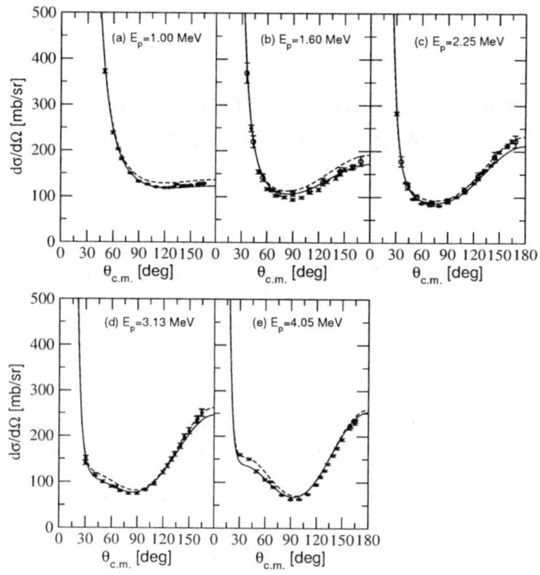

Fig. 1. Differential cross section for p-^{3}He elastic scattering at the indicated energies. The solid (dashed) curve is the HH calculation for the AV18+UR (AV18) potential. The experimental points are from Ref.11 (solid circles) and Ref.12 (open circles).

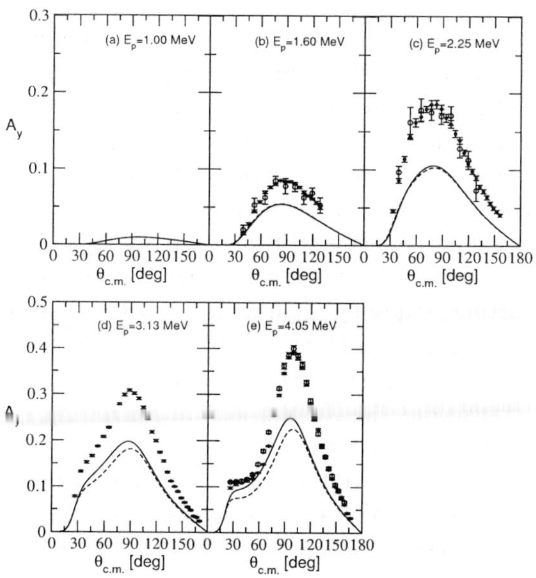

Fig. 2. The analyzing power A_y for p-^{3}He elastic scattering at the indicated energies. The solid (dashed) curve is the HH calculation for the AV18+UR (AV18) potential. The experimental points are from Ref.11(solid circles), Ref.13 (open circles pannels b and c) and Ref.14 (open circles pannel e).

THE NUMBER SELF-CONSISTENT RENORMALIZED RANDOM PHASE APPROXIMATION

A. MARIANO

Departamento de Física, Universidad Nacional de La Plata, cc.67
La Plata, Buenos Aires (1900), Argentina
mariano@venus.fisica.unlp.edu.ar

RPA and its quasiparticle generalization (QRPA) have been widely used to study electromagnetic transitions and beta decays in medium and heavy nuclei, being the pn-QRPA charge exchange mode extensively employed in the description of single and double beta decays in vibrational nuclei. However develops a collapse, i.e. it presents imaginary eigenvalues for strengths beyond a critical value of the force. Extensions called renormalized QRPA (RQRPA) do not develop any collapse going beyond the simplest quasiboson approximation, however they present several drawbacks which will be analyzed.

Keywords: QRPA; collapse; renormalization

1. Introduction

A whole family of extensions of the QRPA, called renormalized QRPA (RQRPA) are known that do not develop any collapse by implementing the Pauli principle in a consistent way, pursuing the quasiboson approximation.[1–4] However, still in its simplest versions there is a violation of the non energy weighted Ikeda sum rule. Calculations to determine the amount of the violation and some improvements to the RQRPA, in order to restore the sum rule, have been presented.[5] It has been shown that treating simultaneously BCS and QRPA equations one can fulfil the Ikeda sum rule for the Fermi case when a schematic model is used.[6] In the present work we go a step further, studying ^{76}Ge with a RQRPA (SRQRPA) where at the same time the mean field is changed by minimizing the energy and fixing the number of particles in the correlated ground state. In spite of this, the Ikeda sum rule is violated for Fermi transitions due to the lack of scattering terms in the phonon operators. Quite recently these terms have been introduced in the structure of the excitation operators, being the QRPA matrixes not trivially renormalized[7] obtaining the so called full RQRPA (FRQRPA). In this case the sum rule is fulfilled, but it has been shown[8] for a schematic Hamiltonian that the FQRPA cannot surpass the collapse if one imposes consistency between the approximation and the structure of the vacuum. In this work we show for the case of ^{76}Ge, that the FRQRPA is very close to the usual QRPA in the collapse region, for the Fermi transitions.

2. Different Approximations

In this section we summarize the different approximations, in a quasiparticle scheme, to treat states of an open-shell nucleus. The Hamiltonian is

$$H = H_p + H_n + H_{p,n}, \tag{1}$$

where $p \equiv$proton, $n \equiv$neutron and the last term corresponds to the proton-neutron interaction. After the Bogoliubov transformation

$$\alpha_{\mathbf{t}}^\dagger = u_t a_{\mathbf{t}}^\dagger - v_t a_{\bar{\mathbf{t}}}, \tag{2}$$

where $\mathbf{t} \equiv \mathbf{p}, \mathbf{n} = t, m_t$, with $t \equiv \{n_t, l_t, j_t\}$ and $m_t \equiv m_j$ we get

$$H = U + \sum_p 2\Omega_p \epsilon_p \hat{\mathcal{N}}_p + \sum_n 2\Omega_n \epsilon_n \hat{\mathcal{N}}_n + H_{22} + H_{40} + H_{04} + H_{13} + H_{31}, \tag{3}$$

where H_{nm} destroy m quasiparticles and creates n quasiparticles, being $\hat{\mathcal{N}}_t$ the quasiparticle number operators, and where U and the quasiparticle energies ϵ_t are defined as usually.[9] With $|0\rangle$ we will indicate the nuclear ground state.

2.1. *BCS*

Here $|0\rangle \equiv |BCS\rangle$, which satisfies $\alpha_t|0\rangle = 0$, and the quasiparticle mean field occupations v_i are obtained asking that

$$\frac{\partial}{\partial v}\langle 0|H|0\rangle = 0, \qquad \langle 0|\hat{N}|0\rangle = N, \qquad \langle 0|\hat{Z}|0\rangle = Z, \qquad u_i^2 + v_i^2 = 1, \tag{4}$$

getting the BCS equations

$$2(\overline{e_t} - \lambda)v_t u_t - (u_t^2 - v_t^2)\Delta_t = 0, \tag{5}$$

where $\overline{e}_t$ are the dressed single-particle energies and Δ_t are the gaps defined as usually.[9]

2.2. *QRPA*

Residual interactions between protons and neutrons in H_{nm}, can be included using the QRPA, where $|0\rangle \equiv |QRPA\rangle$ and the nuclear excited states $|\lambda JM\rangle$ are constructed as

$$\Omega(\lambda JM)|0\rangle = 0, \qquad |\lambda JM\rangle \equiv \Omega^\dagger(\lambda JM)\ |0\rangle, \tag{6}$$

with

$$\Omega^\dagger(\lambda JM) = \sum_{pn} \left[X_{pn}(\lambda J)A_{pn}^\dagger(JM) - Y_{pn}(\lambda J)A_{pn}(\overline{JM})\right], \tag{7}$$

being $\hat{A}_{pn}^\dagger(JM) \equiv [\alpha_p^\dagger \alpha_n^\dagger]^{JM}$, and assuming the Quasiboson approximation

$$\left[A_{pn}, A_{p'n'}^\dagger\right] \approx \langle 0|\left[A_{pn}, A_{p'n'}^\dagger\right]|0\rangle = \delta_{pp'}\delta_{nn'}D_{pn}, \quad D_{pn} \equiv 1 - \langle 0|\hat{\mathcal{N}}_p + \hat{\mathcal{N}}_n|0\rangle \tag{8}$$

in the equation of motion with $|0\rangle \approx |BCS\rangle$ in the expectation values. The drawbacks of the QRPA are that $\langle 0|\hat{N}, \hat{Z}|0\rangle \neq N, Z$, the QRPA equations are independent of the structure of $|0\rangle$, and finally that $min(\omega_{\lambda J}) \rightarrow 0$ when the particle-particle strength in H_{pn} is increased.

2.3. *SRQRPA*

We can overcome the problems mentioned previously within the number-conserving SRQRPA, where now the conditions (4) are imposed for $|0\rangle \equiv |SRQRPA\rangle$ arriving to a renormalized Gap equation,[10] and assuming that $D_{pn} \neq 1$ being calculated selfconsistently, together with $u's, v's, X_{pn}(\lambda J), Y_{pn}(\lambda J)$, and $\omega_{\lambda J}$ using

$$\langle 0|\hat{\mathcal{N}_p}|0\rangle = \sum_{\lambda J n'} \frac{(2J+1)}{2\Omega_p} |Y(J)_{pn',\lambda}|^2, \quad \langle 0|\hat{\mathcal{N}_n}|0\rangle = \sum_{\lambda J p'} \frac{(2J+1)}{2\Omega_n} |Y(J)_{p'n,\lambda}|^2, \tag{9}$$

which can be obtained from the expansion

$$|0\rangle = N_0 e^{\frac{1}{2}\sum \sqrt{(2J+1)} Y(J)^*_{pn,\lambda} X(J)^{*-1}_{\lambda,p'n'} \left[A^\dagger_{pn}(J) A^\dagger_{p'n'}(J)\right]^0} |BCS\rangle. \tag{10}$$

Here we sum over repeated indexes. Now $\hat{A}^\dagger_{pn}(JM)$, $X_{pn}(\lambda J)$ and $Y_{pn}(\lambda J)$ can be renormalized as

$$\hat{A}^\dagger_{pn}, X_{pn}(\lambda J), Y_{pn}(\lambda J) \rightarrow \hat{A}^\dagger_{pn} D^{-1/2}_{pn}, X_{pn}(\lambda J) D^{1/2}_{pn}, Y_{pn}(\lambda J) D^{1/2}_{pn}, \tag{11}$$

arriving at the same eigenvalue problem of the QRPA but with a renormalized interaction,[5] avoiding the collapse. Finally we mention that the Ikeda sum rule for a transition operator $\hat{O}(J=0,1)^{\pm} = \hat{O}(J=0,1)t^{\pm}$, which reads

$$S_- - S_+ = (2J+1)(N-Z), \quad S_\pm = \sum_{\lambda M} |\langle \lambda JM|\hat{O}^\pm|0\rangle|^2, \tag{12}$$

is not fulfilled within the SRQRPA, in spite of having the good number of particles in $|0\rangle$, due to the lack of scattering terms[5] in the phonon operators (7).

2.4. *FRQRPA*

To overcome the mentioned problems with the sumrule, the FRQRPA was developed from the SRQRPA but including the scattering terms in the phonon (7) through the replacement

$$A^\dagger_{pn} \rightarrow \left(A^\dagger_{pn} + \frac{u_n v_n B^\dagger_{pn} - u_p v_p \bar{B}_{pn}}{v_n^2 - v_p^2} \right) D^{-1/2}_{pn}, \quad B^\dagger_{pn}(JM) = [\alpha^\dagger_p \alpha_{\bar{n}}]^{JM}. \tag{13}$$

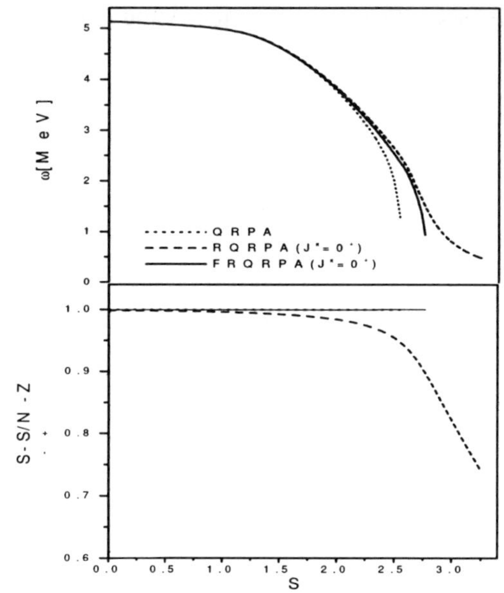

Fig. 1. Lowest $J^{\pi} = 0^{+}$ excitation energies in ^{76}As, calculated from the ^{76}Ge ground state and sumrule, for the QRPA, RQRPA, and FRQRPA approximations, as function of the residual interaction parameter s. $J^{\pi} = 0^{+}$ indicates that we only keep this contribution in the sums of eq.(9).

3. Results

In the present work we study Fermi beta excitations in ^{76}Ge. We adopt a δ-type residual interaction already used previously,[5] and our Hilbert space has six single particle energy levels, including all the single-particle orbitals from oscillator shells $3\hbar\omega$ plus $1g_{9/2}$ and $1g_{7/2}$ from the $4\hbar\omega$ oscillator shell. They were obtained using a Coulomb-corrected Wood-Saxon potential. The QRPA, SRQRPA and the FRQRPA results for the lowest eigenvalue and the sum rule are compared in figure 1. As can be seen the SRQRPA avoids the collapse but does not fulfills the sume rule, while within the FRQRPA the sum rule is fulfilled but the collapse is again present.

References

1. D. J. Rowe, *Phys. Rev.* **175**, 1283 (1968); *Rev. Mod. Phys.* **40**, 153 (1968) .
2. F. Catara, N. Dinh Dang and M. Sambataro, *Nucl. Phys.* **A579**, 1 (1994).
3. J. Toivanen and J. Suhonen, *Phys. Rev. Lett.* **75**, 410 (1995).
4. J. Dukelsky, P. Schuck, *Phys. Lett.* **B387**, 233 (1996).
5. F. Krmpotić, T.T.S. Kuo, A. Mariano, E.J.V. de Passos, A.F.R. de Toledo Piza, *Nucl. Phys.* **A612**, 223 (1996).
6. D.S. Delion, J. Dukelsky, P. Schuck, *Phys. Rev.* **C55**, 2340 (1997).
7. L. Pacearescu, V. Rodin, F. Simkovic, and A. Faessler, *Phys. Rev.* **C68**, 064310 (2003).
8. O. Civitarese, M. Reboiro and J.G. Hirsch, *Phys. Rev.* **C71**, 014318 (2005).
9. M.K. Pal, Y.K. Gambhir, and Ram Raj, *Phys. Rev. Lett.* **155** (1967) 1144.

10. R.V. Jolos and W. Rybarska-Nawrocka, *Z. Phys.* **A296** (1980) 73.

COUPLED-CLUSTER THEORY FOR NUCLEI

T. PAPENBROCK

Department of Physics and Astronomy, University of Tennessee,
Knoxville, TN 37996, USA, and
Physics Division, Oak Ridge National Laboratory,
Oak Ridge, TN 37831, USA
tpapenbr@utk.edu

D. J. DEAN

Physics Division, Oak Ridge National Laboratory,
Oak Ridge, TN 37831, USA

J. R. GOUR

Department of Chemistry, Michigan State University,
East Lansing, MI 48824, USA

G. HAGEN

Department of Physics and Astronomy, University of Tennessee,
Knoxville, TN 37996, USA, and
Physics Division, Oak Ridge National Laboratory,
Oak Ridge, TN 37831, USA

M. HJORTH-JENSEN

Department of Physics and Center of Mathematics for Applications, University of Oslo
N-0316 Oslo, Norway

P. PIECUCH and M. WŁOCH

Department of Chemistry, Michigan State University,
East Lansing, MI 48824, USA

This presentation focuses on some of the recent developments in low-energy nuclear structure theory, with emphasis on applications of coupled-cluster theory. We report on results for ground and excited states in ^{4}He and ^{16}O, and about extensions of coupled-cluster theory to treat three-body forces.

Keywords: Nuclear structure, light nuclei, coupled-cluster theory

1. Introduction

In recent years, much progress has been made in *ab-initio* nuclear structure calculations. The description of light nuclei can now be based on realistic nuclear interactions and powerful methods that solve the quantum many-body problem. The

Green's Function Monte Carlo (GFMC) calculations have established the structure of light nuclei up to mass $A = 12$,[1,2] while the No-Core Shell Model (NCSM) approach has extended *ab-initio* structure calculations to heavier p-shell nuclei.[3] These approaches have led to two main results, namely (i) a unified description of light nuclei starting from nucleons and realistic interactions, and (ii) a much improved understanding and determination of the effective interaction itself. We would like to extend this approach to heavier systems. The coupled cluster method[4,5] is a very promising candidate for the endeavor, as it scales more favorably with increasing system size than the GFMC or the NCSM.

The progress in *ab-initio* nuclear structure calculations goes hand in hand with progress regarding the effective nuclear interactions. For a recent review, we refer the reader to Ref. 6. Several realistic nucleon-nucleon potentials fit the two-nucleon phase shifts with a chi-square of one per degree of freedom. These potentials agree in their long-range part, and differ in the modeling of the short-range part. Note that the latter is presently neither constrained by experimental data on the phase shifts nor by theoretical arguments. As a consequence, different realistic potentials yield different results for many-nucleon systems.

There are at least two possible ways to resolve this dilemma. The first consists of an approach to nuclear interactions that is based on effective field theory.[7,8] Within chiral effective field theory, the effective Hamiltonian is constructed systematically up to a given order of the relative momentum, and the coupling coefficients are fit to reproduce nucleon-nucleon phase shifts and the deuteron bound state.[6] Within this approach, three-body forces naturally enter at higher order. The second way consists of an approach based on the renormalization group. Starting from a realistic potential, one might construct its low-momentum approximation by integrating out momentum modes above a cutoff parameter. This approach leaves the two-nucleon phase shifts unchanged for momenta below the cutoff, and thereby constructs a family of cutoff-dependent potentials that are phase shift equivalent. Interestingly, this approach[9] shows that the realistic potentials collapse onto a "universal" low-momentum potential $V_{\mathrm{low}-k}$. Within this approach, the form of the (cutoff-dependent) three-body force is determined by fit to the three- and four-nucleon system. As in the chiral potential, the occurrence of three-body forces seems unavoidable. This is not surprising: Nucleons are not elementary particles, and the omission of substructure, i.e. the removal of degrees of freedom, is accompanied by the occurrence of many-body forces.

This makes it interesting to test and to further explore the nuclear interaction through *ab-initio* nuclear structure calculations in heavier nuclei. The remainder of this article is divided as follows. In the next section, we briefly describe coupled cluster theory. This is followed by a presentation of results for ^{4}He and ^{16}O. In the fourth section, we describe the coupled-cluster theory with effective three-body forces. We finally give a summary.

2. Coupled-Cluster Theory

The coupled-cluster theory allows one to formulate accurate approximations of the quantum many body problem with relatively low computational cost. It originated in nuclear physics,[4] and saw applications in quantum chemistry a few years later.[5] For reviews, we refer the reader to Refs. 10–15. In recent years, the theory has again been applied to nuclear structure calculations. Guardiola *et al.* performed studies of the center-of-mass problem.[16] Mihaila and Heisenberg[17] employed coupled-cluster theory for the ground state calculation of ^{16}O. They obtained a theoretical description of the electron scattering structure function that agrees very well with experimental data.[18] These calculations used a bare nucleon-nucleon interaction, and the density-dependent part of the three-body interaction. The approach followed in Refs. 19–21 is more conventional in the sense that it utilizes a G-matrix as the interaction, and it employs approximations that have become standards in numerous applications in quantum chemistry.

In coupled-cluster theory, the ground state of the A-body system is represented as

$$|\psi\rangle = e^{\hat{T}}|\Phi\rangle. \tag{1}$$

Here, $|\Phi\rangle$ is a single-particle product state (Slater determinant), and $\hat{T}$ is the cluster operator. It has the form of a particle-hole excitation operator

$$\hat{T} = \sum_{ia} t_i^a \hat{a}_a^\dagger \hat{a}_i + \frac{1}{4}\sum_{ijab} t_{ij}^{ab} \hat{a}_b^\dagger \hat{a}_a^\dagger \hat{a}_i \hat{a}_j + \ldots, \tag{2}$$

and creates $1p-1h$, $2p-2h$,...,$Ap-Ah$ excitations in the product state $|\Phi\rangle$. Here, $\hat{a}_p^\dagger$ and $\hat{a}_q$ are fermionic creation and annihilation operators, respectively, and we have used the following convention: the labels $i, j, k, \ldots$ refer to single-particle orbitals that are occupied in $|\Phi\rangle$, while $a, b, c, \ldots$ refer to unoccupied orbitals. If the expansion (2) is carried out up to $Ap - Ah$ excitations, coupled-cluster theory is exact. The success of coupled-cluster theory, however, relies on the finding that a truncation of Eq. (2) at relatively low order already yields quite accurate results for interacting many-body systems. In what follows, we will restrict ourselves to the truncation after $2p-2h$ clusters (CCSD), and will occasionally include $3p-3h$ clusters approximately.

To obtain the coupled cluster equations within the CCSD approximation, one inserts the ansatz (1) into the Schrödinger equation, and left-multiplies with $\langle\Phi_{i_1\ldots i_k}^{a_1\ldots a_k}|\exp(-\hat{T})$ for $k = 0, 1, 2$. This yields the CCSD equations

$$\begin{aligned} \langle\Phi|\overline{H}|\Phi\rangle &= E, \\ \langle\Phi_i^a|\overline{H}|\Phi\rangle &= 0, \\ \langle\Phi_{ij}^{ab}|\overline{H}|\Phi\rangle &= 0. \end{aligned} \tag{3}$$

Here

$$\overline{H} \equiv \exp\left(-\hat{T}\right) H \exp\left(\hat{T}\right) \tag{4}$$

is the similarity transformed Hamiltonian, and $\langle\Phi_{i_1\ldots i_k}^{a_1\cdots a_k}|$ is the adjoint of the $kp-kh$ excited state $|\Phi_{i_1\ldots i_k}^{a_1\cdots a_k}\rangle = \hat{a}_{a_k}^\dagger \ldots \hat{a}_{a_1}^\dagger \hat{a}_{i_1} \ldots \hat{a}_{i_k}|\Phi\rangle$. The second and third lines of the Eqs. (3) determine the unknown cluster amplitudes t_i^a and t_{ij}^{ab}, and they constitute a coupled set of nonlinear equations. These equations can be solved iteratively, at the relatively low computational cost $\mathcal{O}(n_o^2 n_u^4)$, where n_o and n_u refer to the number of occupied and unoccupied orbitals, respectively. The energy is obtained by inserting the solution of these equations into the first of the Eqs. (3).

Note that the similarity-transformed Hamiltonian (4) is non-Hermitian. This renders coupled-cluster theory a non-variational approach. However, this "disadvantage" is more than compensated by the advantage that $\overline{H}$ can be determined exactly without any further approximation. In a diagrammatic language, only fully connected diagrams enter the expression for $\overline{H}$. For the computation of excited states,[22,23] one diagonalizes the similarity-transformed Hamiltonian (4) in the space of all $1p-1h, \ldots, kp-kh$ excitations of the product state $|\Phi\rangle$. Within the CCSD approximation one has $k=2$, and the excited states can thus be computed for large model spaces where a full matrix diagonalization would be prohibitively expensive. In principle, one could test the quality of the CCSD approximation by comparing the results with a more expensive calculation that also includes $3p-3h$ cluster excitations. Such a calculation scales as $\mathcal{O}(n_o^3 n_u^5)$, and is thus much more expensive. For this reason, we include the $3p-3h$ cluster only approximately and follow the method that is described in Ref. 24.

3. Results for ^{4}He and ^{16}O

Our calculations on ^{4}He are reported in Ref. 20. They employ a G-matrix based on the Idaho-A potential.[25] We worked in a spherical harmonic oscillator basis with oscillator frequency Ω, and controlled the center-of-mass problem by adding the center-of-mass Hamiltonian $\frac{\beta}{2}(P^2/M + M\Omega^2R^2 - 3\hbar\Omega)$ to the Hamiltonian. The Lagrange multiplier β is chosen such that the expectation value of the center-of-mass Hamiltonian is extremal. The calculation was performed in a rather small model space consisting of four oscillator shells. This allowed us to check the quality of the CCSD approximation and to compare with results from exact diagonalizations of the same Hamiltonian. The results are displayed in Table 1. Clearly, CCSD is a good approximation, particularly when compared to CISD – an exact diagonalization in a space of $1p-1h$ and $2p-2h$ excited states which is computationally as expensive as CCSD. This clearly demonstrates that the similarity-transformed Hamiltonian (4) contains much more relevant physics than its Hermitian counterpart. The $3p-3h$ corrections denoted by "CR-CCSD(T)" improve the CCSD result and are close to the exact result. The coupled cluster approximation works equally well for a reference state $|\Phi\rangle$ that is a product of $0s$ oscillator states as for a product state based on the Hartree-Fock (HF) approximation for ^{4}He. For the oscillator basis, both the $1p-1h$ and the $2p-2h$ cluster amplitudes yield considerable contributions to the total binding energy. This is different for the HF basis, where most of the correlation energy stems from the $2p-2h$ cluster amplitudes. The calculations of Ref. 20 also showed that excited states of nuclei can be reliably computed within the CCSD approximation.

Table 1. The ground-state energies of ^{4}He calculated using the oscillator (Osc) and Hartree-Fock (HF) basis states. Units are MeV. The reference energies $\langle\Phi|H'|\Phi\rangle$ are -7.211 (Osc) and -10.520 (HF) MeV. Taken from Ref. 20.

Method	Osc	HF
CCSD	-21.978	-21.385
CR-CCSD(T),c	-22.630	-22.450
CR-CCSD(T),c/$\Delta_0 = 1$	-23.149	-22.783
CISD	-20.175	-20.801
Exact	-23.484	-23.484

Our calculations for ^{16}O employ G-matrices based on the Idaho-A and the chiral N3LO potential.[26] The results are reported in Ref. 21. One of the main questions concerns the convergence of numerical results with respect to the increasing size of the model space. The results of our calculations (See, e.g. Fig. 1 of Ref. 21) show that the ground-state energy converges well for a model space consisting of seven to eight oscillator shells; the difference in binding energy being about 0.5MeV for both model spaces. As previously found for ^{4}He, $3p-3h$ corrections are very small and account for only 0.7MeV additional binding. The binding energies for ^{16}O are -109MeV and -112MeV for the Idaho-A and N3LO potentials, respectively. This suggests that the difference to the experimental value of -128MeV must be attributed to deficiencies of the Hamiltonian itself, and it points to the need of three-body forces.

The computation of the excited states of ^{16}O also yielded interesting results. For the Idaho-A potential, we found a $J^\pi = 3^-$ state at about 12MeV in excitation energy, and a $J^\pi = 0^+$ state at about 22MeV. Calculations with the N3LO potential yielded similar results. The 3^- state exhibits only small corrections due to $3p-3h$ cluster amplitudes, while the corrections are considerable for the 0^+ state. This is no surprise. The excited 3^- state is known to be mainly a single-particle excitation, while the first excited 0^+ state is thought to be an α-particle excitation and thus of $4p-4h$ character. Within the CCSD approximation, it is impossible to accurately reproduce such a state. However, this argument does not apply to the ground state, as the reference Slater-determinant already can be viewed as a cluster of α-particles. Note also that Nature puts the excitation energy of the 3^- state at about 6.16MeV. Again, the deviation of our theoretical result points to deficiencies of the employed Hamiltonian. This motivates us to consider coupled-cluster theory with effective three-body forces.

4. Outlook: Three-Body Forces

The precise description of light nuclei cannot be accomplished with local two-body potentials alone, and the inclusion of three-body forces remedies the situation. Three-body forces affect overall binding as well as level ordering. This occurrence of three-body forces is not surprising. Nucleons, the fundamental degrees of freedom in low-energy nuclear structure theory, are not point particles. Thus, their internal excitations (e.g., the Δ-resonance) are excluded from the model space. This approach is justified by the observation that those excitations are much higher in energy (typically hundreds of MeV) than typical excitations in nuclei (of the order of 1MeV). At low excitation energies, however, the presence of subnuclear degrees of freedom is indistinguishable from three-body forces. This argument is well known from renormalization group theory: integrating out high-momentum (or high-energetic) degrees of freedom creates effective low-momentum Hamiltonians with interactions that have a higher rank than two.

This makes it necessary to use three-body forces in coupled cluster theory. As a first step, one has to derive coupled cluster equations for three-body Hamiltonians. Formally, these are given by Eqs. (3). However, the similarity-transformed Hamiltonian (4) has to be evaluated for three-body terms. This introduces a number of new terms to the coupled cluster equations. In practice, $\overline{H}$ can most conveniently be evaluated in a diagrammatic form. The CCSD equations for the energy, and the cluster amplitudes t_i^a and t_{ij}^{ab} consist of two, fifteen, and 51 diagrams, respectively. As an example, Fig. 1 shows the CCSD diagrams of the three-body force that contribute to the energy.

Fig. 1. A diagrammatic equation for the contributions of three-body forces to the coupled cluster energy within the CCSD approximation.

Here, the long bars denote the three-body Hamiltonian, while the shorter bars denote the $1p-1h$ and $2p-2h$ cluster operators. The energy is obtained by fully contracting the Hamiltonian with the cluster operators. The algebraic expression corresponding to Fig. 1 is

$$E = \frac{1}{4}\sum_{klmcde} \langle klm||cde\rangle t_k^c t_{lm}^{de} + \frac{1}{6}\sum_{klmcde} \langle klm||cde\rangle t_k^c t_l^d t_m^e. \tag{5}$$

Here, $\langle klm||cde\rangle$ denotes the three-body matrix elements. For the efficient numerical implementation, we write the CCCD equations in factorized form,[27] such that more complex diagrams result from successive contractions of simple sub-diagrams with individual cluster amplitudes. This approach allows us to re-use many sub-diagrams, and to order the contractions such that the computational cycle count is minimal. At present, we have derived and implemented coupled-cluster theory for three-body

Hamiltonians. Near-future calculations for ^{16}O and ^{40}Ca will unravel the role that three-body forces play for medium-mass nuclei, and will also help us to better constrain the three-body force itself. The pending results of these calculations will be reported elsewhere.

5. Summary

We employed coupled-cluster theory for nuclear structure calculations, and obtained an *ab-initio* description of light nuclei. Within the two-particle, two-hole cluster approximation, we obtained results for the ground and excited states of ^{4}He and ^{16}O that are fully converged with respect to the size of the model space, and corrections due to three-particle, three-hole clusters are small. The difference between our results for oxygen and experimental values suggest that the microscopic Hamiltonian we employed is deficient, and that three-body forces might be the missing ingredient. We have extended coupled-cluster theory to treat three-body interactions, and derived the corresponding coupled cluster equations. This will enable us to explore the role of these forces in medium-mass nuclei.

Acknowledgments

Research supported by the U.S. Department of Energy under Contracts No. DE-FG02-96ER40963 (University of Tennessee), No. DE-AC05-00OR22725 with UT-Battelle (Oak Ridge National Laboratory), No. DE-FG02-01ER15228 (Michigan State University), the National Science Foundation (Michigan State University), and the Research Council of Norway (University of Oslo).

References

1. S. C. Pieper and R. B. Wiringa, *Ann. Rev. Nucl. Part. Sci.* **51**, 53 (2001).
2. R. B. Wiringa and S. C. Pieper, *Phys. Rev. Lett.* **89**, 182501 (2002).
3. P. Navrátil and W. E. Ormand, *Phys. Rev.* **C68**, 034305 (2003); *Phys. Rev. Lett.* **88**, 152502 (2002).
4. F. Coester, *Nucl. Phys.* **7**, 421 (1958); F. Coester and H. Kümmel, *Nucl. Phys.* **17**, 477 (1960).
5. J. Čížek, *J. Chem. Phys.* **45**, 4256 (1966); *Adv. Chem. Phys.* **14**, 35 (1969).
6. R. Machleidt and I. Slaus, *J. Phys.* **G27**, R69 (2001).
7. S. Weinberg, *Phys. Lett.* **B363**, 288 (1990).
8. U. van Kolck, *Prog. Part. Nucl. Phys.* **43**, 337 (1999).
9. S. K. Bogner, T. T. S. Kuo, and A. Schwenk, *Phys. Rep.* **386**, 1 (2003).
10. H. Kümmel, K. H. Lührmann, and J. G. Zabolitzky, *Phys. Rep.* **36**, 1 (1978).
11. R. J. Bartlett, *J. Phys. Chem.* **93**, 1697 (1989); J. Paldus and X. Li, *Adv. Chem. Phys.* **110**, 1 (1999).
12. R. F . Bishop, in *Microscopic Quantum Many-Body Theories and Their Applications*, eds.: J. Navarro and A. Polls, Lecture Notes in Physics, Vol. **510** (Springer, Berlin, 1998), p. 1.
13. T. D. Crawford and H. F. Schaefer III, *Rev. Comput. Chem.* **14**, 33 (2000).

14. H.G. Kümmel, in *Recent Progress in Many-Body Theories*, eds.: R. F. Bishop *et al.* (World Scientific, Singapore, 2002), p. 334.
15. P. Piecuch, K. Kowalski, Iso Pimienta, P.-D. Fan, M. Lodriguito, M. J. McGuire, S. A. Kucharski, T. Kus, and M. Musial, *Theor. Chem. Acc.* **112**, 349 (2004).
16. R. Guardiola, P. I. Moliner, J. Navarro, R. F. Bishop, A. Puente, and N. R. Walet, *Nucl. Phys.* **A609**, 218 (1996).
17. J. H. Heisenberg and B. Mihaila, *Phys. Rev.* **C59**, 1440 (1999); B. Mihaila and J. H. Heisenberg, *Phys. Rev.* **C61**, 054309 (2002).
18. B. Mihaila and J. H. Heisenberg, *Phys. Rev. Lett.* **84**, 1403 (2000).
19. D. J. Dean and M. Hjorth-Jensen, *Phys. Rev.* **C69**, 054320 (2004).
20. K. Kowalski, D.J. Dean, M. Hjorth-Jensen, T. Papenbrock, and P. Piecuch, *Phys. Rev. Lett.* **92**, 132501 (2004).
21. M. Włoch, D. J. Dean, J. R. Gour, M. Hjorth-Jensen, K. Kowalski, T. Papenbrock, and P. Piecuch, *Phys. Rev. Lett.* **94**, 212501 (2005).
22. H. Monkhorst, *Int. J. Quantum Chem. Symp.* **11**, 421 (1977); K. Emrich, *Nucl. Phys.* **A351**, 379 (1981).
23. J. F. Stanton and R. J. Bartlett, *J. Chem. Phys.* **98**, 7029 (1993).
24. M. Włoch, J. R. Gourm K. Kowalski and P. Piecuch, *J. Chem. Phys.* **122**, 214107 (2005).
25. D. R. Entem and R. Machleidt, *Phys. Lett.* **B524**, 93 (2002).
26. D. R. Entem and R. Machleidt, *Phys. Rev.* **C68**, 041001 (2003).
27. S. A. Kucharski and R. J. Bartlett, *Theor. Chim. Acta* **80**, 387 (1991).

CORRELATIONS IN HOT ASYMMETRIC NUCLEAR MATTER

A. POLLS, A. RIOS, A. RAMOS

Department d'Estructura i Constituents de la Matèria, Universitat de Barcelona, E-08028 Barcelona,Spain
artur@ecm.ub.es

H. MÜTHER

Institut für Theoretische Physik, Universität Tübingen D-72076 Tübingen, Germany

The single-particle spectral functions in asymmetric nuclear matter are computed using the ladder approximation within the theory of finite temperature Green's functions. The internal energy and the momentum distributions of protons and neutrons are studied as a function of the density and the asymmetry of the system. The proton states are more strongly depleted when the asymmetry increases, whereas the occupation of the neutron states is enhanced compared to the symmetric case. Preliminary results for the entropy and the free energy are also presented.

Keywords: spectral functions; correlations; asymmetric nuclear matter.

1. Introduction

The equation of state (EOS) of asymmetric nuclear matter is a necessary ingredient in the description of astrophysical environments such as supernova explosions or the structure of neutron stars.[1] The study of asymmetric nuclear matter is also relevant to understand stable nuclei because they themselves are asymmetric nuclear systems. In addition, the recent availability of data concerning nuclei far from the stability valley has revitalized the interest in the study of asymmetric systems. Therefore, the EOS of nuclear matter in a wide range of densities and temperatures is with no doubt one of the challenging open problems in nuclear physics.

The microscopic evaluation of this EOS starting from realistic models of the nucleon-nucleon (NN) interactions requires a rigorous treatment of the nucleon-nucleon (NN) correlations and the use of very sophisticated many-body techniques.[2–4] In fact, the strong short-range and tensor components, which are needed in realistic NN interactions to fit the NN scattering data, lead to important modifications of the nuclear wave function. A very simple indication of the importance of NN correlations is provided by the observation that a simple Hartree-Fock calculation for nuclear matter at the empirical saturation density using such realistic interactions typically results in positive energies rather than the empirical value of -16 MeV per nucleon.

Correlations do not only affect the bulk properties but also modify the single-particle properties in a substantial way. Several recent calculations have shown without ambiguity how the NN correlations produce a partial occupation of the single-particle states which would be fully occupied in a mean field description. This depletion has been experimentally corroborated in the analysis of $(e, e'p)$ reactions on ^{208}Pb, which led to the conclusion that the occupation numbers for deeply bound proton states were depleted by about 15-20 %.[5]

Several theoretical methods have been developed to describe these correlations in nuclear systems. These include the Brueckner hole-line expansion and also variational approaches using correlated basis functions.[6] Recently, enormous progress has been achieved in using the self-consistent Green's function technique (SCGF).[3,4] This method gives direct access to the single-particle spectral function, i.e., the distribution of strength in energy when one adds or removes a particle of the system with a given momentum. The knowledge of the single-particle spectral functions allows to study all single-particle properties of the system, and also permits to evaluate the internal energy by means of the Koltun sum-rule. At the same time, the modification of the single-particle properties affects the effective interactions among the nucleons in the medium and both quantities, the single-particle propagator and the effective interactions in the medium, should be self-consistently determined.

Most of the microscopic calculations have been addressed to study symmetric nuclear matter (SNM) and pure neutron matter (PNM). The study of asymmetric nuclear matter is technically more involved and only a few calculations are available.[7-9] In this paper we want to report our last contributions in the field of SCGF theory[9,10] applied to asymmetric nuclear matter. We will also present some strategies and preliminary results concerning the calculation of the entropy and the free energy of nuclear matter, which allow one to extract the thermodynamic properties at finite temperature.

2. Asymmetric nuclear matter within SCGF approach

2.1. *Spectral functions*

For a given Hamiltonian H, the Green's function for a system at finite temperature can be defined in a grand-canonical formulation for both real and imaginary times:

$$ig(\mathbf{k}t; \mathbf{k}'t') = Tr\{\rho \mathbf{T}\left[a_{\mathbf{k}}(t) a^{\dagger}_{\mathbf{k}'}(t')\right]\}. \tag{1}$$

$\mathbf{T}$ is the time ordering operator that acts on a product of Heisenberg operators $a_{\mathbf{k}}(t) = e^{itH} a_{\mathbf{k}} e^{-itH}$ in such a way that the operator with largest time argument t (or it in the case that t is imaginary) is put to the left. The trace is to be taken over all energy eigenstates and all particle number eigenstates of the many-body system, weighted by the statistical operator

$$\rho = \frac{1}{Z} e^{-\beta(H-\mu N)}, \tag{2}$$

where β and μ denote the inverse temperature and the chemical potential, respectively, N is the particle number operator and Z is the grand partition function, $Z = Tr\left\{e^{-\beta(H-\mu N)}\right\}$. Notice that in this theoretical introduction we omit the isospin quantum numbers.

For a homogeneous system, the Green's function is diagonal in momentum space and depends only on the modulus of $\mathbf{k}$ and on the time difference $\Delta t = t' - t$. We start with the case $\Delta t > 0$, and introduce the following definition of the so called correlation function:

$$g^{<}(k, \Delta t) = Tr\{\rho e^{i\Delta t H} a_{\mathbf{k}}^{\dagger} e^{-i\Delta t H} a_{\mathbf{k}}\}\,. \tag{3}$$

$g^{<}(k, \Delta t)$ can be expressed as a Fourier integral over all frequencies,

$$g^{<}(k, \Delta t) = \int_{-\infty}^{+\infty} \frac{d\omega}{2\pi} e^{i\omega\Delta t} A^{<}(k, \omega) \tag{4}$$

if $A^{<}(k, \omega)$ is defined by

$$A^{<}(k, \omega) = 2\pi \sum_{nm} \frac{e^{\beta(E_m - \mu N_m)}}{Z} \mid \langle \Psi_n \mid a_k \mid \Psi_m \rangle \mid^2 \delta\left(\omega - (E_m - E_n)\right)\,. \tag{5}$$

Notice that $\mid \Psi_m\rangle$ are simultaneous eigenstates of both the number operator and the Hamiltonian. A similar procedure can be performed for $\Delta t < 0$, yielding a function

$$A^{>}(k, \omega) = e^{\beta(\omega-\mu)} A^{<}(k, \omega)\,. \tag{6}$$

The spectral function at finite temperature is defined as the sum of the two positive functions, $A^{<}$ and $A^{>}$, $A(k,\omega) = A^{<}(k,\omega) + A^{>}(k,\omega)$, and thus $A^{<}(k,\omega) = f(\omega)A(k,\omega)$ and $A^{>} = (1 - f(\omega))A(k,\omega)$, where $f(\omega) = \left[1 + e^{\beta(\omega-\mu)}\right]^{-1}$ is the Fermi function.

Expression (5) for $A^{<}$, can be compared to the result for the hole spectral function at zero temperature,

$$A_h(k, \omega) = 2\pi \sum_{n} \mid \langle \Psi_n^{A-1} \mid a_k \mid \Psi_0^A \rangle \mid^2 \delta\left[\omega - (E_0^A - E_n^{A-1})\right]\,, \tag{7}$$

where $\mid \Psi_0^A\rangle$ is the ground state of a A particle system and $\mid \Psi_n^{A-1}\rangle$ labels the excited energy eigenstates of a system that contains $A-1$ particles. The physical interpretation of the hole spectral function in a system at zero temperature is quite simple: $A_h(k,\omega)$ is the probability to remove a particle from the ground state of the A-body system, such that the residual system is left with an excitation energy $E_n^{A-1} = E_0^A - \omega$, being E_0^A the ground state energy of the A particle system. The fact that the lowest possible energy of the final state is the ground state energy of the $A-1$ particle system limits the energy domain of the hole spectral function to a maximum value of $\omega = E_0^A - E_0^{A-1} = \mu$. In a similar way, the particle spectral function A_p can be defined as the probability to attach a further nucleon with momentum k to the system in such a way that the excitation energy of the $A+1$ system with respect to the ground state energy of the initial system is $\omega = E^{A+1} -$

E_0^A with a minimum value of μ. Therefore, at zero temperature the particle and hole spectral spectral functions are defined in a separated energy range.

The situation is different at finite temperature. Since thermally excited states $| \Psi_m\rangle$ are always included in the grand-canonical ensemble average, one can take out a particle from a thermally excited state and end up in a state close to the ground state of the residual system. This leads to a contribution to $A^<$ for an energy ω larger than μ. Similar arguments allow to conclude that $A^<$ extends to the energy region below μ. In any case, there is no longer an energy separation between $A^<$ and $A^>$, and the maxima of both functions can even coincide. A simple example, is provided by the free Fermi gas at finite temperature. In this case, $A^<$ and $A^>$ are δ peaks located at the same energy, $\epsilon(k) = k^2/2m$, but have different strengths, $f(\epsilon(k))$ and $[1 - f(\epsilon(k))]$, respectively.

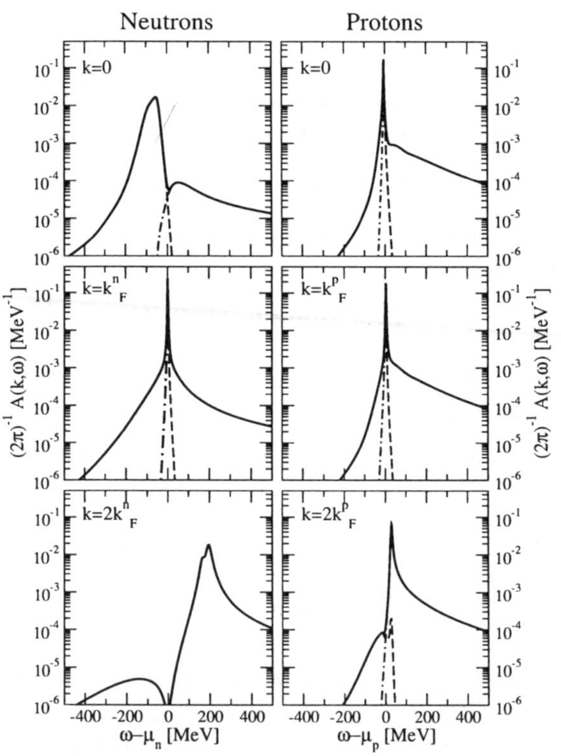

Fig. 1. Neutron (left panels) and proton (right panels) single-particle spectral functions at $\rho = 0.16\ \mathrm{fm}^{-3}$, $T = 5$ MeV and proton fraction $x_p = 0.04$ for three different momenta. Both $A_\tau^<$ (dashed lines) and $A_\tau^>$ (dot-dashed lines) are displayed, together with the the total spectral function A_τ (solid lines).

To illustrate these changes, the full spectral function $A(k,\omega)$ (solid lines), as well as $A^<(k,\omega)$ (dashed lines) and $A^>(k,\omega)$ (dash-dotted lines) are shown in Fig. 1, for neutrons (left panels) and protons (right panels) for three different momenta, $k=0$, k_F^τ and $2\ k_F^\tau$, with k_F^τ the Fermi momentum of each nucleon species, either a proton ($\tau=+1/2$) or a neutron ($\tau=-1/2$), at a density $\rho=0.16$ fm^{-3}, a proton fraction of $x_p=0.04$ and a temperature of $T=5$ MeV.

For the case of neutrons at $k=0$, the peak of the spectral function is provided by $A_n^<$. This is due to the fact that the position of this peak, which can be identified with the neutron dynamical quasiparticle energy $\epsilon_{qp}^n(k)$, is well below the neutron chemical potential μ_n, and therefore the value of $f(\epsilon_{qp}^n(0))$ is very close to one. The thermal effects do not fill up completely the minimum of the spectral function at $\omega=\mu_n$ and we still observe some kind of separation in energy between $A_n^<$ and $A_n^>$. However, around $\omega=\mu_n$ there is an energy interval in which both $A_n^<$ and $A_n^>$ are small but different from zero. A similar situation is observed for the neutron spectral function at $k=2k_F^n$. In this case, the quasiparticle energy is well above μ_n, which means that $(1-f(\epsilon_{qp}))$ is very close to one and the peak structure is supplied by $A_n^>$. Again, we observe a clear separation between $A_n^<$ and $A_n^>$. On the contrary, at $k=k_F^n$, the quasiparticle peak is close to the chemical potential, which means that $f(\epsilon_{qp}^n(k))$ is around 0.5. Therefore, both $A_n^<$ and $A_n^>$ have a peak at $\omega=\mu_n$, and the overlap region of $A_n^<$ and $A_n^>$ is enhanced.

In the case of very small proton fraction, like $x_p=0.04$, the Fermi momentum k_F^p is rather small and consequently the quasiparticle energies of the chosen momenta are close to the Fermi energy. Therefore, there are strong overlaps between $A_p^<$ and $A_p^>$ and the peaks are rather narrow.

The spectral function is contained in the imaginary part of the single-particle Green's function, which in turn is determined through a Dyson equation once the self-energy has been calculated. The self-energy is computed in the ladder approximation, in which the pairs of particles or holes in the intermediate states are described with the full spectral functions. In this way, the effective interaction among nucleons in the medium is affected by the single-particle properties, and both things should be defined in a self-consistent way.[3,4,9]

2.2. *Internal energy and momentum distributions*

Once the self-consistent solution for the spectral functions has been obtained, the internal energy per particle is given by

$$\frac{E}{A}=\frac{\nu}{\rho}\sum_\tau\int\frac{d^3k}{(2\pi)^3}\int_{-\infty}^{+\infty}\frac{d\omega}{2\pi}\frac{1}{2}\left(\frac{k^2}{2m_\tau}+\omega\right)A_\tau(k,\omega)f_\tau(\omega)\,. \tag{8}$$

where $\nu=2$ is the spin degeneracy. The BHF approach can be obtained from the previous expression by assuming that the single-particle spectral functions are characterized by only one energy having the full strength accumulated in this energy, $A_\tau(k,\omega)=\delta(\omega-\epsilon_\tau^{BHF})$ with $\epsilon_\tau^{BHF}(k)$ the BHF single-particle energy. Notice also that in the BHF approximation, the two body propagator which appears in the intermediate states of the G-matrix, does not contain propagation of holes.

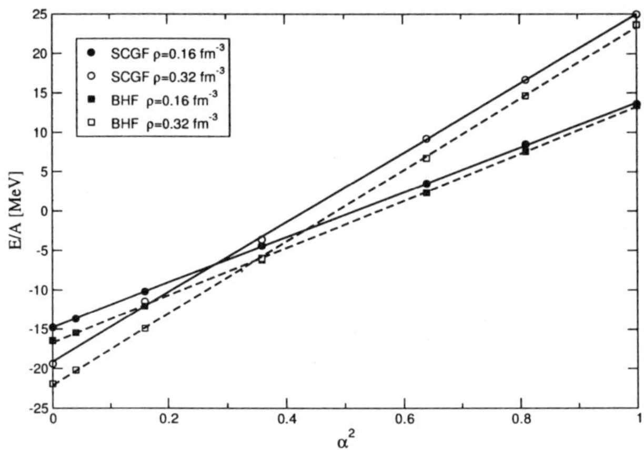

Fig. 2. Energy per particle as a function of α^2 calculated in the SCGF approach (full lines) and in the BHF approach (dashed lines) at two densities, $\rho = 0.16$ fm^{-3} (solid symbols) and $\rho = 0.32$ fm^{-3} (empty symbols).

The asymmetry dependence of the internal energy for both approaches is shown in Fig. 2, for two densities $\rho = 0.16$ fm^{-3} and 0.32 fm^{-3}. The SCGF results (solid lines) are compared with the BHF (dashed lines) ones. All the results discussed in this paper have been computed for the charge-dependent Bonn (CDBONN) potential,[11] which is nonlocal and exhibits a soft tensor component. The temperature that we consider, $T = 5$ MeV, should be low enough for the conclusions on the asymmetry dependence to be valid at $T = 0$ MeV and high enough to avoid the instabilities related to neutron-proton pairing. The first point to notice is the linear dependence of the energy in terms of α^2, the square of the asymmetry $\alpha = (\rho_n - \rho_z)/\rho$, for both types of calculations (BHF and SCGF) in the full range of variation of asymmetry from SNM to PNM. One can then assume a quadratic dependence of the energy per particle in terms of α and the symmetry energy $a_s(\rho)$,

$$e(\rho, \alpha) = e(\rho, 0) + a_s(\rho)\alpha^2 + \dots . \tag{9}$$

A second point is the fact that the propagation of holes and the use of the spectral functions in the intermediate states of the ladder equation results in a repulsive effect with respect to the continuous choice BHF calculation. For a given density, the difference between the SCGF and the BHF results does not depend so much on the asymmetry, although this difference is slightly greater in SNM than in PNM. This leads to a small decrease of the symmetry energy for the SCGF calculation compared to the BHF. At $\rho = 0.16$ fm^{-3}, $a_s = 30.0$ MeV in the BHF approximation whereas the SCGF scheme provides $a_s = 28.6$ MeV.

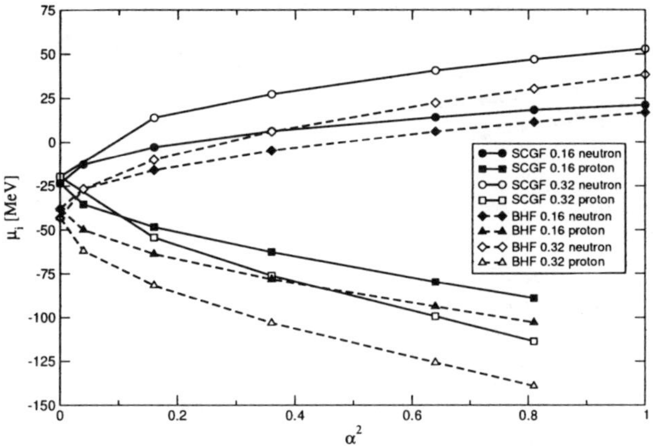

Fig. 3. Neutron (upper curves) and proton (lower curves) chemical potentials as a function of the asymmetry calculated in the SCGF approach (solid lines) and in the BHF approach (dashed curves) at two densities, $\rho = 0.16$ fm^{-3} (solid symbols) and $\rho = 0.32$ fm^{-3} (empty symbols).

The momentum distribution of each component is calculated from an integral over the energy of the spectral function,

$$n_\tau(k) = \int_{-\infty}^{\infty} \frac{d\omega}{2\pi} A_\tau(k,\omega) f_\tau(\omega) . \tag{10}$$

It is worth to remind that, for a given total density, the partial fraction $x_\tau = \rho_\tau/\rho$ of the respective particle species is given by:

$$x_\tau = \frac{\nu}{\rho} \int \frac{d^3k}{(2\pi)^3} n_\tau(k) . \tag{11}$$

By considering a fixed composition, Eq. (11) can be used to determine the chemical potential of each species. The dependence of these chemical potentials on the asymmetry is shown in Fig.3 for the same previously stated densities and a temperature of $T = 5$ MeV. The chemical potentials of neutrons (upper curves) and protons (lower curves) have been calculated in the SCGF approach (solid lines) and in the BHF approach (dashed curves). When the asymmetry increases, the neutron chemical potential increases and becomes positive, whereas the proton chemical potential becomes more and more negative. The dependence on α^2 is no longer lineal. It is important to notice that the SCGF approach is thermodynamically consistent, in the sense that the chemical potential derived from the normalization condition of the partial density Eq. (11) coincides with the ones obtained from the derivative of the free energy. In contrast, it is well known that, in the BHF approach, the chemical potential derived from the normalization condition substantially differs from the one obtained by the thermodynamical relation.

The momentum distributions provide a clear measurement of the effects of correlations. In Fig. 4 we show the occupation of the zero-momentum state as a function

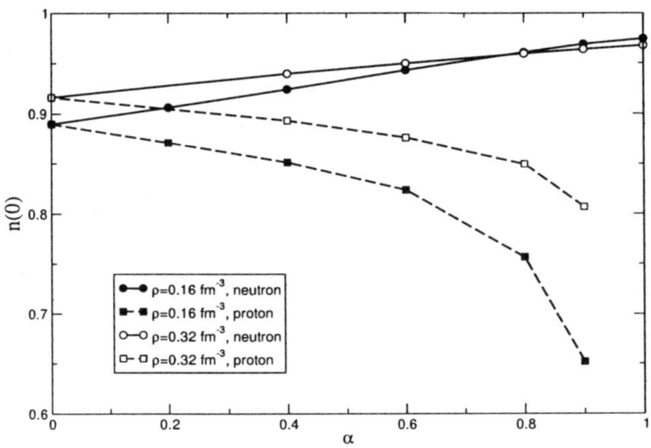

Fig. 4. Dependence of the occupation of the $k = 0$ single-particle state on the asymmetry for neutrons (full lines) and protons (dashed lines) at two given densities, $\rho = 0.16$ fm^{-3} (solid symbols) and $\rho = 0.32$ fm^{-3} (empty symbols).

of the asymmetry. In the symmetric case, we observe an unexpected behavior on the occupation. Namely, at higher densities there is a lower depletion than at lower densities. For a given density, the proton depletion increases with the asymmetry, indicating the importance of the neutron-proton correlations. In contrast, the depletion of the neutrons becomes smaller. From that point of view, neutron matter is a less correlated system. Also it is worth to notice that for PNM one recovers the expected behavior of having a larger depletion for higher densities.

The full momentum distribution of protons and neutrons at $\rho = 0.16$ fm^{-3} and $\alpha = 0.2$ is presented in Fig. 5, for the same $T = 5$ MeV. Firstly, one observes that the BHF momentum distributions do not contain correlation effects and are thus close to a normal thermal Fermi distribution. The momentum distributions obtained within the framework of the SCGF contain, besides thermal effects, important short-range and tensor correlations. These are reflected in the depletion of the occupation at low momentum and in a larger occupation at large momenta compared to BHF results. Notice also that the proton momentum distribution is more depleted than the neutron one. This is in agreement with the fact that the protons (i.e., the less abundant particle in natural systems) are more affected by correlations, mainly due to their interaction with neutrons.

3. Calculation of the entropy of correlated SNM

The computation of the entropy of a correlated system of fermions is a long-standing problem in many-body quantum physics. It can be calculated either within a Fermi liquid theory, using the concept of Landau quasiparticles, or within a more microscopic formalism, in which the dynamical quasiparticle energies are defined as the poles of dressed propagators. Carneiro and Pethick discussed the computation of

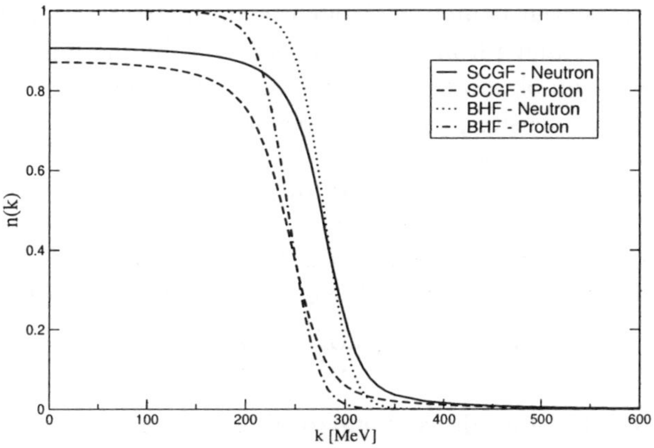

Fig. 5. Momentum distribution for the SCGF approach (full line for neutrons, dashed line for protons) and the BHF approach (dotted line for neutrons, dot-dashed line for protons) at a density $\rho = 0.16$ fm^{-3}, an asymmetry $\alpha = 0.2$ and a temperature $T = 5$ MeV.

the entropy within both formalisms.[12,13] In these two references it is shown that a good approximation to the entropy per particle at low temperatures is provided by the following expression:

$$s_b = -\frac{1}{\rho}\int \frac{d^3k}{(2\pi)^3}\int_{-\infty}^{\infty}\frac{d\omega}{2\pi}B(k,\omega)\left\{f(\omega)\ln f(\omega) + [1-f(\omega)]\ln[1-f(\omega)]\right\}\ , \quad (12)$$

where $B(k,\omega)$ has the properties of a spectral function and can be extracted from the single-particle Green's function.[12,13] Taking $B(k,\omega) = \delta(\omega - \epsilon_{qp}(k))$ one recovers the quasiparticle expression for the entropy, which is normally used in the BHF calculations at finite temperature.

One can also give some *naive* estimations for the entropy of correlated systems. In Eq. (10), for instance, one can see that the main difference between the momentum distribution within BHF and within SCGF comes both from the inclusion of the spectral function and the additional integration over energies. Following the same philosophy, one can give the following estimate of the entropy:

$$s_a = -\frac{1}{\rho}\int \frac{d^3k}{(2\pi)^3}\int_{-\infty}^{\infty}\frac{d\omega}{(2\pi)}A(k,\omega)\left\{f(\omega)\ln f(\omega) + [1-f(\omega)]\ln[1-f(\omega)]\right\}\ , \quad (13)$$

which, in the approximation $A(k,\omega) = \delta(\omega - \epsilon_{qp}(k))$, leads also to the quasiparticle expression for the entropy. Finally, another "reasonable" estimation of the entropy would be given by using the quasiparticle expression with the fully correlated SCGF momentum distribution:

$$s_k = -\frac{1}{\rho}\int \frac{d^3k}{(2\pi)^3}\left\{n(k)\ln n(k) + [1-n(k)]\ln[1-n(k)]\right\}\ . \quad (14)$$

On the other hand, one can also think in a thermodynamical procedure to calculate the entropy. In fact, once the SCGF calculation is performed, one has access to the chemical potential computed from the normalization condition Eq. (11). Since μ can also be obtained from the free energy via the thermodynamical relation $\mu = \partial F/\partial N \mid_T$, one can obtain the variation of the free energy at constant T, from an arbitrary reference density ρ_1 up to a given density ρ, by

$$\frac{F}{V} - \frac{F_1}{V} = \int_{\rho_1}^{\rho} d\rho\, \mu(\rho) \,. \tag{15}$$

The knowledge of the exact internal energy thanks to Koltun's sum rule, Eq. (8), allows us to calculate the variation of the entropy of the system through:

$$\frac{\Delta S_T}{V} = \frac{1}{T}\left(\frac{\Delta E}{V} - \frac{\Delta F}{V}\right) \,. \tag{16}$$

On the left panel of Fig. 6, we compare these estimations of the entropy as a function of the density for the case of SNM at $T = 10$ MeV. Since we want to represent the entropy (and not its variation), we take the reference density to be $\rho_1 = 0.1$ fm^{-3}, i.e. at this density s_T is taken to be equal to what we consider to be the best estimation, s_b. Except for s_k, one can consider that all the approximations are in a reasonable agreement and the dependence with density is very similar. The reason that s_k gives larger values for the entropy is due to the fact that, by using the fully-depleted $n(k)$, we are taking the correlation effects on $n(k)$ as thermal ones. In other words, the large depletion due to correlations mimics, in this case, a large effective temperature and thus one gets a large entropy.

On the right panel of Fig. 6, we present the free energy together with the chemical potential as a function of the density for the same fixed temperature $T = 10$ MeV. We observe a good agreement between all the approaches, except for the one associated with s_k. Notice also that the chemical potential crosses the free energy per particle at the density for which the free energy presents a minimum. This confirms the fullfilment of the Hugenholtz-van Hove theorem and supports the thermodynamical consistency of the different approaches.

Acknowledgments

Useful discussions with Drs. Tobias Frick, Isaac Vidaña, Marcello Baldo and Hans Schulze are gratefully acknowledged. This work has been supported by the Europäische Graduirtenkolleg Tübingen-Basel (DGF-SNF), the Grant No. FIS2005-03142 from MEC (Spain) and FEDER, and the Generalitat de Catalunya Project No. 2005SGR-00343. A. Rios acknowledges the support of DURSI and the European Social Funds.

References

1. H. Heiselberg and M. Hjorth-Jensen, *Phys. Rep.* **328**, 237 (2000).
2. M. Baldo, *Nuclear Methods and the Nuclear Equation of State, Int. Rev. of Nucl. Phys*, Vol. 9 (World Scientific, Singapore, 1999). (McGraw-Hill, New York, 1964).
3. H. Müther and A. Polls, *Prog. Part. Nucl. Phys.* **45**, 243 (2000).

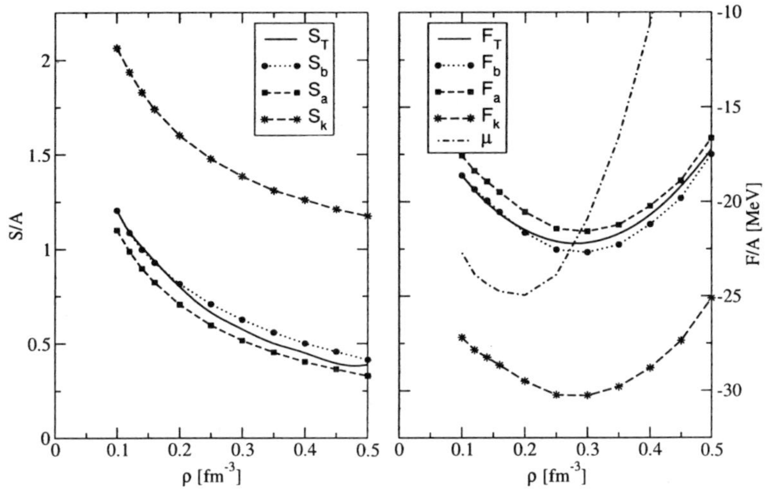

Fig. 6. Left Panel: different approximations (see text) to the entropy per particle of SNM as a function of density for a fixed temperature $T = 10$ MeV. Right Panel: different approximation (see text) to the free energy per particle of SNM as a function of density for a fixed temperature $T = 10$ MeV.

4. W. H. Dickhoff and C. Barbieri, *Prog. Part. Nucl. Phys.* **52**, 377 (2004).
5. M.F. van Batenburg, Ph.D. Thesis, Univ. of Utrecht, Utrecht, 2001.
6. S. Fantoni and A. Fabrocini, in *Microscopic Quantum Many-Body Theories and Their Applications*, ed. J. Navarro and A. Polls (Springer, New York, 1998).
7. A.E.L. Dieperink, Y. Dewulf, D. Van Neck, M. Waroquier, and V. Rodin, *Phys. Rev.* **C68**, 064307 (2003).
8. I. Vidaña and I. Bombaci, *Phys. Rev.* **C66**, 045801 (2002).
9. T. Frick, H. Müther, A. Rios, A. Ramos, and A. Polls, *Phys. Rev.* **C71**, 014313 (2005).
10. T. Frick, H. Müther, and A. Polls, *Phys. Rev.* **C69**, 054305 (2004).
11. R. Machleidt, F. Sammarruca, and Y. Song, *Phys. Rev.* **C53**, R1483 (1996).
12. C.J. Pethick and G.M. Carneiro, *Phys. Rev.* **A7**, 304 (1973).
13. G.M. Carneiro and C.J. Pethick, *Phys. Rev.* **B11**, 1106 (1975).

AUTHOR INDEX

SUBJECT INDEX